LARGE-SCALE
SOLAR POWER
SYSTEM DESIGN

McGRAW-HILL'S GREENSOURCE SERIES

Attmann
Green Architecture: Advanced Technologies and Materials

Gevorkian
Alternative Energy Systems in Building Design
Large-Scale Solar Power System Design: An Engineering Guide for Grid-Connected Solar Power Generation
Solar Power in Building Design: The Engineer's Complete Design Resource

GreenSource: The Magazine of Sustainable Design
Emerald Architecture: Case Studies in Green Building

Haselbach
The Engineering Guide to LEED–New Construction: Sustainable Construction for Engineers, 2d ed.

Luckett
Green Roof Construction and Maintenance

Melaver and Mueller (eds.)
The Green Building Bottom Line: The Real Cost of Sustainable Building

Nichols and Laros
Inside the Civano Project: A Case Study of Large-Scale Sustainable Neighborhood Development

Winkler
Green Facilities: Industrial and Commercial LEED Certification
Recycling Construction & Demolition Waste: A LEED-Based Toolkit

Yellamraju
LEED–New Construction Project Management

Yudelson
Green Building Through Integrated Design
Greening Existing Buildings

LARGE-SCALE SOLAR POWER SYSTEM DESIGN

AN ENGINEERING GUIDE FOR GRID-CONNECTED SOLAR POWER GENERATION

PETER GEVORKIAN, PH.D., P.E.

New York Chicago San Francisco Lisbon London Madrid
Mexico City Milan New Delhi San Juan Seoul
Singapore Sydney Toronto

The McGraw·Hill Companies

Cataloging-in-Publication Data is on file with the Library of Congress

Copyright © 2011 by The McGraw-Hill Companies, Inc. All rights reserved. Printed in the United States of America. Except as permitted under the United States Copyright Act of 1976, no part of this publication may be reproduced or distributed in any form or by any means, or stored in a data base or retrieval system, without the prior written permission of the publisher.

2 3 4 5 6 7 8 9 0   QVR/QVR   1 7 6 5 4 3 2

ISBN 978-0-07-176327-1
MHID 0-07-176327-9

Sponsoring Editor
Joy Evangeline Bramble

Editing Supervisor
Stephen M. Smith

Production Supervisor
Richard C. Ruzycka

Acquisitions Coordinator
Alexis Richard

Project Manager
Anupriya Tyagi,
Glyph International

Copy Editor
Erica Orloff

Proofreader
Amy Rodriguez

Indexer
Robert Swanson

Art Director, Cover
Jeff Weeks

Composition
Glyph International

Printed and bound by Quad Graphics .

McGraw-Hill books are available at special quantity discounts to use as premiums and sales promotions, or for use in corporate training programs. To contact a representative, please e-mail us at bulksales@mcgraw-hill.com.

The pages within this book were printed on acid-free paper containing 100% postconsumer fiber.

About the Author

 **Peter Gevorkian, Ph.D., P.E.,** is President of Vector Delta Design Group, Inc., an electrical engineering and solar power design consulting firm located in La Canada Flintridge, California. He holds a B.S. in electrical engineering, an M.S. in computer science, and a Ph.D. in electrical engineering. He has been the recipient of numerous awards for engineering merit, design achievement, solar power design, and renewable energy systems.

Dr. Gevorkian has taught computer science, automation control, and renewable energy systems engineering, and has written many technical papers for national and international symposiums. He is the author of the books *Sustainable Energy Systems in Architectural Design, Sustainable Energy Systems Engineering, Solar Power in Building Design,* and *Alternative Energy Systems in Building Design,* all published by McGraw-Hill.

CONTENTS

FOREWORD

With the publication of *Large-Scale Solar Power System Design,* Peter Gevorkian once again shows his mastery of a subject matter that, on the face of it, appears arcane and unremittingly esoteric. In Dr. Gevorkian's hands, this encyclopedic examination of solar power system design and implementation comes alive as an indispensable tool in our quest for sustainable, energy-efficient solutions to many of today's most pressing environmental challenges.

This work is a testament to Dr. Gevorkian's power as a communicator. He has a uniquely inspired gift: the facility to take a comprehensive body of knowledge and, without compromise, synthesize its essence for an extraordinarily mixed audience comprising sustainability adherents, engineers, mathematicians, physicists, architects, public officials, building developers, and owners—even lay end users.

Dr. Gevorkian has a big-picture vision of the aspects of this subject matter that appeals to a variety of sectors, including persons with a primarily technical bent, those readers who are conscious of the contemporary public will toward alternative-energy strategies, and those people for whom the business case of solar technology is all-important. Dr. Gevorkian is beyond literate on the subject of solar energy.

His ability to weave a commonsense logic that provides simplified explanations of fundamentally complex issues makes this work eminently readable. Dr. Gevorkian tells a good story, over and over again. What makes this book so important is the nexus of the known and unknown that he continually revisits. Dr. Gevorkian is a teacher throughout this book. He continually defines and redefines, clarifies and reclarifies, and not so much advocates for you to become one of the converted as he offers you the validation—and space—that you need to proclaim what you have already decided to be. He nudges you to feel the impulse in your viscera and to express it with a roar like the New Age intellect that you purport to be.

Any book that evolves from an initial assessment of relevant definitions of subject matter to an exercise in clarifying Max Planck and Einstein, from a repository of how-to applications to a work that ultimately makes a nearly unassailable financial case for large-scale solar energy investment, is a unique and valued instrument. *Large-Scale Solar Power System Design* has offered Dr. Gevorkian the opportunity to craft a masterwork.

Dr. Gevorkian is poised on the precipice of sustainability's future. It is the future of people of common sense, post-excess practitioners of efficiency and moderation. In the coming years, when the practical application of solar energy to every aspect of the collective life experience is an afterthought, this delightful work will stand the test of time.

Dr. Lance A. Williams

Dr. Lance A. Williams is Executive Director of the U.S. Green Building Council's Los Angeles chapter. He is LEED Accredited Professional who has traveled extensively and viewed sustainability in a wide variety of settings. His primary focus is the relationship between culture and sustainability.

INTRODUCTION

This book is intended to be a comprehensive design reference guide for professionals vested in PV solar power generation system designs, solar power system integration, project management, and system tests and evaluations of large-scale grid-connected solar power systems.

The book may be used as a graduate or postgraduate textbook in alternative energy systems and electrical engineering studies, as well as a professional extension course for individuals interested in pursuing a career in the solar power system industries.

As a design reference manual, this book is intended to set a standard for engineering design and construction guidelines for large-scale solar power system projects. Design methodologies reflected in the book have been based on my years of pragmatic solar power and electrical engineering design experiences.

Currently, there is an acute shortage of engineering design and construction personnel worldwide who possess an in-depth knowledge of large-scale grid-connected solar power systems. It is my hope that people in the field and considering entering the field find this book a valuable resource.

Furthermore, in the past several years, skilled professionals with design and installation experience have been (and, in the near future, will be) in great demand. However, our higher-learning institutions (universities and technical schools) seriously lack teachers and instructors with real hands-on experience. It is these types of educators who are required to respond to the demand created by ongoing large-scale solar power projects worldwide.

Peter Gevorkian, Ph.D., P.E.

ACKNOWLEDGMENTS

I would like to thank my colleagues and other individuals who have encouraged me and assisted me in writing this book. I am especially grateful to all agencies and organizations that provided photographs and allowed use of some textual material, and to my colleagues who read the manuscript and provided valuable insight. Thanks to Arlen Gharibian for the preliminary edit of the manuscript.

My thanks go to Dr. Lance Williams, Executive Director, U.S. Green Building Council, Los Angeles Chapter; Ken Touryan, Ph.D.; Gabriel Paoletti, EATON Corporation; Dr. Vahan Garboushian, President, AMONIX Inc.; Robert McConnell, Ph.D., Sr. V.P., AMONIX Inc.; Nancy Hartsoch, V.P., SolFocus; Hagob Panossian, Ph.D.; Dr. Subhendu Guha, United Solar Ovonics; Dr. William Nona, Architect, National Council of Architectural Registration Boards; Behzad Eghtesady, P.E., Chief Electrical Engineer, Los Angeles Department of Building and Safety; and Eddie Alahverdian, P.E., Sr. Engineer, Solar Power Systems Design, Los Angeles Department of Water and Power.

My thanks also go to AMONIX Inc., Torrance, California; Atlantis Energy Systems, Inc., Sacramento, California; California Energy Commission, Sacramento, California; Museum of Water & Life, Center for Water Education, Hemet, California; Solar Integrated Technologies, Los Angeles, California; U.S. Green Building Council, Los Angeles Chapter; U.S. Department of Energy, National Renewable Energy Laboratories; Wikipedia, for coverage of the fuel cell and hydrogen industries; Sandia National Laboratories; and Solargenix Energy, Newport Beach, California.

DISCLAIMER NOTE

This book examines large-scale solar power system design, with the sole intent to familiarize the reader with the design guidelines of existing solar photovoltaic technologies in order to encourage engineers, educators, solar power integrators, and management personnel to promote deployment of solar power energy systems.

The principal objective of the book is to emphasize solar power co-generation design, application, and economics. Also, Chap. 11 covers passive solar systems that have in the past few decades undergone notable improvements throughout the industrialized world.

Neither the author, the publisher, nor individuals, organizations, or manufacturers referenced or credited in this book make any warranties, express or implied, or assume any legal liability or responsibility for the accuracy, completeness, or usefulness of any information, products, and processes disclosed or presented.

Reference to any specific commercial product, manufacturer, or organization does not constitute or imply endorsement or recommendation by the author.

LARGE-SCALE
SOLAR POWER
SYSTEM DESIGN

SOLAR POWER SYSTEM TECHNOLOGIES

Introduction

Solar or photovoltaic (PV) cells are electronic devices that essentially convert the solar energy of sunlight into electric energy or electricity. The physics of solar cells is based on the same semiconductor principles as diodes and transistors, which form the building blocks of the entire world of electronics.

Solar cells convert energy as long as there is sunlight. In the evenings and during cloudy conditions, the conversion process diminishes. It stops completely at dusk and resumes at dawn. Solar cells do not store electricity, but batteries can be used to store the energy.

One of the most fascinating aspects of solar cells is their ability to convert the most abundant and free forms of energy into electricity without moving parts or components. Also, they do not produce any adverse forms of pollution that affect the ecosystem, as is associated with most known forms of nonrenewable energy production methods, such as fossil fuels, hydroelectric power, or nuclear energy plants.

In this chapter, we will review a number of solar power PV system technologies, manufacturing processes, and the intercellular connectivity of basic flat-panel technologies.

Solar Cell Electronics

An electrostatic field is produced at a PN junction of a solar cell by impinging photons that create 0.5 to 0.6 V of potential energy, which is characteristic of most silicon-based PN-junction photovoltaic technologies. Voltage potential generated by the cells is analogous in function to a small battery. When connecting the positive and negative leads in parallel or series combination, similar to conventional batteries, the PN-interconnected cells could generate higher currents and voltages.

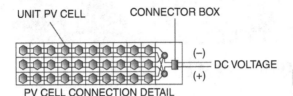

Figure 1.1   **Photovoltaic cell interconnection.**

In general, conventional PV panels are configured from serial interconnected sub-groupings of cells, which include multiple hundreds of PN junctions that are connected in parallel, forming a cell module. The module produces 0.5 to 0.6 V and a current of several amps, which is proportional to the number of parallel connections.

In photovoltaic panels, the serial interconnection of multiple cells connected in a series produces voltages proportional to the number of series-connected unit cells. For example, when connecting a 48-cell module in a series, PV panel assembly will produce a nominal current characterized by the cell module and a voltage proportional to the number of serially interconnected cell modules multiplied by 0.5 to 0.6 V. This will yield a panel that produces a nominal voltage, or 24 V. Figure 1.1 depicts solar cell module interconnection within a PV panel.

After interconnecting photovoltaic panels in a series or parallel, the resulting electrical current and voltages of the solar power string produce desired voltages and currents that are compatible for inverters that convert direct current (DC) to alternating current (AC). Figure 1.2 depicts photovoltaic solar string interconnection.

In typical commercial and industrial solar power system installations, string voltages may vary within 300 to 1000 V and currents of 5 to 10 A. Commercially available PV panels used in most solar power generation systems are available at 12, 24, 48, and 96 V.

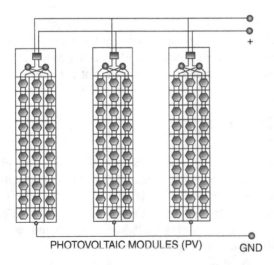

PHOTOVOLTAIC MODULES (PV)   GND

Figure 1.2   **Photovoltaic solar string interconnection.**

Solar Cell Technologies, Manufacturing, and Packaging

Currently, solar cell technologies fall into three main categories: monocrystalline (single-crystal construction), polycrystalline (semi-crystalline), and amorphous silicon thin film.

Currently, solar cells are manufactured from monocrystalline, polycrystalline amorphous, and thin-film-based materials. A more recent undisclosed solar technology, known as organic photovoltaics, is also under commercial development. Each of the technologies has unique physical, chemical, manufacturing, and performance characteristics, and is best suited for specialized applications.

In this section, we will discuss the basic manufacturing principles. In subsequent chapters, we will review the production and manufacturing processes of several solar power cell technologies.

MONOCRYSTALLINE AND POLYCRYSTALLINE SILICON CELLS

The heart of most monocrystalline and polycrystalline photovoltaic solar cells is a crystalline silicon semiconductor. This semiconductor is manufactured by a silicon purification process, ingot fabrication, wafer slicing, etching, and doping, which finally forms a PNP junction that traps photons. This results in the release of electrons within the junction barrier, thereby creating a current flow.[1]

The manufacturing of a solar photovoltaic cell itself is only one part of the process of manufacturing a whole solar panel product. To manufacture a functionally viable product that will last over 25 years requires that the materials be specially assembled, sealed, and packaged to protect the cells from natural climatic conditions and to provide proper conductivity, electrical insolation, and mechanical strength.

One of the most important materials used in sealing solar cells is the fluoropolymer manufactured by DuPont, Elvax. This chemical compound is manufactured from *ethylene vinyl acetate* resin. It is then extruded into a film and used to encapsulate the silicon wafers that are sandwiched between tempered sheets of glass to form the solar panel. One special physical characteristic of the Elvax sealant is that it provides optical clarity while matching the refractive index of the glass and silicon, thereby reducing photon reflections. Figure 1.3 depicts various stages of the monocrystalline solar power manufacturing process. Figure 1.4 depicts frame assembly of a typical solar photovoltaic module.

Another chemical material manufactured by DuPont, Tedlar, is a polyvinyl fluoride (PVF) film. It is extruded with polyester film and applied to the bottom of silicon-based photovoltaic cells as a backplane that provides electrical insolation and protection against climatic and weathering conditions. Other manufacturing companies, such as Mitsui Chemical and Bridgestone, also manufacture products comparable to Tedlar, which are widely used in the manufacture and assembly of photovoltaic panels.

Another important product manufactured by DuPont chemical is Solamet, a silver metallization paste used to conduct electric currents generated by individual solar

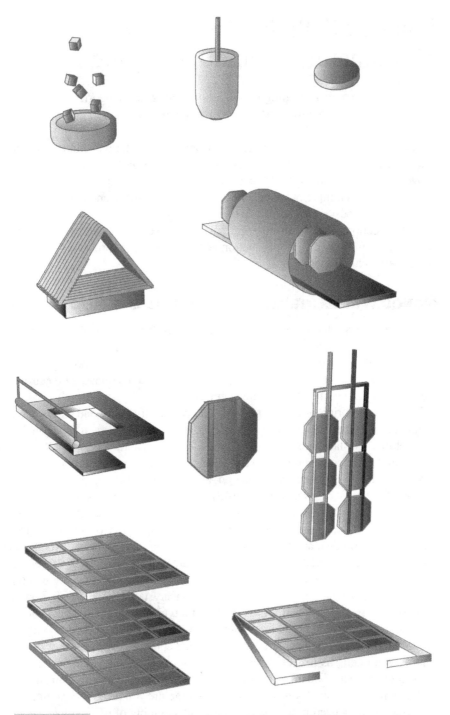

Figure 1.3   Monocrystalline solar panel manufacturing process. Solar panel frame assembly.

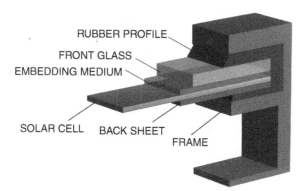

Figure 1.4 **Frame assembly of typical PV module.**

silicon cells within each module. Solamet appears as micronwide conductors that are so thin, they do not block solar rays.

A *dielectric silicon-nitride* product used in photovoltaic manufacturing creates a sputtering effect that enhances silicon, trapping sunlight more efficiently. Major fabricators of polycrystalline silicon are Dow Corning and General Electric in the United States, and Shin-Etsu Handotai and Mitsubishi Material in Japan.

Because of the worldwide silicon shortage, the driving cost of solar cells has become a limiting factor for lowering the manufacturing costs. Currently, silicon represents the largest cost component of manufactured solar panels. To reduce silicon costs, the current industrial trend is to minimize the wafer thickness from 300 to 180 μm.

Note that the process of ingot slicing results in 30 percent wasted material. To minimize this material waste, General Electric is currently developing a technology to cast wafers from silicon powder. Cast wafers thus far have proven to be somewhat thicker and less efficient than the conventional sliced silicon wafers. However, they can be manufactured faster and avoid the 30 percent waste produced by wafer sawing.

Crystalline solar photovoltaic module production In this section, we will review the production and manufacturing process cycle of a crystalline-type photovoltaic module. The product manufacturing process presented here is specific to SolarWorld Industries. However, it is representative of the general fundamental manufacturing cycle for the monosilicon class of commercial solar power modules currently offered by a large majority of manufacturers.

The manufacture of monocrystalline photovoltaic cells starts with silicon crystals, which are found abundantly in nature in the form of flint stone. The word *silicon* is derived from the Latin *silex,* meaning hard stone, which is an amorphous substance found in nature. It consists of one part silicon and two parts oxygen (SiO_2). Silicon (Si) was first produced in 1823 by Jöns Jacob Berzelius, when he separated the naturally occurring ferrous silica (SiF_4) by heat exposure with potassium metal. The commercial production of silicon commenced in 1902 and resulted in an iron-silicon alloy with

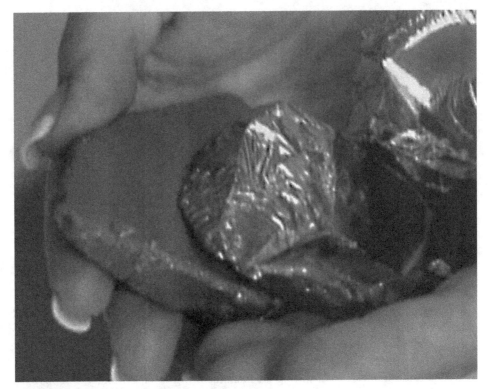

Figure 1.5   Silicon crystals.

an approximate weight of 25 percent iron, which was used in steel production as an effective deoxidant. Today, more than 1 million tons of metallurgical-grade 99 percent pure silicon is used by the steel industry. Approximately 60 percent of the referenced silicon is used in metallurgy, 35 percent in the production of silicones, and approximately 5 percent for the production of semiconductor-grade silicon.[1]

In general, common impurities found in silicon are iron (Fe), aluminium (Al), magnesium (Mg), and calcium (Ca). The purest grade of silicon used in semiconductor applications contains about one part per billion (ppb) of contamination. Silicon purification involves several different types of complex refining technologies such as chemical vapor deposition, isotopic enrichment, and a crystallization process. Figure 1.5 depicts mined silicon crystals prior to the ingot manufacturing process.

Chemical vapor deposition   One of the earlier silicon-refining processes, known as chemical vapor deposition, produced a higher grade of metallurgical silicon. This consisted of a chemical reaction of silicon tetrachloride ($SiCl_4$) and zinc (Zn) under high-temperature vaporization conditions, and yielded pure silicon through the following chemical reaction:

$$SiCl_4 + 2(Zn) \text{ results into } Si + 2\ ZnCl_2$$

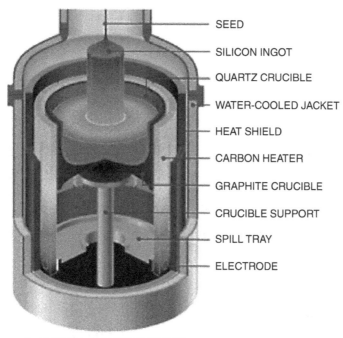

SEED

SILICON INGOT

QUARTZ CRUCIBLE

WATER-COOLED JACKET

HEAT SHIELD

CARBON HEATER

GRAPHITE CRUCIBLE

CRUCIBLE SUPPORT

SPILL TRAY

ELECTRODE

Cz CRYSTAL PULLING FURNACE

Figure 1.6   Czochralski crystallization furnace chamber.

The main problem with this process was that $SiCl_4$ always contained boron chloride (BCl_3) when combined with zinc-produced boron, a serious contaminant. In 1943, a chemical vapor deposition was developed that involved replacing the zinc with hydrogen (H). This gave rise to pure silicon since hydrogen, unlike zinc, does not reduce the boron chloride to boron. Further refinement involved replacing silicon tetrachloride with trichlorosilane ($SiHCl_3$), which is readily reduced to silicon. Figure 1.6 depicts silicon crystallization melting and an ingot manufacturing chamber.

Czochralski crystal growth   In 1916 a Polish metallurgist, Jan Czochralski, developed a technique to produce silicon crystallization which bears his name. The crystallization process involved inserting a metal whisker into molten silicon and pulling it out with increasing velocity. This allowed for the formation of pure crystal around the wire and was thus a successful method of growing single crystals. The process was further enhanced by attaching a small silicon crystal seed to the wire rod. Further production efficiency was developed by attaching the seed to a rotatable and vertically movable spindle. Incidentally, the same crystallization processing apparatus is also equipped with special doping ports where P- or N-type dopants are introduced into the crystal for generation of PN- or NP-junction-type crystals. These are used in the construction of NPN or PNP transistors, diodes, light-emitting diodes, solar cells, and virtually all high-density, large-scale integrated circuitry used in electronic technologies.[1]

Figure 1.7   A formed silicon ingot cylinder.

The chemical vaporization and crystallization process described here is energy-intensive and requires a considerable amount of electric power. To produce purified silicon ingots at a reasonable price, silicon ingot production plants are located within the vicinity of major hydroelectric power plants, which produce an abundance of low-cost hydroelectric power. Ingots produced from this process are either circular or square in form, and they are cleaned, polished, and distributed to various semiconductor manufacturing organizations. Figure 1.7 is a photograph of a formed silicon ingot cylinder.

SOLAR PHOTOVOLTAIC CELL PRODUCTION

The first manufacturing step in the production of photovoltaic modules involves incoming ingot inspection, wafer cleaning, and quality control. After completing the incoming process, in a clean-room environment the ingots are sliced into millimeter-thick wafers and both surfaces are polished, etched, and diffused to form a PN junction. After being coated with antireflective film, the cells are printed with a metal-filled paste and fired at high temperature. Each individual cell is then tested for 100 percent functionality and is made ready for module assembly.

The photovoltaic module production process involves robotics and automatic controls in which a series of robots assemble the solar cells step by step, laying the modules, soldering the cells in a predetermined pattern, and then laminating and framing the assembly as a finished product. After framing is completed, each PV module is tested under artificial insolation conditions, and the results are permanently logged and serialized. The last step of production involves a secondary module test, cleaning, packaging, and crating. In general, the efficiency of the PV modules produced by this technique ranges from 15 to 18 percent.

Photovoltaic module life span and recycling   To extend the life span of solar power photovoltaic modules, PV cell assemblies are laminated between two layers of protective covering. In general, the top protective cover is constructed from $1/4$- to $5/8$-in tempered glass, and the lower protective cover either from a tempered glass or a hard plastic material. A polyurethane membrane is used as a gluing membrane, which holds the sandwiched PV assembly together. In addition to acting as the adhesive agent, the membrane hermetically seals the upper and lower covers, preventing water penetration or oxidation. As a result of hermetical sealing, silicon-based PV modules are able to withstand exposure to harsh atmospheric and climatic conditions.

Even though the life span of silicon-based PV modules is guaranteed for a period of at least 20 years, in practice it is expected that the natural life span of the modules will exceed 45 years without significant degradation.

In order to minimize environmental pollution, SolarWorld has adopted a material recovery process whereby obsolete, damaged, or old PV modules (including the aluminium framing, tempered glass, and silicon wafers) are fully recycled and reused to produce new solar photovoltaic modules. Figure 1.8 depicts a solar panel lamination robotic arm. Figure 1.9 depicts a solar panel inspection station.

Figure 1.8   **A robotic solar power laminator.** *Photo courtesy of Martifer.*

Figure 1.9   A solar panel inspection station. *Photo courtesy of Martifer.*

CONCENTRATOR TECHNOLOGIES

Concentrator-type solar technologies are a class of photovoltaic systems that deploy a variety of lenses to concentrate and focus solar energy on semiconductor material used in the manufacture of conventional PV cells.

The advantage of these types of technologies is that for a comparable surface area of silicon wafer, it becomes possible to harvest considerably more solar energy. Because silicon wafers used in the manufacture of photovoltaic systems represent a substantial portion of the product cost, by using relatively inexpensive magnifying concentrator lenses it is possible to achieve a higher-efficiency product at a lower cost than conventional PV power systems.

One of the most efficient solar power technologies commercially available for large-scale power production is a product manufactured by Amonix. This concentrator technology has been specifically developed for ground installation only and is suitable only for solar farm-type power cogeneration. The product efficiency of this unique PV solar power concentrator technology, under field test conditions in numerous applications in the United States (determined by over half a decade of testing by the Department of Energy, Arizona Public Service, Southern California Edison, and the University of Nevada, Las Vegas), has exceeded 26 percent. This is nearly twice that of comparable conventional solar power systems. Currently, Amonix is in the process of developing a multijunction concentrating cell that will augment solar power energy production efficiency to 36 percent.

THIN-FILM SOLAR CELL TECHNOLOGY

The last decade has witnessed tremendous progress in the science and technology of thin-film silicon that uses amorphous and nanocrystalline photovoltaic materials. The key factor that will determine the wide-scale acceptance of thin-film solar photovoltaic products is attributed to the lower cost of solar electricity achieved using this technology. Two factors, namely efficiency of solar modules and throughput of production equipment, will play a key role in making the thin-film technology a viable cost-effective way of generating inexpensive solar power energy.

In the past five years, the world market for photovoltaic has been growing at an annual rate of about 40 percent. In spite of the economic downturn in the global economy, the shipment for 2009 exceeded 6000 MW, a gain of 10 percent over 2008. Until recently, the PV technology mix was dominated by single-crystal and polycrystalline silicon PV technologies. Thin-film silicon amorphous and nanocrystalline technologies accounted for about 3 percent of the total market. The low material cost and ease of large-scale manufacturing of thin-film technologies have resulted in further expansion in thin-film silicon photovoltaic system deployment. Current projections are that thin-film silicon technology may within the next few years capture up to 30 percent of the global share of the PV markets.

The key driver for large-scale deployment of PV is the levelized cost of electricity that depends on the energy yield in terms of kWh produced per kW installed and the installed system cost. Thin-film silicon alloy solar cells are less sensitive to temperature than the crystalline counterpart, and they produce more kWh/kW under real-world conditions. Flexible solar laminates discussed further in this chapter have lower installation costs for rooftop applications that reduce the overall system cost. In order to reduce module cost, thin-film technologies must strive to increase the power-output efficiency while lowering the cost of production. As in competing technologies, success of thin-film technologies will be to achieve solar electricity production at grid parity.

The core material of thin-film solar cell technology is amorphous silicon. This technology, instead of using solid polycrystalline silicon wafers, uses *silane gas,* which is a chemical compound that costs much less than crystalline silicon. Solar-cell manufacturing involves a lithographic-like process in which the silane film is printed on flexible substrates, such as stainless steel or *Plexiglas* material, on a roll-to-roll process.

Silane (SiH_4), also called *silicon tetrahydride, silicanel,* and *monosilane,* is a flammable gas with a repulsive odor. It does not occur in nature. Silane was first discovered in 1857 by F. Wohler and H. Buffy by reacting hydrochloric acid (HCL) with an Al-Si alloy. Silane is principally used in the industrial manufacture of semiconductor devices for the electronic industry. It is used for *polycrystalline deposition,* interconnection, masking, growth of epitaxial silicon, chemical vapor deposition of silicon diodes, and the production of amorphous silicon devices such as photosensitive films and solar cells.

Even though thin-film solar power cells have about 5 to 9 percent efficiency in converting sunlight to electricity when compared to the 15 to 20 percent efficiency of polysilicon products, they have an advantage in that they do not need direct sunlight to produce electricity. As a result, they are capable of generating electric power over a longer period of time.

United Solar Ovonic (USO) technology   The three key components of USO technology consist of roll-to-roll production, multijunction thin film silicon solar cell structure, and flexible solar laminates. As shown in Fig. 1.10 innovation has played a key role in taking the laboratory results to production. In 1981, USO built its first roll-to-roll machine where the web was transported through just a single chamber. Subsequently, a pilot plant producing same-gap amorphous silicon (a-Si:H/a-SiH) double-junction cells was installed in 1991. With advances in the laboratory demonstrating the advantages of the triple-junction cell technology, in 1996 the company built its first triple-junction processor, which had an annual production capacity of 5 MW. Recognizing the advantage of flexible PV products for the rooftop market, the company introduced its first building-integrated photovoltaic (BIPV) product in 1997. With increasing acceptance of its products, production capacity has been augmented to 150 MW of annually.

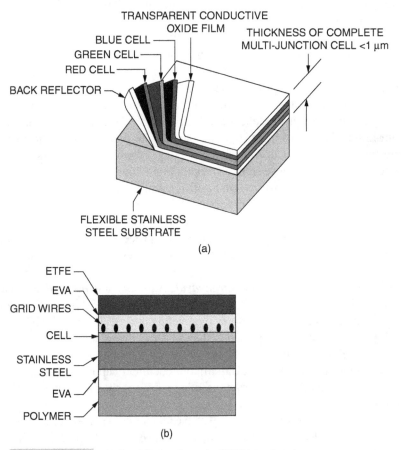

Figure 1.10   United Solar Ovonic (USO) technology.

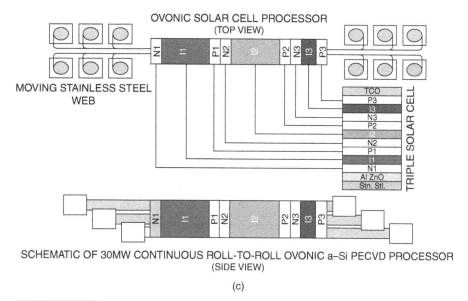

SCHEMATIC OF 30MW CONTINUOUS ROLL-TO-ROLL OVONIC a–Si PECVD PROCESSOR
(SIDE VIEW)

(c)

Figure 1.10   *(Continued)*

According to USO, the commercial laminate fabrication process consists of three basic steps. In the first is a proprietary roll-to-roll deposition technology, which is used to deposit an amorphous silicon/amorphous silicon-germanium/amorphous silicon-germanium (a-Si:H/a-SiGe:H/a-SiGe:H) forming a triple-junction solar cell on a flexible and lightweight stainless steel substrate, using a radio frequency (RF) glow-discharge system. Six rolls of stainless steel, each 1.5 miles long, are loaded into the triple-junction processor, and 9 miles of solar cells are produced in 62 hours. The bottom subcell absorbs the red light, the middle cell the yellow/green light, and the top cell the blue light.

An Al/ZnO back reflector at the bottom of the solar cell improves the reflectivity and texture of the substrate, which results in improved light trapping and enhancement in conversion efficiency. The second fabrication step consists of cutting the roll of solar cell into smaller pieces, and processing them for cell delineation, short passivation, and top and bottom current-collection bus bar application. The third and final step consists of interconnecting the individual solar cells into a series string and encapsulating the string in UV-stabilized and weather-resistant polymers to form the final product. Figure 1.10 shows schematics of (*a*) spectrum-splitting triple-junction solar cell structure, (*b*) cross section of the module, and (*c*) roll-to-roll a-Si:H alloy processor and triple-junction structure formation.

The UNI-SOLAR laminate shown in Fig. 1.11 offers unique qualities and advantages compared to conventional sold crystalline PV technologies that include being lightweight and flexible, with an adhesive and a release paper at the back that can be easily adhered and integrated with the roof. This reduces the installation cost significantly. Figure 1.11 offers a comparison between conventional rigid solar panels and USO flexible solar laminate.

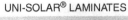

CONVENTIONAL SOLAR CELLS   UNI-SOLAR® LAMINATES

Figure 1.11   Conventional rigid solar panels versus USO flexible solar laminate.

The current laminate has a total-area efficiency of 6.7 percent and an aperture-area efficiency of 8.2 percent. The manufacturer claims to have an aggressive plan that will increase the aperture-area efficiency first to 10 percent and later to 12 percent within the next several years.

Measures considered for improving the product efficiency in the near future include use of two parallel approaches. One of the methods used will be replacement of the Al/ZnO back reflector (BR) with a superior Ag/ZnO BR; the other will be use of an improved deposition process to develop high-quality a-Si:H, a-SiGe:H, and nanocrystalline silicon (nc-Si:H) alloys. The improved deposition process is expected to result in higher throughput from the deposition machine as well.

Back reflector   To improve photon absorption, the thin-film cell structure is deposited on the back surface referred to as the specular textured surface, which reflects the photons back, thus improving photon absorption (Fig. 1.12a and b). Random back scattering is increased by the path length that photons travel. The back reflection

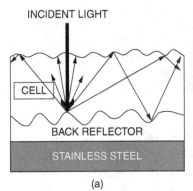
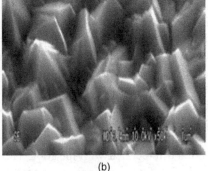

(a)   (b)

Figure 1.12   (a) Schematic diagram of a textured back reflector. (b) Atomic microscope photographs of back reflectors' texture.

property of the material is assigned a refractive index for various types of materials. For instance, the refractive index of a-Si:H alloy is about 25 reflections.

Other methods of efficient light trapping include the use of optical confinement by a grating structure that uses nanoparticles.

Laminate manufacturing improvements   As mentioned earlier, the current USO thin-film product has an aperture-area efficiency of 8.2 percent and a total-area efficiency of 6.7 percent. In order to improve product efficiency, the company plans in the near future to reduce the inactive area in our laminate to obtain higher total-area efficiency for the same aperture area. The introduction of superior background reflector in the production line will also increase the efficiency. Figure 1.13 depicts a cell footprint

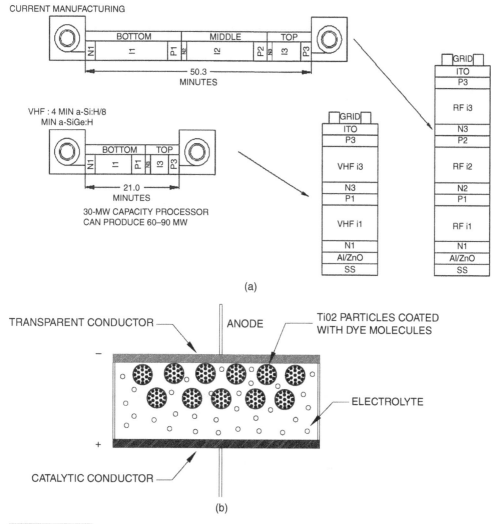

(a)

(b)

Figure 1.13   Reduction in cell footprint using high-frequency deposition method.

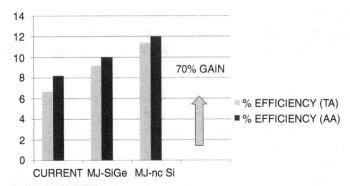

Figure 1.14   Near-term improvement in total-area (TA) and aperture-area (AA) efficiency.

TABLE 1.1 CELL PARAMETERS NEEDED TO OBTAIN INITIAL 25 PERCENT CELL EFFICIENCY FOR DIFFERENT STRUCTURES				
	CURRENT	**OPTION 1**	**OPTION 2**	**OPTION 3**
Cell structure	a-Si/a-SiGe/ nc-Si	a-Si/a-Si/ nc-Si	a-Si/a-SiGe/ nc-Si	a-Si/nc-Si/ nc-Si
V_{oc} (V)	2.24	2.50	2.30	2.15
J_{sc} (mA/cm^2)	9.13	11.8	12.3	12.6
FF	0.75	0.85	0.85	0.85
Eff (%)	15.4	25.1	24.0	23.0

using the high-frequency deposition method. Figure 1.14 is bar chart representation of USO near-term improvement in total-area and aperture-area efficiency.

UNI-SOLAR has already demonstrated a small-area initial cell efficiency of 15.4 percent using a triple-junction structure incorporating nc-Si:H. Table 1.1 represents cell parameters that must be achieved to obtain 25 percent cell efficiency for different cell structures. Some of the major improvements needed are in fill factor and short-circuit current density; these are expected to be achieved by deposition methods, such as rapid thermal annealing, epitaxy using hot-wire CVD, hollow cathode discharge, etc. Use of a multijunction cell fabrication approach to increase the fill factor from 0.75 to 0.85 could be a great challenge.

Reaching Grid Parity

The cost of solar electricity has decreased significantly over the last decade. The incentives offered by several countries for the deployment of PV resulted in

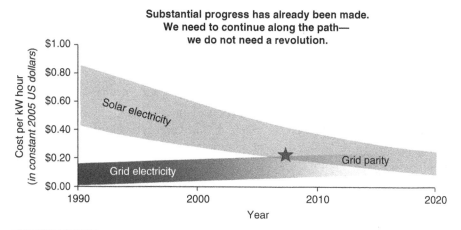

Figure 1.15 **Path toward grid parity.** *Source: Solar America Initiative*

addition of manufacturing capacity. This allowed the industry to take advantage of economy of scale. In the past decade, there have been numerous innovations in every PV technology, which have resulted in cost reduction across the entire value chain from module to deployment. Figure 1.15 represents a chart that was first provided by DOE in 2007 showing the historical decline in the price of solar electricity and the goals to achieve.[2] The installed cost of rooftop systems in the United States today is between $3.50 and $4/W, and using the Solar Advisor Model[3] (SAM discussed elsewhere in this book) provided by DOE and the investment tax credit that is available today, solar electricity production costs have already reached lower than 10 c/kWh in sunny places like California and Arizona.

SOLAR INTEGRATED TECHNOLOGIES

Solar Integrated Technologies has developed a flexible solar power technology specifically for use in roofing applications. The product meets the unique requirements of applications in which the solar power cogeneration also serves as a roofing material. This particular product combines solar film technology overlaid on a durable single-ply polyvinyl chloride roofing material. It offers an effective combined function as a roof covering and solar power cogeneration that can be readily installed on a variety of flat and curved roof surfaces. Even though the output efficiency of this particular technology is considerably lower than that of conventional glass-laminated mono- or polycrystalline silicon photovoltaic systems, its unique pliability and dual-function (it is used as both a solar power co-generator and a roof-covering system) makes it indispensable in applications in which roof material replacement and coincidental renewable energy generation become the only viable options.

Until the introduction of this product, the installation of solar panels on large-area flat or low-slope roofs was limited because of the heavy weight of traditional, rigid crystalline solar panels. This lightweight solar product overcomes this challenge and eliminates any related roof penetrations.

The Solar Integrated Technologies BIPV roofing product is installed flat as an integral element of the roof and weighs only 12 oz/ft^2, allowing installation on existing and new facilities. Application of this technology offsets electric power requirements of buildings, and where permitted in net metering applications, excess electricity can be sold to the grid. Figure 1.16 depicts Solar Integrated Technologies product configuration.

In addition to being lightweight, this product uses unique design features to increase the total amount of sunlight converted to electricity each day, including better performance in cloudy conditions.

Both the single-ply PVC roof material and BIPV solar power system are backed by an extensive 20-year package operations and maintenance service warranty. Similar to all solar power cogeneration systems, this technology also offers a comprehensive real-time data acquisition and monitoring system, whereby customers are able to monitor exactly

Figure 1.16   Manufacturing process of film technology. *Photo courtesy of Integrated Solar Technologies, Los Angeles, CA.*

the amount of solar power being generated with real-time metering for effective energy management and utility bill reconciliation.

Custom-Fabricated BIPV Solar Cells

Essentially, BIPV is a term commonly used to designate the custom-made assembly of solar panels specifically designed and manufactured to be used as an integral part of building architecture. These panels are used as architectural ornaments, such as windows, building entrance canopies, solariums, curtain walls, and architectural monuments.

The basic fabrication of BIPV cells consists of laminating mono- or polycrystalline silicon cells, which are sandwiched between two specially manufactured tempered-glass plates. This is referred to as a glass-on-glass assembly. A variety of cells arranged in different patterns and spacing are sealed and packaged in the same process as described previously in this chapter. Prefabricated cell wafers used by BIPV fabricators are generally purchased from major solar power manufacturers.

The fabrication of BIPV cells involves complete automation, whereby the entire assembly is performed by special robotic equipment that can be programmed to implement solar cell configuration layout, lamination, sealing, and framing in a clean-room environmental setting without any manual labor intervention. Some solar power fabricators, such as Sharp Solar of Japan, offer a limited variety of colored and transparent photovoltaic cells for aesthetic purposes. Cell colors, which are produced in deep marine, sky blue, gold, and silverfish brown, usually are somewhat less efficient and are manufactured on an on-demand basis.

Because of their lower performance efficiency, BIPV panels are primarily used in applications in which there is the presence of daylight, such as in solariums, rooms with skylights, or sunrooms. In these cases, the panels become an essential architectural requirement. Figures 1.17 and 1.18 depict custom BIPV modules manufactured by Atlantis Energy Systems.

Figure 1.17   **BIPV modules in a window glazing.**
Photo courtesy of Atlantis Energy Systems.

Figure 1.18   BIPV modules. *Photo courtesy of Atlantis Energy Systems.*

Polycrystalline Photovoltaic Solar Cells

In the polycrystalline process, the silicon melt is cooled very slowly under controlled conditions. The silicon ingot produced in this process has crystalline regions, which are separated by grain boundaries. After solar cell production, the gaps in the grain boundaries cause this type of cell to have a lower efficiency compared to that of the monocrystalline process described in this text. Despite the efficiency disadvantage, a number of manufacturers favor polycrystalline PV cell production because of the lower manufacturing cost.

Amorphous Photovoltaic Solar Cells

In the amorphous process, a thin wafer of silicon is deposited on a carrier material and doped in several process steps. An amorphous silicon film is produced by a method similar to the monocrystalline manufacturing process and is sandwiched between glass plates, which form the basic PV solar panel module.

Even though the process yields relatively inexpensive solar panel technology, it has disadvantages such as a larger installation surface, lower conversion efficiency, and inherent degradation during the initial months of operation that continues over the life span of the PV panels.

The main advantages of this technology are a relatively simple manufacturing process, lower manufacturing costs, and lower production energy consumption.

Thin-Film Cadmium Telluride Cell Technology

In this process, thin crystalline layers of cadmium telluride (CdTe, of about 15 percent efficiency) or copper indium diselenide (CuInSe$_2$, of about 19 percent efficiency) are deposited on the surface of a carrier base. This process uses very little energy and is very economical. It has simple manufacturing processes and relatively high conversion efficiencies.

Gallium-Arsenide Cell Technology

This manufacturing process yields a highly efficient PV cell, but as a result of the rarity of gallium deposits and the poisonous qualities of arsenic, the process is very expensive. The main feature of gallium-arsenide (GaAs) cells, in addition to their high efficiency, is that their output is relatively independent of the operating temperature, and is primarily used in space programs.

Dye-Sensitized Solar Cells

Dye-sensitized solar cells (DSC) are a class of solar cells which are formed by placing a semiconductor between a photo-sensitized anode and an electrolyte. They have photochemical properties that provide charge separation by absorbing solar energy. This class of cells is also referred to as *Grätzel cells,* the name of the original inventor. Figure 1.19 depicts a dye-sensitized solar epitaxial configuration.

BASIC PRINCIPLES OF DYE-SENSITIZED SOLAR CELLS

Dye-sensitized solar cells essentially separate the two functions provided by silicon in a conventional semiconductor type solar cell design. Under normal conditions, the silicon in a semiconductor cell acts as both the source of photoelectrons and also forms the potential barrier that allows for the separation of the charges that create current.

However, in the dye-sensitized solar cell the semiconductor is used for charge separation only; the photoelectrons are provided from a separate *photosensitive dye.*

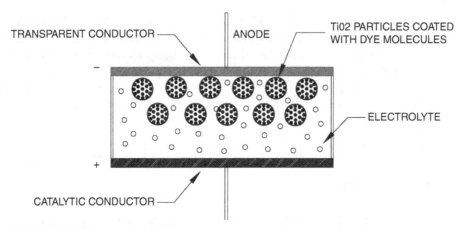

TRANSPARENT CONDUCTOR

ANODE

Ti02 PARTICLES COATED
WITH DYE MOLECULES

ELECTROLYTE

CATALYTIC CONDUCTOR

Figure 1.19   **Dye-sensitized solar cell epitaxial configuration.**

Furthermore, charge separation is not provided only by the semiconductor, but works in concert with a third element of the cell (an electrolyte that is kept in contact with both the semiconductor and the dye). Figure 1.20 depicts a dye-sensitized flexible solar power cell manufactured by Nano Solar.

Since the dye molecules are quite small in size, in order to effectively capture a reasonable amount of incoming solar rays or sunlight, the layer of dye molecules are made much thicker than the molecules themselves. To resolve this problem, a *nano-material* in the form of a scaffold is used to hold or bundle large numbers of the dye

Figure 1.20   **A dye-sensitized flexible solar power cell manufactured by Nano Solar.** *Photo courtesy of Nano Solar.*

molecules in a 3D matrix. Bundles of large numbers of molecules thus provide a large cell surface area. Currently, this scaffolding is fabricated from semiconductor material that effectively serves double duty.

A dye-sensitized solar cell, described above, has three primary parts. The top portion of the cell, termed anode, is constructed from a glass coated with a layer of a transparent material made of fluorine-doped tin oxide (SnO_2:F). On the back is a thin layer of *titanium dioxide* (TiO_2) that forms into a highly porous structure with an extremely large surface area.

The titanium dioxide (TiO_2) plate is in turn immersed in a mixture of a photosensitive dye substance, called *ruthenium-polypyridine*, and a solvent. After immersing the film in the dye solution, a thin layer of the dye is covalently bonded to the surface of the TiO_2. Another layer of electrolyte, referred to as iodide, is spread thinly over the above conductive sheets. Finally, a backing material, typically a thin layer of platinum metal, is placed as the lowest layer. Upon the formation of these layers, the front and back parts are then joined and sealed together to prevent the electrolyte from leaking.

Although this technology makes use of some costly materials, the amounts used are so small that they render the product quite inexpensive when compared to the silicon needed for the fabrication of conventional semiconductor cells. For instance, titanium oxide (TiO_2) is an inexpensive material widely used as a white paint base.

Dye-sensitized solar cells operate when sunlight enters the cell through the transparent SnO_2F, striking the dye on the surface of the TiO_2. Highly energized photons striking the dye are absorbed and create an excited state in the dye, which, in turn, injects electrons into the conduction band of the TiO_2 (which are moved by a chemical diffusion gradient to the clear *anode* located on top). In the meantime, the dye molecule loses an electron, which could result in the decomposition of the substance if another electron is not provided.

In the process, the dye strips one electron from iodide in the electrolyte below the TiO_2, oxidizing it into a substance called tri-iodide. This reaction takes place very quickly when compared to the time that it takes for the injected electron to recombine with the oxidized dye molecule, thus preventing the recombination reaction that would cause the solar cell to short-circuit. The tri-iodide then recovers its missing electron by mechanically diffusing to the bottom of the cell, where the counter electrode reintroduces the electrons after flowing through the external circuit load. Figure 1.21 depicts multijunction solar cell epitaxial layer.

In this measurement, the electrical power is expressed as the product of short circuit current (Isc) and open voltage (Voc). Another solar cell efficiency measurement, defined as q*uantum efficiency,* is used to compare the chance of one photon of impacted solar energy resulting in the creation of a single electron.

In quantum efficiency terms, dye-sensitized solar cells (DSSc) are extremely efficient. Because of their nanostructured configurations, there is a strong chance that a photon will be absorbed. Therefore, they are considered highly effective in converting solar rays into electrons.

Most of the power conversion losses in DSSc technology result from conduction losses in the TiO_2 and the clear electrode, as well as optical losses in the front electrode. Their overall quantum efficiency is estimated to be approximately 90 percent.

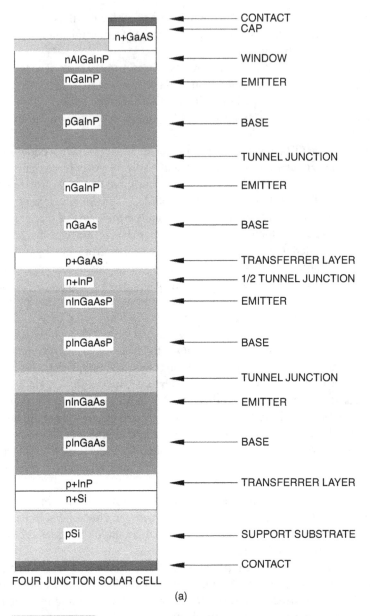

FOUR JUNCTION SOLAR CELL

(a)

Figure 1.21   Multijunction solar cell epitaxial layer.

The maximum voltage generated by DSSc is simply the difference between the Fermi level of the TiO_2 and the potential of the dye electrolyte, which is about 0.7 V total (V_{oc}). This is slightly higher than semiconductor-based solar cells, which have a maximum voltage of about 0.6 V.

Even though DSSc are highly efficient in turning photons into electrons, it is only those electrons with enough energy to cross the TiO_2 band gap that result in current

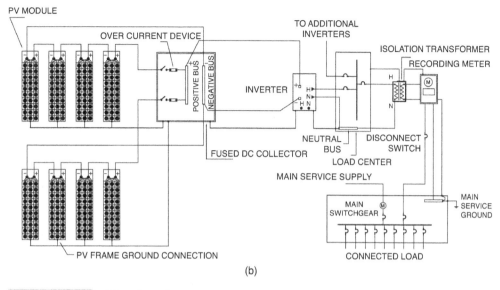

Figure 1.21   (*Continued*)

being produced. Also, since the DSSc band gap is slightly larger than it is in silicon, fewer photons in sunlight become usable for generating solar power.

Furthermore, the electrolyte in DSSc limits the speed at which the dye molecules can regain their electrons and become available for a renewed cycle of photo-excitation.

As a result of these factors, the maximum current output limit for a DSSc is 20 mA/cm² compared to 35 mA/cm² generated by silicon-based solar cells. This in turn is translated into a conversion efficiency of about 11 percent. Semiconductor-based solar cells, on the other hand, operate between 12 and 15 percent efficiency. In comparison, flexible thin-film cells operate at a maximum of 8 percent efficiency.

Another approach used to enhance the efficiency of DSSC involves injecting an electron directly into the TiO_2, where the electron is boosted within the original crystal. In comparison, the injection process used in the DSSc does not introduce a hole into the TiO_2, only an extra electron. Although in doing so, electrons have a better possibility of recombining when back in the dye. The probability that this will occur is so low that the rate of electron-hole recombination efficiency becomes insignificant.

In view of this characteristic of low losses, they can perform more efficiently under cloudy skies, whereas traditional designs would cut out power production at lower limits of illumination. This operational characteristic of DSSC renders the cell ideal for indoor applications.

A significant disadvantage of DSSc technology is the use of a liquid electrolyte, which has temperature stability problems. At low temperatures, the electrolyte can freeze, ending power production and potentially leading to physical damage. Higher temperatures can also cause the liquid to expand, making sealing of the panels very difficult.

ADVANCEMENTS IN DYE-SENSITIZED SOLAR CELL TECHNOLOGY

The early experimental version of DSSc had a narrow bandwidth that functioned in the high-frequency end of the solar UV and blue spectrum. Subsequently, due to the use of an improved dye electrolyte, a wider response to the low-frequencies range of red and infrared was achieved.

At present, with the use of a special dye with a deep brown-black color (referred to as black dye), the conversion rate of photons into electrons has significantly improved to almost 90 percent. There is only a 10 percent loss attributed to optics and the top electrode.

A critically significant characteristic of the black dye is that over millions of cycles of simulated exposure to solar irradiance, the output efficiency of DSSc has shown outstanding performance with no discernable decrease in efficiency.

In recent tests using an improved electrolyte, the thermal performance of the solar cell has been pushed to 60°C with remarkable conversion efficiency.

An experiment recently conducted in New Zealand used a wide variety of organic dyes such as *porphyrin* (a natural *hemoprotein* present in *hemoglobin*) and chlorophyll to achieve an efficiency of 7.1 percent.

Currently, dye-sensitized solar cell technology is still in the infantile stages of its production cycle. It is expected that in the near future, with the use of newer dye electrolytes and quantum dots, there will be significant improvement in efficiency gains.

Multijunction Photovoltaic Cells

Multijunction solar cells were first developed and deployed for satellite power applications, where the high cost was offset by the net savings offered by the higher efficiency.

Multijunction photovoltaic cells are a special class of solar or photovoltaic cells that are fabricated with multiple levels of stacked, thin film semiconductor PN junctions. Their production technique is referred to as *molecular beam epitaxy* or the *metal-organic vapor phase process*. Each type of semiconductor is designed with a characteristic valance band gap that allows for the absorption of a certain bandwidth or spectrum of solar electromagnetic radiation.

In a single-layered or band-gap solar cell, valance-band efficiency is limited because of the inability of the PN junction to absorb a broad range of electromagnetic rays or photons in the solar spectrum. Photons below the band gap in the blue spectrum either pass through the cell, or because of molecular agitation are converted within the material. Energy in photons above the band gap in the red spectrum are also lost, since only the energy necessary to generate the hole-electron pair is utilized. The remaining energy is converted into heat. Multijunction solar cells, which have multiple layers and therefore junctions with several band gaps, cause different portions of the solar spectrum to be converted by each junction at a greater efficiency.

MULTIPLE-JUNCTION SOLAR CELL CONSTRUCTION

Multijunction photovoltaic cells use many layers of film deposition, or epitaxy. By using differing alloys within the eighth column of the periodic table, the band gap of each layer is tuned to absorb a specific band of solar electromagnetic radiation. The efficiency of multiple-junction solar cells is achieved by the precise alignment of respective superimposed band gaps.

To achieve maximum output efficiency all epitaxial layers that are in series are optically aligned from top to bottom, such that the first junction receives the entire spectrum. Photons above the band gap of the first junction are absorbed in the first layer (red spectrum photons). Green and yellow photons below, which pass through the first layer, are absorbed by the second band gap, and finally, the third band gap absorbs the high-energy blue spectrum photons.

Most commercialized cells utilize a tandem PN electrical connection, which allows series cumulative or composite current output through positive and negative terminals. An inherent design constraint of tandem cell configuration is that in series material ohmic resistance limits connection the current through each junction. Since the point current of each junction is not the same, the efficiency of the cell is reduced.

MULTIJUNCTION SOLAR CELL MATERIALS

In general, most multijunction cells are categorized by the substrate used for cell manufacture. Depending on the band-gap characteristics, multijunction solar cell substrates are constructed from various epitaxial layers that make use of different combinations of semiconductors, metals, and rare Earth metal alloys such as germanium, gallium arsenide, and indium phosphide.

Gallium arsenide substrate   Twin junction cells with indium gallium phosphide and gallium arsenide are constructed on gallium arsenide wafers. Alloys of $In_{0.5}Ga_{0.5}P$ through $In_{0.53}Ga_{0.47}P$ are used as high-band-gap alloys. This alloy range provides band gaps with voltages ranging from 1.92 to 1.87 eV (eV stands for electronvolt). Gallium arsenide (GaAs), on the other hand, allows for the fabrication of junctions with a lower band gap of 1.42 eV.

Since the solar spectrum has a considerable quantity of photons in its lower regions than does the GaAs band gap, a significant portion of the energy is lost as heat, limiting the efficiency of the GaAs substrate cells.

Germanium substrate   Triple-junction cells, consisting of indium gallium phosphide, gallium arsenide, (or indium gallium arsenide), and germanium are fabricated on germanium wafers.

Much like GaAs, because of the large band-gap difference between GaAs (1.42 eV) and Ge (0.66 eV), the current match becomes very poor. As a result, current throughput suffers limiting output efficiency. Currently, efficiencies for InGaP, GaAs, and Ge cells are within the 25 to 32 percent range. In a recent laboratory test of cells, using additional junctions between the GaAs and Ge junction produced efficiencies that exceeded 40 percent.

Indium phosphide substrate   Indium phosphide is also used as a substrate to fabricate cells with band gaps between 1.35 and 0.74 eV. Indium phosphide has a band gap of 1.35 eV. Indium gallium arsenide ($In_{0.53}Ga_{0.47}As$), lattice matched to indium phosphide, has a band gap of 0.74 eV. An alloy composed of three elements—indium, gallium arsenide, and phosphide—has resulted in the fabrication of optically matched lattices that perform with greater efficiency.

Recently, multijunction cell efficiencies have been improved through the use of concentrator lenses. This has resulted in significant improvements in solar energy conversion, creating price reductions that have made the technology competitive with silicon flat-panel arrays.

MULTIJUNCTION CELL TECHNOLOGY

This process employs two layers of solar cells, such as silicon (Si) and GaAs components, one on top of another, to convert solar power with higher efficiency. The staggering of two layers allows the trapping of a wider bandwidth of solar rays, thus enhancing the solar cell solar energy conversion efficiency, as shown in Fig. 1.21.

Even though these types of cells have lower conversion efficiency than the existing thin-film and solid-state semiconductor-based technologies, in the near future they are expected to offer a price–performance ratio large enough to replace a significant amount of electricity generated by fossil fuels (because of the lower cost of materials and inexpensive production).

Polymer Solar Cells

Polymer solar cells, also referred to as plastic cells, are a relatively new technology that covert solar energy to electricity through the use of polymer materials. This class of solar cells, unlike conventional semiconductors, is based on the photovoltaic system technologies described above. Neither silicon nor any alloy material is used in their fabrication.

At present, polymer solar cells are being researched by a number of universities, national laboratories, and several companies around the world.

Compared to silicon-based devices, polymer solar cells are lightweight, biodegradable, and inexpensive to fabricate. The use of polymer substances renders the cells flexible, facilitating greater design possibilities and diverse applications.

Because Fullerene (a plastic-based material) is inexpensive and readily available, solar cell manufacturing is extremely easy to mass produce at a cost of approximately one-third that of traditional silicon solar cell technology.

Some potential uses of polymer solar cells include applications in a wide variety of commercial products, including small televisions, cell phones, and toys.

Comparative Analysis

In order to compare differences between existing solid-state semiconductors and dye-synthesized solar cells, it would be important to review the construction and operational characteristics of both technologies.

As discussed in earlier chapters, conventional solid-state semiconductor solar cells are formed from two doped crystals, one doped with an impurity that forms a slightly negative bias (which is referred to as an N-type semiconductor and has a free electron), and the other doped with an impurity that provides a slight positive bias (which is referred to as P-type semiconductor and lacks free electrons). When placed in contact to form a PN junction, some of the electrons in the N-type portion will flow into the P-type to fill in the gap, or electron hole.[1]

Eventually, enough electrons flow across the boundary to equalize what is called the *Fermi levels* of the two materials. The resulting PN junction gives rise to the location where charge carriers are depleted or accumulated on each side of the interface. This transfer of electrons produces a potential barrier for electron flow, which typically has a voltage of 0.6 to 0.7 V.

Under direct exposure to solar rays, photons in the sunlight strike the bound electrons in the P-type side of the semiconductor and elevate their energy, a process that is referred to as *photo-excitation*.

HIGH-ENERGY CONDUCTION BAND

In a high-energy conduction band the electrons in the conduction band are prompted to move about the silicon, giving rise to electron flow, or electricity. When electrons flow out of the P-type and into the N-type material, they lose energy while moving or circling through an external circuit. Eventually, when they enter back into the P-type material, they recombine with the valence-band (at lower-energy potential) hole they left behind, therefore permitting sunlight energy to be converted into electrical current.

Some of the disadvantages of the conventional solar power technology discussed in this chapter include a large bandgap difference in energy between the valence and conduction bands. The *bandgap* creates a situation that only photons with that amount of energy can overcome, the potential difference that contributed to producing a current.

Another shortcoming of conventional semiconductor-based solar power technology is that higher-energy photons, at the blue and violet ends of the solar-ray spectrum, have more than enough energy to cross the bandgap. Even though a small fraction of this energy is transferred into the electrons, a much larger portion of it becomes wasted as heat, reducing cell efficiency because of ohmic drop.

Another issue is that in order to have a reasonable chance of capturing a photon in the P-type layer, it has to be fairly "thick." This in turn promotes the recombination of electrons and holes within the gap material before reaching the PN junction. These phenomena produce an upper limit on the efficiency of silicon solar cells, which is currently about 12 to 15 percent for production-type solar cells, and somewhat closer to 40 percent under ideal laboratory test conditions.

Besides these physical impediments, the most important setback of the semiconductor solar cells is their production cost, since production requires a thick layer of silicon in order to have reasonable photon capture rates. That necessary silicon is a somewhat rare and very expensive commodity.

Some of the measures undertaken to reduce semiconductor-based solar cells have resulted in the development of thin-film approaches, which involve the use of P and N type semiconductor power paste in a type of lithographic printing process. As of today, because of the loss of electrons and molecular decomposition, film-based solar power technologies have had limited applications.

Another design approach that shows great improvement in efficiency and promise is the *multijunction* approach. This process involves stacking several layers of junctions of solar cells, which capture a much wider spectrum of solar energy. However, this type of cell is very expensive to produce and will mainly be marketed for large commercial applications.

Bio-Nano Generators

Bio-nano generators are biological cells that function like a fuel cell at a nanoscale, molecular level. Bio-nano cells are essentially electrochemical devices, which function like galvanic cells. They use blood glucose, drawn from living cells, as a reactant or fuel. This is similar to how the body generates energy from food. The bio-nano generation process is achieved by means of special enzymes that strip electrons from glucose, freeing them to generate electrical current, much like in fuel cells.

It is estimated that the average person's body could generate 100 W of electricity using a bio-nano generator. The electricity generated by bio-nano processes could perhaps some day power body-embedded devices, such as pacemakers and blood circulation pumps. It is also suggested that the future development of bio-nano generator robots, fueled by sugar from glucose, could be embedded to perform various bodily functions. At present, research conducted on bio-nano generators is still experimental; however, progress in the field holds significant prospects for advancing the technology.

Concentrator Photovoltaic Systems

This section is intended to explore the principle of concentrator photovoltaic (CPV) solar power systems. Chapter 8 of this book provides detailed coverage of various CPV technologies.

Solar photovoltaic systems are a class of technology that use special lenses or reflectors that focus sunlight onto solar cell modules. In general, concentrator lenses

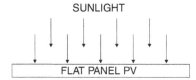

Figure 1.22   Graphic of solar ray impact on the surface of a flat-plate PV module.

deployed frequently have concentration ratios of 10 to 500 or more, are mostly made of inexpensive plastic or glass materials, and are engineered with refracting features that direct sunlight onto the small, narrow PN-junction area of cells. Module efficiencies of most PV cells (discussed previously) normally range from 10 to 18 percent, whereas concentrator-type solar cell technology efficiencies can exceed 30 percent.

CONCENTRATOR OPTICS

Refractive optics is used to concentrate the sun's irradiance onto a solar cell (further discussed in Chap. 8) (see Figs. 1.22 and 1.23). In Fig. 1.23, a square Fresnel lens, incorporating circular facets, is used to turn sun rays to a central focal point. A solar cell is mounted at this focal point and converts the sun power into electric power. A number of Fresnel lenses are manufactured as a single piece, or parquet.

The solar cells are mounted on a plate, at locations corresponding to the focus of each Fresnel lens. A steel C-channel structure maintains the aligned positions of the lenses and cell plates.

WHY CONCENTRATION?

Before photovoltaic systems can provide a substantial part of the world's need for electric energy, there needs to be a large reduction in their cost. Studies conducted by the Department of Energy (DOE), Electrical Power Research Institute (EPRI), and others

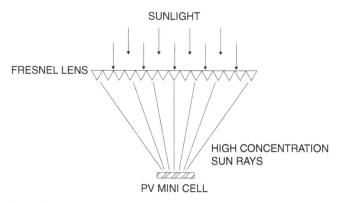

Figure 1.23   Diagram of Fresnel lens concentration principle.

show that concentrating solar energy systems can eventually achieve lower costs than conventional PV power systems. The lower cost results from the following element.

Less expensive material   Because the semiconductor material for solar cells is a major cost element of all photovoltaic systems, one approach to cost reduction is to reduce the required cell area by concentrating a relatively large area of solar insolation onto a relatively small solar cell.

An Overview of Solar Power System Applications

The following is an overview of basic types and classifications of solar power system deployment methodologies. A comprehensive guide for designing large-scale solar power system design is covered in subsequent chapters.

SOLAR POWER SYSTEM CONFIGURATION AND CLASSIFICATIONS

In general, solar power systems consist of the following configurations:

Directly connected DC solar power system

Stand-alone DC solar power systems with battery backup

Stand-alone hybrid solar power systems with generator and battery backup

Grid-connected solar power system

Directly connected DC solar power   As depicted in Fig. 1.24, positive and negative outputs are connected to a DC pump motor via a two-pole single throw switch. This type of solar power system configuration is typically used in agricultural applications, where either regular electrical service is unavailable or the cost is prohibitive. A floating or submersible DC pump connected to the PV array provides a constant stream of well water that is accumulated in a reservoir for farm or agricultural use.

Stand-alone DC solar power systems with battery backup   The solar power photovoltaic array configuration shown in Fig. 1.25 is a DC system with battery backup. Such configuration PV arrays are connected in series to obtain the desired DC voltage, such as 12, 24, or 48 V. Outputs of that are in turn connected to a DC collector panel equipped with specially rated overcurrent devices, such as ceramic-type fuses. The positive lead of each PV array conductor is connected to a dedicated fuse, and the negative lead is connected to a common neutral bus. All fuses as well are connected to a common

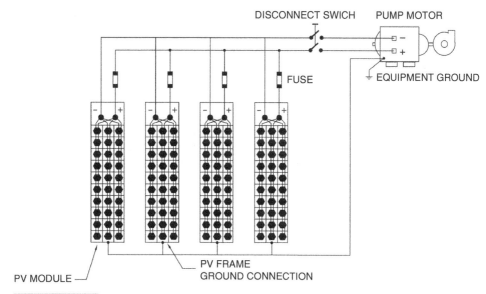

Figure 1.24   Directly connected solar power DC pump diagram.

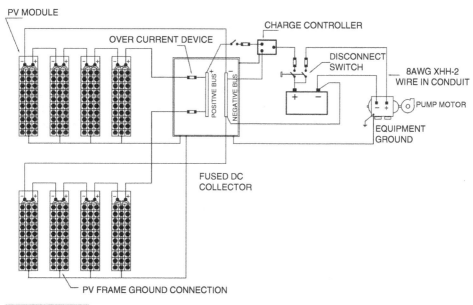

Figure 1.25   A battery-backed solar power-driven DC pump.

positive bus. The output of the DC collector bus, which represents the collective amperes and voltages of the overall array group, is connected to a DC charge controller. This regulates the current output and prevents the voltage level from exceeding the maximum needed for charging the batteries.

The output of the charge controller is connected to the battery bank by means of a dual DC cut-off disconnect. The cut-off switch, when turned off for safety measures, disconnects the load and the PV arrays simultaneously. Under normal operation during the daytime when there is adequate solar insolation, the load is supplied with DC power while simultaneously charging the battery. When sizing the solar power system, the DC power output from the PV arrays should be taken into account to ensure it is adequate to sustain the connected load and the battery trickle charge requirements.

Battery storage sizing depends on a number of factors, such as the duration of an uninterrupted power supply to the load when the solar power system is inoperative, which occurs at nighttime or during cloudy days. It should be noted that battery banks inherently produce a 20 to 30 percent power loss caused by heat when in operation.

When designing a solar power system with a battery backup, the designer must determine the appropriate location for the battery racks and room ventilation, to allow for dissipation of the hydrogen gas generated during the charging process. Sealed-type batteries do not require special ventilation.

All DC wiring calculations discussed take into consideration losses resulting from solar exposure, battery cable current derating, and equipment current resistance requirements, as stipulated in NEC 690 articles.

Stand-alone hybrid solar power system with standby generator   A stand-alone hybrid solar power system with standby generator is essentially identical to the DC solar power system just discussed, except that it incorporates two additional components, as shown in Fig. 1.26. The first component is an inverter. Inverters are electronic power equipment that is designed to convert direct current into alternating current. The second component is a standby emergency DC generator. The principal function of inverters is to convert DC to AC current. This is achieved by chopping the DC current waveforms into equal segments, referred to as square waves. These are in turn filtered and shaped into sinusoidal AC waveforms.

In general, DC-to-AC inverters are intricate electronic power conversion equipment designed to convert direct current to a single- or three-phase current that replicates the regular electrical services provided by utilities. Special electronics within inverters, in addition to converting direct current to alternating current, are designed to regulate the output voltage, frequency, and current under specified load conditions. As discussed here, inverters also incorporate special electronics that allow them to automatically synchronize with other inverters when connected in parallel. Most inverters, in addition to PV module input power, also accommodate auxiliary input power to form a standby generator. This is used to provide power when battery voltage is dropped to a minimum level.

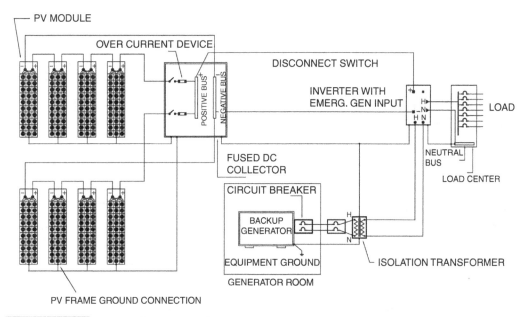

Figure 1.26   **A stand-alone hybrid solar power system with standby generator.**

Special designed inverters, referred to as the *grid-connected* type, incorporate synchronization circuitry that allows the production of sinusoidal waveforms in unison with the electrical service grid. This type of inverter, when connected to the electrical service grid, effectively acts as an AC auxiliary power generation source. Design and performance characteristics of grid-type inverters are required to meet specific international design and performance standards, and are strictly regulated by utility agencies. As mentioned, some inverters incorporate an internal AC transfer switch that isolates allows connection from an auxiliary AC input from an AC standby generator.

Grid-connected solar power cogeneration system   Figure 1.27 depicts a typical grid-connected solar power system diagram. In such configuration the power cogeneration system is similar to the hybrid system just described. The essence of a grid-connected system is *net metering*. Standard service meters are odometer-type counting wheels that record power consumption at a service point by means of a rotating disc, which is connected to the counting mechanism. The rotating discs operate by an electro-physical principle called an *eddy current*, which consists of voltage and current measurement sensing coils that generate a proportional power measurement.

New electric meters make use of digital electronic technology that registers power measurement by solid-state current- and voltage-sensing devices that convert analog-measured values into binary value. These values are displayed on the meter bezels by liquid-crystal display (LCD) readouts.

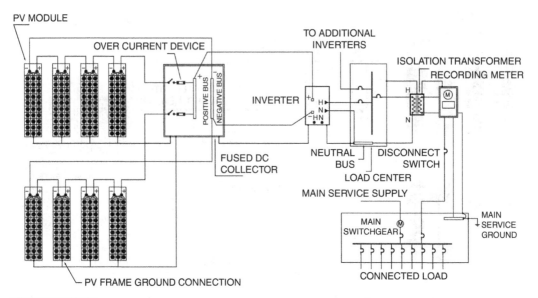

Figure 1.27 **A typical grid-connected hybrid solar power system with standby generator.**

In general, conventional meters only display power consumption; that is, the meter counting mechanism is unidirectional.

Net metering The essential difference between a grid-connected system and a stand-alone system is that inverters, which are connected to the main electrical service, must have an inherent line frequency synchronization capability to deliver the excess power to the grid.

Net meters, unlike conventional meters, have a capability to record consumed or generated power in an exclusive summation format. That is, the recorded power registration is the net amount of power consumed and the total power used minus the amount of power that is produced by the solar power cogeneration system. Net meters are supplied and installed by utility companies that provide grid-connection service systems. Net-metered solar power cogenerators are subject to specific contractual agreements and are subsidized by state and municipal governmental agencies.

Grid-connection isolation transformer In order to prevent spurious noise transfer from the grid to the solar power system electronics, a delta-y isolation transformer is placed between the main service switchgear disconnects the inverters. The delta winding of the isolation transformer, which is connected to the service bus, circulates noise harmonics in the winding and dissipates the energy as heat.

Isolation transformers are also used to convert or match the inverter output voltages to the grid. In commercial installations, inverter output voltages range from 208 to

230 V (three-phase), which must be connected to an electric service grid that supplies 277/480 V power.

Some inverter manufacturers incorporate output isolation transformers as an integral part of the inverter system, which eliminates the use of external transformation and ensures noise isolation.

References

1. Gevorkian, Peter, *Alternative Energy Systems in Building Design,* McGraw-Hill, New York, 2010.
2. Solar America Initiative, http://www1.eere.energy.gov/solar/initiatives.html.
3. Solar Advisor Model, https://www.nrel.gov/analysis/sam/.

SOLAR POWER SYSTEM PHYSICS
AND EFFECTS OF AMBIENT
PARAMETER VARIATION

Introduction

In order to acquire proficiency in solar power system design, the engineer or the designer must be familiar with the atmospheric, climatic, and environmental factors that affect output power performance of solar power cells and PV modules. This chapter will discuss the basics theory of solar cell, solar physics, and ambient climatic and atmospheric changes affecting performance characteristics of PV cells and modules.

Physics of Solar Cells
Background: Photons

The study of solar PV physics requires a fundamental understanding of photons. In physics, a photon is an elementary particle, the quantum of the electromagnetic interaction and the basic unit of light and all other forms of electromagnetic radiation. It is also the electromagnetic force carrier. The effects of the electromagnetic force can be observed at both the microscopic and macroscopic levels; because the photon has no rest mass, this allows for interactions at long distances. Like all elementary particles, photons are governed by quantum mechanics and exhibit wave–particle duality; in other words, they exhibit properties of both waves and particles.

In 1900, Max Planck was working on black-body radiation and suggested that the energy in electromagnetic waves could only be released in "packets" of energy. In his 1901 article in *Annalen der Physik,* he called these packets "energy elements." The word *quanta* (singular *quantum*) was coined before 1900 to mean particles or amounts of different quantities, including electricity. In 1905, Albert Einstein went further by suggesting that electromagnetic waves could only exist in quantum or in discrete *wave packets*. Einstein called such a wave packet the light quantum (or in German, *das Lichtquant*). The name *photon* derives from the Greek word for light, φως transliterated *phôs*, and was coined in 1926 by the physical chemist Gilbert Lewis, who proposed a theory that photons were "un-creatable and indestructible." Even though the theory was contradicted by many experiments, photon was adopted as a name reference by physicists. In physics, a photon is usually denoted by the Greek letter gamma, γ.

PHYSICAL PROPERTIES OF PHOTONS

The photon is massless energy that has no electric charge and that does not decay when traveling in empty space. Photons have two polarization states and are described by components of their wave vector, which determine their wavelength, λ, and their direction of propagation.

Photons are emitted in processes. For instance, when a charge is accelerated it emits synchrotron radiation. Photon emission also occurs during a molecular, atomic, or nuclear transition of an electron to a lower energy level. In ideal vaccum space, photons move at the speed of light (c). Photon energy and momentum are related by the equation $E = pc$, where p stands for the magnitude of the momentum of vector **p**. The equation is derived from the relativity theory where the mass, $m = 0$.

$$E^2 = p^2c^2 + m^2c^4$$

The energy and momentum of a photons are proportional to their frequency (v) or the inverse of their wavelength (λ):

$$E = \hbar w = hv = \frac{hc}{\lambda}$$

$$p = \hbar k$$

In the above formula, **k** stands for the wave vector, the wave number where $k = |\mathbf{k}| = 2\pi/\lambda$, $\omega = 2\pi v$ represents the angular frequency. The $\hbar = h/2\pi$ is referred to as the reduced Planck constant.

The **p** being a vector point defines the direction of the photon's propagation. Magnitude of the momentum is defined as

$$p = \hbar k = \frac{hv}{c} = \frac{h}{\lambda}$$

Photons also carry an angular momentum spin, which is not dependent on their frequency. The equation described above is used to represent the classical relationship of photon energy and momentum, which are the basis of electromagnetic radiation. It should be noted that the pressure of electromagnetic radiation on various objects results in the transfer of photon momentum on an object, which forces displacement of electrons in the area of impact. The above electromagnetic force is defined as the fundamental principle of photoelectric effect, which gives rise to electron mobility or displacement in photovoltaic PN junctions. This will be discussed further in this chapter.

In the latter part of the nineteenth century, physicists discovered a new phenomenon. When light is incident on liquids or metal cell surfaces, electrons are released. However, no one had an explanation for this bizarre occurrence. At the turn of the century, Albert Einstein provided a theory for this, which won him the Nobel Prize in physics and laid the groundwork for the theory of the *photoelectric effect.* Figure 2.1 shows the photoelectric effect experiment. When light is shone on metal, electrons are released. These electrons are attracted toward a positively charged plate, thereby giving rise to a photoelectric current.

Einstein explained the observed phenomenon by a contemporary theory of *quantized energy levels,* which was previously developed by Max Planck. The theory described light as being made up of miniscule bundles of energy called *photons.* Photons impinging on metals or semiconductors knock electrons off atoms. In the 1930s, these theorems led to a new discipline in physics called *quantum mechanics,* which consequently led to the discovery of transistors in the 1950s, and to the development of semiconductor electronics.[1]

The principle operation of solar cells and PV modules is caused by a physical phenomenon called *photogeneration of charge carriers.* In essence, solar energy is photons in sunlight, which impacts solar panels and which are absorbed by semiconducting materials, such as silicon.

Electrons (negatively charged) within semiconductor elements such as silicon, when hit by photons, are knocked loose from their atoms. This allows them to flow through the material to produce electricity. Due to the special construction and composition of solar cells (discussed in this chapter), the electrons are allowed to move in one direction, which constitutes useable electrical DC current. Effectively, solar cells, function can be described as devices that convert solar energy into electricity. Figure 2.2*a, b, c,* and *d* depict PN junction configuration and charge transfer through the junction barrier.

When a photon impacts a piece of silicon, the following scenarios develop. Lower energy photons pass straight through the silicon, and some photons, after colliding with the

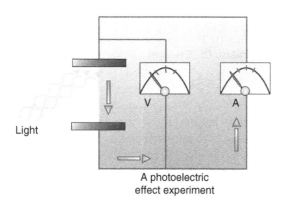

Light

A photoelectric
effect experiment

Figure 2.1   **Photoelectric effect.**

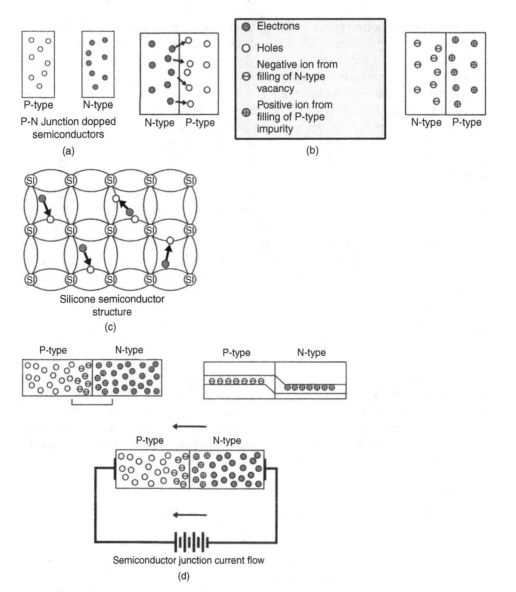

Figure 2.2   (*a*) Semiconductor PN doping; (*b*) PN junction electron crossover; (*c*) Semiconductor electron-hole displacement; (*d*) Semiconductor junction current flow.

semiconductor, are reflected off the surface. In other circumstances, photons are absorbed by the semiconductor. In some instances when the photon energy is higher than the silicon's so-called absorption capacity, the energy is transformed and dissipated as heat.

When a photon is absorbed, its energy is given to an electron within the valence band of the crystal lattice. Electrons within the valence band are tightly bound in covalent bonds between neighboring atoms, and under normal circumstances are unable to move

SHORT-CIRCUIT CURRENT (Isc)

Short-circuit current (Isc) of solar cells is defined as the maximum current flow under no-load, zero-voltage conditions, and when output leads are shorted. Short-circuit current measurement characterizes the maximum circuit design current of a solar array. Isc characteristics are directly proportional to solar irradiance and are reduced when the cell PN junction temperature rises.

When in the field, short-circuit current measurement can be measured by connecting a clamp-on meter to the short-circuited cable loop of solar array, by connecting positive (+) and negative (−) terminal lugs. Likewise, similar measurement can be done with an in-line multimeter, by clamping the positive (+) and negative (−) terminals of the solar array terminal to input current of the meter. Typical short-circuit current measurement of flat-panel PV modules may vary from 1 to 10 VA. Since solar array or string is formed by serial interconnection of several PV modules, the output current remains identical to Isc of a single PV module.

As PV module specifications are generally recorded in degrees Celsius, temperature values can be converted to Fahrenheit with the following formula:

$$F = (9/5 \times C) + 32$$
$$C = 5/9 \times (F - 32)$$

where F = temperature in degree F°
$\quad C$ = temperature in degree C°

Charge Carrier Separation

The process of electron excitation and its separation from atomic structures is referred to as *charge carrier separation*. There are two main modes for charge carrier separation in a solar cell, namely *drift* and *diffusion*.

In drift mode, electron flow or displacement driven by an electrostatic field is established across the device. In diffusion mode, electrons are displaced from a lower carrier concentration zone to a higher carrier concentration zone, which has higher carrier concentration.

In solar cell technologies that are based on PN junctions, the principle mode of charge carrier separation is by drift. However, in non–PN-junction-type solar cells such as *dye-sensitized* or *polymer solar cells,* the principle mode of separation electron is via charge carrier diffusion.

PN-Junction Solar Cell Technology

Most flat-panel photovoltaic solar cell technologies are configured from crystalline semiconductor PN junctions that are fabricated from silicon. For simplification purposes, these types of solar cells could be considered as a layer of N-type silicon being brought into close contact with P-type silicon. However, the fabrication of PN-junction solar cells involves the diffusion of N-type and P-type dopants into adjacent sides of silicon wafers.

In practice, a piece of P-type silicon is placed in intimate contact with a piece of N-type silicon, where the diffusion electrons takes place from the N region of the high-electron-concentration side into the region of low-electron concentration on the P side of the junction. When the electrons diffuse across the PN junction, they recombine with holes on the P-type side.

The diffusion of carriers creates charge build-up on either side of the junction, which results in the build-up of an electric field. The electric field creates a diode effect that promotes charge flow, or a drift current, that opposes and eventually balances out the diffusion of electrons and holes. This region where electrons and holes diffuse across the junction is called the *depletion region* or *space-charge region* because it no longer contains any mobile charge carriers.

SOLAR CELL CONNECTION TO AN EXTERNAL LOAD

Manufacturing solar cells involves fabricating ohmic metal-semiconductor contacts that sandwich both sides of the N-type and P-type solar cell, forming electrodes that connect the cell to an external load. Under solar irradiance, electrons created on the N-junction flow onto the P-type side as DC current that travels through wire, powering external connected loads. The electrons continue flowing through the load until they reach the P-type semiconductor-metal contact where they recombine with holes that were created as an electron-hole pair on the P-type side of the solar cell, or holes that were swept across the junction from the N-type side after being created there.

Solar Cell Equivalent Circuit

Performance characteristics of PN-type solar cells, are best defined as the electrical equivalent of a diode. An ideal solar cell, from an electrical engineering point of view, may be modeled by a current source in parallel with a diode, a shunt resistance, and a series-resistance component, *s*, as in the schematic diagram in Fig. 2.3*a* and *b*.

CHARACTERISTIC EQUATION

From the equivalent circuit, the current produced by the solar cell equals that produced by the current source, minus the current that flows through the diode, minus that which flows through the shunt resistor, as defined by the following formula:

$$I = I_L - I_D - I_{SH}$$

where I = output current in amperes
       I_L = photo-generated current in amperes
       I_D = diode current in amperes
       I_{SH} = shunt current in amperes

The voltage governs the current through circuit elements across them:

$$V_J = V + IR_S$$

where V_j = voltage across both diode and resistor R_{SH} in volts
V = voltage across the output terminals in volts
I = output current in amperes
R_S = series resistance in ohms

SERIES RESISTANCE

When series resistance increases in solar cells, the voltage drop between the junction voltage and the terminal voltage becomes greater for the flow of current, which results in sagging of the current-controlled portion of the I-V curve. This results in a significant reduction of the terminal voltage, V, and a slight reduction in the short-circuit current, I_{SC}. Likewise, high values of R_S also produce a significant reduction in I_{SC}. As a result, series resistance significantly effects the power output performance of the solar cell. Figure 2.4 shows the effect of increased ohmic series resistance for crystalline silicon solar cells.

Losses of series resistance can be computed by the quadratic equation $P_{Loss} = V_{Rs} \times I = I^2 R_S$, which produces an approximated value of the cell power output.

SHUNT RESISTANCE

A decreased value of shunt resistance current diverted through the shunt resistor increases for a given level of junction voltage. The net result is that the voltage-controlled portion of the I-V curve begins to sag, which produces a significant decrease in the terminal current I and a slight reduction in open circuit voltage, V_{OC}. Lower values of the shunt resistor R_{SH} result in a significant reduction in V_{OC}. Figure 2.5 depicts effect of shunt resistance on the current-voltage characteristics of a solar cell.

REVERSE SATURATION CURRENT

When the output current of a solar cell (I_o) increases, it results in the reduction of open-circuit voltage, V_{OC}, a phenomenon that is directly related to PN junction reverse current saturation, which increases cell junction temperature. Reverse saturation current is effectively a "leakage current" that results from carrier recombination in the neutral regions on either side of the junction. Figure 2.6 depicts the effect of reverse saturation current on the current-voltage characteristics of a solar cell.

IDEALITY FACTOR

The *ideality factor,* also called the *emissivity factor,* is a cell functional performance parameter multiplier that describes how closely the solar PN junction diode's behavior

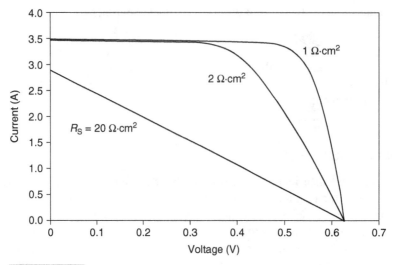

Figure 2.5   Effect of shunt resistance on the current-voltage characteristics of a solar cell.

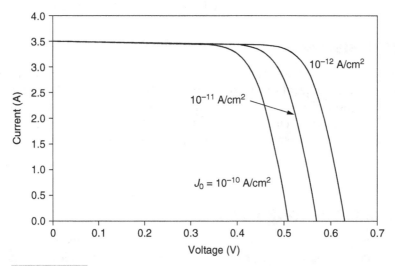

Figure 2.6   Effect of reverse saturation current on the current-voltage characteristics of a solar cell.

matches the equivalent diode circuit described previously. Under perfect analogous performance, the performance factor is assigned a value of $n = 1$. In actual situations, recombinations of silicon crystalline cells in the space-charge region are not ideal and I-V measurements as shown in Fig. 2.7 do change, which yield n values greater than one.

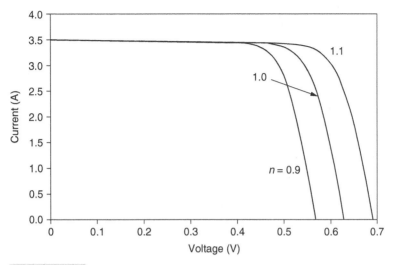

Figure 2.7 Effect of ideality factor on the current-voltage characteristics of a solar cell.

Energy Conversion Efficiency

Solar cell *energy conversion efficiency,* designated by Greek letter eta, η, is the percentage of power converted from absorbed solar irradiance when connected to an electrical load. The efficiency is calculated using the ratio of the maximum power point, P_m, divided by the impacted solar irradiance, E, measured in W/m², on the surface of a PV module Ac in m² under standard test conditions (STC).

$$\eta = \frac{P_m}{E \times A_c}$$

Standard tests are conducted by PV module manufacturers under the specific temperature of 25°C and an irradiance of 1000 W/m² with an air mass 1.5 (AM 1.5) spectrum discussed earlier in this chapter. In the United States, STC test conditions correspond to the irradiance and spectrum of sunlight incident on a clear day upon a sun-facing 37°-tilted surface with the sun at an angle of 41.81° above the horizon. This test condition represents approximate solar noon near the spring and autumn equinoxes in the continental United States with the surface of the cell aimed directly at the sun. Under STC test conditions, a solar cell of 16 percent efficiency with a 100 cm² (0.01 m²) surface area must produce approximately 1.6 W of power.

The efficiency of a solar cell is an aggregation of reflectance efficiency, thermodynamic efficiency, charge carrier separation efficiency, and conductive efficiency. As such, the overall efficiency of a solar cell is the product of each of the named individual efficiencies.

Cell Temperature

Figure 2.8 represents temperature effects on output performance of solar cells. As seen in Fig. 2.8, output voltage and current of the cell exhibit lower energy output inversely proportional to the ambient temperature. In other words, as the ambient temperature increases, the output voltage decreases. From an operational point of view, the net effect of the rise in temperature minimally affects the output current. However, the net effect translates into a reduction of open-circuit voltage, Voc. Likewise, a decrease in ambient temperature increases the Voc voltage.

For most crystalline silicon solar cells, the power-output production is reduced by about 0.50%/°C. However, the rate of reduction for the high-efficiency crystalline silicon cells is around 0.35%/°C. For amorphous silicon solar cells, it is is 0.20 to 0.30%/°C, depending on how the cell is made.

However, the amount of photogenerated current, I_L, due to thermally generated carriers, is slightly increased. This effect is very marginal and is about 0.065%/°C for crystalline silicon cells and 0.09%/°C for amorphous silicon cells.

The overall effect of temperature on cell efficiency can be computed using these factors in combination with the characteristic equation. However, because the change in voltage is much stronger than the change in current, the overall effect on efficiency tends to be similar to that on voltage. Most crystalline silicon solar cells decline in efficiency by 0.50%/°C, and most amorphous cells decline by 0.15 to 0.25%/°C. The *I-V* profiles shown in the *I-V* curves (Fig. 2.8) are typically for a crystalline silicon solar cell at various temperatures.

When designing solar power systems, the design must consider effects of extremities of temperature variation, which occur in polar equatorial regions.

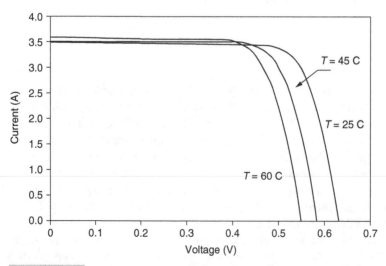

Figure 2.8   Effect of temperature on the current-voltage characteristics of a solar cell.

PV MODULE TEMPERATURE RESPONSE

Almost all PV modules, when operating in high ambient temperatures, exhibit voltage drop and a slight increase in current that results in output deterioration. Under extended exposure to ambient heat, PV modules' encapsulation may undergo gradual deterioration, which can cause permanent premature module degradation. In general, cell temperature refers to the internal temperature at the PN junction. Cell temperature can be affected by wind speed, solar irradiance, humidity, and thermal characteristics of PV lamination.

When designing solar power systems, a cell temperature value multiplier, referred to as a *temperature-rise coefficient* (shown in PV module specification sheets), is used to calculate compensation for solar irradiance under various ambient temperature conditions. The formula used to estimate the temperature compensation is:

$$T_{CELL} = T_{AMB} + (T_{RISE} \times E)$$

where T_{CELL} = cell temperature (in °C)
T_{AMB} = ambient temperature (in °C)
T_{RISE} = temperature-rise coefficient (in °C)
E = solar irradiance (in kW/m^2)

In order to mitigate such occurrences, PV arrays must be configured in a manner that allows natural convection. In practice, PV arrays must be configured with the back of the module facing the wind; therefore, module installation laid flat and close to a surface must be avoided.

The following calculation demonstrates cell-temperature change for a roof-mount flat panel around 32°C ambient temperature, a cell-temperature coefficient of 26°C/kW/m^2, and solar irradiance of 1100 kW/m^2;

$$T_{CELL} = 32 + (26 \times 1.1) = 60.6$$

Cell-temperature rise, T_{RISE}, likewise can be calculated by using the following formula:

$$T_{RISE} = (T_{CELL} - T_{AMB})/E$$

Note that to achieve a degree of accuracy in estimating cell-temperature rise, it is necessary to calculate the value under various ambient temperature conditions. Temperature-rise coefficient is used to predict PV-module power output performance under various ambient conditions.

Temperature-coefficient parameter The temperature-coefficient parameter is defined as the rate of change in voltage and current caused by temperature change. A negative coefficient implies that the parameter decreases with cell temperature increase, whereas positive coefficient means that the parameter increases with increasing cell temperature. Temperature coefficients are expressed as unit changes per degree of temperature change or percentage change per degree of temperature.

Temperature coefficients are specific to each PV module and vary from one manufactured device to another. Typical temperature coefficients for silicone cells are as follows:

$$\text{Voltage} = -0.00225 \text{ V/°C} +/-0.10\%/°C$$

$$\text{Current} = 0.0000037 \text{ A/°C} +/-0.0010\%/°C$$

Solar-cell-module temperature coefficients are calculated by the following formula:

Temperature coefficient calculation for voltage $C_v = C_{Vcell} \times N_s$

where C_v = PV-module absolute temperature coefficient for voltage (in V/°C)
C_{Vcell} = Cell absolute temperature coefficient for voltage (in V/°C/cell)
N_s = Number of series connected cells within a module

The temperature coefficient calculation for current is:

$$C_I = C_{Icell} \times N_P \times A$$

where C_I = PV module absolute temperature coefficient (in A/°C)
C_{Icell} = Cell absolute-temperature coefficient for current (in A/°C/cm^2)
N_p = Number of parallel connected cell strings within the PV module
A = Cell area (in cm^2)

The following illustrates temperature-coefficient calculation for voltage and current for a monosilicone-type PV module with 72 cells, each cell having a surface area of 144 cm^2 arranged in four strings of 18 cells each.

$$C_v = C_{Vcell} \times N_s$$

$$C_v = -0.00225 \times 18 = -\mathbf{0.0405} \text{ V/°C}$$

$$C_I = C_{Icell} \times N_P \times A$$

$$C_I = 0.0000037 \times 4 \times 144 = \mathbf{0.00022} \text{ A/°C}$$

In practice, when calculating solar-power string Voc of a solar power system installed in an area where ambient temperature variations swing from 45°C to −30°C, the following Voc voltage adjustment is required:

PV module Voc = 55.2 V at STC, 25°C

PV string = 10 modules

String Voc = 55.2 × 10 = 552 V

Cv = − 0.0405 V/°C

Lowest ambient temperature = −35°C

Temperature swing from STC = −60°C

Voltage adjustment = −0.0405 × −60 = 2.43 V

Adjusted value of PV-string Voc = 552 + 2.43 = 554.43 V

Percent changes in temperature coefficient are relatively standard for crystalline silicon (cSi) solar cells. In general, percentage values of temperature coefficients discussed above used are used in the following formula to calculate adjusted Voc values of a PV module for different ambient temperatures:

$$Cv = V_{REF} \times C_{\%V}$$

$$C_I = I_{REF} \times C_{\%I}$$

$$C_P = P_{REF} \times C_{\%P}$$

where Cv = Absolute temperature coefficient (in V/°C)
  V_{REF} = Rated or reference voltage (in V)
  $C_{\%V}$ = Temperature coefficient for voltage (in °C)

For example, a monosilicon PV module with V_{OC} = 50.9 V will have a unit change temperature coefficient value of:

$$Cv = 50.9 \times -0.00405 = -0.206 \text{ V/°C}$$

THERMODYNAMIC EFFICIENCY LIMIT

From a physical point of view, solar cells are rare quantum energy conversion devices, and they are subject to a "thermodynamic efficiency limit." Thermodynamic efficiency refers to a condition when photons with energy below the band gap PN junction cannot generate a hole-electron pair. As a result, energy absorbed is not converted to useful DC current output and is instead converted to heat. For photons with energies above the band gap, a small fraction of the energy is converted to useful output. Photons of greater energy absorbed above the bandwidth are converted to kinetic energies, which results in carrier combinations. As discussed previously, such excess kinetic energies are converted to heat through phonon interactions as the kinetic energy slowing down the carriers' equilibrium velocity.

In multijunction solar cells, multiple band gaps are capable of absorbing wider solar irradiance spectrums and have greater conversion efficiency.

FILL FACTOR

Fill factor (FF) is a ration that defines the overall power performance characteristics of a solar cell. The ration defines the maximum power point divided by the open-circuit voltage (V_{oc}) and the short-circuit current (I_{sc}), and is expressed by the following formula:

$$FF = \frac{P_m}{V_{oc} \times I_{sc}} = \frac{\eta \times A_c \times E}{V_{oc} \times I_{sc}}$$

As implied by the formula, the fill factor of a PV module or a solar cell is directly affected by the values of the cell's series and shunt resistance. Increasing shunt resistance (*Rsh*) and decreasing the series resistance (*Rs*) results in a higher fill factor and greater efficiency.

Comparative Analysis of Solar Cell Energy Conversion Efficiencies

As discussed previously, energy conversion efficiency of solar cells depends on multiple factors such as solar-irradiance spectrum, ambient temperature, and atmospheric factors. An international solar cell test condition referred to as IEC standard 61215 is used to compare the performance of cells designed for deployment under terrestrial, temperate conditions. It uses its standard temperature and conditions (STC) at an irradiance of 1 kW/m^2, a spectral distribution close to solar radiation through air mass of AM 1.5, and a cell temperature of 25°C. Under this test condition, the resistive load connected to the output of PV modules or a solar cell is varied until the peak or maximum power point (MPP) is achieved. The power at this point is recorded as watt-peak (Wp).

As discussed earlier in this chapter, air mass has an effect on power output. In space, where there is no atmosphere, the spectrum of the sun is relatively unfiltered; however, on the surface of the earth, the solar spectrum is diminished. To account for the spectral differences, a system was devised to calculate this filtering effect. The efficiency of Silicon solar cells under air mass (AM) of 1.5 is reduced by approximately two-thirds that of AM 0.

Currently, multijunction silicon-based solar cells tested in a laboratory environment reach 42.8 percent efficiency, whereas amorphous cells exhibit 6 to 8 percent performance efficiency. Solar cell energy conversion efficiencies for commercially available *multicrystalline silicone solar cells* are around 14 to 19 percent.

It should be noted that producing high-efficiency cells with performance efficiencies of about 30 percent involves using expensive exotic materials such as gallium arsenide, indium selenide, and multijunction manufacturing methodologies that substantially increase cell production cost.

Atmospheric and Climate Effects on Photovoltaic Modules

SOLAR ENERGY

The sun is composed of mostly a gaseous body of hydrogen and small amounts helium. The gaseous cloud under the influence of gravity and electromagnetic fields swirls, which forms the heated core. This results in nuclear fusion of the hydrogen atoms. Within its core, intense gravitation fuses light hydrogen atoms together transforming them into heavier element helium, which results in the release of enormous amounts of energy that radiate outward. The radiated energy, in the form of waves or particles, travels the sun's visible surface or the *photosphere* and escapes into space in the form of light and heat radiation.

The distance between the sun and earth measures 93 million miles. This distance is established as an astronomical unit (AU) and is used as an interstellar measuring unit. Since light travels at 186,000 miles per second, radiation from the sun reaches earth's surface in eight minutes. In addition to radiated energy, hydrogen fusion reaction within the sun's core also releases high-energy photons. A photon, which is a fundamental unit of energy, takes thousands of millions of years to travel to the surface of the sun and escape as visible light.

At any instant, earth receives approximately 170 gigawatts (GW) of power; a small fraction, if harvested, could exceed all present and future energy requirements of mankind.

SOLAR RADIATION

The fusion process within the sun's core converts the differential mass between atomic weights of fused hydrogen atoms and helium into energy that radiates from the surface of the sun in all directions. Solar radiation traveling toward the earth, because of various atmospheric and environmental factors, loses its energy along the way and eventually a portion of it is absorbed and converted into electrical energy by PV devices.

Solar radiation outside the earth's atmosphere is referred to as *extraterrestrial radiation,* also called *top-of-atmosphere* (TOA) *radiation,* which is used in the design of solar power systems for orbiting satellites.

SOLAR IRRADIANCE

Solar irradiance is the *intensity* of solar energy impacting an imaginary unit surface. Solar irradiance is expressed as watts per square meter (W/m^2) or kilowatts per square meter (kW/m^2). Solar irradiance is an instantaneous value of energy and does not represent cumulated energy over a period of time. Solar irradiance is used to measure *instantaneous peak power output performance* of any solar power energy device or a PV module.

Solar irradiance constantly fluctuates up and down with the rise-and-fall cycle of the sun, but more so because of the orbital variation of distance between the earth and the sun.

The diminishing value of irradiance due to distance is explained by the inverse square law of physics, which states that the amount of radiation (Ra) is proportional to the inverse distance (d) from the source (Rs), expressed as $Ra = Rs/d^2$. Figure 2.9 depicts solar irradiance absorption and reflection.

Solar energy, commonly referred to as solar irradiance, is the aggregate amount of accumulated solar energy over a time period. The period of energy accumulation time could be an hour, a day, a month, a year, or a life cycle of a solar power system. Solar irradiance is measured in watts per hour (W/h^2) or kilowatts per hour (kW/h^2). As such, irradiance determines the amount of solar power energy production capacity or power output performance measure of a photovoltaic system.

Solar irradiance on the surface of the earth rises in the morning, reaches a peak at noon, and falls back to zero after dusk.

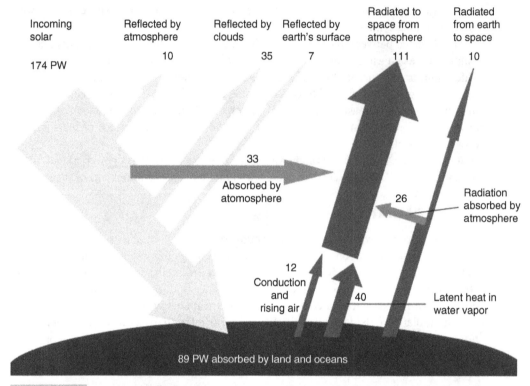

Figure 2.9   Solar irradiance absorption and reflection.

Solar irradiance is calculated by the following formula:

$$H = E \times t$$

where H = Solar irradiation (intensity in W/m²)
E = Solar irradiance (power in W/m²)
T = Time (in hours)

For example, assuming an average solar irradiance of 800 W/m² over a time period of 8 hours will yield the following:

$$H = 800 \text{ (Wh/m}^2) \times 8 \text{ (hr)} = 6400 \text{ (Wh/m}^2) \text{ or } 6.6 \text{ (kWh/m}^2)$$

SOLAR CONSTANT

Solar radiation outside the earth's atmosphere, because of the absence of atmospheric pollutants, clouds, and water particles, which scatter and absorb the solar energy, is greater than the energy that reached the surface of the earth. The amount of solar power or irradiance measured at AU1 (astronomic unit 1) is approximately 1366 W/m². This irradiance value is relatively constant and varies insignificantly over a long period of time. This constant measure of solar power can only be used on orbiting satellites and cannot be applied at earth's surface. At earth's surface at sea level, solar radiation measures 1000 W/m².

SOLAR ENERGY SPECTRUM

Energy emanated from the sun is *electromagnetic radiation* in the form of varying lengths of waves, which have electromagnetic properties. The waveform lengths define the energy content of the solar power spectrum, which ranges from millionths of a meter (gamma rays) to several kilometers (radio waves). The solar electromagnetic spectrum, therefore, includes a wide range of wavelengths, some of which fall within ultraviolet and infrared visible light band.

EFFECTS OF ATMOSPHERE ON SOLAR RADIATION

Solar radiation, when entering the earth's atmosphere, is partially absorbed and scattered by ozone, water vapor, carbon dioxide, dust particles, and gases. Other factors that diminish the solar radiation strength are clouds, dust storms, atmospheric pollution, and volcanic eruptions.

Radiation impacting the earth's surface is classified into two categories: *direct radiation* and *diffused radiation.* Aggregation of direct and diffused solar radiation is referred to as *total global radiation.* Yet another source of diffused radiation, called *albedo radiation* or *albedo reflectance*, results when sun's direct radiation is reflected back into the atmosphere.

Direct radiation is unobstructed solar radiation that travels directly from the sun and impacts the surface of the earth without scattering. Rays of direct solar radiation travel in parallel and cause shading when obstructed by various objects. On the other hand, diffused solar radiation is dispersed and scattered and is received at the surface of the earth from many directions. Global radiation, an accumulated direct and diffused radiation, varies from 10 to 100 percent of strength during daylight hours.

In general, flat-PV panels absorb total global radiation, whereas concentrating-type PV (HCPV) modules and solar thermal absorption collecting devices absorb direct radiation.

AIR MASS

As discussed, the amount of solar power radiation reaching the surface of the earth is directly related to the amount of energy scattered through the atmosphere. When the sum is at its apex, or the zenith, the atmospheric mass has the smallest distance and mass. The zenith angle (θz) is the angle between the sun and the zenith.

As the zenith angle increases, the sun's rays pass through the atmosphere's mass, which reduces the rays' intensity in proportion to the atmospheric air mass.

Air mass is assigned a value of 1 (AM1.0) when the sun is directly overhead of sea level. At the outer atmosphere, the value of air mass because of the lack of any irradiance impediment is AM 0.

At any location on the surface of the earth, the air mass is calculated with the following formula:

$$AM = 1/Cos\ \theta z$$

where AM = air mass value
θz = zenith angle in degrees

For example, at an azimuth angle of $\theta z = 60$ degrees, Cos 60 = 0.5, therefore AM_{LOCA} = 1/ 0.5 = AM 2.

Note that air mass value depends on the time of the day, the year, and the altitude of a specific location. The air mass barometric pressure value at the sea level is 1013 millibars or mbar.

Air mass of any location on the surface of the earth is calculated by the following formula:

$$AM_{LOCAL} = AM \times (P_{LOCAL}\ /\ 1013)$$

where AM_{LOCAL} = the local air mass
AM = air mass at sea level
P_{LOCAL} = local atmospheric pressure (in millibars)
1013 = atmospheric pressure at sea level (in millibars)

In practice, the air mass in any part of the earth's surface can be calculated by using the following procedure:

1 Set a measuring ruler at length of (L_R in inches) in vertical position.
2 Measure the shadow length from the base of the ruler (L_S in inches).
3 Calculate the zenith angle $\theta z = Arctan(L_S / L_R)$.

For example, air mass value calculated at zenith angle of 50° and atmospheric pressure of 800 mbar will yield the following air mass value:

$$AM = 1/Cos\ 30° = 1/0.866 = 1.155$$

$$AM_{LOCAL} = AM \times (P_{LOCAL}/1013) = 1.55 \times (800/1013) = 1.22$$

The average air mass value in the United States is 1.5. It should be noted that the solar irradiance value, at any point on the earth's surface, is calculated by dividing the solar constant irradiance AU (1366 W/m^2), by the AM_{LOCAL} value.

PEAK SUN HOUR

As discussed previously, when solar irradiance reaches the earth's surface, it loses a significant amount of its energy, such that the value of the solar constant 1366 W/m^2 is reduced by about one-third to a standard value measuring 1000 W/m^2 at sea level. The standard irradiance level increases in clear, unpolluted areas, such as higher altitudes. Figure 2.10 depicts the daily solar sun profile.

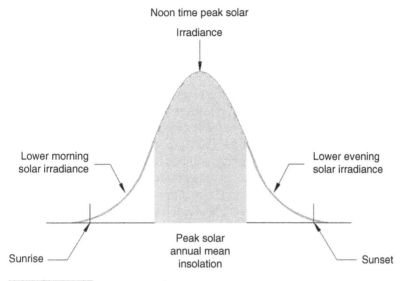

Noon time peak solar
Irradiance

Lower morning
solar irradiance

Lower evening
solar irradiance

Peak solar
annual mean
insolation

Sunrise

Sunset

Figure 2.10   **Daily solar rise and fall profile.**

Peak sun hours are the number of hours required during the day for a solar power system to accumulate energy at the *peak sun condition*. For example, in a specific location where the solar irradiance is measured to be 800 W/m^2 over a period of 7 hours, the total accumulated irradiance is converted to peak sun hours by the following calculation:

$$\text{Accumulated energy} = 800 \text{ W/m}^2 \times 7 \text{ hrs} = 5600 \text{ Wh/m}^2$$

$$\text{Peak sun hours} = 5600 \text{ Wh/m}^2/1000 \text{ Wh/m}^2 = 5.6 \text{ hours}$$

INSOLATION

Insolation is the solar energy that reaches the earth's surface over the course of the day and is expressed as kW/m^2/day, which effectively is equal to the energy produced during peak sun hours discussed previously.

Solar Radiation Measurement Devices

In view of the fact that solar irradiation varies moment to moment and is instantaneous, solar energy measurements are conducted over long periods. Therefore, in order to calculate an accurate energy-output performance, it is important to develop a historical database of the solar radiation. The energy measurement database is quite important in benchmarking PV solar power-system performance over an extended period of time.

Such solar power output measurements are essential during integration and final acceptance of a commissioning of solar power systems.

PYRANOMETER

A pyranometer is an instrument used for measuring *total global solar irradiance* within the solar aperture or field of view. The instrument is used to measure direct and diffused radiation incident on the plain of the PV modules. In general, pyranometer(s) are mounted on support brackets, which have the same plane as solar arrays. Measurements of solar irradiance are recorded in predetermined intervals and stored in a remote data acquisition and monitoring system for further information processing.

Precision-type pyranometers (from the Greek *pyr*—fire; *ano*—sky, *metron*—measure) use thermocouple sensors that measure heat and produce current that is proportional to solar irradiance. Less expensive pyranometers use silicon (cSi) cells or *photodiodes* to measure solar irradiance.

PYRHELIOMETER

A pyrheliometer (from the Greek *pyr*—fire; *helio*—sun; *metron*—measure) is an instrument that measures *direct solar radiation* in the field of view. The instrument does not measure diffused component of radiance, and it is always secured to a sun-tracking mechanism that points the device directly at the sun. The device is used in dual-axis solar power-tracking systems and solar thermal technologies.

REFERENCE CELLS

Reference cells are solar cells encapsulated in aluminium housing covered with an optical glass. When exposed to solar rays, reference cells produce small amounts of electrical current that is linearly proportional to solar irradiance. The amount of current in milliamperes is displayed or expressed in $mA/kW/m^2$ or $A/kW/m^2$. Reference cells' housing also includes a thermocouple, which is attached to the back of the solar cell and provides temperature measurement that is used to correct and compensate the output against temperature variations.

Solar Array Orientation

As referenced earlier in this chapter, seasonal solar path and geographic location of the solar platform, and climatic and atmospheric conditions, significantly impact the amount of solar radiation received at the surface of the earth. Likewise, orientation of solar arrays as well as their tilt and azimuth angle determine the incident angle of solar irradiance, which is of prime design consideration for harvesting the maximum amount of solar power. Figure 2.11 depicts world renewable energy production.

In general, the optimal angle for harvesting the maximum amount of solar power energy is the local latitude. Variation from the latitude angle results in lesser power output.

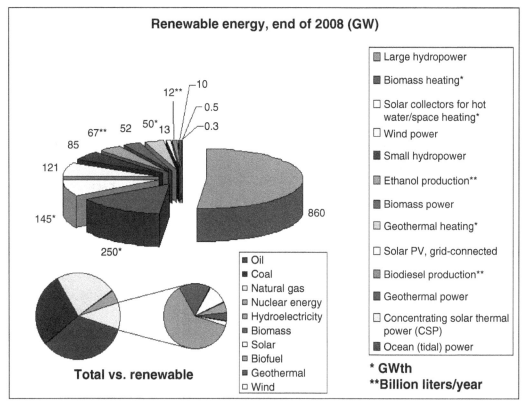

Figure 2.11   World renewable energy production.

Depending on the location of the solar power platform and the low-peak and high-peak power electrical power tariffs (usually much higher in summertime), in some instances it is preferable to reduce the tilt angle of the solar array to a few degrees. The sun's travel path and the azimuth angle (which is higher in May to October in the northern hemisphere) means smaller solar array tilt angles will have optimum irradiance angles, which will yield higher amounts of electrical power at times when the electrical power tariff charges are at the their highest.

Note that to harvest maximum solar arrays in lower latitudes the tilt angle must be very small, whereas in higher altitudes closer to the North Pole, tilt angles may be installed in a vertical position.

Physics of Solar Intensity

The amount of solar intensity of light that impinges upon the surface of solar PV panels is determined by an equation referred to as Lambert's cosine law, $I = k \times$ cosine A, as depicted in Fig. 2.12. Lambert's cosine law states that intensity of light falling on a plane

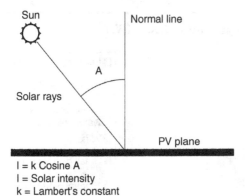

I = k Cosine A
I = Solar intensity
k = Lambert's constant
A = Solar angle

Figure 2.12   The basic physics of the solar intensity equation diagram.

is directly proportional to the cosine of the angle made by the direction of light source to the normal of the plane. In other words, when in summertime the angle of the sun is directly overhead, the magnitude of intensity is at its highest, since the cosine of the angle is zero; therefore cosine 0 = 1. This implies $I = K$ or equals Lambert's constant. Figure 2.12 depicts the basic physics of the solar intensity equation diagram.

Array Azimuth Angle

In view of the sun's seasonal travel path in the Northern Hemisphere, the optimal azimuth angle of solar arrays is due south. Therefore, to harvest the maximum amount of solar energy, PV support structures must be installed in a southward direction (surface of the solar array north facing south). In some instances when building roof pitch angles do not permit north-to-south orientation, solar arrays may be installed with azimuth angles facing east or west. In such installations, +45° to −45° azimuth angular variation could reduce solar energy output production by about 10 percent.

Insolation

As discussed previously, the amount of energy received from the sun's rays that strike the surface of our planet is referred to as insolation (I). The amount of energy that reaches the surface of the earth is largely subject to climatic conditions, such as seasonal temperature changes, cloudy conditions, and the angle at which solar rays strike the ground.

As our planet rotates around the sun at an axis tilted at approximately 23.5°, the *solar declination angle, I* (shown in Fig. 2.13), constantly varies throughout its revolution around an oval-shaped orbit and changes from +23.5° on June 21/22, when the earth's axis is tilted toward the sun, to −23.5° by December 21/22, when the earth's axis is tilted away from the sun. The earth's axis at these two seasonal changes, referred to as the summer and winter equinoxes, is zero degrees.

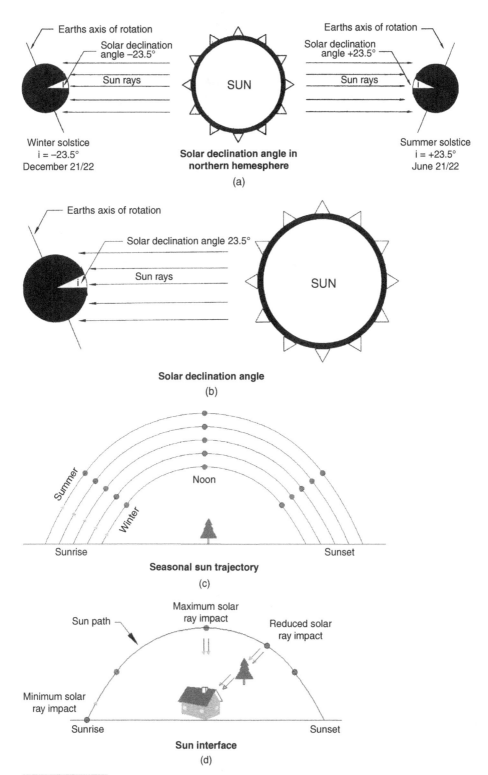

Figure 2.13 (*a*) A solar declination angle in Northern Hemisphere; (*b*) a solar declination angle; (*c*) sun trajectory, *courtesy of Solmetric;* (*d*) sun interfacing, *courtesy of Solmetric.*

The solar declinations (shown in Figure 2.13a, b, c, and d) result in seasonal cyclic variations in solar insolation. For the sake of discussion, if we consider earth a sphere of 360°, within a 24-hour period it will have rotated 15° around its axis each hour (commonly referred to as the hour angle). This daily rotation of the earth around its axis gives the notion of sunrise and sunset.

The *hour angle, H* (shown in Fig. 2.14), is the angle that the earth has rotated since midday, or solar noon. At noon, when the sun is exactly above our heads and does not cast any shadow on vertical objects, the hour angle equals zero degrees.

By knowing the solar declination angle and the hour angle, we could apply geometry and find the angle of the observer's zenith point looking at the sun, which is referred to as the zenith angle, *Z* (shown in Figure 2.15). Figure 2.14 depicts solar angle.

The amount of average solar energy striking the surface of the earth is established by measuring the sun's energy rays that impact perpendicular to a square meter area, referred to as the solar constant *(S)*. The amount of energy on top of the earth's atmosphere, measured by satellite instrumentation, is 1366 W/m². Due to the scattering and reflection of solar rays upon entering the atmosphere, solar energy loses 30 percent of its power. As a result, on a clear, sunny day, the energy received on the earth's surface is reduced to about 1000 W/m². The net solar energy received on the surface of the earth is also reduced by cloudy conditions, as well as being subject to the incoming angle of radiation. Figure 2.15 depicts solar zenith angle.

Calculation of solar insolation is determined as follows:

$$I = S \times \text{cosine } Z$$

where

$$S = 1000 \text{ W/m}^2$$
$$Z = (1/\text{Cosine}) \times (\text{Sine L} \times \text{Sine i} + \text{Cosine L} \times \text{Cosine I} \times \text{Cosine H})$$
$$L = \text{Latitude}$$
$$H \text{ (hour angle)} = 15 \text{ degrees} \times (\text{Time} - 12).$$

Time in the above formula is the hour of the day from midnight.

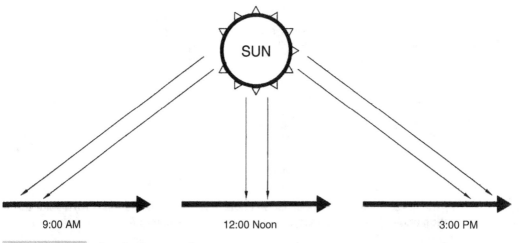

9:00 AM   12:00 Noon   3:00 PM

Figure 2.14   A solar hour angle.

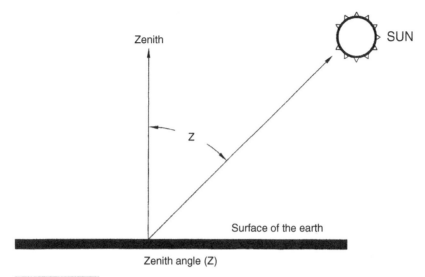

Figure 2.15 Solar zenith angle.

Figure 2.16 is solar array tilt angle annual insolation table for Los Angeles, California. Figures 2.17a through f are corresponding insolation profiles of the insolation table that show corresponding solar hours per day impacting photovoltaic panels over a period of one year. Figure 2.18 is photovoltaic NREL solar resource map of the United States.

Location los angeles   LATITUD 33.93
                        LONGIT −118.4

TILT DEGREES		JAN.	FEB.	MAR.	APR.	MAY	JUNE	JULY	AUG.	SEPT.	OCT.	NOV.	DEC.	YEAR AV.
0	AV.	2.80	3.60	4.80	6.10	6.40	6.60	7.10	6.50	5.30	4.20	3.20	2.60	4.93
	MIN.	2.30	3.00	4.00	5.50	5.70	5.60	6.40	6.10	4.40	3.80	2.10	4.70	4.47
	MAX.	3.30	4.40	5.60	6.80	7.20	7.70	8.00	7.00	5.80	4.50	3.00	5.10	5.70
LAT. −15	AV.	3.80	4.50	5.50	6.40	6.40	6.40	7.10	6.80	5.90	5.00	4.20	3.60	5.47
	MIN.	2.90	3.60	4.50	5.80	5.70	5.40	6.30	6.30	4.70	4.40	3.40	2.70	4.64
	MAX.	4.60	5.70	6.40	7.30	7.30	7.30	7.90	7.20	6.60	5.60	4.90	4.30	6.26
LAT.	AV.	4.40	5.00	5.70	6.30	6.10	6.00	6.60	6.60	6.00	5.40	4.70	4.20	5.58
	MIN.	3.30	3.80	4.70	5.60	5.40	5.00	5.90	6.10	4.80	4.70	3.70	3.00	4.67
	MAX.	5.40	6.40	6.70	7.20	6.80	6.70	7.30	7.00	6.70	6.00	5.60	5.00	6.40
LAT. +15	AV.	4.70	5.10	5.60	5.90	5.40	5.20	5.80	6.00	5.70	5.50	5.00	4.50	5.37
	MIN.	3.40	3.80	4.50	5.20	4.80	4.40	5.20	5.50	4.50	4.70	3.90	3.10	4.42
	MAX.	5.90	6.60	6.60	6.70	6.10	5.80	6.30	6.40	6.50	6.10	6.00	5.40	6.20
90	AV.	4.10	4.10	3.80	3.30	2.50	2.20	2.40	3.00	3.60	4.20	4.30	4.10	3.47
	MIN.	2.90	3.00	3.10	2.90	2.30	2.10	2.30	2.80	2.90	3.50	3.20	3.30	2.86
	MAX.	5.20	5.40	4.50	3.60	2.70	2.30	2.50	3.20	4.10	4.70	5.20	3.70	3.93

Solar insolation for flat plate collector facing south
at fixed tilt (kWh/m*m/day)

Figure 2.16   Solar array tilt angle annual insolation profile for Los Angeles, CA.

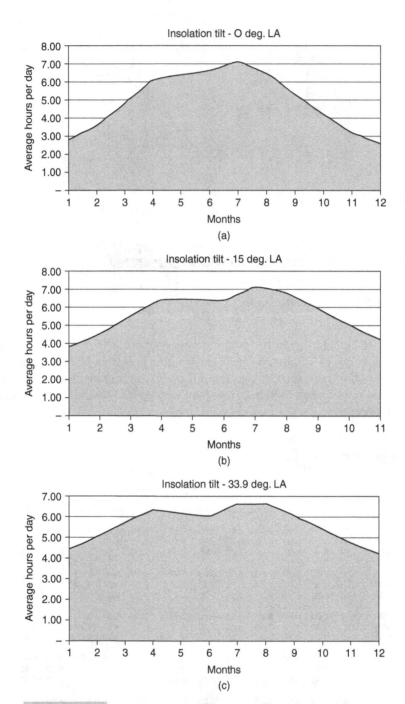

Figure 2.17 (*a*) Annual insolation profile for 0° tilt angle;
(*b*) annual insolation profile for 15° tilt angle; (*c*) annual insolation profile for 33.9° tilt angle; (*d*) annual insolation profile for +15° tilt angle; (*e*) – annual insolation profile for +90° tilt angle.

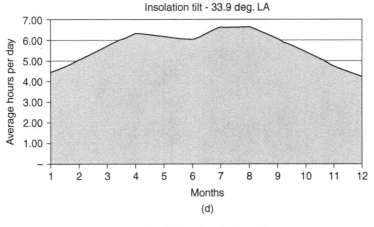

Insolation tilt - 33.9 deg. LA

(d)

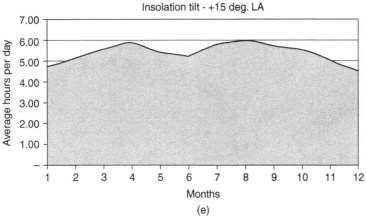

Insolation tilt - +15 deg. LA

(e)

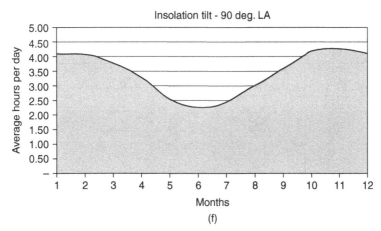

Insolation tilt - 90 deg. LA

(f)

Figure 2.17   (*Continued*)

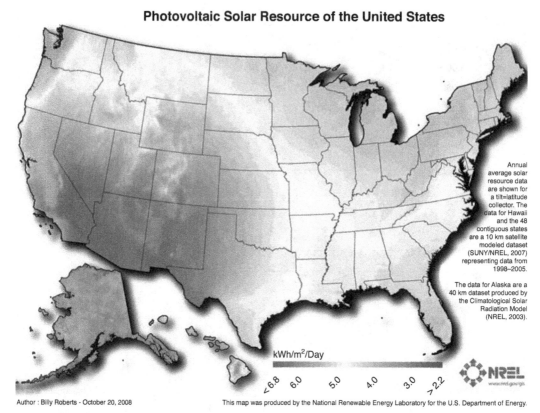

Figure 2.18   **Photovoltaic solar resource map of the United States.** *Courtesy of NREL.*

Reference

1. Gevorkian, Peter, *Alternative Energy Systems in Building Design,* McGraw-Hill, New York, 2010.

3

SOLAR PHOTOVOLTAIC POWER
SYSTEM COMPONENTS

Introduction

The photovoltaic modules discussed in Chapter 1 represent only one of the basic elements of a solar power system. They work in conjunction with complementary components, inverters, solar-tracker isolation transformers, power distribution panels, and storage battery systems that are essential to the solar power energy conversion process. This chapter will explore these other components.

Inverters

As described previously, PV panels generate direct current that can only be used by a limited number of devices. Most residential, commercial, and industrial devices and appliances are designed to work with alternating current. Inverters are devices that convert direct current into alternating current. Although inverters are usually designed for specific application requirements, the basic conversion principles remain the same. Essentially, the inversion process consists of the following.

THE WAVE FORMATION PROCESS

This process is direct current characterized by a continuous potential of positive and negative references (bias). It is essentially chopped into equidistant segments, which are then processed through circuitry that alternately eliminates positive and negative portions of the chopped pattern, which then results in a waveform pattern called a square wave. Figure 3.1 shows an inverter single line diagram.[1]

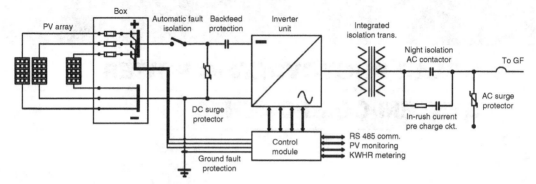

Figure 3.1   **Inverter single-line diagram.** *Courtesy of SatCon Canada.*

WAVESHAPING OR FILTRATION PROCESS

A square wave, when analyzed mathematically (by Fourier series analysis), consists of a combination of a very large number of sinusoidal (alternating) wave patterns called harmonics. Each wave harmonic has a distinct number of cycles (rise-and-fall patterns within a time period).

An electronic device referred to as a choke (magnetic coils) filters or passes through 60-cycle harmonics, which form the basis of sinusoidal current. Solid-state inverters use a highly efficient conversion technique known as envelope construction. Direct current is sliced into fine sections, which are then converted into a progressive rising (positive) and falling (negative) sinusoidal, 60-cycle waveform pattern. This chopped sinusoidal wave is passed through a series of electronic filters that produce an output current, which has a smooth sinusoidal curvature.[1]

PROTECTIVE RELAYING SYSTEMS

In general, most inverters used in photovoltaic applications are built from sensitive, solid-state electronic devices that are very susceptible to external stray spikes, load short circuits, and overload voltages and currents. To protect the equipment from harm, inverters incorporate a number of electronic circuitry:

- Synchronization relay
- Undervoltage relay
- Overcurrent relay
- Ground trip or overcurrent relay
- Overvoltage relay
- Overfrequency relay
- Underfrequency relay

Most inverters designed for PV applications allow simultaneous paralleling of multiple units. For instance, to support a 60-kW load, outputs of three 20-kW inverters may be connected in parallel. Depending on the power system requirements, inverters can produce single- or three-phase power at any required voltage or current capacity. Standard outputs available are single-phase 120-V AC and three-phase 120/208- and 277/480-V AC. In some instances, step-up transformers are used to convert the output of 120/208-V AC inverters to higher voltages.

INPUT AND OUTPUT POWER DISTRIBUTION

To protect inverters from stray spikes, resulting from lightning or other such high-energy spikes, DC inputs from PV arrays are protected by fuses housed at a junction boxes located in close proximity to the inverters. Additionally, inverter DC input ports are protected by various types of semiconductor devices that clip excessively high voltage spikes resulting from lightning activity.

To prevent damage resulting from voltage reversal, each positive (+) output lead within a PV cell is connected to a rectifier, a unidirectional (forward-biased) element. AC output power from inverters is connected to the loads by means of electronic or magnetic-type circuit breakers. These serve to protect the unit from external overcurrent and short circuits.[2]

TYPES OF PHOTOVOLTAIC INVERTER SYSTEMS

In general, photovoltaic inverter systems are classified in two categories: stand-alone or *grid independent* and utility-interactive or grid connected. As the classification names suggest, the main difference between them is whether the inverters are connected to batteries or are connected to the electrical grid.

Stand-alone inverter systems   Stand-alone inverters are usually connected to battery banks, which provide the DC power. In PV systems, PV modules provide the DC source, which is transformed to CA power. In such applications, the principle functions of the solar power PV modules are connected to the inverter via a charge controller to the batteries. The inverters must be sized to meet the AC demand load requirements, and PV modules are sized to provide adequate charge to support the battery system. In stand-alone type systems, the loads are directly impacted by the charge storage capacity of the battery system.

Utility-interactive inverters   Utility-interactive inverters are designed to operate in parallel with the electrical utility grid systems. In grid-connected systems, PV modules provide DC power, which is directly connected to the inverters. The AC power produced by the inverters is either directly used by the load, or in the case of surplus power, will feed the grid system in synchronous manner. Interactive AC power output is directly proportional to the photovoltaic DC source.

In grid-connected applications, PV systems are connected to the utility at the point-of-service entrance, on the supply- or load-distribution side of the panel. In essence, the grid acts as an infinitely large electrical energy storage reservoir that can store surplus energy and supply it back whenever the need arises. This forward and backward transaction of electrical energy is realized by means of a net-metering system, which adds or subtracts the amount of energy used or delivered to the grid.

Bimodal inverters   Bimodal inverters include specific electronic circuitry that allows them to operate in either interactive or stand-alone mode. Bimodal inverters are generally deployed in applications in which, under solar irradiance conditions, the PV module provides the DC current directly to the inverter for conversion. In the meantime, it delivers sufficient charge to the battery charge controller to keep the battery fully charged. In the absence of solar irradiance, the AC load is sustained by the battery backup system. Bimodal inverters usually are used in systems that require relatively small on-site loads. Regardless of the size or type, inverters are constructed with sophisticated electronic and electrical power technologies. Large grid-connected inverters use sophisticated design features, such as *anti-islanding protection, maximum power point tracking,* and *sine wave output* generation circuitry.

The anti-islanding protection feature is specifically designed to turn off the inverter during grid outages. This feature prevents accidental electrical shock hazard to linespersons working on the power line during blackout periods.

DC TO AC SWITCHING MECHANISM

In large inverters, DC to AC power conversion takes place by utilizing solid-state electronic circuits and switching devices. The principle electronic device used in solid-state-type inverters is a thyristor, which is also know as a silicon-controlled rectifier (SCR), shown in Fig. 3.2. SCRs are essentially electronic current valves or gates that allow

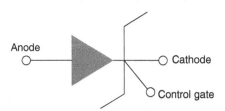

Silicon-controlled rectifier
thyristor

Figure 3.2   **Silicon-controlled rectifier.**

current flow in one direction only when a control signal is applied to the gate. Similar to a wall-mount light switch, an SCR can be in either an on or off position. SCRs are in essence heavy-duty electronic power switches that are capable of controlling thousands of amperes of current flow. In general, all large-capacity inverters designed to operate above 10 kW, such as120/208-V, or 277/480-V three-phase systems, use SCR-switching devices.

On the other hand, smaller-size inverters use metal-oxide semiconductor field-effect transistors (MOSFET) or insulated gate bipolar transistors transistors (IGBT) to for power switching or commutation. MOSFET-switching devices are particularly well suited for high-speed switching that produces clean AC sine wave output. MOSFET inverter power-output capacities are usually limited to a maximum of 10 kW. However, IGBT-switching devices (which operate at lower commutation speed) are used in high-voltage inverter designs that have capacities exceeding 100 kW.

In each of the described switching devices used in inverters to convert DC to AC power, control circuitry creates sequential commutation or switching functions. In some instances, switching is accomplished by an external signal derived from the grid, referred to as line commutation synchronization, and in some instances, an internal microprocessor generates timing signals that control the switching process (self-commutation).

LINE COMMUTATED INVERTERS

In self-commutation-type inverters, switching is controlled automatically by an external source such as the grid power. This class of inverter commutation takes place by alternately turning the switches on and off by positive and negative half-cycles to the utility voltage sinusoidal waveforms, which automatically synchronize the inverter output waveform and frequency with the grid. The main drawback of line-commutated inverters is that they are grid-dependent and cannot operate in stand-alone mode. Figure 3.3 depicts line-commutated inverter grid synchronization using external line voltage envelop for triggering the switching devices.

SELF-COMMUTATED INVERTERS

Self-commutated inverters use specially designed timing control circuitry that controls switching device activation and deactivation. The main advantage of this type of inverter is that it can function with or without external grid synchronization signals. As mentioned, the switching function is accomplished by an internal microprocessor timing-control system that, in addition to external-line synchronization, allows the inverters to have superior control of the external output waveform, as well as additional features such as power factor correction and harmonic suppression.

Self-commutated inverters are also classified as either a voltage or a current-output source. Voltage-source inverters use the input DC voltage to generate a voltage source and deliver AC voltage output that has constant amplitudes with variable width. On the other hand, current-output source inverters use the DC-input voltage and deliver AC-current output at constant amplitude and variable width. Stand-alone and bimodal inverters are generally voltage-source type.

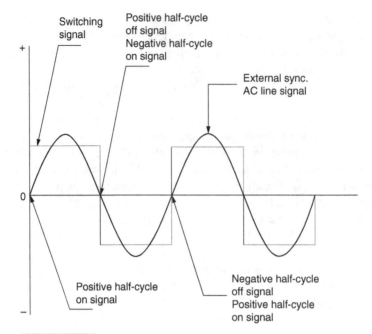

Figure 3.3   Line-commutated inverter DC-to-AC switching.

SQUARE WAVE INVERTERS

Square wave inverters are a class of equipment that uses multiple stages of switching process to convert the input DC into a sine wave AC power. Square wave output waveforms are quite inefficient; they contain significant amounts of harmonics, which can be very harmful to connected loads. To transform the square wave in to a sinusoidal waveform, inverters use sophisticated wave-shaping circuitry and filters that modify the wave into a sinusoidal form.

Wave-shaping control circuits use specific timing mechanisms that symmetrically and sequentially control gates of the switching devices, which convert the DC input into positive and negative half-waves of AC output. Types of square wave conversion and filtering technologies include H-bridge systems and push-pull systems.

H-bridge inverter circuits   H-bridge inverter circuits are similar in function to full-wave rectifier circuits, however, they operate in reverse. An H-bridge inverter, using a set of controlled switches described previously, switches the DC input power into square AC output power. As shown in Fig. 3.4, switching occurs when one pair of switches opens while the other two close in a back-and-forth alternating manner. This results in the transformation of DC to AC power.

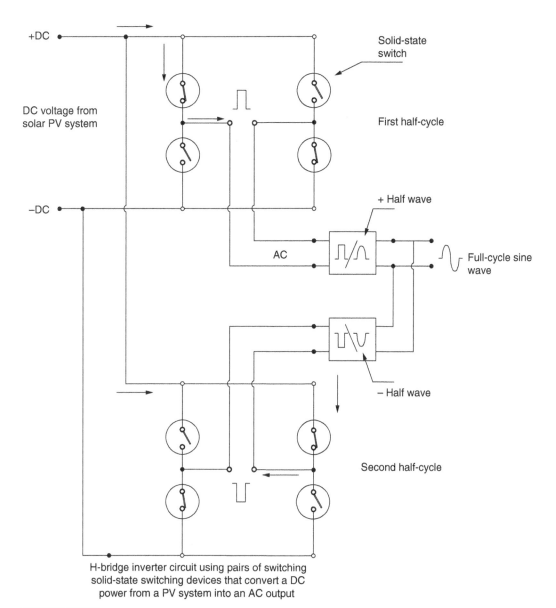

+DC

DC voltage from
solar PV system

−DC

Solid-state
switch

First half-cycle

AC

+ Half wave

Full-cycle sine
wave

− Half wave

Second half-cycle

H-bridge inverter circuit using pairs of switching
solid-state switching devices that convert a DC
power from a PV system into an AC output

Figure 3.4   H-bridge inverter system diagram.

Push-pull inverters circuits   As shown in Fig. 3.5, in push-pull inverter systems, DC-to-AC power conversion takes place by switching current flow in a center-tapped transformer. The switching process occurs when the top switch closes, allowing flow of current from the DC source through the upper half of the transformer in the first half-cycle. The reverse occurs in the second half-cycle, which results in the production of full-cycle AC power.

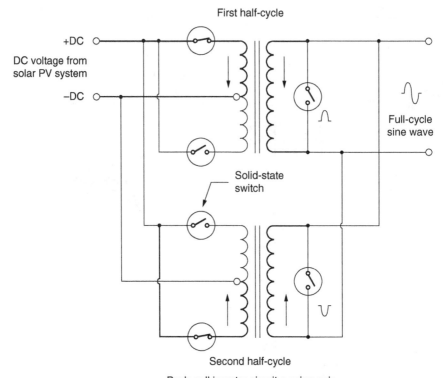

Figure 3.5 **Push-pull inverter system diagram.**

HIGH- AND LOW-FREQUENCY WAVEFORM SWITCHING

In order to generate a sine wave, the square wave output of inverters is refined to remove the excess harmonics through a process referred to an improved sine approximation. The approximation is realized by adjusting the duration of the square-wave pulses, which track a sinusoidal waveform envelope called square-wave modification. Output voltage-level control is achieved by deploying step-up transformers. In order to generate a multistep square wave, as shown in Fig. 3.6*a*, *b*, and *c*, the square waves are superimposed on each other, forming a close match to a sine wave. Inverters that use 50- or 60-cycle frequency sine-wave generation are referred to as low-frequency inverters.

Another type of sine-wave generation involves sine-wave generation by a technique called pulse-width modulation (PWM) control. Sine-wave generation in this type of inverter is accomplished by switching the commutation switches via pulses that have a variable length of time (Figs. 3.6*a*–*c*).

High-frequency pulse-width modulated inverters have an advantage in that the power converted has the least amount of harmonics in the current. Additionally, because of the smaller size of the transformer, they are lighter and more efficient. These types of

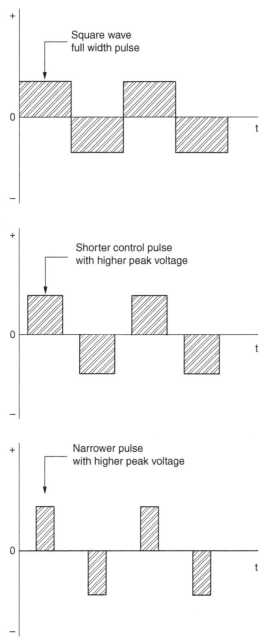

Square wave
full width pulse

Shorter control pulse
with higher peak voltage

Narrower pulse
with higher peak voltage

Square wave duration and magnitude control
by gate pulse duration control

(a)

Figure 3.6 (*a*) Square-wave
generation.

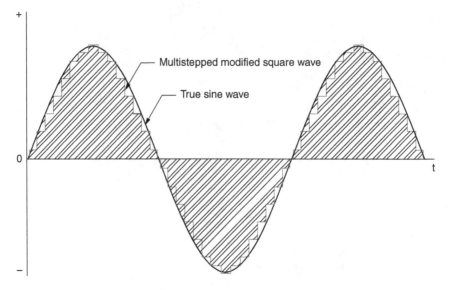

Sine wave generation by combination of multiple modified
square wave with varying magnitudes and duration

(b)

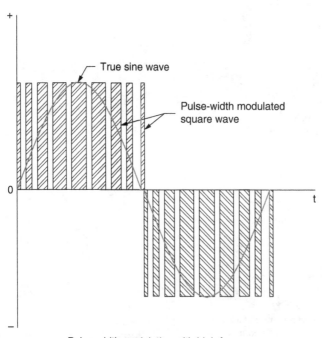

Pulse-width modulation with high frequency
since wave generation by approximation

(c)

Figure 3.6   (*b*) Sine-wave generation formed from modified square
waves. (*c*) Sine-wave generation by high-frequency pulse modulation.

inverters also use a DC-DC converter to step up the input DC voltage swings to 1000 V, which allows longer PV string circuits that require less material and manpower.

INVERTER GROUNDING SYSTEMS

We now discuss the application of ungrounded DC photovoltaic systems in the United States and give guidance on their use within the National Electric Code (NEC).

Traditionally, the United States has used only grounded DC systems, whereas Europe and Japan have used ungrounded systems. The Europeans have demonstrated that ungrounded systems result in higher system efficiencies. Now there is a trend in the United States to begin to use ungrounded systems to take advantage of these higher efficiencies. We explore where the improvement of efficiency comes from, and the need for using ungrounded systems, and how the National Electric Code (NEC) is addressing ungrounded systems.

Ground isolated inverters   Inverter topology determines the application of a grounded versus an ungrounded system. In the United States (with grounded systems), inverters incorporate an isolation transformer in their topology. Figure 3.7*a* and *b* show

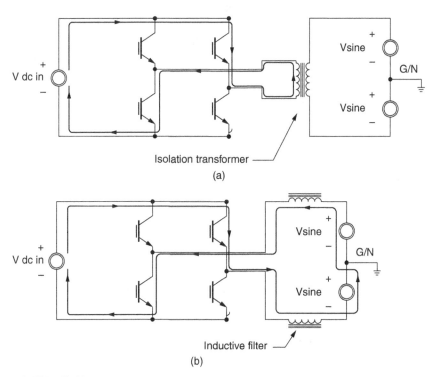

Figure 3.7   (*a*) Grounded system inverter. (*b*) Ungrounded system inverter.

typical single-phase grid-tie inverters used in the United States (PV powered). The isolation transformer represents a significant part of the overall volume and cost of the inverter. The isolation transformer provides two features to the inverter:

Output Filter   The transformer, being a reactive component, helps filter the inverter's pulse-width modulation signal. The transformer is generally not the only reactive component in the filter design.

An inductor and a capacitor provide additional filtering so that a pure sine wave is generated at the inverter's AC output.

Voltage Step-Up   The maximum voltage that an inverter can output is approximately 10 percent less than the maximum voltage that can be produced on the DC side of the system. In the line isolation system (LIS), the typical maximum DC voltage is 600 VDC, with the actual operating voltage as low as 330 VDC.

The output voltage of the inverter, on the other hand, must match the grid's maximum voltage. For example, in a 240-V AC installation, the peak grid voltage can be as high as 373 V. For a 480-V three-phase AC-system maximum-peak grid, voltage can be as much as 747 V. An isolation transformer permits stepping-up the voltage to match the grid voltage for a given input DC voltage.

As shown in Fig. 3.7a and b, when decoupling between the inverter AC and DC with grounded systems, the DC section of the inverter system requires isolation from the AC system so that each system's grounds are not coupled through the source circuits.

Isolation transformers typically generate system losses of approximately 1 to 2 percent and lower the overall system efficiency with the same proportion. To improve the efficiency of inverter, the inverter isolation transformers should be eliminated. Eliminating the isolation transformers requires special compensation circuits that provide the same features listed previously. Some of the circuits required to fulfill these features include:

1 *Output Filter.* The output filter circuits are designed to include filter elements, by addition of an inductor and/or capacitor.
2 *Output Voltage Scale-Up.* The output voltage scale-up is achieved by using higher-input DC voltage. For small single-phase inverters, the inverter input swing voltage band ranges 330-V to 6001-V DC. On the other hand, for 480-V AC three-phase systems, the inverter input DC voltage swing band is extended to 10-DV DC.
3 *Grounding System.* Figure 3.7a depicts an inverter (as input and output current sources) schematic with and without an isolation transformer. In both diagrams a and b, the DC system is grounded. The right-hand side of the graphic a shows an isolation transformer with no direct path between the DC and AC grounds. However, in the right-hand side of graphic b, the transformer is replaced with line inductors that provide a direct short through the coupled grounds.

In order to avoid the coupled ground problem, either the AC or the DC ground must be removed. The removal of the AC ground is not possible because the NEC does not permit it. The removal of the DC ground (ungrounded system) is presently addressed by the NEC.

I short; an uncomplicated way to improve the efficiency of an inverter is to eliminate the isolation transformer. In general, the ungrounded inverters which do not have isolation transformer allow the input DC voltages to be higher.

The NEC, in order to harmonize with the European standards (IEC), has recently upgraded the code permitting ungrounded DC systems. In order to use ungrounded systems, the Underwriters Laboratories (UL) standard mandates use of double-insulated wire (referred to as PV wire), which has identical specifications to wire used in Europe.

NEC requirement for installing an ungrounded PV system   The solar power component interconnection is addressed in NEC under section 690.32. The requirements include the following:

- Disconnects. All photovoltaic source and output circuit conductors shall have disconnects complying with 690, Part III.
- Overcurrent Protection. All photovoltaic source and output circuit conductors shall have overcurrent protection complying with 690.9.
- Ground-Fault Protection. All photovoltaic source and output circuits shall be provided with a ground protection device or system that complies with: detects a ground fault; indicates that a ground fault has occurred; automatically disconnects all conductors of causes the inverter or charge controller connected to the fault circuit to automatically cease supplying power to output circuits.
- Source Circuit Conductors. Photovoltaic source circuit conductors shall consist of the following: non-metallic jacketed multi-conductor cable; conductors installed in a raceway; or conductors listed and identified as photovoltaic (PV) wire installed as exposed single conductors.

Underwriters Laboratories (UL) standards   For product safety, the industry in the United States has worked with Underwriters Laboratories (UL) to develop *UL1741, Standard for Static Inverter and Charge Controller for Use in Independent Power Systems.* This has become the safety standard for inverters being used in the United States. Standard UL1741 covers many aspects of inverter design, including enclosures, printed circuitboard configurations, interconnectivity requirements (such as the amount of direct current the inverters can inject into the grid), total harmonic distortion (THD) of the output current, inverter reaction to utility voltage spikes and variations, reset and recovery from abnormal conditions, and reaction to islanding conditions when the utility power is disconnected.

Islanding is a condition that occurs when the inverter continues to produce power during a utility outage. Under such conditions, the power produced by a PV system becomes a safety hazard to utility workers who could be inadvertently exposed to hazardous electric currents. Because of this, inverters are required to include anti-islanding control circuitry to cut the power to the inverter and disconnect it from the grid network.

Anti-islanding also prevents the inverter output power from getting out of phase with the grid when the automatic safety interrupter reconnects the inverter to the grid (which could result in high-voltage spikes that can cause damage to conversion and utility equipment). Figures 3.8 and 3.9 depict electronics of a SOLECTRIA Renewables inverter and a typical installation.

Institute of Electrical and Electronics Engineers   The Institute of Electrical and Electronics Engineers (IEEE) provides suggestions for customers and utilities alike regarding the control of harmonic power and voltage flickers that frequently occur on

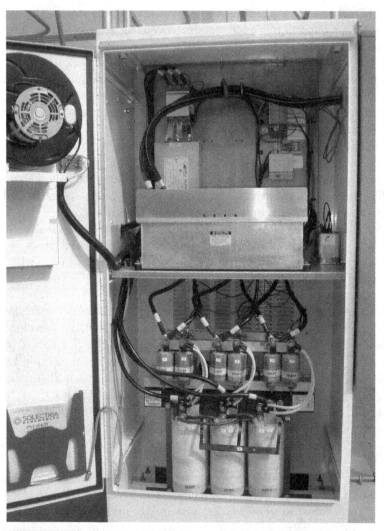

Figure 3.8   **View of inverter electronics.** *Photo courtesy of Solectria Renewables.*

Figure 3.9   View of SATCON 500-kW inverter. *Photo courtesy of Solectria Renewables.*

utility buses in its IEEE 929 guideline (not a standard), *Recommended Practice for Utility Interface of Photovoltaic (PV) Systems.* Excessive harmonic power flow and power fluctuation from utility buses can damage a customers' equipment. Therefore, a number of states including California, Delaware, New York, and Ohio specifically require that inverters be designed to operate under abnormal utility power conditions.

POWER LIMIT CONDITIONS

The maximum size of a PV power cogeneration system is subject to limitations imposed by various states. Essentially, most utilities are concerned about large sources of private grid-connected power generation, since most distribution systems are designed for unidirectional power flow. The addition of a large power cogeneration system, on the other hand, creates bidirectional current flow conditions on the grid, which in some instances can diminish utility network reliability. However, it is well known that small amounts of cogenerated power do not usually create a grid disturbance significant enough to be a cause for concern. To regulate the maximum size of a cogeneration system, a number of states have set various limits and caps for systems that generate in excess of 100 kW of power.

UTILITY-SIDE DISCONNECTS AND ISOLATION TRANSFORMERS

In the states of California, Delaware, Florida, New Hampshire, Ohio, and Virginia, utilities require that visible and accessible disconnect switches be installed outside for grid service isolation. It should be noted that several states, including California, require that customers open the disconnect switches once every four years to check that the inverters are performing the required anti-islanding.

In other states, such as New Mexico and New York, grid isolation transformers are required in order to reduce noise created by private customers that could be superimposed on the grid. However, this requirement is not a regulation that is mandated by the UL or the Federal Communication Commission (FCC).

PV POWER COGENERATION CAPACITY

In order to protect utility companies' norms of operation, a number of states have imposed a cap on the maximum amount of power that can be generated by photovoltaic systems. For example, New Hampshire limits the maximum to 0.05 percent, and Colorado to 1 percent of the monthly grid network peak demand.

INVERTER SURGE WITHSTAND CAPABILITY

In most instances, power distribution is undertaken through a network of overhead lines that are constantly exposed to climatic disturbances such as lightning, which results in power surges. Additional power surges could also result from switching capacitor banks used for power factor correction, power conversion equipment, or during load shedding and switching. The resulting power surges, if not clamped, could seriously damage inverter equipment by breaking down conductor insulation and electronic devices.

To prevent damage caused by utility spikes, the IEEE has developed nationally recommended guidelines for inverter manufacturers to provide appropriate surge protection. A series of tests devised to verify IEEE recommendations for surge immunity are performed by the UL as part of equipment approval.

PV SYSTEM TESTING AND MAINTENANCE LOG

States including California, Vermont, and Texas require that comprehensive commission testing be performed on PV system integrators to certify that the system is operating in accordance with expected design and performance conditions. It is interesting to note that for PV systems installed in the state of Texas, a log must be maintained of all maintenance performed.

EXAMPLE OF A UL1741 INVERTER

The following is an example of a UL1741-approved inverter manufactured by SatCon, Canada.

An optional combiner box, which includes a set of special ceramic overcurrent protection fuses, provides accumulated DC output to the inverter. At its DC input, the inverter is equipped with an automatic current fault isolation circuit, a DC surge protector, and a DC backfeed protection interrupter. In addition to the preceding, the inverter has special electronic circuitry that constantly monitors ground faults and provides instant fault isolation. Upon the conversion of DC to AC, the internal electronics of the inverter provide precise voltage and frequency synchronization with the grid.

An integrated isolation transformer within the inverter provides complete noise isolation and filtering of the AC output power. A night isolation AC contactor disconnects the inverter at night or during heavy cloud conditions. The output of the inverter also includes an AC surge isolator and a manual circuit breaker that can disconnect the equipment from the grid.

A microprocessor-based control system within the inverter includes waveform envelope construction and filtering algorithms, and a number of program subsets that perform anti-islanding, voltage, and frequency control.

As an optional feature, the inverter can also provide data communication by means of an RS-485 interface. This RS-485 interface can transmit equipment operational and PV-measurement parameters such as PV-output power, voltage, current, and totalized kW-hour metering data for remote monitoring and display.

Storage Battery Technologies

One of the most significant components of solar power systems consists of battery backup systems that are frequently used to store electric energy harvested from solar PV systems for use during the absence of sunlight (such as at night and during cloudy conditions). Because of the significance of storage battery systems, it is important for design engineers to have a full understanding of the technology since this system component represents a notable portion of the overall installation cost. More importantly, the designer must be mindful of the hazards associated with handling, installation, and maintenance. To provide an in-depth knowledge about the battery technology, this section covers the physical and chemical principles, manufacturing, design application, and maintenance procedures of the storage battery. In this section, we also attempt to analyze and discuss the advantages and disadvantages of different types of commercially available solar power batteries and their specific performance characteristics.

The battery is an electric energy storage device that in physics terminology can be described as a device or mechanism that can hold kinetic or static energy for future use. For example, a rotating flywheel can store dynamic rotational energy in its wheel and releases the energy when the primary mover, such as a motor, no longer engages the connecting rod. Similarly, a weight held at a high elevation stores static energy embodied in the object mass, which can release its static energy when dropped. Both of these are examples of energy storage devices or batteries.

Energy storage devices can take a wide variety of forms, such as chemical reactors and kinetic and thermal energy storage devices. It should be noted that each energy storage device is referred to by a specific name; the word battery, however, is solely used for electrochemical devices that convert chemical energy into electricity by a process referred to as galvanic interaction. A galvanic cell is a device that consists of two electrodes, referred to as the anode and the cathode, and an electrolyte solution. Batteries consist of one or more galvanic cells.

It should be noted that a battery is an electrical storage reservoir and not an electricity-generating device. Electric-charge generation in a battery is a result of chemical interaction, a process that promotes electric charge flow between the anode and the cathode in the presence of an electrolyte. A recharging process that can be repeated numerous times resurrects the electro-galvanic process that eventually results in the depletion of the anode and cathode plates. When delivering stored energy, batteries incur energy losses as heat when discharging or during chemical reactions when charging.

MAJOR BATTERY TYPES

Solar power backup batteries are divided into two categories based on what they are used for and how they are constructed. The major applications where batteries are used as solar backup include automotive systems, marine systems, and deep-cycle discharge systems.

The major manufactured processes include flooded or wet construction, gelled electrolyte, and absorbed glass mat (AGM) types. AGM batteries are also referred to as "starved electrolyte" or "dry" type, because instead of containing wet sulphuric acid solution, the batteries contain a fiberglass mat saturated with sulphuric acid that has no excess liquid. Figure 3.10 depicts storage battery operation principle.

Common flooded-type batteries are usually equipped with removable caps for maintenance-free operation. Gelled-type batteries are sealed and equipped with a small vent valve that maintains a minimal positive pressure. Absorbed glass mat batteries are also equipped with a sealed regulation-type valve that controls the chamber pressure within 4 lb/in^2.

As discussed earlier, common automobile batteries are built with electrodes that are grids of metallic lead containing lead oxides that change in composition during charging and discharging. The electrolyte is diluted sulphuric acid. Lead-acid batteries, even

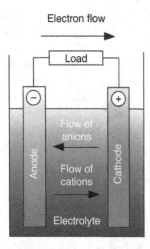

Figure 3.10   Storage battery operation principle.

though invented nearly a century ago, are still the battery of choice for solar and backup power systems. With improvements in manufacturing, batteries can last as long as 20 years.

Nickel-cadmium or alkaline storage batteries, in which the positive active material is nickel oxide and the negative material contains cadmium, are generally considered very hazardous because of the cadmium. The efficiency of alkaline batteries ranges from 65 to 80 percent compared to 85 to 90 percent for lead-acid batteries. Their nonstandard voltage and charging current also make them very difficult to use.

Deep-discharge batteries used in solar power backup applications generally have lower charging and discharging rate characteristics and are more efficient.

In general, all batteries used in PV systems are lead-acid type batteries. Alkaline-type batteries are used only in exceptionally low temperature conditions of below 5°F. Alkaline batteries are expensive to buy and due to the hazardous contents are very expensive to dispose of.

BATTERY LIFE SPAN

The life span of a battery will vary considerably depending on how it is used, how it is maintained and charged, the temperature, and other factors. In extreme cases, it can be damaged within 10 to 12 months of use when overcharged. On the other hand, if the battery is maintained properly, the life span could be extended over 25 years. Another factor that can shorten the life expectancy by a significant amount is when the batteries are stored uncharged in a hot storage area. Even dry-charged batteries when sitting on a shelf have a maximum life span of about 18 months; as a result, most are shipped from the factory with damp plates. As a rule, deep-cycle batteries can be used to start and run marine engines. When starting, engines require a very large inrush of current for a very short time. Regular automotive starting batteries have a large number of thin plates for maximum surface area. The plates, as described previously, are constructed from impregnated lead-paste grids similar in appearance to a very fine foam sponge. This gives a very large surface area, and when deep-cycled, the grid plates quickly become consumed and fall to the bottom of the cells in the form of sediment. Automotive batteries will generally fail after 30 to 150 deep-cycles, if they are indeed deep-cycled; they may last for thousands of cycles in normal starting use discharge conditions. Deep-cycle batteries are designed to be discharged down time after time and are designed with thicker plates.

The major difference between a true deep-cycle battery and regular batteries is that the plates in a deep-cycle battery are made from solid lead plates and are not impregnated with lead oxide paste. Figure 3.11 is a single-line diagram of a battery-backed solar power system. Figure 3.12 shows a typical solar battery bank system.

The stored energy in batteries, in general, is discharged rapidly. For example, short bursts of power are needed when starting an automobile on a cold morning, which results in high amounts of current being rushed from the battery to the starter. The standard unit for energy or work is the joule (J), which is defined as 1 watt-second (W-s) of mechanical work performed by a force of 1 newton (N) or 0.227 lb pushing or moving

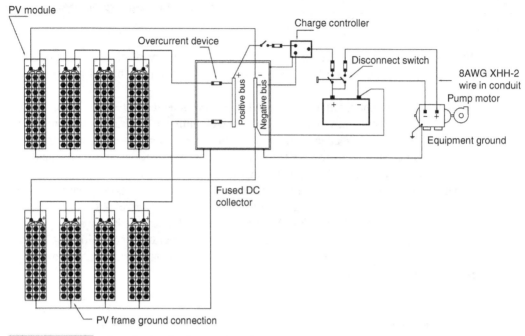

Figure 3.11   A typical solar battery bank system.

Figure 3.12   **Deep-cycle battery packs.** *Photo courtesy of Solar Integrated Technologies.*

a distance of 1 m. Since 1 hour has 3600 seconds, 1 watt-hour (Wh) is equal to 3600 J. The stored energy in batteries is either measured in milliampere (mA) hours if small, or ampere-hours (Ah) if large. Battery ratings are converted to energy if their average voltages are known during discharge. In other words, the average voltage of the battery is maintained relatively unchanged during the discharge cycle. The value in joules can also be converted into various other energy values as follows:

Joules divided by 3,600,000 yields kilowatt-hours.

Joules divided by 1.356 yields English units of energy foot-pounds.

Joules divided by 1055 yields British thermal units.

Joules divided by 4184 yields calories.

BATTERY POWER OUTPUT

In each instance, when power is discharged from a battery, the battery's energy is drained. The total quantity of energy drained equals the amount of power multiplied by the time the power flows. Energy has units of power and time, such as kilowatt-hours or watt-seconds. The stored battery energy is consumed until the voltage and current available to levels of the battery are exhausted. Upon depletion of stored energy, batteries are recharged over and over again until they deteriorate to a level where they must be replaced by new units. High-performance batteries generally have the following notable characteristics. First, they must be capable of meeting the power demand requirements of the connected loads by supplying the required current while maintaining a constant voltage. Second, they must have sufficient energy storage capacity to maintain the load power demand as long as required. Finally, they must be as inexpensive and economical as possible and be readily replaced and recharged.

BATTERY INSTALLATION AND MAINTENANCE

Unlike many electrical apparatuses, standby batteries have specific characteristics that require special installation and maintenance procedures. If not followed correctly, these can impact the quality of the battery performance.

As mentioned previously, the majority of today's emergency power systems make use of two types of batteries: lead-acid and nickel-cadmium (NiCad). Within the lead-acid family, there are two distinct categories, namely flooded or vented (filled with liquid acid) and valve-regulated lead acid (VRLA, immobilized acid).

Lead-acid and NiCad batteries must be kept dry at all times and in cool locations (preferably below 70°F), and must not be stored for long in warm locations. Materials such as conduits, cable reels, and tools must be kept away from the battery cells.

BATTERY INSTALLATION SAFETY

What separates battery installers from the layperson is the level of awareness and respect for DC power. Energy stored in the battery cell is quite high and sulphuric acid (lead-acid batteries) or potassium hydroxide (a base used in NiCad batteries) electrolytes could be very harmful if not handled professionally. Care should always be exercised when handling these cells. The use of chemical-resistant gloves, goggles, a face shield, and protective sleeves is highly recommended. The battery room must be equipped with an adequate shower or water sink to provide for rinsing of the hands and eyes in case of accidental contact with the electrolytes. Stored energy in a single NiCad cell of 100-Ah capacity can produce about 3000 A if short-circuited between the terminal posts. A fault across a lead-acid battery can send shrapnel and terminal post material flying in any direction, which can damage the cell and endanger workers.[2]

RACK CABINET INSTALLATION

Stationary batteries must be mounted on open racks or on steel or fiberglass racks or enclosures. The racks should be constructed and maintained in a level position and secured to the floor and must have a minimum of 3 feet of walking space for egress and maintenance.

Open racks are preferable to enclosures since they provide a better viewing of electrolyte levels and plate coloration. They also have easier access for maintenance. For multistep or bleacher-type racks, batteries should always be placed at the top or rear of the cabinet to avoid anyone having to reach over the cells. The manufacturer-supplied connection diagram should always be used to ensure the open positive and negative terminals when charging the cells. If possible, it is important to delay delivery in the event of installation scheduling delays.

BATTERY SYSTEM CABLES

Appendix A provides code-rated DC cable tables for a variety of battery voltages and feed capacities. The tables provide American wire gauge (AWG) conductor gauges and voltage drops calculated for a maximum of a 2 percent drop. Whenever larger drops are permitted, the engineer must refer to NEC tables and perform specific voltage-drop calculations.

BATTERY CHARGER CONTROLLER

A charge controller is essentially a current-regulating device that is placed between the solar panel array output and the batteries. These devices are designed to keep batteries charged at peak power without overcharging. Most charge controllers incorporate special electronics that automatically equalize the charging process.

DC FUSES

All fuses that are used as overcurrent devices that provide a point of connection between PV arrays, and collector boxes must be DC-rated. Fuse ratings for DC branch circuits, depending on wire ampacities, are generally rated from 15 to 100 A. The DC-rated fuses familiar to solar power contractors are manufactured by a number of companies, such as Bussman, Littlefuse, and Gould. They can be purchased from electrical suppliers. Various manufacturers identify the fuse voltage by special capital letter designations.

As a rule, PV output must be protected with extremely fast-acting fuses. The same fuses can also be utilized within solar power control equipment and collector boxes. Some of the fast-acting fuses used are manufactured by the same companies just listed.

Junction boxes and equipment enclosures   All junction boxes utilized for interconnecting raceways and conduits must be of waterproof construction and be designed for outdoor installation. All equipment boxes, such as DC collectors, must either be classified as MENA 3R or NEMA 4X (outdoor use enclosures).

Lightning Protection

In geographic locations such as Florida, where lightning is a common occurrence, the entire PV system and outdoor-mounted equipment must be protected with appropriate lightning-arrestor devices and special grounding. This will provide a practical mitigation and a measure of protection from equipment damage and burnout.

LIGHTNING EFFECT ON OUTDOOR EQUIPMENT

Lighting surges are comprised of two elements: voltage and the quantity of charge delivered by lightning. The high voltage delivered by lightning surges can cause serious damage to equipment since it can break down the insulation that isolates circuit elements and the equipment chassis. The nature and the amount of damage are directly proportional to the amount of current resulting from the charge.

In order to protect equipment damage from lightning, devices known as surge protectors or arrestors are deployed. The main function of a surge arrestor is to provide a direct conduction path for lightning charges to divert them from the exposed equipment chassis to the ground. A good surge protector must be able to conduct a sufficient current charge from the stricken location and lower the surge voltage to a safe level quickly enough to prevent insulation breakdown or damage.

In most instances, circuits have a capacity to withstand certain levels of high voltages for a short time; however, the thresholds are so narrow that if charges are not removed or isolated in time, the circuits will sustain an irreparable insulation breakdown.

The main purpose of a surge arrestor device is therefore, to conduct the maximum amount of charge and reduce the voltage in the shortest possible time. Reduction of a voltage surge is referred to as voltage clamping, as shown in Figures 13.13a through c. In general, clamping depends on device characteristics, such as internal resistance, the response speed of the arrestor, and the point in time at which the clamping voltage is measured.

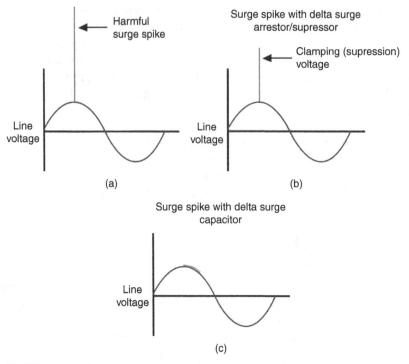

Figure 3.13   (*a*) Effect of lighting surge spike. (*b*) Effect of lightning surge spike clamping. (*c*) Effect of lightning surge spike suppression.

When specifying a lightning arrestor, it is necessary to take into account the clamping voltage and the amount of current to be clamped (for example, 500 V and 1000 A). Let us consider a real-life situation in which the surge rises from 0 to 50,000 V in 5 ns. At any time during the surge (say at 100 ns), the voltage clamping would be different from the lapsed time (say at 20 ns), where the voltage could have been 25,000 V. Nevertheless, the voltage is arrested with a high current rating with conductivity and will remove the surge current from the circuit more rapidly, therefore providing better protection. Figure 3.14 is a graphic diagram showing deployment of a lightning surge arrestor in a rectifier circuit.

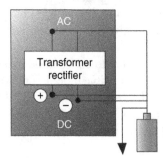

Figure 3.14   Deployment of a lightning surge arrestor in a rectifier circuit.

Central Monitoring and Logging System

In large commercial solar power cogeneration systems, power production from the PV arrays is monitored by a central monitoring system, which provides a log of operation performance parameters. The central monitoring station consists of a PC-type computer that retrieves operational parameters from a group of solar power inverters by means of an RS-232 interface, a power line carrier, or wireless communication system. Upon receipt of the performance parameters, a supervisory software program processes the information and provides data in display or print format. Supervisory data obtained from the file can also be accessed from distant locations through Web networking.

Some examples of monitored data are:

- Weather-monitoring data
- Temperature
- Wind velocity and direction
- Solar power output
- Inverter output
- Total system performance and malfunction
- DC power production
- AC power production, accumulated daily, monthly, and annually

The following solar power monitoring system represents an example of an integrated data acquisition system, which has been designed to acquire and display real-time performance parameters regarding electrical power performance and atmospheric data.

The system hardware configuration of a data acquisition system consists of a desktop computer and data-logging software (which processes and displays measured parameters from the following sensors and equipment):

- Anemometer (for measuring meteorological data)
  - Ambient air temperature sensor
  - Wind speed
  - Outdoor air temperature sensor
- Pyrometer for measuring solar insolation
- Photovoltaic power output performance measurement sensors
  - AC current and voltage transducer
  - DC current and voltage transducer
  - Kilowatt hour meter transducer
  - Optically isolated RS-422 or RS-232C Modem

DISPLAY AND SUN SERVER MONITORING SOFTWARE

This software provides acquisition and display of real-time data every second and displays the following on a variety of display monitors:

- DC current
- DC voltage
- AC current
- AC voltage
- AC kilowatt hours
- Solar plane of array irradiance
- Ambient temperature
- Wind speed

Calculated parameters displayed include:

- AC power output
- Sunlight conversion efficiency to AC power
- Sunlight conversion efficiency to DC power
- Inverter DC-to-AC power conversion efficiency
- Avoided pollutant emissions of CO_2, SOX, NOX gases

The above information and calculated parameters are displayed on monitors and updated once every second. The data is also averaged every 15 minutes and stored in a locally accessible database. The software also includes a "Virtual Array Tour" that allows observers to analyze the component of the PV array and monitoring system. The software also provides an optional portal Web capability whereby the displayed data could be monitored from a remote distance over the internet.

The monitoring and display software could also be customized to incorporate descriptive text, photographs, schematic diagrams, and user-specific data. Some of the graphing capabilities of the system include the following:

- Average plots of irradiance, ambient temperature, and module temperature (which are updated every 15 minutes and averaged over one day).
- Daily values or totals of daily energy production, peak daily power, peak daily module temperature, and peak daily irradiance plotted over a specified month.
- Monthly values of monthly energy production, monthly incident solar irradiance, and avoided emission of CO_2, NOX, and SOX plotted over a specified year.

Displayed information   A standard display will usually incorporate a looping background of pictures from the site, graphical overlays of the power generation in watts and watt-hours for each building, and the environmental impact from the solar system. The display also shows current meteorological conditions. Displayed data in general should include the following combination of items:

- Project location (on globe coordinates—zoom in and out)
- Current and historic weather conditions
- Current positions of the sun and moon, with the date and time
- Power generation from the total system and/or the individual solar power arrays
- Historic power generation

- Solar system environmental impact
- Looping background solar system photos and video
- Educational PowerPoint presentations
- Installed solar electric power overview
- Display of renewable energy system environmental impact statistics

The display should also be programmed to periodically show additional information related to the building's energy management or the schedule of maintenance relevant to the project.

Transmitted data from the weather-monitoring station should include air temperature, solar cell temperature, wind speed, wind direction, and sun intensity measured using a pyrometer.

Inverter monitoring transmitted data must incorporate a watt-hour transducer that will measure voltage (DC and AC), current (DC and AC), power (DC and AC), AC frequency, watt-hour accumulation, and inverter error codes and operation.

The central supervising system must be configured with an adequate CPU processing power and storage capacity to permit future software and hardware upgrading. The operating system should preferably be based on Windows XP or an equivalent system operating software platform.

The data communication system hardware must be such to allow switch selectable RS-232/422/485 communication transmission protocols. It must also have software selectable data transmission speeds. The system must also be capable of frequency hopping from 902 to 928 MHz on the FM bandwidth and be capable of providing transparent multipoint drops.

Animated video and interactive programming requirements   A graphical program builder must be capable of animated video and interactive programming and have an interactive animation display feature for customizing the measurements listed earlier. The system must also be capable of displaying various customizable chart attributes such as labels, trace colors and thickness, axis scale, limits, and ticks.

Ground-Mount PV Module Installation and Support Structures

Ground-mount outdoor PV array installations can be configured in a wide variety of ways. The most important factor when installing solar power modules is the PV module orientation and panel incline.

In general, maximum power from a PV module is obtained when the angle of solar rays impinge directly perpendicular (at a 90-degree angle) to the surface of the panels. Since solar ray angles vary seasonally throughout the year, the optimum average tilt angle for obtaining the maximum output power is approximately the local latitude minus 9 or 10 degrees.

In the northern hemisphere, PV modules are mounted in a north-south tilt (high end north), and in the southern hemisphere in a south-north tilt. Appendix A also includes U.S. and world geographic location longitudes and latitudes.

To attain the required angle, solar panels are generally secured on tilted prefabricated or field-constructed frames that use rustproof railings (such as galvanized UNISTRUT or commercially available aluminium or stainless-steel angle channels) and fastening hardware (such as nuts, bolts, and washers). Prefabricated solar power support systems are also available from UniRac and several other manufacturers.

When installing solar support pedestals, also known as stanchions, attention must be paid to structural design requirements. Solar power stanchions and pedestals must be designed by a qualified, registered professional engineer. Solar support structures must take into consideration prevailing geographic and atmospheric conditions such as maximum wind gusts, flood conditions, and soil erosion.

Typical ground-mount solar power installations include agricultural grounds, parks and outdoor recreational facilities, carports, and large commercial solar power-generating facilities, also known as solar farms. Most solar farms are owned and operated by electric energy-generating entities. Prior to installing solar power system, local electrical service authorities, such as building and safety departments, must review structural and electrical plans.

Roof-Mount Installations

Roof-mount solar power installations are made of either tilted or flat-type roof support structures (or a combination of both). Installation hardware and methodologies also differ depending on whether the building already exists or is a new construction. Roof attachment hardware materials also vary for wood-based and concrete constructions. Figure 3.15 depicts a prefabricated PV module support railing system used for roof-mount installations.

WOOD-CONSTRUCTED ROOFING

In new constructions, the PV module support system installation is relatively simple. This is because locations of solar array frame pods, which are usually secured on roof rafters, can be readily identified. Prefabricated roof-mount stands that support railings and associated hardware (such as fasteners) are commercially available from a number of manufacturers. Solar power support platforms are specifically designed to meet physical configuration requirements for various types of PV module manufacturers. Figure 3.16 shows graphic diagram of a typical solar power support railing installation detail.

Some types of PV-module installations, such as the one shown in Fig. 3.16a to c, have been designed for direct mounting on roof-framing rafters, without the use of specialty railing or support hardware. When installing roof-mount solar panels, care

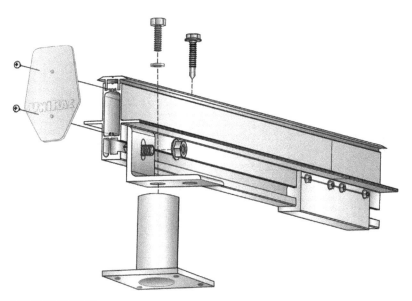

Figure 3.15 Prefabricated PV module support railing for roof-mount system. *Drawing courtesy of UniRac.*

must be taken to meet the proper directional tilt requirement. Another important factor to be considered is that solar power installations, whether ground- or roof-mounted, should be located in areas free of shade caused by adjacent buildings, trees, or air-conditioning equipment. In the event of unavoidable shading situations, the solar power PV module location, tilt angle, and stanchion separations should be analyzed to prevent cross-shading.

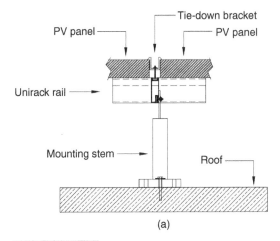

(a)

Figure 3.16 (*a*) Typical roof-mount solar power installation detail.

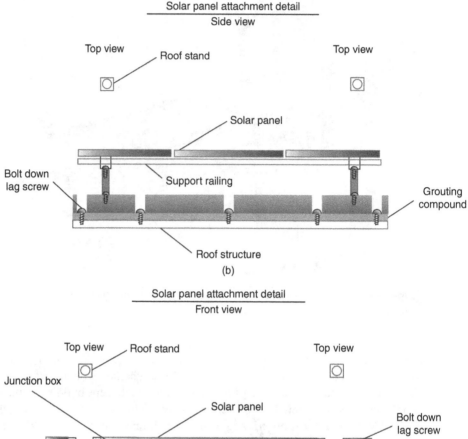

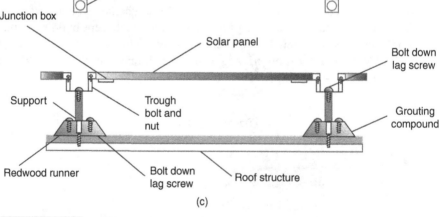

Figure 3.16   (*b*) **Front view of the solar power railing attachment. (*c*) Side view of the solar power railing attachment.**

LIGHTWEIGHT CONCRETE ROOFING

Solar power installation PV module support systems for concrete roofs are config-ured from prefabricated support stands and railing systems, similar to the ones used in wooden roof structures. Stanchions are anchored to the roof by means of rust-resistant expansion anchors and fasteners.

In order to prevent water leakage resulting from roof penetration, both wood and concrete standoff support pipe anchors are thoroughly sealed with waterproofing

Figure 3.17   Structural concrete roof-mounted solar railing system assembly, using double-sided UNISTRUT railing. *Photo courtesy of Vector Delta Design Group, Inc.*

compounds. Each standoff support is fitted with thermoplastic boots that are in turn thermally welded to roof cover material, such as single-ply PVC.

PHOTOVOLTAIC STANCHION AND SUPPORT STRUCTURE TILT ANGLE

As discussed earlier, in order to obtain the maximum output from the solar power systems, PV modules or arrays must have an optimum tilt angle that will ensure a perpendicular exposure to sun rays. When installing rows or solar arrays, the spacing between stanchions must be such that there should not be any cross-shading. In the design of a solar power system, the available roof area is divided into a template format that compartmentalizes rows or columns of PV arrays. Figure 3.17 is a photograph of UNISTRUT railing support system.

Solar-Tracking Systems

Tracking systems are support platforms that orient solar PV module assemblies by keeping track of the sun's movement from dusk to sundown, thus maximizing solar energy power generation efficiency. Trackers are classified as passive or active and may be constructed to track in single or dual axis. Single-axis trackers usually have a single-axis tilt movement, whereas dual-axis systems trackers also move in regular intervals adjusting for an angular position.

In general, single-axis trackers, when compared to fixed stationary-tilted PV support systems, increase solar power capture by about 20 to 25 percent. Dual-axis trackers, on the other hand, can increase solar power production from 30 to 40 percent. Solar power concentrators, which use Fresnel lenses to focus the sun's energy on a solar cell, require a high degree of tracking accuracy to ensure that the concentrated sunlight is focused precisely on the PV cell.

Fixed-axis systems orient the PV modules to optimize power production for a limited time performance and generally have a relatively low annual power production. On the other hand, single-axis trackers (even though less accurate than dual-axis tracker applications) produce strong power in the afternoon hours and are deployed in applications, such as grid-connected solar power farms, that enhance power production in morning and afternoon hours.

Compared to the overall cost of PV systems, trackers are relatively inexpensive devices that significantly increase the power output performance efficiency of PV panels. Even though some tracker systems operate with degree of reliability, they usually require seasonal position adjustments, inspection, and periodic lubrication.

POLAR TRACKERS

Polar trackers are designed to have one axis rotate in the same pattern as the earth. Essentially, polar trackers are aligned perpendicular to an imaginary ecliptic disc that represents the apparent mathematical path of the sun. To maintain relative accuracy, these types of trackers are manually adjusted to compensate for the seasonal ecliptic shifts that occur in autumn, winter, spring, and summer. Polar trackers are usually used in astronomical telescope mounts, in which high-accuracy solar tracking is an absolute requirement.

HORIZONTAL-AXLE TRACKERS

Horizontal trackers are designed to orient a horizontal axle by either passive or active mechanisms. Essentially, a long tubular axle is supported by several bearings that are secured to some type of wooden, metallic, or concrete pylon structure frames. The tubular axles are installed in north-south orientation, whereas PV panels are mounted on the tubular axle that rotates on an east-west axis that tracks the apparent motion of the sun throughout daylight hours. It should be noted that single-axis trackers do not tilt toward the equator, and as a result their power-tracking efficiency is significantly reduced in mid-winter. However, their productivity increases substantially during spring and summer seasons when the sun path is directly above the sky. Horizontal-axle single-axis trackers, because of the simplicity of their mechanism, are considered to be very reliable, easy to clean and maintain, and are not subject to self-shading.

VERTICAL-AXLE TRACKERS

Vertical-axle trackers are constructed in a manner as to allow pivotal movement of PV panels mounted about a vertical axis. These types of trackers have a limited use and are usually deployed in high latitudes where the solar path travels in a long arc. PV panels mounted on a vertical axis system are suitable for operation during long summer days in northern territories, where there are extended solar days. Figures 3.18 and 3.19 depict the rendering of dual-axis solar power tracker installations.

Figure 3.18 **Self-contained dual-axis solar power tracking system.** *Photo courtesy of MARTIFER.*

Figure 3.19 **T20 single-axis solar power tracking system.** *Photo courtesy of SunPower.*

References

1. Gevorkian, Peter, *Alternative Energy Systems in Building Design,* McGraw-Hill, New York, 2010.
2. Gevorkian, Peter, *Solar Power in Building Design,* McGraw-Hill, New York, 2008.

4

PHOTOVOLTAIC POWER SYSTEM
FEASIBILITY STUDY

Introduction

The essential steps required for solar power systems engineering design include site evaluation, feasibility study, site-shading analysis, photovoltaic mapping or configuration analysis, DC-to-AC power conversion calculations, PV module and inverter system selection, and total solar power array electrical power calculations.

In order to have a holistic understanding of solar power cogeneration, system designers must have a basic appreciation of insolation concepts, shading analysis, and various design parameters that affect output performance and efficiency of the overall system. It should be kept in mind that the responsibility of a professional solar power designer is to ensure that the engineering design will offer utmost system performance efficiency.

In general, all solar power systems (from inception of the project to completion) ascend through several phases, each requiring specific effort. The following are various phases of solar power system lifecycle:

1 Preliminary project assessment
2 Project site survey
3 Feasibility study or preliminary design
4 Feasibility study report
5 Proposal
6 Contract award negotiations
7 Final design
8 Project integration
9 System integration
10 System test and verification
11 Final test and acceptance or performance verification
12 System maintenance

Feasibility Study

Undoubtedly, solar power feasibility study is the most fundamental engineering effort required for assessing and planning any type of solar power system design. Feasibility study is the cornerstone of solar power design because it provides an in-depth, meaningful assessment of the energy potential of solar project platforms, such as rooftop, carport, or ground-mount solar power systems.

Solar feasibility study is also of paramount importance to any investment in solar power systems, since it provides detailed assessments of solar energy production potential, as well as establishes a fundamental platform for future engineering design.

The following are fundamental tasks involved in conducting site solar power platform evaluations and assessments of energy production potential:

- Roof-top, carport, and ground mount systems platform topology study
- Shading analysis
- Study and evaluation of the site electrical system and grid connectivity configuration
- Electrical power demand loading analysis for the entire site
- Solar platform topology mapping design and infrastructure study, which includes the following:
  - Preparation of solar platform topology diagrams
  - Study of alternative technologies that will be best suited for each type of solar platform
  - Study of PV module support structures for each platform, including carport structures
  - Preliminary study of underground conduit runs and solar interplatform interconnectivity
  - Econometric analysis of the solar power project, which may include solar-projected construction cost estimates for each type of solar power platform by use of analytical software modeling methods
  - Preparation of a comprehensive feasibility study report that incorporates results of all of the above engineering studies

To conduct a solar feasibility study, the engineer or the designer must obtain the following customer-supplied documentation:

- Complete electrical engineering documents, which will include as-built single-line diagrams, switchgear riser diagrams, panel schedules, and the location of all electrical and power distribution rooms and enclosures
- Electrical bills for the last 12 months
- Site plans
- Roof plans (preferably showing roof-mount mechanical equipment and vents)
- Site underground conduit plans
- Any recent energy audit documentation and reports

Solar power feasibility studies usually involve several site visits and a close collaborative effort with the client.

Site Survey

Fundamentally, medium- and large-scale solar PV systems (as referenced in Chap. 3) fall into three main categories: roof or canopy mounts [also referred to as building-integrated solar power system (BIPV)], carport canopies, and ground-mount systems. Most often commercial projects include a composite of all three types of solar power systems.

The following forms embody specific site survey information for each type of solar power system.

Type A—Site survey for roof-mount BIPV systems

1 Customer information log sheet information
   - Primary contact
   - Telephone number
   - E-mail address
2 Project name
3 Site address
4 Project type
   - Private
   - Municipal
   - Institutional
   - Federal
   - State
   - Public agency
   - Other
5 Type of engineering tasks required
   - Solar power system feasibility study
   - Feasibility study and construction supervision
   - Design supervision only
   - Design supervision and project coordination
   - Design only
   - Design and construction supervision
   - Design build
   - Turnkey design build
6 Building type
7 Number of buildings assigned as solar platform
8 Existing building _____ New building _____
9 Area coordinates
   - Longitude
   - Latitude
   - Site elevation from sea level

10 Area climatic data
- Minimum temperature
- Maximum temperature
- Maximum humidity
- Annual rain precipitation
- Pollution index
- Maximum wing sheer
- Local barometric pressure (millibars)

11 Solar energy information
- Solar irradiance in watts per square meter _____ W/m^2
- Average nominal daily insolation hours

12 Roof type
- Wood
- Lightweight concrete
- Structural concrete
- Corrugated steel

13 Other

14 Roof covering material
- Asphalt _____ Age _____
- Single-ply PVC _____ Age _____
- Rock asphalt _____ Age _____
- Composite tile _____ Age _____
- Other _____ Age _____

15 Roof surface
- Flat _____
- Inclined _____ Angle of inclination_____
- Curved _____

16 Roof support structure

17 Roof load-bearing capacity
- PV support mounting system
- Penetrating type
- Nonpenetrating-type ballast
- Railing system type
- Other

18 Roof parapet type and height

19 Roof orientation and azimuth angle

20 Visual shading obstructions

21 Roof-mount equipment and objects

22 Roof photograph

23 Roof sketch

24 Roof architectural and mechanical plans

25 Roof shading
- Tall buildings
- Trees
- Equipment

26 Pathfinder or Solametric shading evaluation

27 Availability of roof-mount utilities
- Water line
- Electricity

28 Location of the main electrical service

29 Feeder-chase location to the main electrical service

30 Approximate distance from mid-sections and roof corners to the electrical chase

31 Electrical documents
- Service provider
- Service voltage _____ Phase _____ Wire _____
- Calculated electrical demand load in amperes
- Nominal electrical peak calculated power demand
- Electrical single line diagram(s)
- Electrical service equipment/room layout plan
- Electrical metering/switchgear bus ampacity
- Electrical bus short-circuit current availability Isc = _____
- 12 consecutive months of electrical bill
- Availability of spare breaker for solar power grid connection
- Spare power bus-loading capacity
- Future power demand requirements

32 Present cost electrical energy tariff and tariff class

33 Last energy audit date

34 Feeder chase distance to the electrical service room (ft)

35 Electrical room space availability for mounting inverter(s), AC combiner box(s), AC disconnect and solar meter

36 Net unobstructed roof area

37 Roof access location and type
- Stairwell
- Outdoor wall-mount ladder
- Indoor ladder and hatch door
- Other

38 Roof height from ground

39 Building number of stories

40 Material storage area

41 Special observations

42 Solar power platform energy production potential study
- Available net unobstructed surface area _____ SF
- Percent of useable surface area _____ SF
- Type of PV technology proposed
- PV manufacturer _____ Model No._____
- Physical dimensions
- Length _____ inches
- Width _____ inches
- Depth _____ inches
- PV support platform type

- PV module specification
- STC power _____ watts
- PTC power _____ watts
- Isc _____ Amps
- Voc _____ volts
- Surface area _____ SF
- PV-module quantity _____
- PV Watts V.2 or California Solar Initiative (CSI) power output calculated energy profile
- Inverter system configuration
- Centralized
- Distributed
- Inverter power requirements
- Inverter manufacturer
- Model
- Quantity
- Inverter and DC/AC combiner box platform configuration strategy
- Inverter pad configuration
- Feeder conduit configuration
- Financing
  - Private financing
  - Private lease option to buy
  - Municipal lease
  - Design, build and construction finance
  - Power purchase agreement (PPA)
- Contract lifecycle
- Warrantee
- PV module
- Inverters
- Manpower

1 Maintenance
   - Self maintained
   - Maintenance contract duration
   - Contract lifecycle maintenance duration
2 Financial subsidy
   - Federal percent contribution
   - State/province percent contribution
   - Municipal percent contribution
3 Econometric analysis
   - Solar power/electrical systems engineering design
   - Shop drawing design documentation
   - Site system engineering design (civil, site survey, soils, environmental)
   - Site system integration supervision and project management
   - Hardware bill of materials

4 Data acquisition system
  - Integrated system
  - Third-party
5 Site preparation
  - System installation, integration, and test
  - Project logistics
  - Test and commissioning
  - Customer training
  - Spare parts
  - System lifecycle maintenance
  - Travel and lodging expenses
  - Other
6 State and federal contributions
  - State _____ Percent contribution _____
  - Municipal _____ Percent contribution _____
7 Econometric analysis
  - Contract lifecycle in years
  - Hardware bill of materials
8 Data acquisition system
  - Integrated system
  - Third party
9 Site preparation
10 System installation and integration
11 Logistics (material transportation, storage, etc.)
12 Test and commissioning
13 Other observations

When conducting field surveys of the existing roofs or solar platforms, special care must be taken to ensure that there are no mechanical, construction, or natural structures that could cast a shadow on the solar panels. Shade from trees and sap drops could create an unwanted loss of energy production. One of the solar PV modules in a chain, when shaded, could act as a resistive element that will alter the current and voltage output of the whole array.

The solar power system engineer or designer should also consult with the architect to ensure that the installation of solar panels will not interfere with the roof-mount solar window, vents, and air-conditioning unit ductwork. The architect must also consider roof penetrations, installed weight, anchoring, and the system must meet seismic requirements.

Type B—Site survey form for solar power parking canopy systems

1 Customer information
  - Primary contact
  - Telephone number
  - E-mail address

2 Project name
3 Site address
4 Project type
- Private
- Municipal
- Institutional
- Federal
- State
- Public agency
- Other

5 Type of engineering services
- Solar power system feasibility study
- Feasibility study and construction supervision
- Design supervision only
- Design supervision and project coordination
- Design only
- Design and construction supervision
- Design build
- Turnkey design build

6 Financing
- Private financing
- Private lease option to buy
- Municipal lease
- Design, build, and construction finance
- Power purchase agreement (PPA)
- Contract lifecycle

7 Building type
8 Number of buildings assigned as solar platform
9 Existing building _____ New building _____
10 Area coordinates
- Longitude
- Latitude
- Site elevation from sea level (ft)

11 Area climatic data
- Minimum temperature
- Maximum temperature
- Maximum humidity
- Annual rain precipitation
- Pollution index
- Maximum wing sheer
- Local barometric pressure (millibars)

12 Solar energy information
- Solar irradiance in watts per square meter _____watts/m^2
- Average nominal daily insolation hours _____

13 Nominal parking stall dimensions
- Width _____ ft
- Length _____ ft

14 Parking canopy height restrictions

15 Ground cover
- Asphalt
- Concrete
- Paver

16 Documentation requirements
- Parking plan
- Site utility plan
- Site drainage and water flow lines

17 Construction restriction conditions

18 Solar canopy design strategic considerations
Solar power canopy power production potential

19 Available net unobstructed canopy surface area _____ SF

20 Percent of useable parking surface area _____ SF

21 Type of PV technology proposed

22 PV manufacturer _____ Model No. _____

23 Physical dimensions
- Length _____ inches
- Width _____ inches
- Depth _____ inches

24 PV canopy support platform type

25 PV module specification
- STC power _____ watts
- PTC power _____ watts
- Isc _____ Amps
- Voc _____ volts
- Surface area _____ SF

26 PV module quantity _____

27 PV watts V.2 power output calculated energy profile

28 Inverter system configuration
- Centralized
- Distributed

29 Inverter power requirements
- Inverter manufacturer
- Model
- Quantity

30 Inverter and DC/AC combiner box platform configuration strategy

31 Inverter pad configuration

32 Feeder conduit configuration

33 Location of the main electrical service

34 Feeder chase location to the main electrical service

35 Approximate distance from mid-sections and roof corners to the electrical chase (ft)

36 Electrical documents
- Service provider
- Service voltage _____ Phase _____ Wire _____
- Bus bar ampacity
- Nominal electrical peak calculated power demand
- Electrical single line diagram(s)
- Electrical service equipment/room layout plan
- Electrical metering/switchgear bus ampacity
- Electrical bus short circuit current availability Isc = _____
- 12 consecutive months of electrical bill
- Availability of spare breaker for solar power grid connection
- Available spar power bus loading capacity
- Future power demand requirements

37 Current cost of electrical energy per kW/hr _____
- Electrical tariff class _____

38 Last energy audit date

39 Chase distance to the electrical service room (ft)

40 Electrical room space availability for mounting inverter(s), AC combiner box(s), AC disconnect, and solar meter

41 Financing
- Private financing
- Lease option to buy
- Municipal lease
- Design, build, and construction finance
- Power purchase agreement (PPA)
- Contract lifecycle

42 Warrantee
- PV module
- Inverters
- Manpower

43 Maintenance
- Self-maintained
- Maintenance contract duration
- Contract lifecycle maintenance duration

44 Financial subsidy
- Federal percent contribution
- State/province percent contribution
- Municipal percent contribution

45 Econometric analysis
- Solar power/electrical systems engineering design
- Shop drawing design documentation
- Site system engineering design (civil, site survey, soils, environmental)

- Site system integration supervision and project management
- Hardware bill of materials

46 Data acquisition system

47 Integrated system

48 Third party

49 Site preparation
- System installation, integration, and test
- Project logistics
- Test and commissioning
- Customer training
- Spare parts
- System lifecycle maintenance
- Travel and lodging expenses
- Other

Type C—Site survey form for ground-mounted solar power systems

1 Customer information
- Primary contact
- Telephone number
- E-mail address

2 Project name

3 Site address and location

4 Project type
- Private
- Municipal
- Institutional
- Federal
- State
- Public agency
- Other

5 Type of engineering services
- Solar power system feasibility study
- Feasibility study and construction supervision
- Design supervision only
- Design supervision and project coordination
- Design only
- Design and construction supervision
- Design build
- Turnkey design build

6 Financing
- Private financing
- Private lease option to buy
- Municipal lease
- Design, build, and construction finance

- Power purchase agreement (PPA)
- Contract lifecycle

7 Area coordinates
- Longitude
- Latitude
- Site elevation from sea level

8 Site topology map

9 Site survey map

10 Area climatic data
- Minimum temperature
- Maximum temperature
- Maximum humidity
- Annual rain precipitation
- Local barometric pressure (millibars)
- Pollution index
- Maximum wing sheer
- Annual lightning occurrence and frequency

11 Solar energy information
- Solar irradiance in watts per square meter (watts/m^2)
- Local nominal daily insolation hours

12 Ground type
- Natural dirt
- Rocky terrain
- Sandy terrain
- Expansive soil
- Farming area
- Ground-covered vegetation
- Soil erosion or settlement risk
- Water table depth
- Natural ground drainage
- Ground drainage requirement

13 Soil analysis requirement

14 Ground geographic setting description

15 Requirement for environmental negative impact report

16 Ground-mounting height restrictions

17 Available unobstructed site acreage

18 Site survey documents

19 Site-grading requirements

20 Solar power platform type
- Fixed angle
- Single-axis solar-tracking system
- Dual-axis tracking system flat PV
- High-power dual-axis solar concentrator (CPV)

21 Site foundation and civil engineering requirements

22 Site visual shading evaluation study requirement
23 Grid interconnection system
  ■ Existing service line
  ■ Existing transmission line
24 Electrical grid specification
  ■ Grid voltage _____ kV
  ■ Grid power-loading capacity _____ MW
  ■ Projected future grid loading profile over planned lifecycle of the transmission line
25 Solar power configuration
  ■ Available ground use percentage for farming or other use
  ■ Special solar power collection and transformation platform configuration requirements
  ■ Tentative solar power production capacity _____ MW
26 Equipment platform location restrictions
27 Existing underground utility system such as electricity, gas, water, or sewage
28 Solar farm site design considerations
  ■ Solar support system configuration
  ■ DC combiner box location
  ■ Inverter system design approach—centralized or distributed
  ■ Equipment platform design considerations for:
      a. Inverters
      b. AC accumulators
      c. Solar current disconnects
      d. Metering system
      e. Data acquisition and communication system
      f. Site utility system
      g. Site security alarm system
      h. Site fencing system
      i. Site lighting system
      j. Equipment platform canopies
      k. Underground conduit feeders system
      l. Site grounding and lightning protection system
      m. Step-up transformer pad(s)
      n. Grid interconnection equipment platform
29 Site storage facility planning
30 Transportation logistics
31 Electrical/solar system engineering design
32 Civil structural engineering
33 Site installation shop drawings
34 Design and integration project management
35 Site engineering project coordination and management
36 Site mobilization
37 Customer training

38 Financing
- Private financing
- Lease option to buy
- Municipal lease
- Design, build, and construction finance
- Power purchase agreement (PPA)
- Contract lifecycle (years)

39 Warrantee
- PV module
- Inverters
- Manpower

40 Maintenance
- Self-maintained
- Maintenance contract duration (years)
- Contract lifecycle maintenance duration (years)

41 Financial subsidy
- Federal percent contribution
- State/province percent contribution
- Municipal percent contribution

42 Econometric analysis

43 Solar power/electrical systems engineering design

44 Shop drawing design documentation

45 Site system engineering design (civil, site survey, soils, environmental)

46 Site system integration supervision and project management

47 Hardware bill of materials

48 Data acquisition system
- Integrated system
- Third party

49 Site preparation requirements

50 System installation, integration, and test

51 Logistics for material storage and transportation

52 Test and commissioning

53 Customer training

54 Spare parts

55 System lifecycle maintenance

56 Travel and lodging expenses

Solar Power System Preliminary Design Considerations

After establishing solar power area clearances, the solar power designer must prepare a set of electronic templates representing standard array configuration assemblies. Solar array templates then could be used to establish a desirable output of DC power. When laying blocks of PV arrays, consideration must be given to the desirable tilt inclination to avoid cross-shadowing. In some instances, the designer must also consider trading solar power output efficiency to maximize the power output production. As a rule of thumb, the most desirable mounting position for fixed-angle solar PV installations is inclined about the angle of the local latitude minus (−)10 degrees to realize maximum solar insolation.

For example, the optimum tilt angle in New York will be 39°, whereas in Los Angeles it will be about 25° to 27°. To avoid cross-shading, the adjacent profiles of two solar rows of arrays could be determined by simple trigonometry. This could determine the geometry of the tilt by the angle of the associated sine (shading height) and cosine (tandem array separation space) of the support structure incline. It should be noted that flatly laid solar PV arrays may incur about a 9 to 11 percent power loss, but depending upon the number of installed panels, it could exceed 30 to 40 percent on the same mounting space.

An important design criterion when laying out solar arrays is grouping the proper number of PV modules that would provide the adequate series-connected voltages and current required by inverter specifications. Most inverters allow certain margins for DC inputs that are specific to the make and model of a manufactured unit. Inverter power capacities may vary from a few hundred to many thousands of watts. When designing a solar power system, the designer should make decisions about the use of specific PV and inverter makes and models in advance, thereby establishing the basis of the overall configuration.

It is not uncommon to have different sizes of solar power arrays and matching inverters on the same installation. In some instances, the designer may, due to unavoidable occurrences of shading, decide to minimize the size of the array as much as possible. This limits the number of PV units in the array, which may require a small-size power capacity inverter. The most essential factor that must be taken into consideration is ensuring that all inverters used in the solar power system are completely compatible.

When laying out the PV arrays, care should be taken to allow sufficient access to array clusters for maintenance and cleaning purposes. In order to avoid deterioration of power output, solar arrays must be washed and rinsed periodically. Adequately spaced hose bibs should be installed on rooftops to facilitate flushing the PV units in the evening only, when the power output is below the margin of shock hazard.

Upon completing the PV layout, the designer should count the total number of solar power system components. Using a rule of thumb, the designer must arrive at a unit cost estimate, such as dollars per watt of power. That will make it possible to better approximate the total cost of the project. In general, net power output from PV arrays when converted to AC power is subjected to a number of factors that can degrade the output efficiency of the system.

The California Energy Commission (CEC) rates each manufacturer PV module output power performance according to standard ambient temperature conditions. A special laboratory test referred to as PV USA, also called Power Test Condition (PTC), determines typical power output for all PV modules sold in the United States (in particular those that are subject to federal or state rebate). Essentially, PTC is a figure of merit used in all solar power output performance calculations.

When designing solar power systems, the engineer or the designer must consider numerous environmental parameters that were covered in Chap. 2, which include but are not limited to the following:

- Solar platform shading
- Latitude and longitude
- Ambient temperature
- Yearly average insolation
- Temperature variations
- Solar platform orientation and the azimuth angle
- Roof or support-structure tilt angle
- Inverter efficiency
- Isolation transformer efficiency
- DC and AC wiring losses
- Solar power exposure derating
- Feeder cable and wire percent voltage drops

SHADING ANALYSIS AND SOLAR ENERGY PERFORMANCE MULTIPLIER

One of the most significant steps prior to designing a solar power system is investigating a location for a solar installation platform where the solar PV arrays will be located. In order to harvest the maximum amount of solar energy, all panels (in addition to being mounted at the optimum tilt angle) must be totally exposed to the sun's rays without shading that may be cast by surrounding buildings, objects, trees, or vegetation.

To achieve this above objective, solar power mounting terrain or the platform must be analyzed for year-round shading. Note that the seasonal rise and fall of the solar angle has a significant effect on the direction and surface area of shadows cast.

The following factors must be taken into consideration for evaluating a site for a PV installation. Each of the factors discussed may have an impact on optimal energy production. In addition to latitude and longitude, which determine the sun path characteristics, PV panel orientation (tilt and azimuth) determine the field of view that an array is exposed to the sun. Shading from trees, nearby hills, buildings, or other obstructions can cause significant reduction of direct solar radiance and result in degradation in energy production. Furthermore, regional climatic and atmospheric weather patterns may result in hourly and daily fluctuations in solar insolation.

The above factors effectively interact and determine the solar energy that impacts PV arrays and therefore affect power production and financial return on investment. Therefore site shading evaluation and methodology used in site evaluation emphasize shade analysis and optimal solar access.

To adequately conduct site shading measurement the surveyor must take into consideration all obstacles that may cause shading.

Note that shadows cast by a building or tree will vary from month to month, changing in length, width, and the shape of the shade. In order to analyze yearly shading of a

Figure 4.1   **Solar Pathfinder™ and shading graphs.** *Courtesy of Solar Pathfinder.*

solar platform, solar power designers and integrators use commercial shading analysis instruments, such as Solar Pathfinder™ (shown in Fig. 4.1) and Solmetric SunEye, both of which will be discussed further.[1]

SOLAR PATHFINDER™

Solar Pathfinder™ is a device used for shade analysis in areas that are surrounded by trees, buildings, and other objects that could cast shadows on a designated solar platform.

The device is essentially comprised of a semispherical plastic dome and latitude-specific, disposable solar shading graphic inserts (shown in Figs. 4.1 and 4.3).

The disposable semicircular plates (shown in Fig. 4.2) have 12-month imprinted curvatures that show the percentage of daily solar energy intensity from sunrise (around 5 a.m.) to sunset (around 7 p.m.). Each of the solar energy intensity curves, from January to December, is demarcated with vertical latitude lines that denote the separation of daily hours. A percentage number, ranging from 1 to 8 percent, is placed between

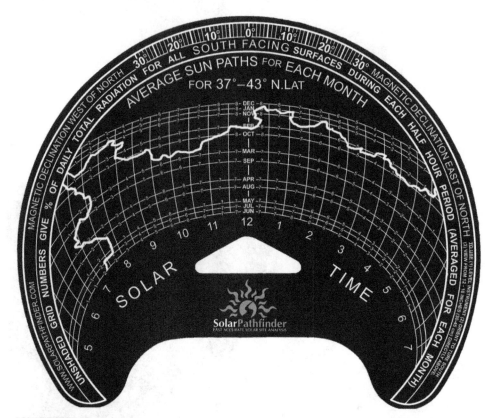

Figure 4.2 **Solarishading graphic insert.** *Photo courtesy of Solar Pathfinder.*

adjacent hourly latitude lines. Percentage values progress upward from sunrise, from a value of 1 percent to a maximum value of 8 percent at noon. They then drop back down to 1 percent at sunset.

Depending on the inclination angle of the sun, the percentage of solar energy values depicted on the monthly curvatures varies for each month. For instance, the maximum percent value for the months of November, December, and January is 8 percent at solar noon (12:00 p.m.). For the rest of the year, from February through October, the maximum percentage is 7 percent.

The total sum of the percentage points shown on the monthly solar energy curves represents the maximum percent of solar insolation (100 percent) on the platform. For instance, the total energy percent shading multiplier for the month of December, January, or any other month, is summed up to a total 100 percent multiplier.

For example, according to the chart 31 to 37 North Latitude for the month of December when summed up equals:

$$\text{Efficiency Multiplier } \% = 2 + 2 + 3 + 4 + 6 + 7 + 7 + 8 + 8 + 8 + 8 \\ + 7 + 7 + 6 + 5 + 4 + 3 + 2 + 2 + 1 = 100\%.$$

Figure 4.3   **Pathfinder semispherical dome showing reflection of surrounding buildings.** *Photo courtesy of Solar Pathfinder.*

The same summation for the month of June equals:

$$
\begin{aligned}
\text{Efficiency Multiplier}\% = {}& 1 + 1 + 1 + 2 + 2 + 3 + 4 + 5 + 5 + 6 + 6 \\
& + 7 + 7 + 7 + 7 + 7 + 6 + 6 + 5 + 5 + 4 \\
& + 3 + 2 + 2 + 1 + 1 + 1 = 100\%.
\end{aligned}
$$

Note that the insolation angle of the sum increases and decreases for each different latitude, hence each plate is designed to cover specific bands of latitudes for the northern and southern hemispheres.

When placing the earlier-referenced plastic dome on top of the platform that holds the curved solar energy pattern, surrounding trees, buildings, and objects that could cast shadows are reflected in the plastic dome. This clearly shows shading patterns at the site, which are in turn cast on the pattern. The reflected shade on the solar pattern shows distinctly defined jagged patterns of shading that cover the plate throughout the 12 months of the year.

A 180° opening on the lower side of the dome allows the viewer to mark the shading on the solar pattern by means of an erasable pen. To determine the total yearly percentage shading multiplier, each portion of the 12 monthly curves not exposed to shading are totaled. When taking the mean average percentage of all 12 months, a representative solar-shading multiplier is derived. This is applied in DC-to-AC conversion calculations, as will be discussed further in this chapter.

The Solar Pathfinder™ dome and shading pattern assembly are mounted on a tripod, as shown in Fig. 4.1. For leveling purposes, the base of the assembly has a fixed leveling bubble at its center that serves to position the platform assembly on a horizontal level. At the lower part of the platform (which holds the pattern plate), a fixed compass indicates the geographic orientation of the unit (shown in Fig. 4.4). The pattern plate is in turn secured to the platform by a raised triangular notch.

Figure 4.4   A Solar Pathfinder platform showing the removable shading graph, the leveling bubble at the center, the triangular holder, and the compass. *Photo courtesy of Solar Pathfinder.*

To record shading, the platform is placed on level ground and the pathfinder is adjusted for proper magnetic declination in order to orient the device toward the true magnetic pole. A small brass lever, when pulled downward, allows the center triangle to pivot or rotate the shading pattern towards the proper magnetic declination angle. Magnetic declination is the deviation angle of the compass needle from the true magnetic pole. Global magnetic declination angle charts are available through magnetic declination Web sites for all countries.

SOLMETRIC SUNEYE™

The SunEye is a device that measures shading and solar access for a specific location. It captures "skyline" views with a digital camera equipped with a calibrated fish-eye lens. Images are processed using an on-board computer, and shading and obstructions are automatically detected. The SunEye (Fig. 4.5) makes these measurements quick and convenient for a wide variety of applications.

SunEye use in solar panel design and installation   The SunEye is used for conducting solar power shading evaluation and analysis. The instrument, in addition to mapping the shading areas of a solar power platform, is capable of transferring the collected shading information to a computer-based software that allows solar power designers to perform computations for optimizing maximum power production.

Figure 4.5   Solmetric SunEye™ shade measuring meter. *Photo courtesy of Solmetric.*

The SunEye and passive solar house design and green architecture   The SunEye is portable convenient shading measurement tool for accurate shading measurement and analysis. The SunEye optimizes the orientation of a structure by identifying where and when the sun will shine. SunEye provides analysis as to questions like: Will there be enough sun in the winter for passive heating of the house given the site-specific shading? How will removing the large oak tree on the southwest corner affect the warming and cooling of the building?, etc. Furthermore, by collecting data inside an existing structure, it allows the user to identify the amount of direct sunlight that will enter a window or skylight or identify the perfect location for a new window or skylight.

The SunEye and home and property inspection   The SunEye gives professional home and property inspectors a way to provide their clients with important information about the solar access of a property. This information can be useful for:

- Identifying potential sites for solar panels, gardens, or new windows
- Determining the amount of passive solar heating or cooling a particular building will experience
- Determining how much direct sunlight will enter a particular window or skylight

The SunEye's integrated digital camera and fish-eye lens capture an image of the entire horizon in 360°. On-board electronics superimpose the paths of the sun throughout the year based on latitude and longitude, detect shade-causing obstructions, and calculate the annual, monthly, daily, and hourly solar access.

How the SunEye works   Armed with this data, the solar system designer can make informed design choices about the optimum location for the solar panels. For example, the designer can find which section of roof is best for solar energy production. Panels that will be shaded at the same time of day should be grouped together in the same string to maintain the energy production from the other strings. The built-in edit tool can be used to simulate removal or trimming of shade-causing trees. An example screenshot in edit mode is shown in Fig. 4.11. Guesswork and ballpark estimation are replaced with solid data.

The SunEye can store more than 100 site readings, transfer data to a PC for further analysis, or export data into a printable report. The SunEye user can also average multiple skylines together. For example, data from the four corners of the array (or string) is automatically averaged together into a single solar access data set. The associated patent application U.S. 2007/0150198 contains additional details of the technical approach. Figure 4.6 depicts shading interaction of a tree.

Figure 4.7 depicts PV module voltage and current output curve under no shading condition, and Fig. 4.8 depicts solar PV module voltage and a current (IV) output curve under shading conditions.

Displaying data with Sun Eye   The SunEye displays annual, seasonal, and monthly solar-access percentage factors and details about obstructions, such as elevation angle versus azimuth angle of objects that will shade that location. The SunEye can export data to the SunEye desktop software to create solar access and shade reports.

Sun path interaction

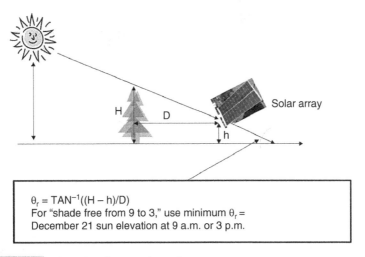

$\theta_r = \text{TAN}^{-1}((H-h)/D)$
For "shade free from 9 to 3," use minimum $\theta_r =$
December 21 sun elevation at 9 a.m. or 3 p.m.

Figure 4.6   **Shading interaction of a tree.**

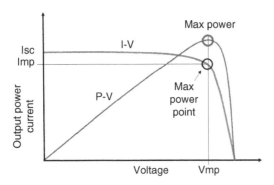

Figure 4.7   **Solar PV module voltage and current output curve under no shading condition.**

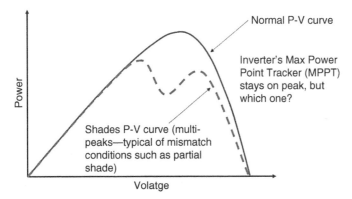

Figure 4.8   **Solar PV module voltage and current (IV) output curve under shading condition.**

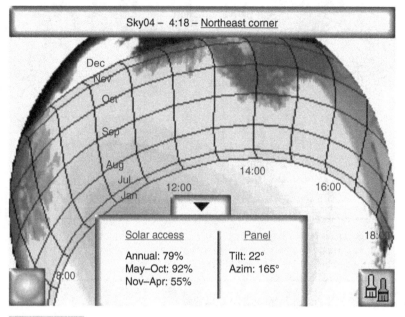

Figure 4.9   Sunpath view.

A special GPS option allows files to be exported and used with Google Earth, so that the user can see exactly where the SunEye data was captured. Figure 4.9 shows the Sunpath view of shading conditions.

Figure 4.10 displays the annual view of the sunpaths drawn on top of the captured skyline. The detected open sky is shown in light gray. The detected shade-causing obstructions are shown in dark gray.

The panel orientation and tilt can be configured to show solar access percentage energy production levels as shown in Fig. 4.10.

The monthly solar access view shows the bar chart of the monthly solar access for the location where the data is captured. The height of the bars and the numbers at the top of each bar indicate the percentage of solar energy available each month for the site-specific shade conditions. If there were no shade obstructions, the bars would all indicate 100 percent. If the location was shaded all year round, the bars would all indicate 0 percent.

Using SunEye data   SunEye data can be transferred to other applications for enhancing calculations of energy production of a solar system. In the case of PV systems, programs such as the Solmetric PV Designer can import the SunEye readings, and combine with weather, orientation, equipment specifications, and layout information to calculate and display energy production in AC kilowatt-hours. Figure 4.11 shows a monitor display of the SunEye software system.

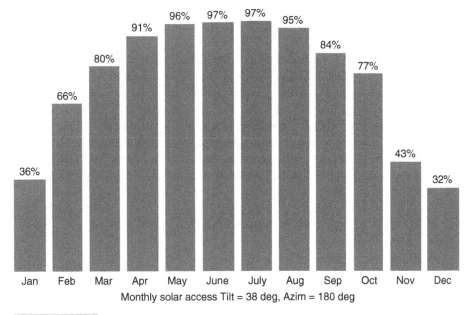

Figure 4.10   Monthly solar access.

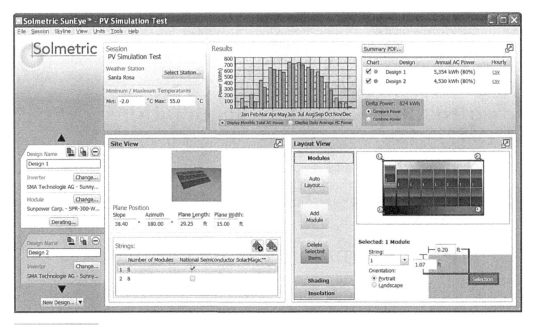

Figure 4.11   Solmetric monitor display.

Meteorological Data

When the design is planned for floor-mount solar power systems, designers must investigate natural calamities, such as extreme wind gusts, periodic or seasonal flooding, and snow precipitation. For meteorological data, contact the NASA Surface Meteorology and Solar Energy Data Set Web site at www.eosweb.larc.nasa.gov/sse/.

To search for meteorological information on this Web site, the inquirer must provide the latitude and longitude for each geographic location. For example, when obtaining data for Los Angeles, California, at latitude 34.09 and longitude 118.4, the statistical data provided will include the following recorded information for each month of the year for the past 10 years:

- Average daily radiation on horizontal surface in $kW/m^2/day$
- Average temperature
- Average wind speed (m/s)

For complete listings of latitude and longitude data, please refer to Appendix A. The following are a few examples for North American metropolitan areas.

Los Angeles, California	34.09 N/118.40 W
Toronto, Canada	43.67 N/–79.38 W
Palm Springs, California	33.7 N/116.52 W
San Diego, California	32.82 N/117.10 W

For local insolation data for the United States, solar power designers may refer to the Web site http://eosweb.lac.nasa.gov/sse.

Structural Design Considerations

In general, one of the most significant issues that a design engineer must be concerned about is the integrity of solar power system support platforms. Whether a roof-mount, carport, or ground-mount system, it is important to ensure that all structural plans are designed and endorsed by a registered structural engineer.

Project Preliminary Cost Estimate

After completing the site survey, the designer must establish the basic design configuration of the solar power system by outlining preliminary topology of the solar array. The designer must also generate drawings and documentation that will assess projected solar power output performance and econometrics associated with the solar power project. In addition to site survey information, the designer must be intimately familiar with all material and have ample experience in estimating manpower, material, and solar power integration costing methodology. The following are typical material and manpower data and information required to complete a preliminary cost estimate:

- Solar PV module specifications
- PV support structure specification for pertinent type of installation
- Inverters
- Combiner boxes
- Electrical disconnects, isolation transformers, and lightning protection devices, conduits system, cables, and miscellaneous parts used in solar power system construction
- Grid connectivity equipment and switchgear components
- Electrical/solar power engineering
- Civil/structural engineering
- Environmental systems engineering
- Material transport and storage
- Possible federal taxes and state sales taxes
- Labor wages (prevailing or nonprevailing) and site supervision (project management)
- Construction drawings and reproduction
- Fencing
- Foundation
- Permit fees
- Maintenance training manuals and instructor time
- Maintenance, casualty insurance, and warranties
- Spare parts and components
- Testing and commissioning
- Customer training
- Construction bond and liability insurance
- Mobilization cost, site office, and utility expenses
- Construction insurance
- Engineering errors and omissions liability insurance
- Construction financing
- Construction bond
- Overhead and profit

Energy Cost Factors

After completing the preliminary engineering study and solar power generation potential, the designer must evaluate the current costs and project the future costs of the electric energy for the entire lifespan of the solar power system. To determine the current value of the electric energy cost for an existing building, the designer must evaluate the actual electric bills for the past two years. Note that the general cost per kilowatt-hour of energy provided by service distributors consists of numerous charges, such as commissioning, decommissioning, bulk purchase, and other miscellaneous cost items that generally appear on electric bills (that vary seasonally) but go unnoticed by consumers.

The most significant of the charges, which is in fact, a penalty, is classified as *peak hour energy*. This charge occurs when the consumer's power demand exceeds the established boundaries of energy consumption as stipulated in tariff agreements. In order to maintain a stable power supply and cost for a unit of energy (a kilowatt-hour), service distributors such as Southern California Electric (SCE) and other power-generating entities generally negotiate a long-term agreement whereby the providers guarantee distributors a set bulk of energy for a fixed sum. Since energy providers have limited power generation capacity, limits are set as to the amount of power that is to be distributed for the duration of the contract. Service providers then use statistics and demographics of the territories served to project power consumption demands, which then form the baseline for the energy purchase agreement. When energy consumption exceeds the projected demand, it becomes subject to a much costlier tariff, which is generally referred to as the *peak bulk energy rate*.

Solar Power System Sizing Estimates

Sizing a solar power system estimate may be accomplished either by detailed analysis of the solar power platform or by using solar power estimating software, which is provided by number of manufacturers and most electric utility providers.

To compute solar power system sizing by detailed analysis, the designer must determine the solar power platform configuration, populate the area with specific solar power modules, and perform computations to determine the overall solar power system power output performance for the duration of one year or the lifecycle of the project.

After determining the solar power system components and material count, the engineer or the designer must account for derate factors, which adversely affect the solar power system output performance. Derate factors include the following:

- PV module nameplate derate factor
- Inverter nameplate conversion efficiency
- PV module impedance mismatch
- Bypass diode loss (current leakage)
- DC wiring losses
- AC wiring losses
- Soiling losses due to dust and grime
- Overall system availability
- Shading losses
- Solar-tracking losses
- Aging degradation losses

Note that a number of the assigned ranges of derate percentages, such as soiling, the overall system reliability, shading losses, and aging values are somewhat arbitrary. For example, soiling loss values will be larger in arid urban and desert-type installations where grime and dust could accumulate on the surface of the PV modules. In such situations, soiling values could be reduced by frequent washdown of the solar power system. Likewise, the mentioned loss factor will also depend on the inclination angle of the PV modules, as well as the amount of yearly rainfall, which automatically cleans and washes the solar power system. The overall system aging can be improved by somewhat oversizing the solar power system.

The overall system reliability, on the other hand, requires a proactive system design strategy. For instance, modularizing the solar power system will significantly minimize chances of failure. An example of such a design strategy can be illustrated in the design of a 1-megawatt (MW) solar power system. In such a design, it is possible to consider a number of options regarding the use of inverters. Options involve using a single 1-MW, 2-500 kW, 4-250 kW, or 10-100 kW inverters. Since inverters have a lower performance life expectancy (5 years) than PV modules (25 years), it becomes obvious that failure of a 1-MW inverter will disable the entire solar power system. On the other hand, failure of one inverter among 10 will only reduce the performance of the solar power system by 10 percent.

Therefore, system reliability entails use of multiple inverters, which in turn may result in a small increase in material and installation cost. When considering long-term power purchase agreement (PPA) contracts, system uptime and reliability becomes of utmost importance.

SOLAR SYSTEM SIZING ESTIMATES USING A PVWATTS II CALCULATOR

PVWatts II is solar power sizing calculator that uses public domain software available on the Web at www.pvwatts.com. The PVWatts calculator is considered the standard solar power sizing methodology for all solar power systems.

The calculator was developed by National Renewable Energy Laboratories. Solar power sizing is relatively simple and requires little effort to compute the size of a solar power system located throughout the United States and abroad. The software system includes large number of libraries for all CEC-approved PV modules, inverters, solar power meters, and up-to-date electric utility rates of every electrical power utility throughout the United States.

The latest revision of PVWatts II provides a display of the map of the United States, which allows the user to zoom in to any region of the country where a solar power system is to be installed. The map window also allows the user to select the site by simply entering the postal zip code.

After entering the zip code, the calculator assigns a unique regional matrix quadrant and subsequently enables the user to make a selection of specific solar power hardware from the CEC's list of certified devices; meanwhile the software selects the local electric utility rate.

After selecting the solar hardware devices, the user is allowed to enter the desired DC power rating, as well as calculated derate factors, referenced previously. In the absence of a specific derate multiplier, the calculator software assigns a default value of 0.77 percent. This accounts for the lowest possible mean efficiency.

After completing the data entry, the calculator provides results of solar power sizing calculations in a tabular form as shown in Fig. 4.12.

The table displays the parameter entry data and additional columns showing year-round (January to December, month to month) values of solar radiation in kW/m²/day, AC energy in kWh (kilowatt hours), and the energy value in dollars, which is a product of the radiation and the utility rate \$/kW/h. The tabulation also displays annual mean solar radiation, totalized annual AC energy production, and annual cost contribution values of the solar power system. The calculator also optionally displays computed derate factor lists.

Note that almost all solar power system proposals throughout the United States that are subject to any kind of rebate program (these include federal, state, public agency, nonprofit organizations, and financed projects) mandate the use of PVWatts II computation. Figure 4.12*a* is PVWatts II DC-to-AC energy and cost-saving calculations and Figure 4.12*b* shows PV Watts II DC-to-AC loss factor calculations.

Customizing PV Watts parameters   As discussed, the PVWatts calculator allows users to substitute its default input parameters with custom values. Solar power designers and engineers can change the PVWatts parameters for:

- DC rating
- DC-to-AC derate factor
- Array type
- Tilt angle
- Azimuth angle
- Electricity cost

The following are narratives describing derating factors that appear in the PVWatts Web site.

DC Rating   The size of a PV system is its nameplate DC power rating. This is determined by adding the PV module power listed on the nameplates of the PV modules in watts and then dividing the sum by 1000 to convert it to kilowatts (kW). PV module power ratings are for standard test conditions (STC) of 1000 W/m² solar irradiance and 25°C PV module temperature. The default PV system size is 4 kW. This corresponds to a PV array area of approximately 35 m² (377 ft²).

Caution: For correct results, the DC rating input must be the nameplate DC power rating described previously. It cannot be based on other rating conditions, such as PVUSA test conditions (PTC). PTC is defined as 1000 W/m² plane-of-array irradiance,

AC energy
&
cost savings

Cautions
for Interpreting
the Results

Station identification	
Cell ID:	0175360
State:	California
Latitude:	34.4°N
Longitude:	118.2°N
PV system specifications	
DC rating:	1000.0 kW
DC to AC derate factor:	0.800
AC rating:	800.0 kW
Arrray type:	Fixed tilt
Array tilt:	34.4°
Array azimuth:	180.0°
Energy specifications	
Cost of electricity:	120.0 ¢/kWh

Results			
Month	Solar radiation (kWh/m²/day)	AC energy (kWh)	Energy value (s)
1	4.63	109310	13089.87
2	5.23	111571	13360.63
3	5.80	135623	16240.85
4	6.24	138452	16579.63
5	6.64	150613	18035.91
6	7.05	150735	18050.52
7	7.02	152632	18277.68
8	7.12	154825	18540.29
9	6.57	139328	16684.53
10	5.86	132329	15846.40
11	5.24	118187	14152.89
12	4.44	104508	12514.83
Year	5.99	1598112	191373.92

(a)

Calculator for overall DC to AC derate factor

Component derate factors	Component derate values	Range of acceptable values
PV module nameplate DC rating	0.96	0.80 – 1.05
Inverter and transformer	0.96	0.88 – 0.98
Mismatch	0.98	0.97 – 0.995
Diodes and connections	0.995	0.99 – 0.997
DC wiring	0.98	0.97 – 0.99
AC wiring	0.99	0.98 – 0.993
Soiling	0.96	0.30 – 0.995
System availability	0.98	0.00 – 0.995
Shading	0.98	0.00 – 1.00
Sun-tracking	1.00	0.95 – 1.00
Age	1.00	0.70 – 1.00
Overall DC to AC derate factor	**0.803**	

(b)

Figure 4.12   (*a*) PV Watts II DC-to-AC energy and cost-saving calculations.
(*b*) PV Watts II DC-to-AC loss factor calculations.

20°C ambient temperature, and 1 m/s wind speed. If a user incorrectly uses a DC rating based on PTC power ratings, the energy production calculated by the PVWatts calculator will be reduced by about 12 percent.

DC-to-AC Derate Factor The PVWatts calculator multiplies the nameplate DC power rating by an overall DC-to-AC derate factor to determine the AC power rating at STC. The overall DC-to-AC derate factor accounts for losses from the DC nameplate power rating and is the mathematical product of the derate factors for the components of the PV system. The default component derating factors used by the PVWatts calculator and their ranges, as well as typical assigned weights, are listed in Table 4.1.

The overall DC-to-AC derate factor is calculated by multiplying the component derate factors. Using aggregated default derate values in PVWatts yields the following:

Overall solar power system DC to AC derate factor:
$$= 0.95 \times 0.92 \times 0.98 \times 0.995 \times 0.98 \times 0.99$$
$$\times 0.95 \times 0.98 \times 1.00 \times 1.00 \times 1.00 = 0.77$$

This calculated derate value of 0.77 means that the AC power rating at STC is 77 percent of the nameplate DC power rating. In most cases, 0.77 will provide a reasonable estimate. However, users can change the DC-to-AC derate factor. The first option is to enter a new overall DC-to-AC derate factor in the provided text box. The second option is to click the *Derate Factor Help* button. This provides the opportunity to change any of the components derate factors. The derate factor calculator then calculates a new

TABLE 4.1 DERATE FACTORS FOR AC POWER RATING AT STC

COMPONENT DERATE FACTORS	PV WATTS WEIGHT	RANGE
PV module nameplate DC rating	0.95	0.80–1.05
Inverter and transformer	0.92	0.88–0.98
Mismatch	0.98	0.97–0.995
Diodes and connections	0.995	0.99–0.997
DC wiring	0.98	0.97–0.99
AC wiring	0.99	0.98–0.993
Soiling	0.95	0.30–0.995
System availability	0.98	0.00–0.995
Shading	1.00	0.00–1.00
Sun-tracking	1.00	0.95–1.00
Age	1.00	0.70–1.00
Overall DC-to-AC derate factor	0.77	0.96001–0.09999

overall DC-to-AC derate factor. The component derate factors are described in the following section.

PV Module Nameplate DC Rating   The PV module nameplate value accounts for the accuracy of the manufacturer's nameplate rating. Field measurements of PV modules may show that they are different from their nameplate rating or that they experience light-induced degradation upon exposure. A derate factor of 0.95 indicates that testing yielded power measurements at STC that were 5 percent less than the manufacturer's nameplate rating.

Inverter and Transformer   This value reflects the inverter's and transformer's combined efficiency in converting DC power to AC power. A list of inverter efficiencies by manufacturer is available from the Consumer Energy Center. The inverter efficiencies include transformer-related losses when a transformer is used or required by the manufacturer.

Mismatch   The derate factors for PV modules mismatch accounts for manufacturing tolerances that yield PV modules with slightly different current-voltage characteristics. Consequently, when connected together electrically, they do not operate at their peak efficiencies. The default value of 0.98 represents a loss of 2 percent because of this mismatch.

Diodes and PV Module Connection Losses   The derate factor accounts for losses from voltage drops across diodes used to block the reverse flow of current and from resistive losses in electrical connections.

DC Wiring Losses   The derate factor for DC wiring accounts for resistive losses in the wiring between modules and the wiring connecting the PV array to the inverter.

AC Wiring Losses   The derate factor for AC wiring accounts for resistive losses in the wiring between the inverter and the connection to the local utility service.

Soiling   The derate factor for soiling accounts for dirt, snow, and other foreign matter on the surface of the PV module that prevent solar radiation from reaching the solar cells. Dirt accumulation is affected by location and weather. Soiling losses are greater (up to 25 percent greater for some California locations) in high-traffic, high-pollution areas with infrequent rain. For northern locations, snow reduces the energy produced, and the severity is a function of the amount of snow and how long it remains on the PV modules. Snow remains longest when subfreezing temperatures prevail, small PV array tilt angles prevent snow from sliding off, the PV array is closely integrated into the roof, and the roof or another structure in the vicinity facilitates snow drift onto the modules. For a roof-mounted PV system in Minnesota with a tilt angle of 23°, snow reduced the energy production during winter by 70 percent; a nearby roof-mounted PV system with a tilt angle of 40° experienced a 40 percent reduction.

System Availability   The derate factor for system availability accounts for times when the system is off because of maintenance and inverter or utility outages. The default value of 0.98 represents the system being off 2 percent of the year.

Shading   The derate factor for shading accounts for situations in which PV modules are shaded by nearby buildings, objects, or other PV modules and arrays. For the default value of 1.00, the PVWatts calculator assumes the PV modules are not shaded. Tools such as Solar Pathfinder can determine a derate factor for shading by buildings and objects. For PV arrays that consist of multiple rows of PV modules and array structures, the shading derate factor should account for losses that occur when one row shades an adjacent row.

   The table below shows the shading derate factor as a function of the type of PV array (fixed or tracking), the ground cover ratio (GCR) (defined as the ratio of the PV array area to the total ground area), and the tilt angle for fixed PV arrays. As shown in the figure, spacing the rows further apart (smaller GCR) corresponds to a larger derate factor (smaller shading loss). For fixed PV arrays if the tilt angle is decreased, the rows may be spaced closer together (larger GCR) to achieve the same shading derate factor. For the same value of shading derate factor, land area requirements are greatest for two-axis tracking, as indicated by its relatively low GCR values compared with those for fixed or one-axis tracking. If you know the GCR value for your PV array, the figure may be used to estimate the appropriate shading derate factor. Industry practice is to optimize the use of space by configuring the PV system for a GCR that corresponds to a shading derate factor of 0.975 (or 2.5 percent loss).

Array Tilt Angle Loss   The optimum tilt angle for PV module performance is the latitude angle of the particular terrain. As discussed earlier in this chapter, irradiance at latitude is perpendicular to the solar PV module. At a latitude angle, annual solar power energy output from PV modules is at its optimum. An increased tilt angle above the latitude will increase power output production in wintertime; however, it will decrease in summertime. Likewise, decreasing the tilt angle from the latitude will increase power production in summertime. The following table relates tilt angle and roof pitch, which is a measure of the ratio of the vertical rise of the roof to its horizontal run.

ROOF PITCH	TILT ANGLE DEGREES
4/12	18.4
5/12	22.6
6/12	26.6
7/12	30.3
8/12	33.7
9/12	36.9
10/12	39.8
11/12	42.5
12/12	45.0

Sun-tracking   The derate factor for sun-tracking accounts for losses in one- and two-axis tracking systems when the tracking mechanisms do not keep the PV arrays at the optimum orientation. For the default value of 1.00, the PVWatts calculator assumes that the PV arrays of tracking systems are always positioned at their optimum orientation and performance is not adversely affected.

Age Factor   The age derate factor accounts for performance losses over time because of weathering of the PV modules. The loss in performance is typically 1 percent per year. For the default value of 1.00, the PVWatts calculator assumes that the PV system is in its first year of operation. For the eleventh year of operation, a derate factor of 0.90 is appropriate.

Note that because the PVWatts overall DC-to-AC derate factor is determined for STC, a component derate factor for temperature is not part of its determination. Power corrections for PV module operating temperature are performed for each hour of the year as the PVWatts calculator reads the meteorological data for the location and computes performance. A power correction of −0.5 percent per degree Celsius for crystalline silicon PV modules is used.

Array Type   The PV array may be fixed, sun-tracking with one axis of rotation, or sun-tracking with two axes of rotation. The default value is a fixed PV array.

Tilt Angle   For a fixed PV array, the tilt angle is the angle from horizontal of the inclination of the PV array (0° = horizontal, 90° = vertical). For a sun-tracking PV array with one axis of rotation, the tilt angle is the angle from horizontal of the inclination of the tracker axis. The tilt angle is not applicable for sun-tracking PV arrays with two axes of rotation. Figure 4.13 shows types of PV array systems.

The default value is a tilt angle equal to the station's latitude. This normally maximizes annual energy production. Increasing the tilt angle favors energy production in the winter, and decreasing the tilt angle favors energy production in the summer.

For roof-mounted PV arrays, the Table 4.2 provides tilt angles for various roof pitches in a ratio of vertical rise to horizontal run.

The default value is an azimuth angle of 180° (south-facing) for locations in the northern hemisphere and 0° (north-facing) for locations in the southern hemisphere. This normally maximizes energy production. For the northern hemisphere, increasing the azimuth angle favors afternoon energy production and decreasing the azimuth angle favors morning energy production. The opposite is true for the southern hemisphere.[2]

The following section discusses the differences between two versions of PVWatts calculators.

PVWATTS VERSION 1

For the United States and its territories, the default value is the average 2004 residential electric rate for the state in which the station is located.[3] For locations outside the United States, the default value is the average 2004 or 2005 residential electric rate for the country in which the station is located.[4] For some countries, no electric cost

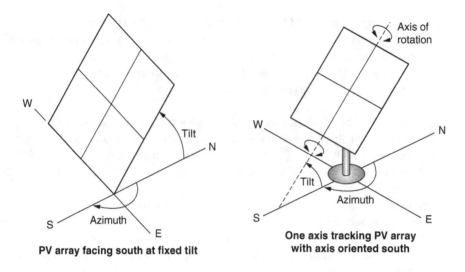

PV array facing south at fixed tilt

One axis tracking PV array
with axis oriented south

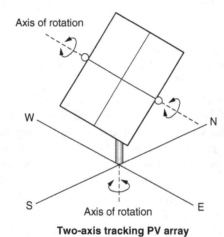

Two-axis tracking PV array

Figure 4.13   Types of PV array systems.

HEADING	AZIMUTH ANGLE (°)
TABLE 4.2 AZIMUTH ANGLES BY HEADING	
N	0 or 360
NE	45
E	90
SE	135
S	180
SW	225
W	270
NW	315

information is available, and the default values are set to zero. For these countries, users should enter a value based on their knowledge. Electric costs are presented in the country's currency.

PVWATTS VERSION 2

The default value is the average 2004 residential electric rate for the 40-km cell chosen. Note that some areas are not covered by any utility provider. For these areas the electric rate for the nearest utility service area is used.[5]

The tables presented here are an example of a PVWatts II calculation for a hypothetical project.

PVWATTS II ENERGY AND COST SAVINGS CALCULATIONS		
Project: ACME		
ZIP code	92392	
Cell ID—National	177360	
State	CA	
Longitude	34.8-N	
Lattitude	117.5-W	
PV system specification		
DC rating in kW	87.4	
DC to AC conversion KW	0.80	
AC power output KW	69.6	
Array type—fixed		
Array tilt	5	
Energy cost		
Derating/loss factors	DERATE	RANGE
PV module nameplate DC rating	0.98	0.8–1.05
Inverter and transformer losses	0.97	0.88–0.98
PV string mismatch	0.98	0.97–0.995
PV cell module diode losses	0.995	0.97–0.997
DC wiring losses	0.98	0.97–0.99
AC wiring losses	0.99	0.98–0.993
Soiling losses	0.95	0.3–0.995
System availability	0.98	0.00–0.995
Shading losses	0.98	0.00–1.00
Sun tracking losses	0.97	0.95–1.00
System aging degradation	1	0.7–1.00
Combined losses	0.80	

COMPUTATIONAL RESULTS

MONTH	SOLAR RADIATION	AC POWER KW/H	ERGY VALUE
1	3.56	7677	$928.89
2	4.3	8674	$1,049.59
3	5.81	15138	$1,831.69
4	7.02	16507	$1,997.33
5	7.91	17984	$2,176.11
6	8.34	17029	$2,060.46
7	8.16	17596	$2,129.14
8	7.64	15943	$1,929.16
9	6.57	14168	$1,714.27
10	5.31	11081	$1,340.81
11	4.15	8660	$1,047.90
12	3.35	7224	$874.10

The software must also perform cumulative lifecycle energy cost analyses for grid-connected and PPA-based solar power projects such as:

■ Totalized energy income from life operation
■ Present value of the projected system salvage amount
■ Federal tax incentive income
■ State tax incentive income
■ Salvage value at the end of the system lifecycle
■ Maintenance cost income
■ Recurring annual maintenance cost
■ Recurring annual data acquisition cost
■ Net income value over the lifecycle of operation

Figures 4.14*a* through *c* are bar charts of annual solar power radiation, annual AC power output, and annual energy values.

Preparing the Feasibility Study Report

As mentioned earlier in this chapter, the key to designing a viable solar power system begins with the preparation of a feasibility report. A feasibility report is essentially a preliminary engineering design report that is intended to inform the end user about significant aspects of a project. The document therefore, must include a thorough definition of the entire project from a material and financial perspective.

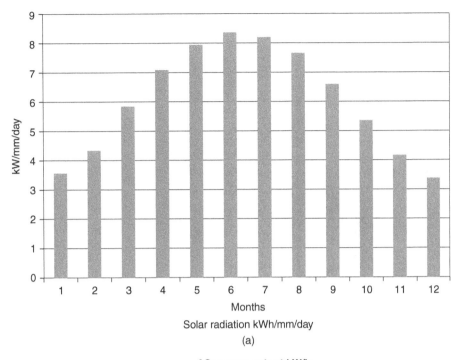

Months

Solar radiation kWh/mm/day

(a)

AC energy output kW/h

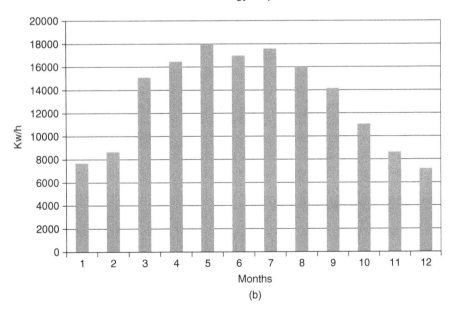

Months

(b)

Figure 4.14   (*a*) Annual solar power radiation. (*b*) Annual AC power output.

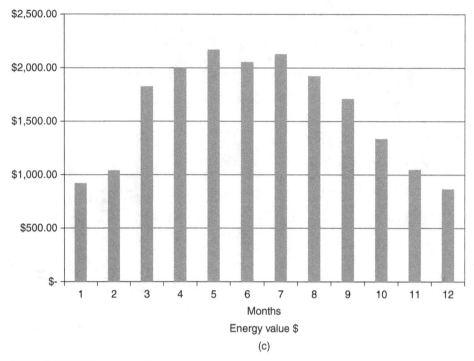

Figure 4.14   (*c*) Annual energy value.

A well-prepared report must inform and educate the client as well as provide a realistic projection of all engineering and financial costs to enable the user to weigh all aspects of a project from start to finish. The report must include a comprehensive technical and financial analysis of all aspects of the project. This includes particulars of local climatic conditions, solar power system installation alternatives, grid-integration requirements, electric power demand, and economic cost projection analysis. The report must also incorporate photographs, charts, and statistical graphs to illustrate and inform the client about the benefits of the solar power or sustainable energy system proposed.

The following are the typical contents of a feasibility study report.

EXECUTIVE SUMMARY

The executive summary is intended to provide an overview of the project feasibility study report which may include the following:

- Solar PV system platform site inspection findings.
- Preliminary solar platform power potential assessment.
- Proposed technology and proposed system configuration overview.
- Overview of solar power system financial analysis, which may also include various financing options such as Energy Savings Performance Contract (ESPC), Utility Energy Savings Contract (UESC), Enhanced Use Lease Contract (EUL), and Direct Funding Contract (DFC).

REPORT FORMAT

In general, the feasibility report format and structure includes the following:

- Executive summary: this part of the report summarizes the overall aspect of the study, its content, and conclusions.
- Site climatic conditions: this section of the report outlines climatic parameters that have been used to calculate the engineering study.
- Site potential platform review: this part of the report includes a roster of solar power platforms that have been used to calculate the solar power potential.
- Service provider electrical bill review for existing buildings: this portion of the report includes an analysis of several months of a customer's electrical bill and concludes the actual mean or average cost of unit of energy cost in $/kWh.
- Design criteria and assumptions: this section of the report outlines design methodology and assumptions made to calculate a project's solar power potential and econometrics.
- Case study: this portion of the report represents a feasibility study of each platform.
- Site evaluation report: this portion of the report includes the results of the site survey and shading analysis of each individual solar power platform.
- Appendix: this section of the report usually includes photographs of typical future solar power system installations such as roof-mount, ground-mount or carport solar power installations.

Example of a Solar Feasibility Study Report

The following is an example of a solar power site feasibility report for a project located in southern California. The report is intended to familiarize the reader with the structure of a typical report, which includes application of methodologies discussed in this chapter. Names of the client and the consulting firm are arbitrary.

The main components of the report include verification of existing and future electrical, mechanical, construction, and architectural drawings, review and analysis of existing utility bills, assessment of local service provider electrical service plans, and shading analysis of the project site. The main objective of feasibility report is to provide clients with a comprehensive assessment of a solar power project's solar energy potential and associated econometrics for executing the project.

It should be noted that a feasibility study, if conducted properly, forms the essential platform for undertaking any future solar power engineering design. The example presented below represents the core content of the report. For the sake of summary, details of three out of eight platforms have been included in the following example.

EXECUTIVE SUMMARY REPORT

This report represents a detailed study of the existing and future solar power platform potentials. The study reflected in this document is based on the following plans and documents provided by Rainbow Disposal (RD) to consultant VDDG:

- Future site electrical
- Existing site electrical plans
- Existing architectural plan Southern California Edison (SCE) as-built plans
- Final site
- SCE electrical bills from June 2009 through June 2010
- Site shading analysis recorded by VDDG

The objective of this report is to evaluate the best possible solar power platform potentials that are comprised of distinct sites within the project's campus. In order to provide accurate evaluation RD campus solar power potential and associated econometrics, each of the following sites has been considered as a stand-alone project. Computation of energy potential for each of the sites has been based on conservative application of parameters. All financial analyses projected in this report are in conformance with California Solar Initiative calculation methodology. The analytical engine used for econometrics has been in full conformance with National Renewable Energy Laboratories (NREL), which is the acceptable standard by all electrical service providers within the jurisdiction of California Energy Commission.

CLIMATIC PARAMETERS

Climatic data parameters applied in solar power potential computations use the NREL reference as follows:

- Reference site nearest to Huntington Beach, CA—Long Beach, CA
- Time zone—GMT-8
- Latitude—33.8167°
- Longitude—118.15°
- Direct normal irradiance—207.6 Wh/m^2
- Diffused horizontal irradiance—76.0 Wh/m^2
- Dry bulb temperature—17.2 C°
- Wind speed—2.7 m/s

SITE DESIGNATION, STATUS, AND POTENTIAL PLATFORM AREA

ELECTRICAL BILL ANALYSIS

Table 4.3 outlines 11 months of electrical bill analysis and variance mean value determination. Cost per kWh is determined by the amount of the actual invoice (which includes

TABLE 4.3 SCE ELECTRICAL BILL EVALUATION

RAINBOW DISPOSAL ELECTRICAL BILL ANALYSIS			7/12/2010
SCE ACCOUNT NO 2-03-649-2809 METERS			
PERIOD	**KWH**	**AMOUNT**	**COST/KWH**
APRIL 30 TO JUNE 1, 10	194668	$27,235.00	$0.140
APRIL 1 TO APRIL 30, 10	184020	$26,969.00	$0.147
MARCH 3 TO APRIL 1, 10	183224	$26,991.00	$0.147
FEBRUARY 1 TO MARCH 3, 10	183072	$23,872.00	$0.130
DEC 2 TO DEC 31, 09	185508	$26,341.00	$0.142
OCT 30 TO DEC 2, 09	202468	$27,008.00	$0.133
SEPT 30 TO OCT 30, 09	182788	$26,308.00	$0.144
JULY 31 TO AUG 31, 09	179264	$42,834.00	$0.239
AUG 31 TO SEPT 30, 09	186344	$45,120.00	$0.242
JULY 1 TO JULY 31, 09	173860	$45,150.00	$0.260
JUNE 2 TO JULY 1, 09	162548	$41,745.00	$0.257
TOTAL 11 MONTHS	2017764	$359,573.00	
MEAN kWh CHARGE OVER 11 MONTHS			$0.18
MEAN kWh-2010			$0.14
MEAN kWh-2009			$0.20
AVERAGE ENERGY CONSUMPTION	183433		

miscellaneous monthly charges) by kilowatt hours consumed. It should be noted that the amount reflected on the bill represents the true cost of the kWh. Bills paid during the month of June to the end of August are indicative of peak power charges that the solar power system will shave. Average value of cost of energy used in economic analysis is $0.18 /kWh.

DESIGN CRITERIA AND ASSUMPTION MADE

In view of the fact that this report represents projection of solar power potential assessment and economic evaluations of each site, in order to complete computation it has been necessary to improvise equipment operational parameters and future solar power platforms, which are not available at the time of preparation of this report.

a. The following are criteria used for computing solar power potential of each site. PV module—Table 4.4 represents footprint calculation of a SANYO Model HIT–210N solar photovoltaic module, which is considered one of the most cost effective and efficient products available in the market.

TABLE 4.4 SOLAR MODULE FOOTPRINT CALCULATION

SOLAR PV MODULE FOOTPRINT
FOR ROOF MOUNT SOLAR POWER INSTALLATION

PV MODEL-SANYO HIT 210N	HIT210N
WATTS	210
LENGTH-INCH-L	62.2
WIDTH-INCH-W	31.4
AREA-SQ.INCH.	1953.08
AREA-FT.	13.6
TILT ANGLE-ϕ-IN DEGREES	5
TILT ANGLE-ϕ-IN RADIANS	0.09
SIN (ϕ)	0.09
r-HYPOTEN	62.2
Y-HEIGHT-Y=SIN(ϕ) X r	5.42
BETA ANGLE β-D-90 DEG. TILT	85
BETA ANGLE β-R	1.48
Tan (β)	11.43
x-FOOT-x = Y/Tan(ϕ)	61.96
x'-INCHES-x' = y X/Tan(β)	0.5
PV SHADE LENGTH-SL = x + x'	62.44
PV SHADE AREA = SL × W, INCH.	1961
FRONT WALKWAY-INCHES	30
FRONT WALKWAY-AREAIPV	942
COMBINED FOOTPRINT/PV-SQ.INCH.	2902.54
PV SHADE FOOTPRINT-SF	20.16

Support structures, whether roof-mount or carport trellises, are assumed to be installed at an angular tilt of 15°. Geometric computation within the table establishes the exact footprint of the PV support structure, which also accounts for the shading at 90° direct natural solar irradiance.

b. PV module is based on 1 percent depreciation.

c. Roof platform useable area is considered 60 percent.

d. PV system derating is based on NREL PVWatts II worst-case 23 percent loss of efficiency or 77 percent system efficiency of operation.

e. Site economic analysis is based on the actual current cost of base energy, $0.18.

f. Mean annual cost of electrical energy escalation is 10 percent, which is in conformance with SCE recent past 12.7 percent escalation.
g. Project lifecycle is assumed to be 25 years.
h. General inflation rate for lifecycle of the project will not exceed 3.5 percent.

PROJECT SUMMARY EVALUATION

Table 4.5 shown below, represents a summary of detail calculation of all project sites
within the project campus. The table is intended to provide summary information at a
single glance. Each of the eight project sites within this report include detailed analysis
of the solar platform energy potential in actual AC kWh, construction cost, associated engineering fees, total aggregate project cost, cost per installed AC watts, and the
payback period. Economic analysis of each site is presented in a spreadsheet detailing
detailed computation throughout 25 lifecycle of the project.

PROJECTED SOLAR POWER OUTPUT CAPACITY

Table 4.6 represents aggregate solar power output capacity of individual buildings within
the Rainbow Disposal campus. Power output generations shown are in DC and AC
kilowatt hours. Solar energy output projections for future building are based upon the
building footprint areas as shown on the final site plan. All roof-mount areas have been
discounted by 40 percent for fire department and roof-mount equipment clearances.

Table 4.7 represents CO_2 pollution that could be averted by deployment of solar photovoltaic systems at each of the eight sites. It should be noted that the amount of CO_2
pollution is based on the assumption that current electricity to be used in California in
the near future will contain a minimal amount of electrical power generated by coal-
fired plants; in actuality the figures could be somewhat higher.

TABLE 4.5   SUMMARY ANALYSIS OF THE SOLAR POWER SYSTEM

RAINBOW DISPOSAL SUMMARY SHEET

SITE	AC KW	CONST. COST	ENG. FEE	TOTAL COST	COST/AC WATT	PAYBACK PERIOD
1	156	$1,097,428.00	$38,410.00	$1,135,838.00	$7.28	11
2	48	$342,866.00	$12,000.00	$354,866.00	$7.32	12
3	633	$4,082,426.00	$142,884.00	$4,225,310.00	$6.67	10
4	210	$1,447,904.00	$50,676.00	$1,498,580.00	$7.13	11
5	190	$1,330,665.00	$46,753.00	$1,377,418.00	$7.25	12
6	407	$2,834,066.00	$99,192.00	$2,933,258.00	$7.21	12
7	100	$690,129.00	$24,435.00	$714,564.00	$7.15	12
8	275	$1,927,421.00	$67,460.00	$1,994,881.00	$7.25	12

TABLE 4.6   PROJECTED SOLAR POWER OUTPUT CAPACITY

PROJECTED SOLAR POWER OUTPUT CAPACITY—RAINBOW DISPOSAL

SITE NO.	SITE DESCRIPTION	KW-DC	KW-AC
1	MAINTENANCE BUILDING	158	121
2A&2B	OFFICE BUILDINGS	48	37
3	TRANSFER STATION NO.1	633	488
4	M.R.F. BUILDING	210	162
5	SECONDARY RECYCLING BUILDING	190	147
6	TRANSFER STATION NO. 2	407	313
7	BIN REPAIR, PAINT BUILDING, AND CANOPY	100	77
8	PARKING LOTS 1 & 2	275	212
TOTALS		2022	1557

To appreciate the significance of CO_2 prevention, note that one pound of dry ice, which is a chilled CO_2 gas, has a dimension of a nearly 8-in. × 8-in. cube. It is hard to estimate what 71,630,082 pounds of CO_2 pollution could do to harm the environment.

POLLUTION PREVENTION

TABLE 4.7 CO_2 PREVENTION

CARBON DIOXIDE EMMISION CALCULATION—RAINBOW DISPOSAL

SITE	PLATFORM POTENTIAL	INSOLATION × 365 DAYS	MULTIPLIER	ANNUAL TOTAL AMOUNT IN TONS	ANNUAL AMOUNT IN POUNDS	25 YEARS AMOUNT IN TONS
1	156	307476	0.009	2767.284	5534568	69182
2	48	94608	0.009	851.472	1702944	21287
3	633	1247643	0.009	11228.787	22457574	280720
4	210	413910	0.009	3725.19	7450380	93130
5	190	374490	0.009	3370.41	6740820	84260
6	407	802197	0.009	7219.773	14439546	180494
7	100	197100	0.009	1773.9	3547800	44348
8	275	542025	0.009	4878.225	9756450	121956
TOTALS					71630082	895376

CONCLUSION

With reference to the above computational results, in view of present and near-future escalation of electrical energy costs, significant peak power charges reflected in the SCE electrical bills and availability of vast solar power platforms within the project campus, it is our opinion that that a payback period of 11 to 12 years should be sufficient justification to consider integration of solar photovoltaic power generation as a hedge against energy cost escalation.

Even though construction of future buildings may not take in the immediate future, however, existing Site 1 (maintenance building), Site 7 (bin repair), and Site 8 (two parking areas) may be good candidates as solar power platforms.

As indicated in Table 4.2, the average monthly electrical power consumption over the past year has averaged 183,433 kWh. This figure translates into an average of 6114 kilowatts or 6.114 MW of electrical usage per day. The theoretical maximum of solar power produced at 50 percent of the available solar power generation potential from various platforms could provide considerable portions of the plant's energy requirements, which in our opinion could provide total peak power shaving during critical summer months.

FEASIBILITY STUDY FOR SITE NO.1

Maintenance Building
Project address: Huntington Beach, California
Climatic data parameters applied in solar power potential computations use the NREL reference as follows:

–Reference site nearest to Huntington Beach, CA—Long Beach, CA
–Time zone—GMT-8
–Latitude—33.8167°
–Longitude—118.15°
–Direct normal irradiance—207.6 Wh/m^2
–Diffused horizontal irradiance—76.0 Wh/m^2
–Dry bulb temperature—17.2C°
–Wind speed—2.7 m/s

Solar power category—Roof mount
Roof area—20,800 square feet
Solar platform useable area—60%
Net available platform area—12,480 square feet
Solar power capacity—156 kW AC
PV module—Sanyo HIT 210N
Total PV count—914 units
Shading multiplier—1
PV Watts II derating multiplier—77%
Estimated PV module installed cost/watt—$3.65
Total PV installed cost—$574,974.00
Estimated cost of inverter per watt—$0.44
Total inverter cost—$69,311.00

Balance of system cost—$310,000.00
Contingency @ 15%—$143,143.00
Total installed cost—$1,097,428.00
Engineering cost @ 3.5% of the total installed cost—$38,410.00
Total project cost—$1,135, 838.00
Cost per AC installed watt—$7.22/ AC watt

Economic analysis summary

Electrical service provider—Southern California Edison (SCE)
Mean utility rate—$0.18
Utility escalation rate—10%
Solar power lifecycle—25 years
CEC rebate—Performance Base Initiative (PBI)
California Solar Initiative (CEC) rebate—$0.25
Rebate period after final acceptance—5 years
Mean general inflation rate over the solar power lifecycle—3.5%
Solar power system performance annual depreciation—1%
Finance class—Private funding
Straight line depreciation—7 years
Payback period—11 years

SOLAR POWER ANALYTICAL GRAPHS

The following graphs are yielded from the NREL, CSI-compliant solar power analytical modeling software engine.

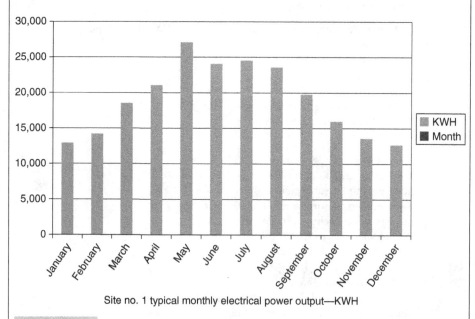

Site no. 1 typical monthly electrical power output—KWH

Figure 4.15   **Typical monthly electrical power output in kilowatts.**

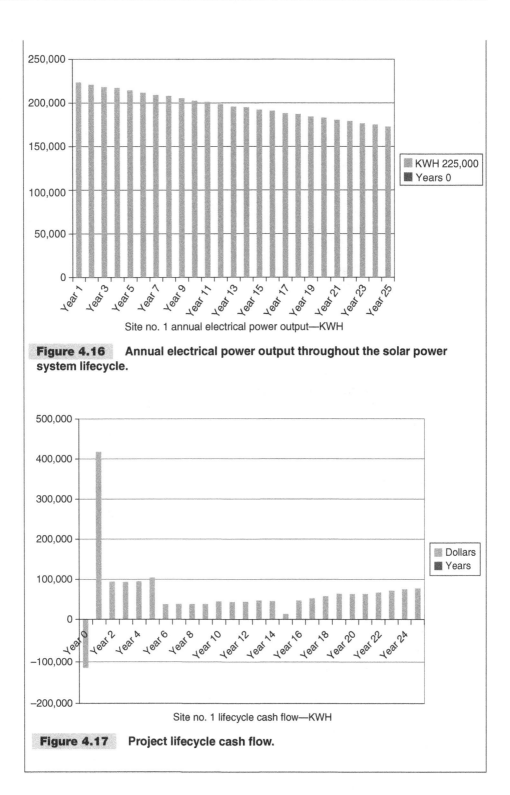

Figure 4.16   Annual electrical power output throughout the solar power system lifecycle.

Figure 4.17   Project lifecycle cash flow.

FEASIBILITY STUDY FOR SITE NO. 2

Office Building

Climatic data parameters applied in solar power potential computations use the NREL reference as follows:

- Reference site nearest to Huntington Beach, CA—Long Beach, CA
- Time zone—GMT-8
- Latitude—33.8167°
- Longitude—118.15°
- Direct normal irradiance—207.6 Wh/m^2
- Diffused horizontal irradiance—76.0 Wh/m^2
- Dry bulb temperature—17.2 C°
- Wind speed—2.7 m/s

Solar Power Category—Roof mount
Roof area—4800 square feet
Solar platform useable area—60%
Net available platform area—2880 square feet
Solar power capacity—48 kW AC
PV module—Sanyo HIT 210N
Total PV count—282 units
Shading multiplier—1
PVWatts II derating multiplier—77%
Estimated PV module installed cost/watt—$3.65
Total PV installed cost—$176,828.00
Estimated cost of inverter per watt—$0.44
Total inverter cost—$21,316.00
Balance of system cost—$100,000.00
Contingency @ 15%—$44,272.00
Total installed cost—$342,866.00
Engineering cost @ 3.5% of the total installed cost—$12,000.00
Total project cost—$354,866.00
Cost per AC installed watt—$7.32/AC watt

Economic analysis summary

Electrical service provider—Southern California Edison (SCE)
Mean utility rate—$0.18
Utility escalation rate—10%
Solar power lifecycle—25 years
CEC rebate—Performance Base Initiative (PBI)
California Solar Initiative (CEC) rebate—$ 0.25
Rebate period after final acceptance—5 years
Mean general inflation rate over the solar power lifecycle—3.5%
Solar power system performance annual depreciation—1%
Finance class—Private funding
Straight line depreciation—7 years
Payback period—12 years

SOLAR POWER ANALYTICAL GRAPHS

The following plotted graphs are yielded from the NREL, CSI-compliant solar power analytical modeling software engine. Figures 4.18 through 4.19 depict graphic representation of monthly power output, the annual power output throughout the solar power system lifecycle and project lifecycle cash flow respectively.

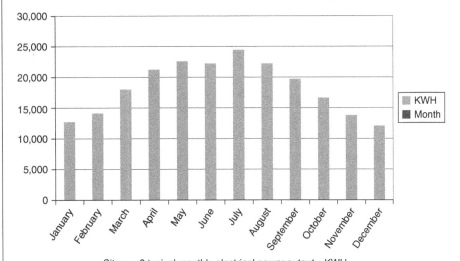

Site no. 2 typical monthly electrical power output—KWH

Figure 4.18   **Typical monthly electrical power output in kilowatts.**

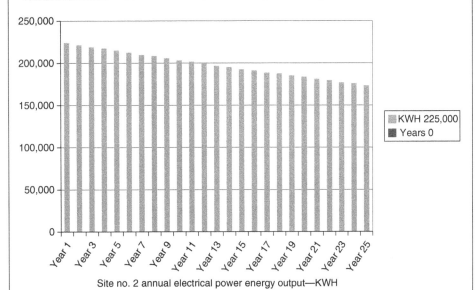

Site no. 2 annual electrical power energy output—KWH

Figure 4.19   **Annual electrical power output throughout the solar power system lifecycle.**

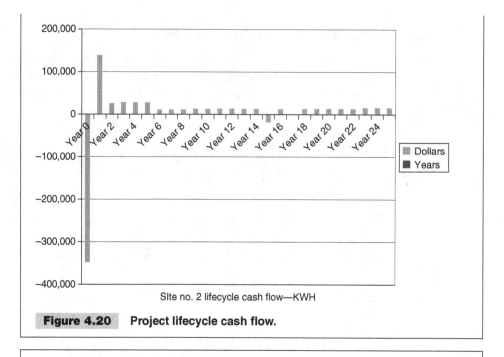

Site no. 2 lifecycle cash flow—KWH

Figure 4.20   **Project lifecycle cash flow.**

FEASIBILITY STUDY FOR SITE NO. 8

Future Parking Lot Solar Canopies

Climatic data parameters applied in solar power potential computations use the NREL reference as follows:

   –Reference site nearest to Huntington Beach, CA—Long Beach, CA
   –Time zone—GMT-8
   –Latitude—33.8167°
   –Longitude—118.15°
   –Direct normal irradiance—207.6 Wh/m^2
   –Diffused horizontal irradiance—76.0 Wh/m^2
   –Dry bulb temperature—17.2 C°
   –Wind speed—2.7 m/s

Solar Power Category—Carport canopy

Number of stalls—161
Car stall area—170 square feet
Net available platform area—27,370 square feet
Solar power capacity—275 kW AC
PV module—Sanyo HIT 210N
Total PV count—1610 units
Shading multiplier—1
PV Watts II derating multiplier—77%
Estimated PV module installed cost/watt—$ 3.65
Total PV installed cost—$1,004,882.00
Estimated cost of inverter per watt—$0.44

Total inverter cost—$121,136.00
Balance of system cost—$550,000.00
Contingency @ 15%—$251,403.00
Total installed cost—$1,927,421.00
Engineering cost @ 3.5% of the total installed cost—$67,460.00
Total project cost—$1,994,880.00
Cost per AC installed watt—$7.25/AC watt

Economic analysis summary
Electrical service provider—Southern California Edison (SCE)
Mean utility rate—$0.18
Utility escalation rate—10%
Solar power lifecycle—25 years
CEC rebate—Performance Base Initiative (PBI)
California Solar Initiative (CEC) rebate—$0.25
Rebate period after final acceptance—5 years
Mean general inflation rate over the solar power lifecycle—3.5%
Solar power system performance annual depreciation—1%
Finance class—Private funding
Straight line depreciation—7 years
Payback period—12 years

SOLAR POWER ANALYTICAL GRAPHS

The following plotted graphs are yielded from the NREL, CSI-compliant solar
power analytical modeling software engine. Figures 4.21 through 4.23 depict
graphic representation of monthly power output, the annual power output throughout the solar power system lifecycle and project lifecycle cash flow respectively.

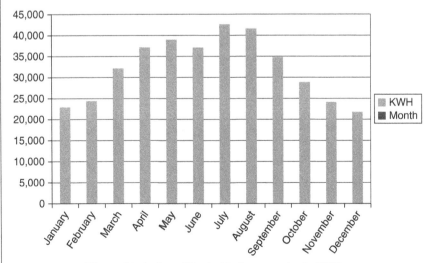

Site no. 8 typical monthly electrical power output—KWH

Figure 4.21   **Typical monthly electrical power output in kilowatts.**

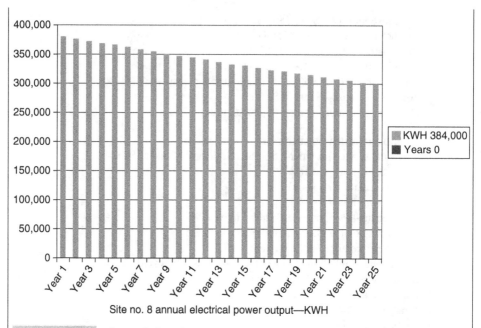

Site no. 8 annual electrical power output—KWH

Figure 4.22 Annual electrical power output throughout the solar power system lifecycle.

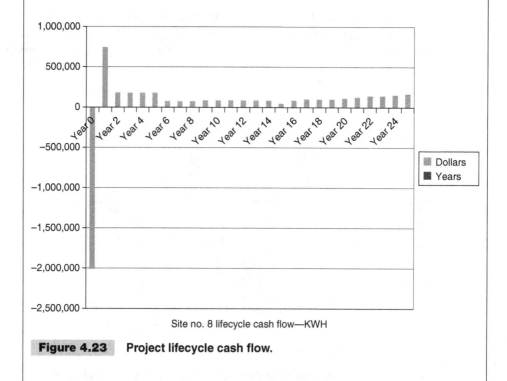

Site no. 8 lifecycle cash flow—KWH

Figure 4.23 Project lifecycle cash flow.

References

1. Gevorkian, Peter, *Alternative Energy Systems in Building Design,* McGraw-Hill, New York, 2010.
2. http://www.nrel.gov/rredc/pvwatts/
3. Energy Information Administration.
4. International Energy Agency Electricity Information 2005; International Energy Agency Energy Prices and Taxes, 4th Quarter 2005; and Eurostat Gas and Electricity Market Statistics 2005.
5. Resource Data International.

SOLAR POWER SYSTEM
COST ANALYSIS

Introduction

Solar power system cost analysis is one of the most important components of Solar power feasibility, since it often establishes cost-benefit analyses and criterion for financial cost justifications of the project. Without accurate financial analysis, solar power projects will undoubtedly be subjected to serious negative consequences.

This chapter will discuss computational performance characteristics of various types of financial analytical methodologies and software that are being used to compute econometrics of solar power systems.

California Solar Initiative Calculator

In addition to PVWatts computations (discussed in Chap. 4), the State of California requires the use of a Web-based calculator referred to as the California Solar Initiative (CSI) calculator. It essentially uses the PVWatts II software engine for calculating the solar power system size.

The CSI calculator also performs project rebate cost estimates for Expected Performance-Based Initiatives (EPBI) as well as Performance-Based Initiatives (PBI). The following are additional computation results provided by the CSI calculator:

- Annual energy production in kilowatt hours
- Summer month power output from May to October in kilowatt hours
- CEC AC power output in kW/h
- Capacity factor in kW
- Prevailing capacity factor in percent
- Design factor in percent

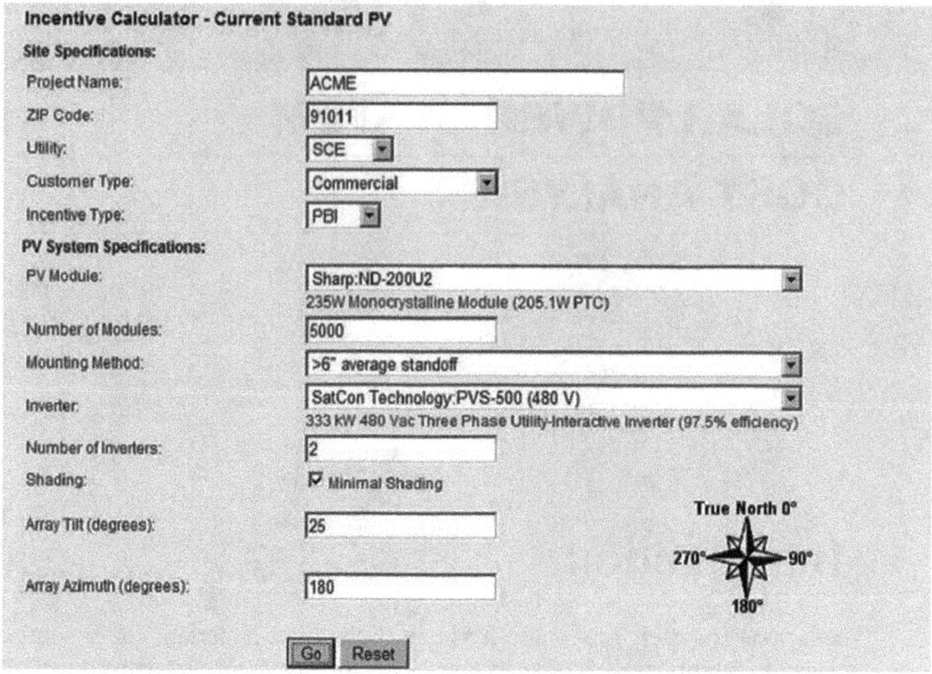

Figure 5.1   CSI solar power calculator data entry page.

- Eligibility annual power output in kilowatt hours
- Incentive rate per kilowatt hours in dollars
- Total CSI incentive contribution in dollars

Figures 5.1 and 5.2 are examples of a hypothetical project located in La Canada, California.

Solar Power Costing Analytical Software

Depending on the type of financing, solar power cost analysis may involve a number of methodologies, which range from public domain solar power cost-estimating engines, such as the National Renewable Energy Laboratory's Solar Advisor Model (SAM), to a number of other commercial software packages. Solar power estimating software packages provide cost approximation based on local electrical utility tariffs, type of solar power system, whether fixed angle, or single- or dual-axis tracking systems. Such software-estimating calculators do not take into consideration the specific project costs. They also do not consider inflation, energy

Incentive Calculator-Current Standard PV

Site Specifications: **Proposed**

Project Name   ACME

ZIP Code   91011

City   La Canada

Utility   SCE

Customer Type   Commercial

Incentive Type   PBI

PV System Specifications:

PV Module   Sharp:ND-200U2
                                                200.0W STC, 173.0WPTC

Number of Modules   5000

Mounting Method   >6" average standoff

DC Rating (kW STC)   1000.0000

DC Rating (kW PTC)   865.0000

Inverter   SatCon Technology:PVS-500 (480 V)

Number of Inverters   2

Inverter Efficiency (%)   96.00

Shading   Minimal Shading

Array Tilt (degrees)   25

Array Azimuth (degrees)   180 True North 0

Results

Annual kWh   1,596.510

Summer Months   May–October

Summer kWh   907,099

CEC-AC Rating   830,400 kW

Capacity Factor   21.947%

Prevailing Capacity Factor   20.000%

Design Factor   109.735%

Eligible Annual kWh   1,596.510

Incentive Rate $0.22/kWh

Incentive $1,756.161

Figure 5.2   CSI solar power calculator data output page.

cost escalation, special labor rates, transportation, and much more. In effect, solar power cost calculators do not provide the accuracy required to conduct a detailed cost estimate. However, most institutional, public, and governmental agencies (as well as public utility commissions) require that all solar power calculations be based on acceptable econometric models. In fact, SAM is considered one of the best estimating software programs available to the public free of charge. It can be accessed at https://www.nrel.gov/analysis/sam/. The following is description of the SAM as it appears in the National Renewable Energy Laboratories.

Solar Advisor Model

SAM combines a detailed performance model with several types of financing (from residential to utility-scale) for most solar technologies. The solar technologies currently represented in SAM include concentrating solar power (CSP) parabolic troughs, dish/Stirling systems, and power tower systems, as well as flat-plate and concentrating PV technologies. SAM incorporates the best available models to allow analysis of the impact of changes to the physical system on the overall economics (including the levelized cost of energy). SAM development continues to add additional financing models and performance models to meet the needs of a growing community of users.

This comprehensive solar technology systems analysis model supports the implementation of the program's Solar America Initiative (SAI) as well as general planning for the Solar Energy Technologies Program (SETP). Use of the SAM software (together with technology and cost benchmarking, market penetration analysis, and other relevant considerations) supports the development of program priorities and direction, and the subsequent investment needed to support solar R&D activities. Most importantly, it promotes the use of a consistent methodology for analysis across all solar technologies, including financing and cost assumptions. SAM allows users to investigate the impact of variations in physical, cost, and financial parameters to better understand their impact on key figures of merit. Figures of merit related to the cost and performance of these systems include, but aren't limited to:

- System output
- Peak and annual system efficiency
- Levelized cost of electricity
- System capital and operating and maintenance (O&M) costs
- Hourly system production

SAM uses a systems-driven approach (SDA) to establish the connection between market requirements and R&D efforts and how specific R&D improvements contribute to the overall system cost and performance. This SDA allows managers to allocate resources more efficiently.[1]

SYSTEM COST DATA

SAM software includes a set of sample files that contain cost data prepared to illustrate its use. The cost data is meant to be realistic, but not to represent actual costs in the marketplace. Actual costs vary depending on the market, technology, and geographic location of a project. Because of price volatility in solar markets, the cost data in the sample files are likely to be out of date.[1]

PHOTOVOLTAIC COST DATA

PV cost input data in SAM is divided into two broad categories: capital and operation/maintenance (O&M) costs. Capital costs are further categorized into direct and indirect costs. Direct costs are costs associated with the purchase of equipment: PV modules, inverter(s), balance of system (BOS), and installation costs. BOS costs are equipment costs that cannot be assigned to either the PV module or the inverter, and may include such costs as mounting racks, junction boxes, and wiring. Installation costs are the labor costs associated with installing the equipment.

Indirect costs may include all other costs that are built into the price of a system such as profit, overhead (including marketing), design, permitting, shipping, etc.

O&M costs are costs associated with a system after it is installed, and are categorized into fixed and variable O&M costs. Fixed O&M costs are costs that vary with the size of the system, and may include the cost of inverter replacements and periodic maintenance checks. Variable O&M costs vary with the output of the system and may be considered to be zero or very small for most PV systems.

SAM uses the total installed cost, which is the sum of direct and indirect costs, to calculate the levelized cost of energy. Since how costs are assigned to each category does not affect the total installed cost, the user can choose to distribute profit, overhead, shipping, and other costs among the component categories. These may consist of modules, inverters, system balance components, and system installation (or it may be considered as a single value and categorized as miscellaneous).

Note that costs in the PV sample files used in the program are based on 2005 costs that are derived from the DOE Multi-Year Program Plan. Therefore, the total installed costs are intended to represent equipment purchase and labor costs plus a margin sufficient to sustain a profitable business with a reasonable return on investment.[1]

Additional sources of PV cost data   The California Energy Commission's *Emerging Renewables Program* Web site provides information about systems installed in California and includes a link to a spreadsheet of total installed system costs for systems installed throughout the states. *SolarBuzz* provides current and historical price data for the United States and around the world based on market studies. The site is a valuable resource that provides detailed statistical information about PV module prices, inverters, and the solar electricity price index.

Guide to using the Solar Advisor Model   SAM consists of the following data entry and display domains:

A. System Summary   This provides a tabulated overview of the computation, which displays the information shown in the table.

Solar power system capacity in kilowatts
Total direct cost in dollars
Total installed cost in dollars
Total installed cost per kilowatts
Solar power system lifecycle in years
Projected inflation rate in percent
Applicable discount rate in percent

B. Climate   This provides information regarding the sites based on postal zip code, which provides the weather and temperature data.

City
State
Time zone in GMT
Elevation in meters
Latitude
Longitude
Direct normal radiance in kW/m^2
Diffused horizontal radiance in kW/m^2
Dry-bulb temperature in degrees centigrade
Wind speed in meters per second (m/s)

C. Utility Rate   This displays the following fields:

Up-to-date cost of electrical power per $/kW
Projected utility inflation rate in percent

D. Financing   This domain allows the user to make a number of project-specific entries for insurance and taxes, federal depreciation options, state depreciation options, project lifecycle, projected inflation rates, and real discount rates. The following are various types of options available in solar power system financing:

- Energy Savings Performance Contract (ESPC). ESPC is partnership between the owner and a federal government agency.
- Energy Service Company (ESCO). An ESCO is a financing system that arranges necessary financing for funding the solar PV plant and guarantees the estimated energy cost savings to the owner. This analysis determines the minimum tariff

at which electric power can be sold from the solar PV plant to the owner's facility.

■ Utility Energy Savings Contract (UESC). In this arrangement, a federal agency enters into partnership with its franchised or serving utilities in order to implement energy improvements at its facilities. The utility arranges financing to cover the capital costs of the project and is repaid by the owner over the contract term, and in turn, provides cost savings to the owner.

■ Enhanced Use Lease Contract (EUL). An EUL program refers to legislative authority that allows owners to lease underutilized land and improvements to a developer or lessee for a term of up to 75 years. In exchange for the EUL, the developer would be required to provide the owner with fair consideration, such as cash or in-kind considerations as determined by the owner.

■ Direct funding. In this option, the owner provides 100 percent funding for the solar power project. No debt financing is assumed.

Factors taken into consideration when calculating the funding also include:

Analysis period (project lifecycle) in years
Projected inflation rate in percent
Real discount rate

Note that discounting is a financial mechanism in which a debtor or the borrower obtains the right to delay payments to a creditor for a defined period of time in exchange for a charge or fee. In other words, the party that currently owes money purchases the right to delay the payment until some future date. The discount, or charge, is the difference between the original amount owed in the present and the amount that has to be paid in the future to meet the debt obligation.

Tax and Insurance   Factors considered in tax and insurance computation include:

Federal tax percent of the project cost
State tax as percent of the project cost
Property tax
Sales tax (in percent)
Solar power system lifecycle in years (in percent)
Insurance cost as percentage of the installed cost

FEDERAL DEPRECIATION

In this domain, the user is allowed to select various choices regarding types of rebate plans, as well as years of equipment depreciation. Choices include the following:

■ No depreciation
■ Modified Accelerated Cost-Recovery System (MACRS) mid-quarter convention

■ Modified Accelerated Cost-Recovery System (MACRS) half-year convention
■ Straight-line depreciation
■ Custom depreciation, which is percent of the installed system cost

In view of the importance of accelerated depreciation, it is important for SAM to fully appreciate depreciation choices that may be best suited for a particular project. The following discussions are intended to familiarize the reader about several depreciation methodologies allowed by the U.S. tax system. It should be noted that selecting a type of depreciation must be always be undertaken by an expert tax accountant.

Modified Accelerated Cost-Recovery System (MACRS)   The Modified Accelerated Cost-Recovery System (MACRS) is a methodology used for recovering capitalized costs of depreciable tangible property other than natural resources. Under this system, the capitalized cost, also referred to as *basis,* is recovered over the lifecycle of a tangible property or asset by annual deductions for depreciation. The lifespan of various types of property are broadly specified in the U.S. Internal Revenue Code. Various classes of asset lifecycles are tabulated and published by the Internal Revenue Service (IRS). The deduction for depreciation is computed in one of two methods: declining balance or straight line.

Depreciable lives by class   As mentioned, MACRS specifies that a taxpayer must compute tax deductions for the depreciation of tangible property or assets using specified lifespans and methods. Assets are divided into classes by type of asset or business in which the asset is used.

For each class, the lifecycles are specified as the general depreciation system (GDS) or alternative depreciation system (ADS). Taxpayers have a choice to use either the GDS or the ADS methodology. Lifecycles of assets may vary from 5 years to 20 years.

FEDERAL ACCELERATED DEPRECIATION FOR SOLAR POWER SYSTEMS

Commercial and industrial systems (upon qualification) can take advantage of special solar power system depreciations (26 USC Sec. 168—MACRS), which allow asset depreciation and amortization over a period of 5 years. Such accelerated depreciation, depending upon combined federal and state tax credits, enables investment recovery of up to 50 percent. It should be noted that the asset value is the total installed less the amount of the rebates received.

In general cash rebates, tax credits, and accelerated depreciation schedules are designed to facilitate short-term returns on investments. This encourages businesses to generate their own solar power. In some instances, such programs can recover up to 40 to 70 percent of the total system cost in a very short period of time.

Note that the 30 percent Federal Investment Tax Credit (ITC) is calculated before any state or utility rebates. The claim is calculated after deducting state rebates, such as the net cost paid by installers, who typically collect the rebate on the owner or client's behalf.

TABLE 5.1 IRS ASSET DEPRECIATION TABLE			
IRS ASSET CLASSES	ASSET DESCRIPTION	ADS CLASS LIFE	GDS CLASS LIFE
00.11	Office furniture, fixtures, and equipment	10	7
00.12	Information systems: computers/peripherals	6	5
00.22	Automobiles, taxis	2	5
00.241	Light general-purpose trucks	4	5
00.25	Railroad cars and locomotives	15	7
00.40	Industrial steam and electric distribution	22	3
01.11	Cotton gin assets	10	7
01.21	Cattle, breeding or dairy	7	5
13.00	Offshore drilling assets	7.5	5
13.30	Petroleum refining assets	16	10
15.00	Construction assets	6	5
20.10	Manufacture of grain and grain mill products	17	10
20.20	Manufacture of yam, thread, and woven fabric	11	7
24.10	Cutting of timber	6	5
32.20	Manufacture of cement	20	15
20.1	Manufacture of motor vehicles	12	7
48.10	Telephone distribution plant	24	15
48.2	Radio and television broadcasting equipment	6	5
49.12	Electric utility nuclear production plant	20	15
49.13	Electric utility steam production plant	28	20
49.23	Natural gas production plant	14	7
50.00	Municipal wastewater treatment plant	24	15
57.0	Distributive trades, and services	9	5
80.00	Theme and amusement park assets	12.5	7

As of December 31, 2009, the IRS has not instituted clear guidance on the issue. Table 5.1 is the IRS MACRS depreciation table showing various asset classifications, asset descriptions, and associated ADS and GDS lifecycles.

DEPRECIATION METHODS

Only the declining balance method and straight-line method of computing depreciation are allowed under MACRS. All solar power systems installed during the current year are considered "placed in service" in the middle of the tax year, referred to as the *half-year convention*. The method and lifecycle used in depreciating an asset is an accounting method, as any deviation or change requires IRS approval.

Alternative depreciation system   The alternative depreciation system also available by IRS pertains to certain assets that must be depreciated under an ADS using the ADS lifecycle, which uses straight-line depreciation methodology. Such depreciation only applies to assets used outside the United States and is not applied to solar power systems.

CALIFORNIA EXEMPTION FROM PROPERTY TAXES

When installing solar power systems in California, note that the state excludes all solar power system installation costs from being added to the property value, which prevents an increase in tax valuation.

Such an exemption means that a solar power system increases the value of property, since the added equity is completely tax-free.

DIRECT CAPITAL COSTS (DCC)

Direct capital cost (DCC) entry sheets allow the user to enter specific installed unit costs ($/kW DC) for the DC component of the solar project, the inverter, storage and transportation, balance of the hardware, and integration costs. Additional fields allow entries for solar power system engineering as percent, non-fixed or fixed cost, and miscellaneous items and sales tax. Note that prior to completing the DCC, the user must complete the PVWatts calculations as discussed previously. Prior to completing the calculations, the designer must complete the solar power system preliminary design and have an estimated value of material and labor costs. Figure 5.3 is an outline of a typical construction manpower and material cost estimate. After computing with PVWatts, the amount of total kW/DC is automatically entered in the system power output capacity field. The remaining entries required include the following:

- A—Direct PV system cost
  - Number of modules
  - DC kW/unit
  - Total system power in kW/DC
  - Combined cost of PV and support structure per $/kW/DC
- B—Inverter system cost. After completing the entries in A, SAM calculates the direct material cost. The row following A calculates the cost of inverter, and has the following entries:
  - Inverter count
  - AC power output rating in kW/AC
  - Inverter cost per $/kW/AC

Typical construction cost estimate

INITIAL COSTS & CREDITS

Engineering Rate — $150.00 per hour

	Hours	Rate		Total	%
Site Investigation	40 $	150.00 $		6,000.00	46.88%
Preliminary Design Coordination	24 $	150.00 $		3,600.00	28.13%
Report Preparation	8 $	150.00 $		1,200.00	9.38%
Travel & Accommodations	1 $	2,000.00 $		2,000.00	15.63%
Other					0.00%
Sub Total				12,800.00	100.00%

DEVELOPMENT

	Hours	Rate		Total	%
Permits & Rebate Applications	8 $	150.00 $		1,200.00	5.41%
Project Management	120 $	150.00 $		18,000.00	81.08%
Travel Expenses	1 $	2,000.00 $		2,000.00	9.01%
Other	1 $	1,000.00 $		1,000.00	4.50%
Sub Total				22,200.00	100.00%

ENGINEERING

	Hours	Rate		Total	%
PV Systems Design	90 $	150.00 $		13,500.00	10%
Architectural Design	90 $	150.00 $		13,500.00	10%
Structural Design	90 $	150.00 $		13,500.00	10%
Electrical Design	420 $	150.00 $		63,000.00	48%
Tenders & Contracting	48 $	150.00 $		7,200.00	5%
Construction Supervision	94 $	150.00 $		14,100.00	11%
Training Manuals	48 $	150.00 $		7,200.00	5%
Sub Total			$	132,000.00	100%

RENEWABLE ENERGY EQUIPMENT

	Hours	Rate		Total	%
PV Modules (per kWh-DC)	255 $	3,900.00 $		994,500.00	92%
Transportation	1 $	5,000.00 $		5,000.00	0%
Other					0%
Tax (Equipment Only)	8.25%			82,046.25	8%
Sub Total				1,081,546.25	100%

INSTALLATION EQUIPMENT

	Hours	Rate		Total	%
PV Module Support Structure (per kWh)	255 $	500.00		127,500.00	18%
Inverter (per kWh)	320 $	488.00		156,160.00	22%
Electrical Materials (per kW)	320 $	250.00		80,000.00	11%
System Installation Labor (per kWh)	320 $	1,000.00		320,000.00	45%
Transportation	1 $	3,000.00		3,000.00	0%
Other	0				0%
Tax (Equipment Only)	8.25%			30,001.95	4%
Sub Total				716,661.95	100%

Figure 5.3   **Typical solar power construction cost estimate.**

- C—Battery system cost. The entries compute the total cost of the inverter system. The third line identical to the inverter computes battery cost (when applicable).
  - Storage battery capacity in kW/h
  - Storage cost kW/AC
  - Cost per $/kW/hr
- D—Balance of other costs
  - Balance of the material costs
  - Fixed installation cost
  - Contingency
- E—Indirect capital costs
  - Engineering, project management, construction supervision either as percentage of the direct cost, variable, or fixed engineering service cost
  - Addition of the above cost, resulting in total installed cost, which is then divided by the total kW/DC, which yields cost per installed watts, $/kW/DC

OPERATION AND MAINTENANCE COSTS

In addition to the DCC, SAM allows users to enter operation and maintenance costs for the duration of the lifecycle of the project, as well as anticipated escalation rates due to inflation. Options include:

- Fixed annual cost   % Escalation rate
- Fixed cost by capacity   % Escalation rate
- Variable cost by generation   % Escalation rate

In general, solar power systems have minimal maintenance requirements. However, to prevent marginal degradation in output performance from dust accumulation, solar arrays require a bi-yearly rinsing with a regular water hose. Since solar power arrays are completely modular, system expansion, module replacement, and troubleshooting are simple and require no special maintenance skills. All electronic DC-to-AC inverters are modular and can be replaced with minimum downtime.

In some instances, a computerized system-monitoring console can provide a real-time performance status of the entire solar power cogeneration system. Installation cost of a software-based supervisory program that features data monitoring and maintenance reporting systems must also be costed in.

ANNUAL SYSTEM PERFORMANCE

This entry allows the user an arbitrary entry for system degradation and system availability or reliability.

The last entry page includes a number of entries specific to types of federal and state investment-based incentives such as:

- Performance-Based Initiative (PBI)
- Capacity-Based Incentive (CPI)
- Investment-Based Incentive (IBI)

SPECIAL COSTING CONSIDERATIONS

As mentioned previously, standard Web-based software calculators are designed to provide rough estimates of solar power production and costing, which are adequate for feasibility studies and rebate applications. However, they lack accuracy and bear no relation to specific requirements of projects.

The following is a description of a comprehensive guide for estimating various categories of solar power systems. In order to have accurate estimating results, the solar power design engineer or the designer must have significant field installation and solar power design experience to account for numerous cost items. These may include but are not limited to the following:

- Site survey and feasibility study
- Engineering design
- Material costs
- Civil and structural design
- Rebate contributions
- Negative environmental impact report
- Field installation labor costs
- Field test and acceptance oversight
- Project management
- Material transportation and storage
- Insurance costs (i.e., errors and omissions, construction insurance, liability insurance)
- Construction loan, construction bond, and long-term financing
- Customer personnel training
- Warrantee and maintenance costs
- System lifecycle profit calculation, such as present value and depreciation cost, etc.

A software program designed for computing econometrics of a solar power system must provide subroutines and computational engines that enable users to insert the numerous variables listed. The software must also compute power performance characteristics, energy production costs, and must include algorithms for dynamic energy cost escalation for the duration of the system lifecycle. As highlighted here, system-costing methodology must account for capital equipment depreciation and return on investment.

Additional cost parameters that influence the cost are system operation performance, dynamic power degradation, present or contractual unit energy cost/kWh, project grid electrical energy cost escalation, rebate profile for Performance-Based Incentives (PBI), initial cost investment, salvage value, and many additional factors. These provide year-to-year solar power income profiles throughout the lifecycle of the contract.

Additional consideration must be given to power purchase agreement (PPA) financial options. In particular, some of the factors that may affect a PPA may include the following:

- Yearly average AC power output and solar power energy value computation
- Dynamic extrapolation of projected unit energy cost escalation for the entire lifecycle of the contract

- Dynamic depreciation of solar power output for system operational lifecycle
- Lifecycle power output potential
- Progressive rebate accumulation for the PBI period based on the available unity energy fund availability
- Cumulative electrical energy cost income from the end-of-rebate period up until the end-of-system contractual lifecycle
- Integrated accumulative income from contractual annual cost escalation factor
- Maintenance income from annual maintenance fees
- Comparative analysis of grid power energy expenditure versus solar power system energy output cost

Costing computation must also include cumulative lifecycle energy cost analysis for grid-connected expenses such as:

- Totalized energy income from life operation
- Present value of the projected system salvage amount
- Federal tax incentive income
- State tax incentive income
- Salvage value at the end of the system lifecycle
- Maintenance cost income
- Recurring annual maintenance costs
- Recurring annual data acquisition costs
- Net income value over the lifecycle of operation

Additional cost factors   In addition to manpower and material costs, there are several additional cost components. These include lifecycle cost variations of utility, operation, and maintenance expenses that are of significant importance for leased or PPA financed projects.

ELECTRICAL ENERGY COST INCREASE

In the past several decades, the cost of electrical energy production and its consistent incline has been an issue that has dominated global economics and geopolitical politics. It has affected our public policies, has been a significant factor in the gross national product (GNP) equation, created numerous international conflicts, and has made more headlines in newsprint and television than any other subject. Electrical energy production not only affects the vitality of international economics, but it is one of the principle factors that determines standards of living, health, and general well-being of the countries that produce it in abundance.

Every facet of the U.S. economy is in one way or another connected to the cost of electrical energy production. Since large portions of global electrical energy production is based on fossil fuel-fired electrical turbines, the price of energy production is therefore, determined by the cost of coal, crude oil, or natural gas commodities. Figure 5.4 depicts the comparative cost inflation of a utility rate at various interest rates during the lifecycle of a project, and Fig. 5.5 is a graphic presentation of a $ 0.13/kW/h utility rate at an inflation rate of 8 percent over the lifecycle of a project.

COMPOUND INTEREST CALCULATION FOR ELECTRICAL ENERGY INFLATION

ENERGY ESCALATION RATE	12%	11%	10%	9%	8%	7%	6%	5%	4%
PRESENT ENERGY COST/kWh	$ 0.13	$ 0.13	$ 0.13	$ 0.13	$ 0.13	$ 0.13	$ 0.13	$ 0.13	$ 0.13
YEARS	COST/kWh	COST/kWh	COST/kWh	COST/kWh	COST/kWh	COST/kWh	COST/kWh	COST/kWh	COST/kWh
1	$ 0.15	$ 0.14	$ 0.14	$ 0.14	$ 0.14	$ 0.14	$ 0.14	$ 0.14	$ 0.14
2	$ 0.16	$ 0.16	$ 0.16	$ 0.15	$ 0.15	$ 0.15	$ 0.15	$ 0.14	$ 0.14
3	$ 0.18	$ 0.18	$ 0.17	$ 0.17	$ 0.16	$ 0.16	$ 0.15	$ 0.15	$ 0.15
4	$ 0.20	$ 0.20	$ 0.19	$ 0.18	$ 0.18	$ 0.17	$ 0.16	$ 0.16	$ 0.15
5	$ 0.23	$ 0.22	$ 0.21	$ 0.20	$ 0.19	$ 0.18	$ 0.17	$ 0.17	$ 0.16
6	$ 0.26	$ 0.24	$ 0.23	$ 0.22	$ 0.21	$ 0.20	$ 0.18	$ 0.17	$ 0.16
7	$ 0.29	$ 0.27	$ 0.25	$ 0.24	$ 0.22	$ 0.21	$ 0.20	$ 0.18	$ 0.17
8	$ 0.32	$ 0.30	$ 0.28	$ 0.26	$ 0.24	$ 0.22	$ 0.21	$ 0.19	$ 0.18
9	$ 0.36	$ 0.33	$ 0.31	$ 0.28	$ 0.26	$ 0.24	$ 0.22	$ 0.20	$ 0.19
10	$ 0.40	$ 0.37	$ 0.34	$ 0.31	$ 0.28	$ 0.26	$ 0.23	$ 0.21	$ 0.19
11	$ 0.45	$ 0.41	$ 0.37	$ 0.34	$ 0.30	$ 0.27	$ 0.25	$ 0.22	$ 0.20
12	$ 0.51	$ 0.45	$ 0.41	$ 0.37	$ 0.33	$ 0.29	$ 0.26	$ 0.23	$ 0.21
13	$ 0.57	$ 0.50	$ 0.45	$ 0.40	$ 0.35	$ 0.31	$ 0.28	$ 0.25	$ 0.22
14	$ 0.64	$ 0.56	$ 0.49	$ 0.43	$ 0.38	$ 0.34	$ 0.29	$ 0.26	$ 0.23
15	$ 0.71	$ 0.62	$ 0.54	$ 0.47	$ 0.41	$ 0.36	$ 0.31	$ 0.27	$ 0.23
16	$ 0.80	$ 0.69	$ 0.60	$ 0.52	$ 0.45	$ 0.38	$ 0.33	$ 0.28	$ 0.24
17	$ 0.89	$ 0.77	$ 0.66	$ 0.56	$ 0.48	$ 0.41	$ 0.35	$ 0.30	$ 0.25
18	$ 1.00	$ 0.85	$ 0.72	$ 0.61	$ 0.52	$ 0.44	$ 0.37	$ 0.31	$ 0.26
19	$ 1.12	$ 0.94	$ 0.80	$ 0.67	$ 0.56	$ 0.47	$ 0.39	$ 0.33	$ 0.27
20	$ 1.25	$ 1.05	$ 0.87	$ 0.73	$ 0.61	$ 0.50	$ 0.42	$ 0.34	$ 0.28
21	$ 2.66	$ 2.04	$ 1.57	$ 1.22	$ 0.95	$ 0.74	$ 0.58	$ 0.46	$ 0.36
22	$ 2.98	$ 2.26	$ 1.73	$ 1.32	$ 1.02	$ 0.79	$ 0.62	$ 0.48	$ 0.38
23	$ 3.33	$ 2.51	$ 1.90	$ 1.44	$ 1.10	$ 0.85	$ 0.65	$ 0.51	$ 0.39
24	$ 3.73	$ 2.79	$ 2.09	$ 1.57	$ 1.19	$ 0.91	$ 0.69	$ 0.53	$ 0.41
25	$ 4.18	$ 3.09	$ 2.30	$ 1.72	$ 1.29	$ 0.97	$ 0.73	$ 0.56	$ 0.42

Figure 5.4 Compound interest table of a $0.13/kW/h electric rate over a period of 25 years.

ENERGY ESCALATION RATE

	12%	3.5%
	GRIG	SOLAR
PRESENT VALUE OF ENERGY/kWh	$ 0.13	$ 0.13

YEARS	COST/kWh	COST/kWh
1	$ 0.15	$ 0.13
2	$ 0.16	$ 0.14
3	$ 0.18	$ 0.14
4	$ 0.20	$ 0.15
5	$ 0.23	$ 0.15
6	$ 0.26	$ 0.16
7	$ 0.29	$ 0.17
8	$ 0.32	$ 0.17
9	$ 0.36	$ 0.18
10	$ 0.40	$ 0.18
11	$ 0.45	$ 0.19
12	$ 0.51	$ 0.20
13	$ 0.57	$ 0.20
14	$ 0.64	$ 0.21
15	$ 0.71	$ 0.22
16	$ 0.80	$ 0.23
17	$ 0.89	$ 0.23
18	$ 1.00	$ 0.24
19	$ 1.12	$ 0.25
20	$ 1.25	$ 0.26
21	$ 2.66	$ 0.32
22	$ 2.98	$ 0.33
23	$ 3.33	$ 0.35
24	$ 3.73	$ 0.36
25	$ 4.18	$ 0.37

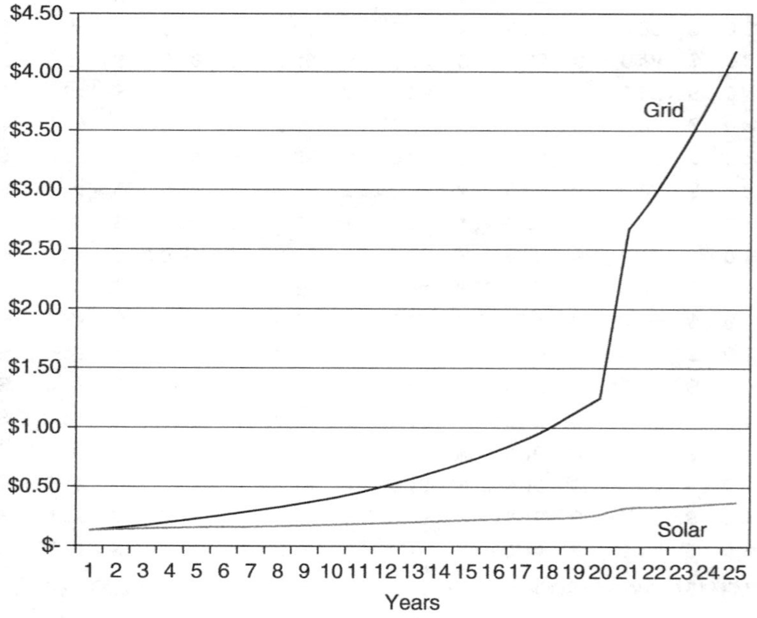

Figure 5.5 Compound interest table of a $0.13/ kW/h electric rate with an 8 percent lifecycle inflation rate.

SYSTEM MAINTENANCE AND OPERATIONAL COSTS

Further, in Chap. 6 we will discuss computerized data acquisition and monitoring systems, which could provide real-time performance status of entire solar power cogeneration systems, as well as features that allow instantaneous indication of system malfunction.

FEDERAL TAX CREDITS FOR COMMERCIAL USE

The Energy Policy Act of 2006, which has been extended to 2016, makes provisions for a 30 percent Investment Tax Credit (26 USC Sec. 48). Note that the *tax credit* is not a deduction; rather, it is a direct, dollar-for-dollar reduction from taxes owed.

The national Solar Energy Industries Association (SEIA) has a document entitled "The SEIA Guide to Federal Tax Incentives," which was compiled with the help of SEIA members and SEIA's tax attorneys, and contains a great deal of specific information and advice regarding the interpretation of the new federal law.[2]

Incentives for Commercial Solar Projects

FEDERAL GRANT

To promote national green energy production, United States Treasury grants are available for commercial solar installations. Instead of taking the 30-percent tax credit, businesses can instead opt to receive a cash grant equal to 30 percent of installed costs of a solar PV system. This grant option has been made possible by the federal stimulus package that was passed in February 2009.

Thirty percent Federal Investment Tax Credit (ITC)   Under U.S. Code Title 26 [Section 48(a)(3)], the federal government extends a corporate tax credit to businesses that invest in renewable power. The types of eligible solar technologies include solar water heat systems, solar space heat, solar thermal electric, solar thermal process heat, and photovoltaic (PV) systems. The credit or the grant is fixed at 30 percent. Note that the credit for businesses is not constrained by a dollar-value cap. So regardless of whether the installation of the solar power system costs $100,000 or $1 million, businesses are permitted to take a 30 percent credit. In October 2008 Congress voted to extend the ITC for eight years, through 2016.

Using the federal renewable energy credit with other incentive programs   Commercial entities planning to take advantage of the federal credit (in conjunction with other incentive programs) should be aware of a few important considerations. In general, most incentives represent income on which federal income taxes are paid. As a result, most incentives do not decrease the basis on which the federal ITC is calculated. For example, if a business receives rebate money from the state government, the business will be required to pay federal income tax on the amount, which does not affect the cost basis used to determine the 30 percent investment tax credit. State rebates (or buy-downs), grants, and other taxable incentives fall into this category.

There are also rare categories of incentives that are not taxable. An example is nontaxable rebates from utilities. Another is a nontaxable grant. When taking these types of incentives, companies are required to reduce the system's cost basis prior to calculating the ITC amount.

For instance, if a business receives $100,000 in nontaxable utility rebates, when determining the ITC amount, the business must subtract the cost of the solar energy system. This would then determine the credit on this adjusted cost basis. The key phrase commonly used is *subsidized energy financing* (SEF), which broadly applies to nontaxable energy incentives. The IRS defines SEF as, "financing provided under a federal, state, or local program, a principal purpose of which is to provide subsidized financing for projects designed to conserve or produce energy."

Note that if a commercial entity pays federal income tax on SEF, additional incentives won't reduce the ITC amount. Further information regarding federal income tax can also be found in the SEIA Web link.[2]

Impact of ITC on depreciation calculations   For federal tax purposes, the Modified Accelerated Cost-Recovery System (MACRS) program (as discussed previously) allows for accelerated depreciation over a period of 5 years. MACRS and the 30-percent investment tax credit are set up to make it easier to purchase a renewable energy system.

It is important to note when calculating the depreciation on a commercial solar energy system that the *tax depreciation basis* (TDB) is distinct from the *tax credit basis* (TCB). Essentially, when counting for a full 30-percent credit, the first-year depreciation value will be 30 percent less. For example, a solar power system costing $100,000 can only be depreciated down to $70,000 ($100,000—30 percent credit). For this reason, the IRS accounts the full value of the credit alongside accelerated depreciation.

Furthermore, IRS rules allow companies to apply half the value of the tax credit when determining the basis on which to calculate depreciation. As such, the tax depreciation basis that a company claims for the solar energy system is reduced by 50 percent of the tax credit amount.

According to this rule, examine a hypothetical company that installs a commercial solar-electric system costing $200,000. The company's tax depreciation basis will be equal to project costs minus half the allowable credit: $200,000 − (50% × $60,000) = $170,000. This example illustrates the difference between the tax credit basis and the tax deprecation basis.

Another scenario concerns earnings and profit. As a rule, gross income is reduced by depreciation to arrive at net income. As such, large allowable depreciation in a given year lowers the net income, which results in lower net income or profits. This in turn, lowers the taxable dividends paid to shareholders. In such a scenario when determining profits, companies may omit the downward basis adjustments and base their calculations on the full-cost basis of the system. A company could lower its tax liability over the short term while investing in a solar energy system that may result in long-term energy savings. These rules have been specifically designed to provide incentive for companies to invest in renewable power. It should also be noted that the tax depreciation basis is used to calculate taxable gains or losses. The 30-percent federal credit does not affect the book depreciation basis.

RETURN ON INVESTMENT (ROI)

Government incentives combined with recent decreases in solar equipment prices make the investment in solar power a good financial decision for businesses. Solar power systems are considered long-term, low-risk, and high-return investments. In general, solar power systems may result in tax-free annual return on investment of 5 to 11 percent. Therefore, solar power systems are considered to be quite competitive with other higher-risk investments, such as stocks and bonds. Moreover, as utility rates increase, the annual return increases. Another attribute of solar power systems is that when installed in commercial projects, such as shopping centers and office buildings, they increase property values and rents, and render properties as environmentally responsive.

PROJECT FINANCING

The following financing discussion is specific to large alternative and renewable energy projects, such as solar, wind, and geothermal projects that require extensive amounts of investment capital. Project financing of such large projects, similar to large industrial projects, involves long-term financing of capital intensive material and equipment.

Since most alternative energy projects in the United States are subject to state and federal tax incentives and rebates, project financing involves highly complex financial structures in which project debt and equity, rebate, federal and state tax incentives, and cash flow generated by grid energy power are used to finance the project. In general, project lenders are given a lien on all of the project assets, including property, which enables them to assume control of a project over the terms of the contract.

Since renewable energy and large industrial projects involve different levels of transactions, such as equipment and material purchase, site installation, maintenance, and financing, a special purpose entity is created for each project. This shields other assets owned by a project sponsor from the detrimental effects of project failure.

As a special purpose joint venture, these types of entities have no assets other than the project. In some instances, capital contribution commitments by the owners of the project company are sometimes necessary to ensure that the project is financially sound.

Alternative energy project financing is often more complicated than alternative financing methods commonly used in capital-intensive projects, such as transportation, telecommunication, and public utility industries.

Renewable energy projects in particular, are frequently subject to a number of technical, environmental, economic, and political risks. Therefore, financial institutions and project sponsors evaluate inherent risks associated with a particular project development and operation, and determine whether projects are financeable.

To minimize risk, project sponsors create special entities that consist of a number of specialist companies operating in a contractual network with each other, and which allocate risk in a way that allows financing to take place.

In general, a project-financing scheme involves a number of equity investors, known as sponsors, which include hedge funds as well as a syndicate of banks that provide loans for the project. The loans are most commonly nonrecourse loans, which are secured by the project itself and paid entirely from its cash flow, rebates, and tax incentives. Projects that involve large risks require limited-recourse financing secured by a surety from sponsors. A complex project finance scheme also may incorporate corporate finance, securitization, options, insurance provisions, or other further measures to mitigate risk.

POWER PURCHASE AGREEMENTS (PPAs)

Power purchase agreements (PPAs) for renewable energy projects are a class of *lease-option-to-buy* financing plans that are specifically tailored to underwrite the heavy cost burden of the project. PPAs, which are also referred to as *third-party ownership* contracts, differ from conventional loans in that they require significant land or property equity, which must be tied up for the duration of the lease. PPAs have the following significant features, which makes them unique as financial instruments.

- They take advantage of federal and state tax incentives, which may otherwise have no value for public agencies, municipalities, counties, nonprofit organizations, or businesses that do not have significant profit margins.
- Properties where renewable energy systems equipment and materials, such as solar PV power support structures, are installed must be leased for the entire duration of the contract agreement, which may exceed 20 years.
- Solar power or the renewable energy system must be connected to the electrical grid.
- Power generated by the renewable energy system must primarily be used by the owner.
- Depending on the lease agreement, excess power produced from the power cogeneration system is accounted toward the third-party owner.
- Equity of the leased property must have liquidity value exceeding the value of the project.

Advantages of PPAs   In general, PPAs have the following significant advantages:

- Projects are financed on equity of properties, such as unused grounds or building rooftops, which otherwise have no value.
- Owners are not burdened with intensive project cost.
- PPAs guarantee owners a hedge against electrical energy escalation costs.

- Energy cost escalation associated with third-party PPAs have significantly less risk than grid-purchased electrical energy.
- The owners assume no responsibility for maintenance and upkeep of the leased equipment or grounds for the duration of the lease period.
- Upon completion of the lease agreement period, owners are offered flexible options for ownership.
- All PPAs intrinsically constitute turnkey *design build contracts*, which somewhat relieve the owners from detailed technical designs.

Disadvantages of PPAs   Since PPAs essentially constitute a contractual rather than an engineering design and procurement agreement, they inherently include a number of undesirable features, which in some instances could neutralize associated benefits discussed above.

Some of the issues associated with PPAs are as follows:

- PPA contracts are extremely complex and convoluted. Contract agreements drafted include legal language and clauses that strongly favor the third-party provider.
- PPA contracts incorporate stiff penalties for premature contract terminations.
- PPA or third-party ownerships generally involve a finance company, an intermediary, such as a sales and marketing organization, a design engineering organization, a general contractor, and in some instances, a maintenance contractor. Considering the fragmented responsibilities and the complexities embodying the collaborative effort of all entities and the lifecycle of the contract, the owners must exercise extreme diligence in executing PPA contracts.
- The owners have no control over the quality of design or materials provided, therefore extra measures of caution should be exercised when evaluating final ownership of the equipment.
- In general, owners who elect to enter into a PPA, such as nonprofit organizations, municipalities, city governments, or large commercial industries, seldom have experienced engineering or legal staff who have had previous exposure to PPA-type contracts.
- The owners tie up the leased grounds or buildings for extended periods of time and assume responsibility for insuring the property against vandalism and damage to property due to natural causes.
- In the event of power outages, third-party ownership agreement contracts penalize the owners for loss of power output generation.
- PPA contracts include yearly energy escalation costs, which represent a certain percentage of the installed cost and must be evaluated with extreme diligence and awareness, as these seemingly small inflationary costs could neutralize the main benefit, which is the hedge against energy-cost escalation.
- PPAs for large, renewable solar power cogeneration contracts are relatively new financial instruments. Therefore, their owners must be careful to take proper measures to avoid unexpected consequences.

SPECIAL FUNDING FOR PUBLIC AND CHARTER SCHOOLS

A special amendment to the CEC mandate, enacted in February 4, 2004, established a Solar Schools Program to provide a higher level of funding for public and charter schools. This is to encourage the installation of PV generating systems at more school sites. Currently, the California Department of Finance has allocated a total of $2.25 million for this purpose. To qualify for the additional funds, the schools must meet the following criteria:

■ Public or charter schools must provide instruction for kindergarten or any of the grades 1 through 12.
■ The schools must have installed high-efficiency fluorescent lighting in at least 80 percent of classrooms.
■ The schools must agree to establish a curriculum tie-in plan to educate students about the benefits of solar energy and energy conservation.

PRINCIPAL TYPES OF MUNICIPAL LEASE

There are two types of municipal bonds. One type is referred to as a "tax-exempt municipal lease," which has been available for many years and is used primarily for the purchase of equipment and machinery that has a life expectancy of seven years or less. The second type is generally known as an "energy-efficiency lease" or a "power purchase agreement." It is used most often on equipment being installed for energy-efficiency purposes, and is used in cases in which equipment has a life expectancy of greater than seven years. Most often this type of lease applies to equipment classified for use as renewable energy cogeneration, such as solar PV and solar thermal systems. The other common type of application that can take advantage of municipal lease plans includes the energy-efficiency improvement of devices, such as lighting fixtures, insulation, variable-frequency motors, central plants, emergency backup systems, energy management systems, and structural building retrofits.

The leases can carry a purchase option at the end of the lease period for an amount ranging from $1.00 to fair market value. They frequently have options to renew the lease at the end of the lease term for a lesser payment than the original payment.

A tax-exempt municipal lease is a special kind of financial instrument that essentially allows government entities to acquire new equipment under extremely attractive terms with streamlined documentation. The lease term is usually for less than seven years. Some of the most notable benefits are:

■ Lower rates than conventional loans or commercial leases
■ Lease-to-own; there is no residual and no buyout
■ Easier application, such as same-day approvals
■ No "opinion of counsel" required for amounts under $100,000
■ No underwriting costs associated with the lease

Entities that qualify for municipal leases   Virtually any state, county, or city municipal government and their agencies (such as law enforcement, public safety, fire, rescue, emergency medical services, water port authorities, school districts, community colleges, state universities, hospitals, and 501 organizations) qualify for municipal leases. Equipment that can be leased under a municipal lease includes essential use equipment and remediation equipment, such as vehicles, land, or buildings. Some specific examples are listed here:

- Renewable energy systems
- Cogeneration systems
- Emergency backup systems
- Microcomputers and mainframe computers
- Police vehicles
- Networks and communication equipment
- Fire trucks
- Emergency management service equipment
- Rescue construction equipment, such as aircraft helicopters
- Training simulators
- Asphalt paving equipment
- Jail and court computer-aided design (CAD) software
- All-terrain vehicles
- Energy management and solid waste disposal equipment
- Turf management and golf course maintenance equipment
- School buses
- Water treatment systems
- Modular classrooms, portable building systems, and school furniture such as copiers, fax machines, closed-circuit television surveillance equipment
- Snow and ice removal equipment
- Sewer maintenance

The transaction must be statutorily permissible under local, state, and federal laws, and must involve something essential to the operation of the project.

Tax-exempt municipal leases   Municipal leases are special financial vehicles that provide the benefit of exempting banks and investors from federal income tax. This allows for interest rates that are generally far below conventional bank financing or commercial lease rates. Most commercial leases are structured as rental agreements with either nominal or fair-market-value purchase options.

Borrowing money or using state bonds is strictly prohibited in all states, since county and municipal governments are not allowed to incur new debts that will obligate payments that extend over multiyear budget periods. As a rule, state and municipal government budgets are formally voted into law; there is no legal authority to bind the government entities to make future payments.

As a result, most governmental entities are not allowed to sign municipal lease agreements without the inclusion of non-appropriation language. Most governments, when using municipal lease instruments, consider obligations as current expenses and do not characterize them as long-term debt obligations.

The only exceptions are bond issues or general obligations, which are the primary vehicles used to bind government entities to a stream of future payments. General obligation bonds are contractual commitments to make repayments. The government bond issuer guarantees to make funds available for repayment, including raising taxes if necessary. In the event that adequate sums are not available in the general fund, "revenue" bond repayments are tied directly to specific streams of tax revenue. Bond issues are very complicated legal documents that are expensive and time-consuming, and in general have a direct impact on the taxpayers and require voter approval. Hence, bonds are exclusively used for very large building projects such as creating infrastructures like sewers and roads.

Municipal leases automatically include a non-appropriation clause; they are readily approved without counsel. Non-appropriation language effectively relieves the government entity of its obligation in the event funds are not appropriated in any subsequent period, for any legal reason.

Municipal leases can be prepaid at any time without a prepayment penalty. In general, a lease amortization table included with a lease contract shows the interest principal and payoff amount for each period of the lease. There is no contractual penalty, and a payoff schedule can be prepared in advance. It should also be noted that equipment and installation can be leased.

Lease payments are structured to provide a permanent reduction in utility costs when used for the acquisition of renewable energy or cogeneration systems. A flexible leasing structure allows the municipal borrower to level out capital expenditures from year to year. Competitive leasing rates of up to 100 percent financing are available with structured payments to meet revenues. This could allow the municipality to acquire the equipment without having current fund appropriation.

The advantages of a municipal lease program include the following:

- Enhanced cash flow financing allows municipalities or districts to spread the cost of an acquisition over several fiscal periods leaving more cash on hand.
- A lease program is a hedge against inflation since the cost of purchased equipment is figured at the time of the lease and the equipment can be acquired at current prices.
- Flexible lease terms structured over the useful lifespan of the equipment can allow financing of as much as 100 percent of the acquisition.
- Low-rate interest on a municipal lease contract is exempt from federal taxation, has no fees, and has rates often comparable to bond rates.
- Full ownership at the end of the lease most often includes an optional purchase clause of $1.00 for complete ownership.

Because of budgetary shortfalls, leasing is becoming a standard way for cities, counties, states, schools, and other municipal entities to get the equipment they need today without spending their entire annual budget to acquire it. It should be noted that municipal leases are different from standard commercial leases because of the mandatory non-appropriation clause. This states that the entity is only committing to funds through the end of the current fiscal year, even if they are signing a multiyear contract.

References

1. National Renewable Energy Laboratories. www.nrel.gov
2. www.seia.com

6

SOLAR POWER SYSTEM DESIGN

Introduction

Previous chapters introduced solar power system technologies, solar physics, special devices and components deployed in solar power systems, project preliminary design requirements (in the context of feasibility studies and costing procedures), and methodology and economics associated with various categories of solar power projects. Those discussions form the essential foundations for undertaking a comprehensive solar power system design.

In this chapter, the essentials of a comprehensive solar power system engineering design are explored. Design procedures discussed reflect pragmatic step-by-step design guidelines that incorporate all the topics included in the previous chapters.

The principle scope of solar power system engineering is to ensure that drawings and documentation produced will meet a multiplicity of objectives. These include ample design information (to allow contractors to provide accurate hardware and manpower system costing) that must be sufficiently detailed to assist contractors to produce site integration shop drawings.

Engineering documents must also include project-specific solar power system specifications that outline the expected installation and system integration requirements.

Solar power engineering design includes a mix of disciplines and technologies, each requiring specific technical expertise. Systems engineering of a solar power system requires an intimate knowledge of all hardware, equipment performance, and application requirements. In general, major system components, such as inverters, batteries, and emergency power generators are available from a wide number of manufacturers. All equipment intended for use in a solar power system design has unique performance specifications that must be carefully evaluated.

The location of a project site, installation space considerations, environmental settings, choice of specific solar power module and application requirements, and numerous other parameters as discussed in previous chapters must be studied for each project. Therefore, the designer must pay special attention to specific solar power platform conditions, and material and component selections (which include PV modules, support structures, wiring, raceways, junction boxes, collector boxes, and inverters).

All materials and equipment selected must be chosen to withstand environmental and atmospheric conditions. In some instances, solar power systems must operate under extreme temperatures, humidity, and wind turbulence or gust conditions. In addition to the preceding environmental adversities, the electrical wiring must withstand degradation under constant exposure to ultraviolet radiation and heat.

Additional factors to consider when designing solar power wiring include PV module characteristic specifications, such as short-circuit current (Isc) values, open-circuit voltage (Voc), specific temperature coefficients, and various performance characteristics that affect power output performance of the solar power system (all of which were discussed in previous chapters).

The solar power system designer, in addition to familiarity with solar power system design, must be intimately familiar with the National Electrical Code, and most specifically, must review Section 690 of the code that specifically addresses solar PV power systems.

Solar Power System Documentation

To meet the engineering design goals, solar power system documents must minimally include a roster of drawings and specifications outlined in this section.

DRAWING ROSTER

In order to distinguish solar electrical plans from solar structural plans, drawing sheets outlined below are assigned arbitrary prefix letters of ES-XX and SS-XX. Designers may choose their proffered drawing designations.

ES-1.0—Title Sheet

ES-2.0—Site Plan

ES-3.0—Electrical Single Line Diagram

ES-4.0—Solar Array Layout

ES-5.0—Solar Power System Feeder Schedule

ES-6.0—Solar Power System Grounding Plan

ES-7.0—Solar Power System Equipment Specification

ES-8.0—System Equipment Safety Plate Designation

SS-1.0 through X.0—Structural plans, provided by PV support structure manufacturers

PV structural drawing numbers and designations are provided by the suppliers or manufacturers of the hardware equipment. PV module system structural installation plans, construction details, and calculations must also be included. The following is an overview of the contents of these drawings and specifications:

ES-1.0—TITLE SHEET

The first page of drawing set is the title sheet. This drawing typically includes the following:

Abbreviations

GND—ground

STC—standard test conditions

IG—isolated ground

CU—copper conductor or electrical

KW—kilowatts

KVA— kilo volt ampere

MCA—minimum circuit amperes

NIC—not included in the contract

NTS—not to scale

SLID—single-line diagram

UON—unless otherwise noted

Voc—open-circuit voltage

WP—weatherproof

PBO—installed by others

G.E.0.—ground electrode conductor, per NEC 250.66 and 250.166

Drawing notes   The following are drawing notes that appear on the title sheet that are intended to instruct solar power system integrators about specific interpretation of the design documents and their pertinent references (in regard to solar power system integration). Typical drawing notes may consist of the following:

■ All negative outputs of the PV modules throughout this solar power system must be grounded at a single point and be connected to the inverter ground.

- All exposed non-current-carrying metal cabinets and parts, including PV module frames, support structures, conduits, and wireways, must be grounded in compliance with NEC articles 2005.134 or 2005.136(a).
- To avoid physical damage, grounding conductors smaller than AWG #6 must be in compliance with NEC article 250.120(c) and shall run through metallic conduits.
- All fasteners, bolts, washers, and nuts used in grounding wires or straps throughout the solar power system must be stainless steel.
- Upon removal or replacement of any metallic equipment chassis or PV module, the grounding connections shall be maintained at all times.
- All inverter, transformer, and equipment platforms must be grounded.
- All combiner boxes and ground buses must be connected to each PV string.
- Grounding wires shall run through metallic conduits.
- All grounding electrodes' bonding conductors must comply to NEC article 690-47(1) and shall be no smaller than the largest grounding electrode conductor.

Electrical notes   Typical electrical notes are sets of electrical system installation requirements that contractors must follow during the integration process. In general, such notes make reference to certain material and equipment specification standards. The following are a number of electrical notes that may be used in the title documentation sheet.

- All DC equipment and devices must be rated for 600 VDC operation.
- All enclosures and conduit hubs must be suitable for grounding per NEC 250.97 and shall be weatherproof.
- All equipment must have an AIC rating greater or equal to that of existing equipment.
- Contractors must consult with inverter manufacturer regarding string configuration, operating and test procedures.
- Inverters shall be compliant to UL 1741 standards and incorporate the following functions:
  - 50/51 phase current
  - 59 overcurrent protection
  - 27 undervoltage protection
  - 810 overfrequency protection
  - 81 underfrequency protection
  - 51 ground fault detection and protection, anti-islanding protection
- All metering and monitoring services provided shall be in accordance with local solar power system ordinances and requirements.
- Use #6 AWG bare CU wire conductor to bond the inverter ground bar to neural DC ground rod electrode per NEC 250.52 and 250.53.

Applicable codes   This section lists all applicable codes to which integrators must adhere. All work performed must also conform to all regulatory codes, laws, and local ordinances.

- 2005 National Electrical Code
- 2007 California Building Code (CBC)
- 2008 California Electrical Code 2007
- California Energy Code
- 2007 California Fire Code

SOLAR POWER EQUIPMENT LABELING REQUIREMENTS

Solar power equipment labeling requirements are important for life safety of maintenance personnel and are mandated by Section 690 of the National Electrical Code (NEC). Solar power system integrators are therefore required to adhere to specific labeling requirements, some of which may be specific to a project. Labeling notes may include the following:

- All solar power equipment must have safety labeling as per local fire marshal requirements.
- All labeling must comply to UL 969 standards
- All labels must be red with white engraved background, weatherproof, and permanently attached to the equipment.
- All disconnect switches must have a label with the following:

> CAUTION
> HIGH VOLTAGE SOLAR DC POWER

- All solar power enclosures, wireways, junction boxes, and conduits shall have permanent labels with corresponding identification as shown on the single-line diagrams.
- Any disconnect switch or a circuit breaker used to disconnect ungrounded circuit conductors within the solar power system shall comply with the NEC.
- All indoor metallic conduits carrying solar DC current shall be securely attached to walls or ceilings by means of unobstructed railings and straps.
- Overcurrent devices carrying DC solar power current shall be DC rated.
- All breakers and disconnect switches deployed in solar power system shall comply with the NEC Article 690.17.

EQUIPMENT NOTES

These notes make reference to equipment installation methodology, which may include the following:

- All equipment installed shall be UL. listed and listed by the California Energy Commission's list of approved equipment.
- All outdoor installed equipment shall have weatherproof enclosures.
- All installed equipment shall be accessible.

■ Primary and secondary sides of all transformers shall be protected in accordance to local NEC and local electrical code requirements.

Scope of work   Scope of work reflects specific instructions regarding PV module type, data acquisition system requirements, and shop drawing requirements.

■ Solar power PV support system shall consist of a penetrating-type PV support railing system.
■ Solar power system shall include a Web-based monitoring system.
■ Solar/electrical diagrams shall consist of a complete set of working drawings including:
  ■ PV array layout diagrams
  ■ Single-line diagram showing all DC combiners, recombiners, boxes, and overcurrent protection devices, such as AC and DC disconnects, and AC accumulators.
  ■ The inverter system
  ■ Conduit and wiring system

Electrical symbols   Electrical symbols shown in Fig. 6.1*a, b,* and *c* are symbols commonly used in solar power system drawing documentation.

ES-2.0—SITE AND VICINITY PLANS

The vicinity plan is the map where the project is located. This map could be obtained from the Google Earth Web site, which shows the roads and the address of a project (Fig. 6.2). The site plan is a proportional scale of a project's footprint overlaid on the Google map, which is used to provide territorial extent of the solar power footprint (Fig. 6.3).

ES-3.0—ELECTRICAL SINGLE-LINE DIAGRAM

The electrical single-line diagram (Fig. 6.4) represents all electrical and solar power system components' interconnections in a single-line pictorial fashion. The interconnected system shown in the figure includes the solar power system array configuration, combiner and recombiners, inverters, solar power system disconnect switches, site switchgear, and grid connectivity points. Single-line diagrams also include equipment riser diagrams and special notes and instructions specific to the project.

ES-4.0—SOLAR ARRAY LAYOUT

Solar array layouts are sets of drawings (shown in Fig. 6.5) or plans that show the topology of solar power system array layout or the footprint of PV module interconnections. The array layout plans reflect the arrangement of a solar power system footprint within nonshaded boundaries of solar platforms. Solar power platforms may be roofs, carport

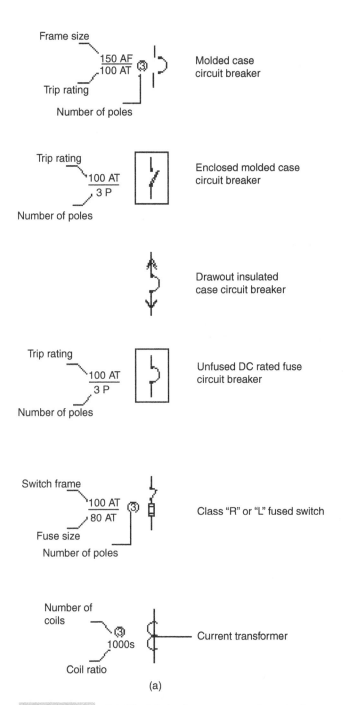

(a)

Figure 6.1   (*a*) Electric/solar power system design symbols.

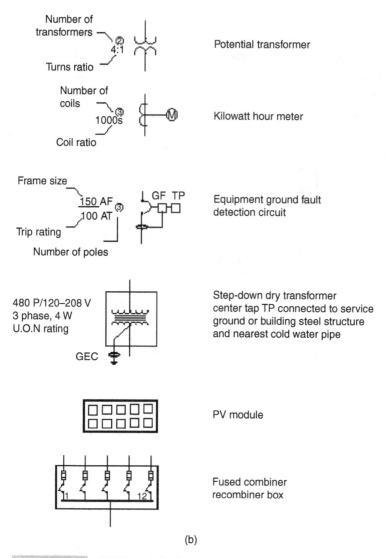

Figure 6.1   (*b*) Electric/solar power system design symbols.

canopies, or open fields where PV modules are to be installed. Topology configurations are usually established during the feasibility study. Essentially solar array layout diagrams are considered to be the foundation of solar system design because they provide the PV module count, array, and subarray configuration, which in turn determine the overall solar power system configuration. Likewise, array configurations determine locations of DC combiner boxes, inverters, conduit runs, and cables. In short, solar array layout plans are fundamental platforms of all categories of solar power system designs. Solar array diagrams also depict solar power string wiring interconnections and string order within each solar array.

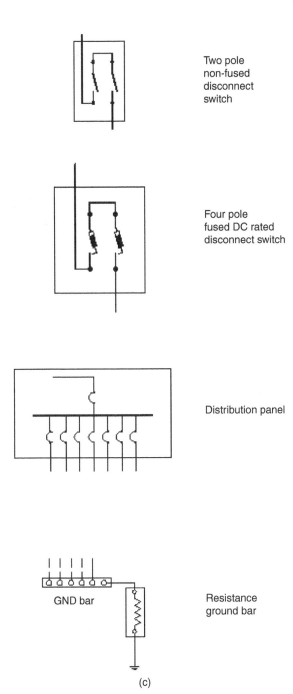

Two pole
non-fused
disconnect
switch

Four pole
fused DC rated
disconnect switch

Distribution panel

GND bar

Resistance
ground bar

(c)

Figure 6.1   (*c*) **Electric/solar power system
design symbols.**

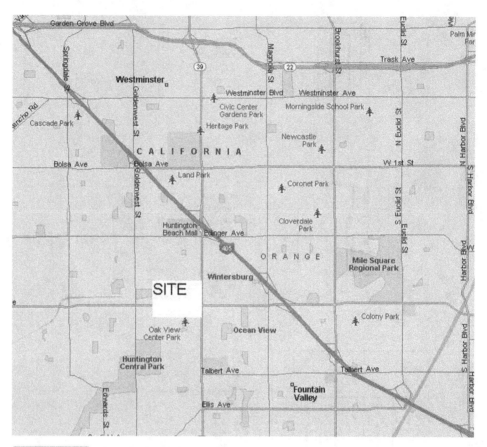

Figure 6.2   A typical vicinity map.

After determining PV module support systems (roof mount, carport, or ground mount), solar power platform topologies are forwarded to the manufacturers who specialize in the fabrication of structural materials and components. After receiving plans, the manufacturers' engineers develop plans that precisely specify exact PV module installation diagrams (which also include PV string assembly details and installation details). Manufacturers' plans also include structural and wind-shear withstands calculations.

Based on documents from the PV support structure system manufacturers, design engineers overlay the string interconnection and identification on the plans. For legibility reasons, solar array layout plans are configured in several layers.

Global solar array topology diagram   The global topology diagram (Fig. 6.6) displays a bird's-eye view of the overall solar system, including highlighted boundaries of all solar power arrays that form the solar power system.

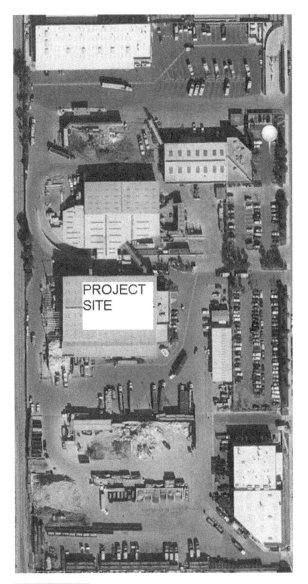

PROJECT
SITE

Figure 6.3   **A typical site map.**

Each of the solar arrays shown within the boundaries bears a unique designation, which is used in subsequent drawings with enlarged scaled proportion corresponding exactly to the physical layout of the PV modules.

The solar power system arrays shown in Fig. 6.7 are used to identify every component within the solar power system. The identification methodology is of significant importance since it is used to cross-reference all system components within a solar power system. The identification systems, as outlined in the legend, are also used to prepare

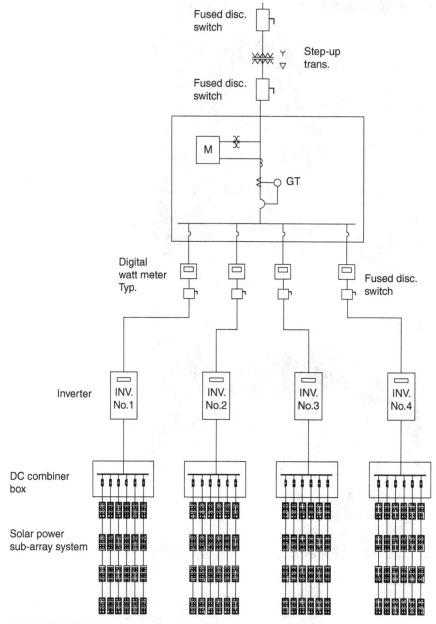

Fused disc.
switch

Step-up
trans.

Fused disc.
switch

M

GT

Digital
watt meter
Typ.

Fused disc.
switch

Inverter

INV.
No. 1

INV.
No. 2

INV.
No. 3

INV.
No. 4

DC combiner
box

Solar power
sub-array system

Figure 6.4   **Solar power system single-line diagram.**

identification tags that are secured to every hardware component within the solar power system. Tagging identification is also used across the board in all drawings, correlating the entire system documentation. Likewise the identification system is also used during integration to tag all system components.

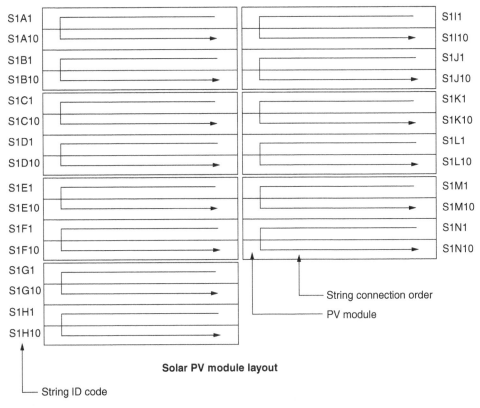

Figure 6.5   Solar array layout.

Each equipment tag must alphanumerically identify complete source-to-destination flow of solar power current that identifies PV string position, the DC combiner box, recombiner boxes, and the destination inverter.

Special requirements for roof-mount solar power systems   The following are highlights of fire marshal mandatory code requirements for all roof-mount solar installations. The fire code addresses specific clearance requirements for walkways, roof access entries, air vents, and air handling and air conditioning equipments. As such, all solar power drawings and documents are required comply with the fire code and must be approved by a local fire department construction services unit prior to the plan check and system installation. At a minimum, the following information shall be presented for approval.

The site plan's solar power arrays drawings must be to scale of the structure. The drawings must show the following:

■ Footprint of the building and north reference point
■ Location of all structures on site

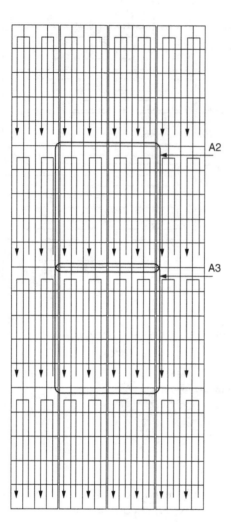

Figure 6.6 Global solar array layout.

- Street address of building
- Access from street to building
- Location of arrays
- Location of disconnects
- Location of required signage
- Location of required access pathways
- Plan and elevation views of the building clearly showing the following:
  - Array placement
  - Roof ridgelines
  - Eave lines
  - Equipment on roof

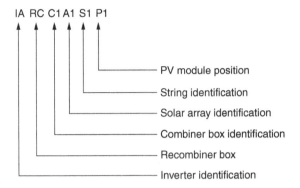

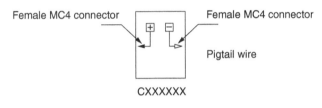

Typical PV module

Figure 6.7   **Solar array designation.**

■ Other objects that may be present on the roof such as vent lines, skylights, and roof hatches
■ Location and verbiage of all markings, labels, and warning signs
■ Building photographs that may be useful in the evaluation of the array placement.

Fire marshal document review   All solar installations on buildings must be approved by the local fire department's construction services unit prior to installation. At a minimum, the site plan (to scale) of the structure, must be presented, on which the PV arrays to be installed are showing the following:

■ Footprint of the building and north reference point
■ Location of all structures on site
■ Street address of building
■ Access from street to building
■ Location of arrays
■ Location of disconnects
■ Location of required signage
■ Location of required access pathways

- Plan and elevation views of the building clearly showing the following:
  - Array placement
  - Roof ridgelines
  - Eave lines
  - Equipment on roof
  - Other objects that may be present on the roof; such as vent lines, skylights, and roof hatches
  - Location and verbiage of all markings, labels, and warning signs

Warning signs and equipment labels   The purpose of warning signs and equipment labels is to warn emergency responders with appropriate warning and guidance so that they can isolate the solar power arrays from the electrical system. This can facilitate identifying energized electrical lines that connect the solar panels to the inverter, as these should not be cut when venting for smoke removal.

Important Note: PV systems exposed to solar rays are always active and produce power at 600 V DC, therefore disconnecting the solar power arrays from the electrical system does not eliminate electrical shock hazards to firefighters.

Later in this chapter, we will discuss solar power system life safety issues and measures that can be mitigated by a central data acquisition and controls system. Detailed fire code and life safety signage have been covered further in this chapter.

Main service disconnect: Residential buildings   In residential buildings (even though not addressed in this book), the markings may be placed within the main service disconnect. The markings are required to be placed on the outside cover of the main service disconnect and must be operable with the service panel closed.

Main service disconnect: Commercial buildings   In commercial buildings, the markings must be placed adjacent to the main service disconnect, and must be clearly visible from the location where the lever is operated.

Access pathways and smoke ventilation   PV array installations on buildings with a hip roof layout must be located in a manner that provides a 3-foot wide clear-access pathway from the eave to the ridge on each roof slope where panels are located. The access pathway must be located at a structurally strong location on the building, such as bearing walls.

Buildings with a single ridge   In buildings with a single ridge, PV panels must be located in a manner that provides two 3-foot wide access pathways from the eave to the ridge on each slope where panels are located. Access pathway clearances do not include eaves or roof overhangs.

Hips and valleys   Solar power installation on roofs with hips and valleys must be located no closer than 1.5 feet to a hip or valley if placed on both sides of the hip or valley.

In cases in which PV panels are to be located on only one side of a hip or valley, the panels may be placed directly adjacent to the hip or valley.

Dead ends In the event that there are two or more access pathways on a roof, pathways must be arranged so that there will be no dead ends greater than 25 feet in length. When access pathways leading to a dead end exceed 25 feet, they must continue on to the next access pathway. At no time shall any access pathway cause a travel distance to exceed 150 feet before arriving at another required access pathway.

Maximum PV array cluster footprint   Uninterrupted sections of PV arrays are not permitted to exceed 150 feet by 150 feet in either of the two axes. All PV arrays must be located no higher than 3 feet below the ridge. Figure 6.8 is a photograph of clearance around roof-mount air-handling equipment.

Roof perimeter clearance requirements   All roof-mount solar power array installations are required to maintain a minimum of 6-foot wide clearance around the perimeter or edges of the roof. The exception to this is in the event that either axis of a building is 250 feet or less. In such a case, a minimum of a 4-feet wide clearance around the roof edges is allowed.

Figure 6.8   A photograph of clearance around roof-mount air-handling equipment.

These clearance requirements also apply to building-integrated photovoltaic (BIPV) systems and skylights, which include the following:

- A minimum of 4-feet clear straight-line pathway must be provided from the access path to skylights and/or ventilation hatches.
- A minimum of 4-feet clear straight-line pathway shall be provided from the access to roof standpipes.
- Not less than 4 feet of clearance is required around roof-access hatches with a minimum of one pathway that is straight and has no less than 4-feet clearance from the parapet or roof edge.
- An access pathway 8 feet or greater in width.
- The access pathways must be 4 feet or greater in width when bordering on the existing roof skylights or ventilation hatches.
- The access pathways must be 4 feet or greater in width with bordering of 4 feet by 8 feet around the venting cut-outs.

Direct current (DC) conductor locations   All conduits, wiring systems, and raceways must be located as close as possible to the ridge, hip, or valley, and from the hip or valley as directly as possible to an outside wall.

All conduit runs between subarrays and DC combiner boxes must use design guidelines that minimize the total amount of conduit used on the roof by taking the shortest path from the array to the DC combiner box.

All DC combiner boxes must be located such that conduit runs are minimized in the pathways between arrays. Figure 6.9 shows a combiner box wiring diagram.

DC wiring requirements   All DC wiring must be run in metallic conduits or raceways when located within enclosed spaces in a building. Whenever possible, DC wiring must run along the bottom of load-bearing members.

Ground-mounted PV array setback requirements   All clearance requirements discussed so far only pertain to roof-mount solar power systems and do not apply to ground-mounted freestanding PV arrays. The only requirement mandated for ground-mount solar power systems is that a minimum of 10 feet of clearance is required around ground-mounted PV systems.

Trellis-type solar power system requirements   The following are minimum requirements for trellis-mount solar power systems:

- Overhead arrays must comply with the same markings, labeling, and warning signs as required of roof-mounted systems.
- All solar power trellis systems must have unobstructed clearance of 7 feet or more between the roof deck surface and the underside of the overhead array.
- At present the city of Los Angeles, California, solar trellis installations must comply with fire marshal code regulations 57.12.03 and 57.138.04. The code has recently been adopted by a number of states.

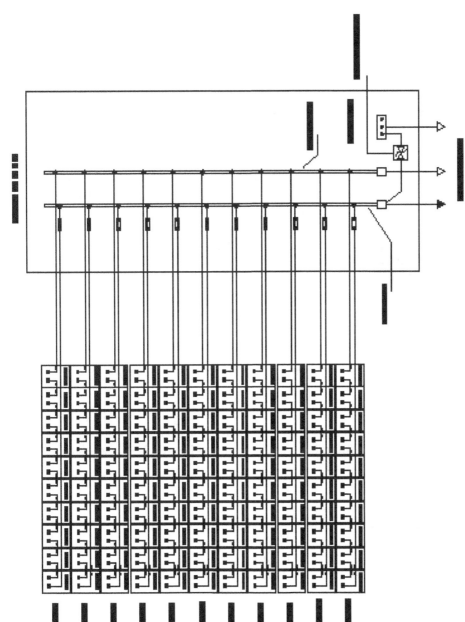

Typical PV string & combiner box wiring

Figure 6.9 Typical solar array wiring.

■ An uninterrupted section of solar PV panels of a trellis must not exceed 150 feet by 150 feet in dimension on either axis.

■ The overhead clearance width between arrays or subarrays must be 4 feet or greater extending from the edge of the array(s) to the roof deck surface. This maintains an unobstructed access pathway and provides for emergency ventilation procedures.

■ Uses of areas below arrays are prohibited.

Equipment installation detail diagrams   In some instances, the solar array drawings outline solar power system installation details for various equipments such as combiner boxes, PV system support structure grounding, solar power system main grounding bus configurations, and special long-run conduit couplings installations. These may be required to either mitigate feeder system expansions and contractions that may occur at building expansion joints or to prevent structural stress during extreme heat and cold ambient temperature conditions. Figures 6.10*a* and *b* are a photographs of typical solar power combiner box integration and an installation diagram. Figure 6.11 is a photograph of a combiner box with an open lead. Figure 6.12 is a diagram of a master grounding bus bar. Figure 6.13 represents a PV module grounding diagram, and Fig. 6.14 shows a PV module assembly ground wiring photograph. Figure 6.15 depicts a diagram of a roof-mount conduit with an expansion joint, and Fig. 6.16 is a photograph of an actual conduit expansion joint installation. Figure 6.17 is a diagram of fused disconnect switches.

ES-5.0—SOLAR POWER SYSTEM FEEDER SCHEDULE

The solar power system feeder schedule drawing (Fig. 6.18) is perhaps the most significant drawing of the entire set of electrical/solar drawing(s). This is because it encompasses all of the solar power system calculations. For the purpose of discussion (in view of the importance of this drawing), the document has been divided in several parts, each of which will be addressed in detail.

A—PV module specifications   The PV module specification shown in Fig. 6.18*b* represents a manufacturer's flat PV panel performance characteristics. The information reflected in the table represents vital information whereby all solar power computations are based. Most of the of the PV module performance parameters were reviewed in Chap. 2 of this book. A number of items required for solar power design are in some instances not listed in the PV module specifications. The following list contains additional specification data that must be obtained from the manufacturer. Supplementary PV module specifications to be obtained include the following:

■ Item A6–Temperature coefficient Pmax
■ Item A7–Temperature coefficient Voc
■ Item A8–Temperature coefficient Isc

(a)

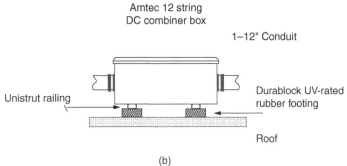

(b)

Figure 6.10   (*a*) Photograph of a roof-mount solar power array installation at Port of Los Angeles, Bert 93 cruise terminal. (*b*) DC combiner box installation detail.

- Item A9–Normal operating temperature NOTC
- Item A10–Lowest ambient temperature
- Item A12–Record low temperature–Rlt
- Item A13–Voltage temperature correction factor VTco
- Item A14–Temperature correction factor Voctf = Rlt × Vtco
- Item A 15–Nameplate of PV modules per string

Figure 6.11   Photograph of a DC combiner box.

Note: For maximum allowable string voltage (Voc × string count) it is recommended that the designer must verify the inverter maximum tracking voltage to ensure that string voltages do not exceed the allowable DC tracking bandwidth.

- Item A16–Corrected string voltage Voc = # of strings × Voctf
- Item A 17–CEC PTC value must be cross verified with California Energy Commission listed PV module rate
- Item A21–Maximum allowable system voltage
- Item A22–Recommended series fuse rating
- Item A23–Performance tolerance warrantee

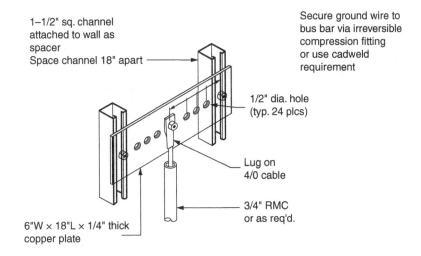

1-1/2" sq. channel attached to wall as spacer
Space channel 18" apart

Secure ground wire to bus bar via irreversible compression fitting or use cadweld requirement

1/2" dia. hole (typ. 24 plcs)

Lug on 4/0 cable

3/4" RMC or as req'd.

6"W × 18"L × 1/4" thick copper plate

Grounding note – Use Thomas & Bets EZ ground compression connector model # BG350–500, for one per conductors, or approved equivalent.

Figure 6.12 Depiction of main grounding bus bar detail.

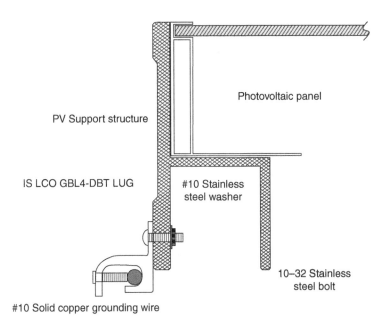

PV Support structure

Photovoltaic panel

IS LCO GBL4-DBT LUG

#10 Stainless steel washer

10–32 Stainless steel bolt

#10 Solid copper grounding wire

Figure 6.13 Depiction of PV support structure grounding.

Figure 6.14   **Photograph of support structure grounding.**

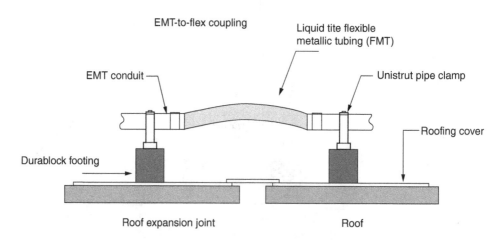

To compensate for thermal expansion for metallic conduits exceeding 100' use
Use liquid tight flexible conduits to connect ends of conduit runs.

Figure 6.15   **Depiction of conduit installation at roof expansion joint.**

Figure 6.16   **Photograph of a roof-mount conduit expansion joint.**

B—Conductor derating   In addition to the specifications in A, the conductor derating values shown in item B are used to calculate combiner box schedules and conductor sizing as shown in Fig. 6.18c.

C—String conductor sizing   The calculation of string conductors involves determining the optimum sizing of wires that will be suitable to operate within given ambient temperature and solar irradiance conditions. The following are itemized analyses of each of the computational steps reflected in Fig. 6.18a through 6.18i.

- Item C1–The string conductor sizing computation shown in Fig. 6.18c determines the maximum number of PV modules allowed per string. Allowed combined string number must not exceed combined multiple of the strings in the corrected value of Voc. For example, if the DC voltage swing of an inverter is 300 to 600 V, the maximum allowed combined or added string Voc value should not exceed 600 V.
- Item C2–Represents Isc value as shown in PV module specification Isc = 5.57 Amps
- Item C3–Applying a derating multiplier of 125 percent for high irradiance, per NEC 690, the is adjusted to Isc = 5.57 × 1.25 = 6.96 Amps

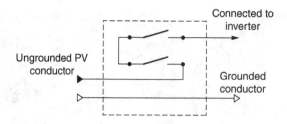

Two-pole non-fused disconnect

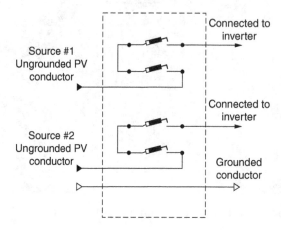

Four-pole fused disconnect switch

Figure 6.17   Depiction of fused disconnect switch diagram.

AC voltage drop calculations

Single-phase   VD = Amps × feet × 2K/C.M
Three-phase   VD = Amps × feet × 2KX.866/C.M
Three-phase   VD = Amps × feet × 2KX.866 × 1.5/C.M
2-pole
A = Amperes
L = Distance from source of supply to load
C.M.- = Cross-sectional area of conductor in circular mills
12 for copper more than 50% loading
11 for copper less than 50% loading
18 for aluminum

DC voltage drop calculation

Percent voltage drop = $((2L × Rc)/1000)) × I)/V$

L = One way conductor length
Rc = Resistance of conductor/1000 feet
I = Current
V = Voltage

Figure 6.18   (a) AC voltage drop calculation formulas.

A PV module specification	Manufacturer	Sanyo-HIT 210
1 Module rated power (Pmax)	210	Watts
2 Maximum power voltage (Vpm)	41.3	Watts
3 Maximum power current (Imp)	5.09	Amperes
4 Open circuit voltage (Voc) @ 25° centigrade	50.9	Volts
5 Short circuit current (Isc)	5.57	Amperes
6 Temperature coefficient (Pmax)	−0.336	Watts/Deg. centigrade
7 Temperature coefficient (Voc)	−0.142	Volts/Deg. centigrade
8 Temperature coefficient (Isc)	1.92	mA/Deg. centigrade
9 Normal operating temperature (NOCT)	46	Centigrade
10 Lowest operational ambient temperature	14 F-10 C	Centigrade
11 STC temperature	25	Centigrade
12 Record low temperature-Rlt	−35	Centigrade
13 Voltage temperature correction factor of VTco	−0.142	Volts/Centigrade
14 Temperature correction factor of Voctf = Rlt × Vtco	4.97	Volts
15 Number of PV modules per string	10	
16 String Voc	509	Volts
17 NEC Table 690.7 voltage correction multiplier	1.13	Volts
18 Corrected string Voc = String Voc × 1.13	575.17	Volts
19 CEC PTC	194.8	Watts
20 Cell efficiency	18.9%	
21 Module efficiency	16.7%	
22 Watts per square foot	15.48	Watts
23 Maximum system voltage	600	Volts
24 Series fuse rating	15	Amps
25 Warrateed tolarence	0 to +10%	

Figure 6.18   (b) PV module specification.

- Item C4–Applying maximum current (MCA) derating factor per NEC 690 results into $Isc = 8.7$ Amps
- Item C5–Applying NEC a conduit fill factor for a maximum of 41 conductors in a raceway (35 percent), yields $Isc = 8.7$ Amps/0.35 = 24.A amps
- Item C6–Temperature derating factor (NEC 310–16) (87 percent) applied to Isc current = 24.9 Amps/0.87 = 28.58 Amps
- Item C7–Specifies the conductor type USE-2 (USE-2 is a conductor that has UV-protected coating)
- Item C8–Specifies conductor size at 90°C as AWG # 10
- Item C9–Specifies cable circular mills as per NEC Table 8, wire gauge table is 10380
- Item C10–Is ampacity of the wire
- Item C11–Specifies the worst-case minimum one-way length of the cable
- Item C12–Is the solar string voltage at maximum power output Vmp

B Conductor derating

	Farenheit	Centigrade
1 Ambient temperature	79	26.1
2 Maximum average high temperature region of the platform	30	17
3 Platform ambient temperature adjustment-2008 NEC 310.15B(2)-C		
4 Adjusted average max. ambient temperature	109	43.1
5 NEC T310.16 ambient temperature correction factor for 90 C	87%	87%

C String conductor sizing

			Multiplier
1 Modules per string	10		
2 Module Isc	5.57	Amps	
3 Hi Irradiance factor derating	6.96	Amps	125%
4 Max current amps (MCA) derating	8.70	Amps	125%
5 Conduit fill-Current carying conductors in raceway (41 or more)	24.9	Amps	35%
6 Temperature derating factor (NEC 310–16)-cable ampacity	28.58	Amps	87%
7 Conductor type	USE-2		
8 Conductor size @ 90° centigrade	AWG#10		
9 Cable circular mills	10380		
10 Wire ampacity	40	Amps	
11 Worst case maximum conductor distance (one way)	200	Feet	
12 String Vmp (module Vmp × Module per string)	413	Volts	
13 String Imp	5.09	Amps	
14 Cable DC resistance (Ohms/1000 ft). NEC Table 8-Rdc	1.26	Ohms	
15 Cable resistance	0.252	Ohms	
16 % Voltage drop = (Imp × dist. × resist)/string voltage	0.31%		

Figure 6.18 *(c)* Conductor derating and solar power string sizing.

C Combiner box feeder schedule calculations

1	Combiner identification		CSAXX-01
2	Number of connected strings		11
3	Derated max current amps (MCA)	Amps	8.70
4	Fuse rating per string	Amps	15
5	String maximum current (MAC) = Isc × PV count × 125%	Amps	76.59
6	String maximum current apers (MOCP) = MCA × Number of strings		95.73
7	Conductor wire fill derate factor-NEC 310–15(B)	Max. OCP	0.8
8	Temperature derate factor (Max. Amb. Tempt F)	4–6 wires	0.87
9	Combined derated amapacit = MOCP	105 to 113 F	137.55
10	Conductor type	Amps	THWN-2
11	Conductor wire guage		#1
12	Conductor ampacity	Amps	150
13	Ground wire-NEC Table 250-122B	OK for 300 Amps	4
14	Conduit size	EMT	1.5"
15	Feeder identification tag-to be the same on the plans		CSAXX-lXX
16	Feeder distance-one way-L	Feet	165
17	Cable resistance/1000 ft-Nec Table 8		0.154
18	Parallel string Imp's per combiner box	11 strings	56.0
19	Percent voltage drop = ((2L × Rc)/1000 × Imp)/string Vmp	Amps	0.69
20	Destination feed recombiner box–Identification		RB-xx-xx

Figure 6.18   (*d*) Combiner box feeder schedule calculations.

D Recombiner box feeder schedule calculations

	RXA-01	RXB-02	RXC-03
1 Recombiner box identification	RXA-01	RXB-02	RXC-03
2 Number of parallel connected incoming circuits	4	4	4
3 PV combiner current Amps	76.59	76.59	76.59
4 PV combiner current Amps	76.59	76.59	76.59
5 PV combiner current Amps	76.59	76.59	76.59
6 PV combiner current Amps	76.59	76.59	76.59
7 Parallel continious output current from the recombiner box-AMPS	306.35	306.35	306.35
8 MCA temperature derating of 125% for 30 centigrade	382.94	382.94	382.94
9 Conductor wire fill derate factor-NEC 310–15(B)-4 wires	478.67	478.67	478.67
10 Temperature derate factor (Max. Amb. Tempt F) 105–113 F	550.20	550.20	550.20
11 Conductor type	THWN-2	THWN-2	THWN-2
12 Conductor size	(2) 250 kcmil	(2) 250 kcmil	(2) 250 kcmil
13 Conductor ampacity-AMPS	580	580	580
14 Ground wire-NEC Table 250-122B	#3	#3	#3
15 Conduit type & size-EMT	3"	3"	3"
16 Feeder tag	RFA-XX	RFB-XX	RFC-XX
17 Approximate feeder distance to inverter-one way distance	150	150	150
18 String Vmp (Module Vmp × Module per string)	413	413	413
19 Imp-Amps = (String Imp × Number of strings)	223.96	223.96	223.96
20 Cable resistance/1000 ft-NEC Table 8-(2)250 kcmil = 0.0535/2	0.02675	0.02675	0.02675
21 Percent voltage drop = [(2L × Rc)/1000 × Imp]/Voltage	0.44%	0.44%	0.44%
22 Recombiner destination disconnect switch identification	RXADS	RXBDS	RXCDS

Figure 6.18 (e) Recombiner box feeder schedule calculations.

E Inverter feeder cable & conduit sizing calculations

	Inverter A	Inverter B
1 Inverter identification	**Inverter A**	**Inverter B**
2 Manufacturer & model-Xantres 500 kw		
3 Output voltage—480 volt, 3 phase, 3 wire	480	480
4 Power output capacity—kw	500	500
5 Continuous output current	601	601
6 Feeder cable derating for 30°C @ 125%	751	751
7 Conductor type	THEWN-2	THWN-2
8 Conductor size	(2) 500 kcmil	(2) 500 kcmil
9 Conductor ampacity—Amps	760	760
10 Ground wire—NEC Table 250-122B	#1/0	#1/0
11 Conduit type & size—EMT	4"	4"
12 Feeder tag	INVA-F	INVB-F
13 Approximate one-way distance to grid—feet	55	55
14 Overcurrent protection device ampacity-Amps	800	800
15 AC voltage drop %	0.105	0.105

Figure 6.18   (*f*) Inverter feeder and cable sizing calculation.

Wire AWG	THHN-2 Ampacity	THWN-2 Ampacity	MCM	Conduit
2000			2016252	
1750			1738503	
1500			1490944	
1250			1245699	
1000	615	545	999424	
900	595	520	907924	
800	565	490	792756	
750	535	475	751581	
700	520	460	698389	4"
600	475	420	597861	4"
500	430	380	497872	4"
400	380	335	400192	4"
350	350	310	348133	3"
300	320	285	299700	3"
250	290	255	248788	3"
4/0	260	230	211600	2-1/2"
3/0	225	200	167000	2"
2/0	195	175	133100	2"
1/0	170	150	105600	2"
1	150	130	83690	1-1/2"
2	130	115	66360	1-1/4"
3	110	100	52620	1-1/2"
4	95	85	41740	1-1/2"
6	75	65	26240	1"
8	55	50	15510	3/4'
10	30	30	10380	1/2"
12	20	20	6530	1/2"

Figure 6.18   (*g*) Conduit and cable table.

Direct-current resistance @ 75°C (167°F)
for copper wire-NEC Table 8

AWG	ohm/1000 ft	AWG	ohm/1000 ft
18	8.08	800	0.0166
16	5.08	900	0.0147
14	3.19	1000	0.0132
12	2.01	1250	0.0106
10	1.26	1500	0.00883
8	0.786	1750	0.00756
6	0.51	2000	0.00662
4	0.321		
3	0.254		
2	0.201		
1	0.16		
1/0	0.127		
2/0	0.101		
3/0	0.0797		
4/0	0.0626		
250	0.0535		
300	0.0446		
350	0.038		
400	0.0331		
500	0.0265		
600	0.0223		
700	0.0189		
750	0.0176		

Figure 6.18   (*h*) NEC Table 8.

- Item C13–String current output at maximum power Imp
- Item C14–Cable DC resistance in ohms/1000 feet of wire (NEC Table 8)
- Item C15–Cable total resistance in ohms
- Item C16–Iv calculation of the percentage voltage drop or percentage VDC

%VDC = (Imp × Cable distance in ft. × resistance /1000 Ft.(/ string voltage)

D—Combiner box feeder schedule calculations   As depicted in Fig. 6.18*d*, the combiner schedule calculations are based on the computed data that were derived in the previous exercises.

- Item C1–Identifies combiner designation, using the same solar array designation nomenclature discussed previously in the chapter.
- Item C2–Specifies number of PV strings in each combiner box, in the example in Fig. 6.18*d* we have 11.
- Item C3–Is the derated MCA value computed previously, which is 8.7 Amps

NEC Table 250.122B   NEC Table 310.15(B)(2)(a)

Grounding wire sizing   Amb. temp adj. factor

OCD Amps	Copper wire AWG	Cond./cond.	Adj. factor AWG
15	14	4–6	80%
20	12	7–9	70%
30	10	10–20	50%
40	10	21–30	45%
60	10	31–40	40%
100	8	Over 41	35%
200	6		
300	4		
400	3		
500	2		
600	1		
800	1/0		
1000	2/0		
1200	3/0		
1600	4/0		
2000	250		
2500	350		
3000	400		
4000	500		
5000	700		
6000	800		

Figure 6.18   (*i*) NEC Tables 250.1228 and 310.15(B)(2)(a).

- Item C4–Specifies the fuse rating capacity as being 15 Amps
- Item C5–Is computation of the string maximum current (MAC) which equals the Isc × Number of PV modules in a string multiplied by NEC 690 derate factor of 125%, which in the example's case is 76.59 Amps
- Item C6–This represents value of MAC multiplied by number of strings connected to the combiner box, which in this case = 8.7 Amps × 11 strings = 95.73 Amps
- Item C7–Specifies derating multiplier factor from NEC Table 310-15(b). Since each string count within the conduit is 4-6, the multiplier = 0.8
- Item C8–Represents NEC derate multiplier factor for temperature which in this case the solar power system is expected to operate at maximum temperature which may vary from 105 to 113°F, which is 0.87
- Item C9–This represents the derated value of the MOC under high temperature operational condition which = 95.7 × 0.8 × 0.87 = 137.55 Amps.
- Item C10–Specifies conductor type (THWN-2 used in solar power applications)
- Item C11–Specifies the proper wire gauge
- Item C12–Specifies the conductor ampacity
- Item C13–Specifies the ground conductor cable for NEC Table 250-122B
- Item C14–Specifies the conduit size

- Item C15–Specifies the conduit identification tag
- Item C16–Specifies the DC feeder maximum one-way distance (L)
- Item C17–Specifies resistance per 1000 feet of cable
- Item 18C–Specifies combined current of 11 strings at Isc = 11 × 4.97 Amps = 54.6 Amps
- Item 19C–Is the computation of percent voltage drop which = $(((2 \times L) \times (\text{ohms}/1000 \text{ Ft.})) \times \text{Imp})/\text{String Vmp}$
- Item 20C–Specifies the destination recombiner box identification.

E—Recombiner box feeder schedule calculations   The recombiner box feeder schedule shown in Fig. 6.18e represents ampacity calculation of confluence of feeder currents accumulated in recombiner boxes. The recombiner box example calculations reflected in the series of figures in 6.18 are based on a combination of four identical combiner boxes which are merged into a single recombiner box.

The following are definitions of the items exhibited:

- Item D1–Identifies tags of three individual recombiner boxes of a solar power system subarray
- Item D2–Indicates the number of combiner boxes connected to the recombiner box
- Items D3–D6–Show value of current feed from each combiner box
- Item D7–Represents sum of the parallel currents of combiner boxes (MCA)
- Item D8–Represents maximum current ampacity (MCA) derated for 30°C ambient temperature
- Item D9–Applies conduit fill derate factor for four wires in the conductor using derate factor shown in NEC Table 310-15(B)
- Item D10–Value of item D is further derated for operation in 105 to 113°F maximum ambient temperature conditions
- Item D11–Indicates type of cables used
- Item D12–Indicates cable size
- Item D13–Indicates cable maximum current-carrying capacity
- Item D14–Shows the appropriate grounding cables as listed by NEC Table 250-122B
- Item D15–Specifies the conduit sizes
- Item D16–Specifies the feeder conduit tags
- Item D17–Indicates the one-way distance of the feeder cable to the destination inverter
- Item D18–Indicates the maximum voltage Vmp of the strings (module Vmp × modules per string)
- Item D19–Represents the maximum combined current (string Imp current × number of strings)
- Item D20–Shows the cable resistance per 1000 feet (NEC Table 8. Note value represented is for 2 × 250 kcmil/2)
- Item D21–Represents calculated value of the percentage of the DC voltage drop
- Item D22–Represents tag of the destination DC disconnect switch

F—Inverter feeder cable and conduit sizing calculation   The inverter feeder calculation shown in Fig. 6.18f represents the calculation of the AC current-carrying capacity of cables, which connect the inverters to the grid connection point, which may be a circuit breaker, an isolation transformer, or an AC combiner box (reverse distribution panel).

- Item E1–Identifies the inverter as the AC current source
- Item E2–Specifies the inverter type and model
- Item E3–Represents the inverter output voltage
- Item E4–Specifies the inverter maximum current output capacity
- Item E5–Specifies the inverter maximum continuous current output
- Item E6–Specifies derated current at 30°C ambient conditions
- Item E7–Specifies the feeder cable type
- Item E8–Specifies the appropriate cable size
- Item E9–Specifies the conductor maximum current-carrying capacity
- Item E10–Specifies the ground-wire size per NEC Table 250-122(B)
- Item E11–Specifies the feeder conduit size
- Item E12–Designates the feeder conduit tag
- Item E13–Shows the approximate one way distance of the feeder cable distance
- Item E14–designated the appropriate current protection device ampacity
- Item E15–Represents percent voltage drop calculation (in this example using Fig. 6.18g–three-phase voltage drop for copper wire)

Solar power system wiring guide   Conductors that are suitable for solar exposure are listed as THW-2, USE-2, and THWN-2 or XHHW-2. All outdoor-installed conduits and wireways are considered to be operating in wet, damp, and UV-exposed conditions. As such, conduits should be capable of withstanding these environmental conditions and are required to be of a thick-wall type such as rigid galvanized steel (RGS), intermediate metal conduit (IMC), thin-wall electrical metallic (EMT), or schedule-40 or -80 polyvinyl chloride (PVC) non-metallic conduits.

For interior wiring where the cables are not subjected to physical abuse, CNM-, NMB-, and UF-type cable are permitted. Care must be taken to avoid installing underrated cables within interior locations, such as attics where the ambient temperature can exceed the cable rating.

Conductors carrying DC current are required to use color-coding recommendations as stipulated in Article 690 of the NEC. Red wire or any color other than green and white is used for positive conductors, white for negative, green for equipment grounding, and bare copper wire for grounding. The NEC allows non-white grounded wires such as USE-2 and UF-2 that are sized #6 or above to be identified with a white tape or marker.

As mentioned earlier, all PV array frames, collector panels, disconnect switches, inverters, and metallic enclosures should be connected together and grounded at a single service grounding point.

ES-6.0—SOLAR POWER SYSTEM GROUNDING PLAN

PV power systems that have an output of 50-V DC under open-circuit conditions are required to have one of the current-carrying conductors grounded. In electrical engineering, the terminologies used for grounding are somewhat convoluted and confusing. In order to differentiate various grounding appellations, it helps to review the following terminologies as defined in NEC Articles 100 and 250.

In the previous paragraph, "grounded" means that a conductor is connected to the metallic enclosure of an electrical device housing that serves as earth.

A grounded conductor refers to a conductor that is intentionally grounded. In PV systems, it is usually the negative of the DC output for a two-wire system or the center-tapped conductor of an earlier bipolar solar power array technology.

An equipment-grounding conductor is a conductor that normally does not carry current and is generally a bare copper wire that may also have a green insulator cover. The conductor is usually connected to an equipment chassis or a metallic enclosure that provides a DC conduction path to a ground electrode when metal parts are accidentally energized. Figure 6.19 is a general note that should appear on the front page of solar power drawing documents. Figure 6.20 shows a solar power array grounding system single line diagram.

ELECTRICAL NOTES

ALL DC EQUIPMENT AND DEVICES SHALL BE RATED FOR 600 VDC OPERATIONS

ALL ENCLOSURES AND COUNDUIT HUBS SHALL BE SUITABLE FOR GROUNDING PER NEC 250.97 AND SHALL BE WEATHER PROOF

ALL EQUIPMENT SHALL HAVE AN AI C RATING GREATER OR EQUAL TO THAT OF EXISITING EQUIPMENT

CONTRACTOR SHALL CONSULT WITH INVERTER MANUFACTURER REGARDINGST RIN CONFIGURATION, OPERATING AND TEST PROCEDURE

INVERTERS SHALL BE COMPLIANT TO UL 1741 STANDARDS AND INCORPORATE THE FOLLOWING FUNCTIONS:

- 50/51 PHASE CURRENT
- 59 OVERCURRENT PROTECTION
- 27 UNDERVOLTAGE PROTECTION
- 810 OVERFREQUENCY PROTECTION
- 81U UNDERFREQUENCY PROTECTION
- 51N GROUND FAULT DETECTION & PROTECTION
- ANTI- ISLANDING PROTECTION

ALL METERING AND MONITORING SERVICES PROVIDED SHALL BE IN ACCORDANCE WITH LOCAL SOLAR POWER SYSTEM ORDINANCES AND REQUIREMENTS.

USE # 6 AWG BARE CU WIRE CONDUCTOR TO BOND THE INVERTER GROUND BAR TO NEURAL DC GROUND ROD ELECTRODE PER NEC 250.52 & 250.53

CAUTION – DC VOLTAGES FROM SOLAR ARRAYS ARE ALWAYS ENERGIZED DURING DAYLIT HOURS. USE EXTREME CAUTION DURING EQUIPMENT INSTALLATION AND MAINTENANCE OPERATION

Figure 6.19   Electrical general note that must be included on the front page of solar power drawing documents.

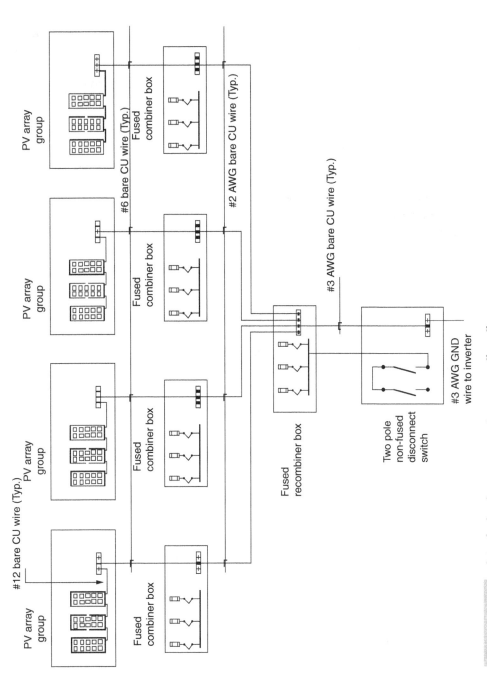

Figure 6.20 A typical solar power system grounding diagram.

A grounding electrode conductor is what connects the grounded conductors to a system-grounding electrode, which is usually located only in a single location within the project site and does not carry current. In the event of the accidental shorting of equipment, the current is directed to the ground, which facilitates actuation of ground-fault devices. A grounding electrode is a grounding rod, or a concrete-encased rebar, (UFR) conductor, and a grounding plate (or simply a structural steel member to which a grounding electrode conductor is connected). As per the NEC, all PV systems, whether grid-connected or stand-alone (in order to reduce the effects of lightning and provide a measure of personnel safety) are required to be equipped with an adequate grounding system. Incidentally, grounding PV systems substantially reduces radiofrequency noise generated by inverter equipment.

In general, grounding conductors that connect the PV module and enclosure frames to the ground electrode are required to carry full short-circuited current to the ground; as such, they should be sized adequately for this purpose. As a rule, grounding conductors larger than AWG #4 are permitted to be installed or attached without special protection measures against physical damage. However, smaller conductors are required to be installed within a protective conduit or raceway. As mentioned previously, all ground electrode conductors are required to be connected to a single grounding electrode or a grounding bus.

Equipment grounding   Metallic enclosures, junction boxes, disconnect switches, and equipment used in the entire solar power system, which could be accidentally energized are required to be grounded. NEC Articles 690, 250, and 720 describe specific grounding requirements. For ground wire sizing refer to NEC Table 250-122B. Equipment grounding conductors similar to regular wires are required to provide 25 percent extra ground-current-carrying capacity and are sized by multiplying the calculated ground current value by 125 percent. The conductors must also be over-sized for voltage drops as defined in NEC Article 250.122(B).

In some installations, bare copper grounding conductors are attached along the railings that support the PV modules. In installations in which PV current-carrying conductors are routed through metallic conduits, separate grounding conductors could be eliminated since the metallic conduits are considered to provide proper grounding when adequately coupled. However, it is important to test conduit conductivity to ensure that there are no conduction path abnormalities or unacceptable resistance values.

Ground-mount solar power system equipment platform grounding   Large-scale solar power systems are usually configured from multiple groups of solar power groups or subsets. Each of the solar subsets are considered as independent virtual power generation cluster zones, configured from a number of solar arrays, combiners, recombiner boxes, and inverters. Output energy of solar power subgroups is eventually merged into a terminal location where AC or DC feeders are accumulated, and eventually connected to the grid. At solar subgroup or the terminal power accumulation centers, equipment such as AC and DC accumulators, disconnect switches, inverters, transformers, switchgear, and communication equipment are secured on reinforced concrete platforms (as shown in Fig. 6.21). To accommodate equipment chassis

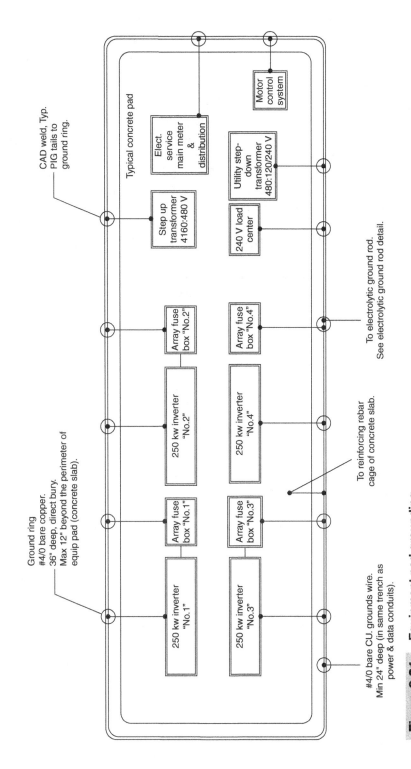

Figure 6.21 Equipment pad grounding.

CAD weld, Typ. PIG tails to ground ring.

Typical concrete pad

Elect. service main meter & distribution

Step up transformer 4160:480 V

Utility step-down transformer 480:120/240 V

Motor control system

240 V load center

Array fuse box "No.2"

Array fuse box "No.4"

250 kw inverter "No.2"

250 kw inverter "No.4"

To electrolytic ground rod. See electrolytic ground rod detail.

Ground ring #4/0 bare copper. 36" deep, direct bury. Max 12" beyond the perimeter of equip pad (concrete slab).

To reinforcing rebar cage of concrete slab.

Array fuse box "No.1"

Array fuse box "No.3"

250 kw inverter "No.1"

250 kw inverter "No.3"

#4/0 bare CU. grounds wire. Min 24" deep (in same trench as power & data conduits).

grounding, the platform is surrounded by a copper wire grounding loop (also referred to as grounding ring) entrenched in the ground. The grounding rings are connected to the concrete platform rebars, and are in turn attached to a number of grounding electrodes by isothermal welding.

At various locations where equipment is to be located, pigtail-grounding wires provide ground attachment to equipment chassis. In arid locations, such as deserts, where ground conductivity is inadequate, electrolytic grounding rods (Fig. 6.22) are used to provide adequate low ground conductivity. Figure 6.21 is a diagram of a solar power equipment platform grounding system. Figure 6.22 shows a detailed diagram of a chemical grounding rod used in poor conductive soil environments.

In addition to grounding solar power system equipment, peripheral fences (Fig. 6.23) surrounding solar power installations, as well as light stands and metallic canopies and enclosures, must all be securely connected to the central grounding system bus bar.

Entrance service power considerations for grid-connected solar power systems   When integrating a solar power cogeneration within existing or new switchgear, it is extremely important to review NEC 690 articles related to switchgear bus capacity.

As a rule, when calculating switchgear or any other power distribution system bus ampacity, the total current-bearing capacity of the bus bars is not allowed to be loaded to more than 80 percent of the manufacturer's equipment nameplate rating. In other words, a bus rated at 600 A cannot be allowed to carry a current burden of more than 480 A.

When integrating a solar power system with the main service distribution switchgear, the total bus current-bearing capacity must be augmented by the same amount as the current output capacity of the solar system. For example, if we were to add a 200-A solar power cogeneration to the switchgear, the bus rating of the switchgear must in fact be augmented by an extra 250 A. The additional 50 A represents an 80 percent safety margin for the solar power output current. Therefore, the service entrance switchgear bus must be changed from 600 to 1000 A, or at a minimum to 800 A.

As suggested previously, the design engineer must be fully familiar with the NEC 690 articles related to solar power design and ensure that solar power cogeneration system electrical design documents become an integral part of the electrical plan check submittal documents.

The integrated solar power cogeneration electrical documents must incorporate the solar power system components such as the PV array systems, solar collector distribution panels, overcurrent protection devices, inverters, isolation transformers, fused service disconnect switches, and net metering within the plans. These elements must also be considered as part of the basic electrical system design.

Electrical plans should incorporate the solar power system configuration in the electrical single-line diagrams, panel schedule, and demand load calculations. All exposed, concealed, and underground conduits must also be reflected on the plans with distinct design symbols and identification that segregate the regular and solar power system from the electrical systems.

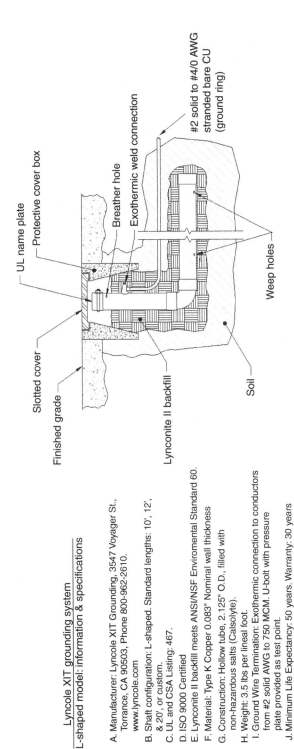

Lyncole XIT grounding system
L-shaped model: information & specifications

A. Manufacturer: Lyncole XIT Grounding, 3547 Voyager St., Torrance, CA 90503, Phone 800-962-2610. www.lyncole.com

B. Shaft configuration: L-shaped. Standard lengths: 10', 12', & 20', or custom.

C. UL and CSA Listing: 467.

D. ISO 9000 Certified

E. Lynconite II backfill meets ANSI/NSF Enviromental Standard 60.

F. Material: Type K Copper 0.083" Nominal wall thickness

G. Construction: Hollow tube, 2.125" O.D., filled with non-hazardous salts (Calsolyte).

H. Weight: 3.5 lbs per lineal foot.

I. Ground Wire Termination: Exothermic connection to conductors from #2 solid AWG to 750 MCM. U-bolt with pressure plate provided as test point.

J. Minimum Life Expectancy: 50 years. Warranty: 30 years

K. L-shaped Model No: K2L-10CS, K2L-12CS, K2L-20CS.

L. GSA Contract pricing Available.

Electrolytic Ground Rod Detail

Labels on figure:
UL name plate
Slotted cover
Finished grade
Protective cover box
Breather hole
Exothermic weld connection
#2 solid to #4/0 AWG stranded bare CU (ground ring)
Lynconite II backfill
Weep holes
Soil

Figure 6.22 **Electrolytic rod grounding detail.**

Figure 6.23   Fence grounding box.

Note that the solar power cogeneration and electrical grounding should be in a single location, preferably connected to a specially designed grounding bus (which must be located within the vicinity of the main service switchgear).

Photovoltaic system ground fault protection   Ground-mounted systems are not required to have the same protection since most grid-connected system inverters incorporate the required GFPD devices.

Ground-fault detection and interruption circuitry perform ground-fault current detection, fault current isolation, and solar power load isolation by shutting down the inverter. Ground-fault isolation technology is currently going through a developmental process, and it is expected to become a mandatory requirement in future installations.

ES-7.0—SOLAR POWER EQUIPMENT SPECIFICATION

This drawing, within the roster, includes pasted equipment specifications and pictorial presentations of all major devices and equipment, such as PV modules, support structures, combiner boxes, disconnect switches, and inverters (as well as a simplified description of the data acquisition system).

ES-8.0—SYSTEM EQUIPMENT SAFETY PLATE SPECIFICATION

Equipment safety specification drawings display actual label designations that must be secured to each piece of equipment within the solar power system that carries solar electrical power. Safety plates are enforceable fire code–mandatory requirements, without which solar power systems will not pass construction inspection or final acceptance. Figures 6.24*a* and *b* depict a sample of equipment safety plates used in solar power system installations.

⚠ **CAUTION**

Photovoltaic system AC disconnect

Do not touch terminal, severe electrical shock hazard

Both ends of incoming and load side terminals may be energized in open position

Maximum current _____ AMPS

Operating voltage _____ VOLTS

⚠ **CAUTION**

Electrical shock hazard

Do not touch terminals without insulated gloves Terminals at both ends may be energized

(a)

SOLAR POWER EQUIPMENT LABELING REQUIREMENTS

1 - ALL SOLAR POWER EQUIPMENT SHALL HAVE SAFETY LABELING
   AS PER LOCAL FIRE MARSHAL REQUIREMENTS

2 - ALL LABELING SHALL COMPLY TO UL 969 STANDARDS

3 - ALL LABELS SHALL BE RED WITH WHITE ENGRAVED BACKGROUND
   WEATHER PROOF AND PERMANENTLY ATTACHED TO THE EQUIPMENT

4 - ALL DISCONNECT SWITCHES SHALL HAVE A LABEL WITH THE FOLLOWING:

CAUTION
HIGH VOLTAGE SOLAR DC POWER

5 - ALL SOLAR POWER ENCLOSURES, WIREWAYS, JUNCTION BOXES,
   AND CONDUITS SHALL HAVE PERMANENT LABELS WITH
   CORRESPONDING IDENTIFICATION AS SHOWN ON THE
   SINGLE LINE DIAGRAMS.

6 - ANY DISCONNECT SWITCH OR A CIRCUIT BREAKER USED TO
   DISCONNECT UNGROUNDED CIRCUIT CONDUCTORS WITHIN THE
   SOLAR POWER SYSTEM SHALL COMPLY WITH THE NATIONAL
   ELECTRICAL CODE (NEC)

7 - ALL INDOOR METALLIC CONDUITS CARRYING SOLAR DC CURRENT
   SHALL BE SECURLY ATTACHED TO WALLS OR CEILINGS BY MEANS
   OF UNISTRUCT RAILING AND STRAPS.

8 - ALL OVERCURRENT DEVICES CARYING DC SOLAR POWER CURRENT
   SHALL BE DC RATED.

9 - ALL BREAKERS AND DISCONNECT SWITCHES DEPLOYED IN SOLAR POWER
   SYSTEM SHALL COMPLY WITH THE NATIONAL ELECTRICAL CODE (NEC)
   ARTICLE 690.17.

(b)

Figure 6.24   (a) Solar equipment warning sign. (b) High voltage caution sign.

SS-1.0—PV SYSTEM STRUCTURAL DRAWINGS

Structural drawings are set of plans that provide exact outline of PV module support structure. Drawings in addition to structural details include structural calculations specific to each type of structure such as ground mount fixed angle system, single axis tracking system, and trellis or roof-mount system. Figure 6.25 below is a typical structural drawing that shows a layout of a roof-mount non-penetrating type ballast system layout. Figure 6.26 is a photograph of an actual ballast installation and Fig. 6.27 is a photograph of a solar string assembly being secured to the ballast railing system.

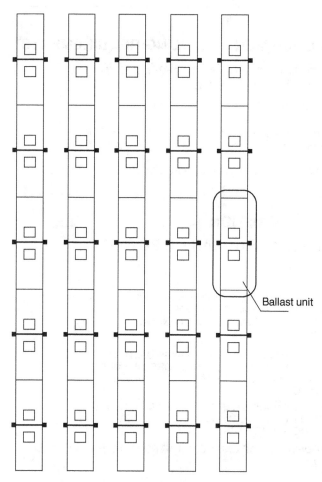

Ballast unit

Non-penetrating type ballast system
for roof-mount solar power system

Figure 6.25   Structural drawing of a roof-mount
PV support structure ballast system layout.

Figure 6.26　Roof-mount nonpenetrating-type solar power support structure ballast system.

Figure 6.27　Solar string array assembly.

7

LARGE-SCALE SOLAR POWER
SYSTEM CONSTRUCTION

Introduction

In view of the multidisciplinary nature of solar PV technology, large-scale solar power system integration (unlike conventional electrical system contracting work) requires considerable experience in a multitude of disciplines. These include electrical systems, solar power system design, civil/structural engineering, and a specific knowledge of solar power system management as outlined in Chap. 8 of this book.

Furthermore, large-scale solar power system installations require a cadre of technical personnel who have fundamental skills in solar power systems and have thorough knowledge and experience in electrical-electronic engineering. In particular, large-scale solar power systems (during the integration phase) involve a series of continuous solar array system and subsystem dynamic tests that require proficiency and somewhat elaborate test and acceptance procedures.

Study and Evaluation of Engineering Plans and Documents

In order to establish a successful large-scale solar power construction enterprise, technical and construction supervisory and construction management personnel must thoroughly evaluate the entire engineering design documentation and specifications. It should be noted that large-scale solar power systems are usually complicated and involve several thousand PV modules and solar power system equipment and support structures. Figure 7.1 shows a summary of a solar power construction cost estimate, which offers an indication of how many hours and personnel are involved in the complexity of a large-scale installation.

INITIAL COSTS

	Hours	Rate/hr or Cost	Total
Site Investigation	40	$ 150.00	$ 6,000.00
Preliminary Design Coordination	24	$ 150.00	$ 3,600.00
Report Preparation	8	$ 150.00	$ 1,200.00
Travel & Accommodation	1	$ 2,000.00	$ 2,000.00
Other	1		
Subtotal			**$ 12,800.00**

DEVELOPMENT

	Hours	Rate/hr or Cost	Total
Permits & Rebate Applications	8	$ 150.00	$ 1,200.00
Project Management	120	$ 150.00	$ 18,000.00
Travel Expenses	1	$ 2,000.00	$ 2,000.00
Other	1	$ 1,000.00	$ 1,000.00
Subtotal			**$ 22,200.00**

ENGINEERING

	Hours	Rate/hr or Cost	Total
PV System Design	90	$ 150.00	$ 13,500.00
Architectural Design	90	$ 150.00	$ 13,500.00
Structural Design	90	$ 150.00	$ 13,500.00
Tenders & Contracting	420	$ 150.00	$ 63,000.00
Construction Supervision	48	$ 150.00	$ 7,200.00
Tenders & Contracting	94	$ 150.00	$ 14,100.00
Custome Training	48	$ 150.00	$ 7,200.00
Subtotal			**$ 132,000.00**

EQUIPMENT & MATERIALS

	Hours	Rate/hr or Cost	Total
PV Modules	255	$ 3,900.00	$ 994,500.00
Transportation	1	$ 5,000.00	$ 5,000.00
Other			
Tax	8.25%		$ 82,046.25
Subtotal			**$ 1,081,546.25**

INSTALLATION EQUIPMENT

	Hours	Rate/hr or Cost	Total
PV Module Structure (per kW)	255	$ 500.00	$ 127,500.00
Inverter (per kW)	320	$ 488.00	$ 156,160.00
Electrical Materials (per kW)	320	$ 250.00	$ 80,000.00
System Installation (kW)	320	$ 1,000.00	$ 320,000.00
Transportation	1	$ 3,000.00	$ 3,000.00
Tax (Equipment Only)	8.25%		$ 30,001.95
Subtotal			**$ 716,661.95**

Figure 7.1   Solar power construction cost estimate summary sheet.

In addition, solar power construction sometimes involves a considerable amount of solar platform preparation, PV support foundation work, logistics, and environmental engineering tasks. All of these form significant cost components of a project. As such, an oversight or negligence in cost accounting could result in serious consequences. Some of the significant tasks that are required for successful project construction include the following:

- Material take-off
- Preparation of shop drawings
- Material and equipment procurement
- Workforce mobilization
- Construction crew training
- Preparation of construction project supervision and management procedures
- Site logistics, which include material storage, site office space, site assembly location, material handling, material transportation, equipment rental, and environmental compliance procedures, etc.
- Establishment of occupational safety and health (OSHA) procedures, project site preparation, which may include grading, solar power support structure foundation works, roof structural reinforcement, roofing material replacement, etc.
- Establishment of site maintenance and security procedures
- Establishment of construction supervision procedures
- Establishment of integration and test procedures
- Establishment of test and commissioning test procedures

Project Superintendent Responsibilities

In addition to the tasks discussed in the previous section, the site superintendent of a large-scale solar power system must be fully familiar with the NEC 690 articles. When starting a project, the site superintendent must become fully familiarized with all solar power system electrical and structural plans and specifications. It should be noted that lack of thorough familiarity may result into costly rework, disputes, and project delays.

FIELD SAFETY PROCEDURES AND MEASURES

As discussed in Chap. 4, all field personnel engaged in solar power construction must be familiarized with imminent danger of exposure to high DC voltages. During the construction period, under exposure to solar irradiance, PV strings continuously produce dangerously high voltage power that may cause serious bodily harm. Therefore, construction safety must be considered to be of paramount importance. To avoid life safety accidents, construction crews must be provided with the following safety procedures and warnings:

- Use modules for their intended purpose only. Follow all module manufacturers' instructions. Do *not* disassemble modules or remove any part installed by the manufacturer.
- Do not attempt to open the diode housing or junction box located on the back side of any factory-wired modules.

- Do not use modules in systems that can exceed 600-V open circuit under any circumstance or combination of solar and ambient temperature.
- Do not connect or disconnect a module unless the array string is open or all the modules in the series string are covered with nontransparent material.
- Do not install during rainy or windy days.
- Do not drop or allow objects to fall on the PV module.
- Do not stand or step on modules.
- Do not work on PV modules when they are wet. Keep in mind that wet modules when cracked or broken can expose maintenance personnel to very high voltages.
- Do not attempt to remove snow or ice from modules.
- Do not direct artificially concentrated sunlight on modules.
- Do not wear jewellery when working on modules.
- Avoid working alone while performing field inspections or repairs.
- Wear suitable eye-protection goggles and insulating gloves rated at 1000 V.
- Do not touch terminals while modules are exposed to light without wearing electrically insulated gloves.
- Always have a fire extinguisher, a first-aid kit, and a hook or cane available when performing work around energized equipment.
- Do not install modules when flammable gases or vapors are present.

Figure 7.2 is a photograph of a roof-mount safety guard railing installation.

Figure 7.2   Safety guard railing.

In addition, the following temporary warning signs and notices must be posted in appropriate locations within the project site.

For solar power enclosures and disconnects:

■ Electric shock hazard—Do not touch terminals—Terminals on both line and load sides may be energized in open position.

For switchgear and metering system:

■ Warning—Electric shock hazard—Do not touch terminals—Terminals on both line and load side may be energized in open position.

For solar power equipment:

■ Warning—Electric shock hazard, dangerous voltages and currents.

For battery rooms or racks:

■ Warning—Electric shock hazard, dangerous voltages and currents, explosive gas, no sparks or flames, no smoking, acid burns. Wear protective clothing.

Construction Supervision

One of the most important tasks involved in large-scale solar power construction is construction supervision and system tests and commissioning. The following is a detailed outline of solar power project management and system acceptance test procedures. Supervision and verification steps outlined are minimal requirements only; additional information and verification may be required to meet specifics of the project, which will vary for every project.

As already mentioned, the integration and testing of large-scale solar power systems and the evaluation of power output performance involve relatively complex and numerous measurements. As such, precise evaluations and measurements of performance parameters are extremely important to ensure the long-term integrity of the project. The following are step-by-step instructions that a site superintendent must follow.

1 Roster of site documentation must include but is not limited to the following:
   ■ Plan check documents
   ■ Construction documents
   ■ Shop drawing documents
   ■ Design and specification documents
   ■ Client's RFP and specification documents
   ■ PV module specification and equipment cut sheets and specifications
   ■ Integrator's test procedure documents
   ■ Rebate application documents (if applicable)
   ■ Record of all minutes of meetings

2 Verify construction schedule and completion milestone correspondence with the original contract documents. If construction schedule exceeds the permissible dateline for completing project construction, undertake measures to ensure that additional manpower, materials, or equipment are assigned to the project to mitigate delays.

3 Verify all major solar power system equipment, such as inverters, transformers, reverse AC distribution panels, and grid connectivity equipment specifications and factory test performance documents.

4 Verify all design modifications to construction drawings resulting from electrical plan check changes. Also, ensure that deviations from the original proposed plans and documents are within the boundaries of system configuration, material content, and expected power output performance potential.

5 Verify solar power array support structures, construction drawings, and documentation. Ensure that array boundaries and clearance comply with local ordinances and fire marshal fire code requirements.

6 Verify all PV support structural attachments to ensure that they comply with seismic requirements. Specific attention must be given to nonpenetrating-type platforms' points of attachment. Verify if point of attachments are properly sealed.

7 Verify electrical PV bonding straps and metallic enclosure bonds. Check for special measures taken when bonding various types of conductive material (e.g., zinc-coated steel, aluminium, and copper). All ground-wire resistance with reference to the service ground must not exceed specified ohmic tolerances (e.g., around 5 to 10 ohms).

8 Verify PV string wiring that connects to DC combiner boxes. Verify that all cables are properly strapped and secured to the platform chassis.

9 Verify all metallic and nonmetallic conduits' attachment to roof surface. Check for soundness of structural support and electrical insulation integrity.

10 Verify DC and AC electrical conduit attachment strapping and roof insulation support blocks.

11 Refer to shop drawings and check if preassembled PV platforms comply with the approved construction design documents.

12 Verify conduit and cable identification labels and tags to make certain that they match with the approved construction document.

13 Verify that PV modules, solar string DC cable labels, tags, and identifications are in place and verify that they match with the final construction documents.

14 Verify all AC and DC safety plates and ensure that they comply with fire code requirements.

15 Establish a regular (preferably frequent) site inspection and verification schedule. Take notes of site meeting discussions. Schedule a weekly meeting with the integrator's site project attendant or supervisor and keep a log of progress made. Take photographs at every stage of integration such as roof-railing installation, PV array assembly, equipment platform installation, panel installation, and take pictures of conversion equipment (such as inverters and transformers).

16 Verify that roof-mount systems have provisions for water hose bibs located at proper distances that would allow periodic PV soling wash-down.

17 Verify that the solar power site (whether roof- or ground-mounted) has adequate illumination (floodlights) and has special provisions for emergency and security alarm systems.

18 Verify that the roofing has been waterproofed and tested for leakage.

19 Ensure that all fasteners used throughout the solar power installation are of stainless steel or noncorrosive materials.

20 Ensure that all metallic structures and PV support structures are constructed from noncorrosive metals such as galvanized steel or aluminium.

Field Performance Test and Final Acceptance Commissioning Procedure

SOLAR POWER SYSTEM INTEGRATION

The following are solar power integration procedures that must be undertaken during the construction phase. These step-by-step instructions are vitally important to solar power installations because they provide important methodology for successful realization of a solar power system. Note that any shortcut in the prescribed testing methodology might result in noncompliance with final inspection and acceptance of the solar power system. The following are step-by-step procedures and instructions for construction and integration:

1 Inspect all solar power systems, subsystem components, and equipment to see if they are installed and secured properly.

2 Verify that the PV support platforms and structural components of the system are built according to the plan-checked construction documents.

3 Verify that the roof has been properly waterproofed and tested and inspected after the support structure installation.

4 Verify that the roof-mount solar power system arrays' layout has adequate walkways, roof access, and equipment clearances required by the National Electrical Code and local fire marshal code requirements.

5 In the event of roofs without parapets, verify that life safety and security railings have been properly installed and comply with Uniform Building Code (UBC) requirements.

6 Verify that all metallic structures, enclosures, and conductive surfaces are properly grounded.

7 Verify that all conduit runs are installed according to the plan-checked drawings.

8 Verify that all the PV racks and fasteners used throughout the solar power system are constructed from rust-proof materials, such as hot-dipped galvanized or stainless steel.

9 Verify that dissimilar metallic structures attachments have been isolated from each other to prevent galvanic corrosion.

10 Verify that all exposed conduits used throughout the system are either electrical metallic tubing (EMT) or rigid metal conduit (RMC).

11 Verify that DC and AC cables, conduits, equipment, and boxes are labeled according to the approved plans.

12 Verify that the system has been connected to the electric utility grid.

13 Verify that the (Web-based) data acquisition system has been connected and is operational.

14 Verify that the Utility Service Company (USC) has signed off the grid-connectivity permits.

During the construction the phase, project superintendents should keep a log of daily progress and take photographs of all major system equipment, components, PV arrays, and subarrays, and equipment installation. Each photograph should be dated and identified. The daily project log must also include commentaries regarding project progress and associated problems that arise during the construction.

Avoiding Construction Negligence

One of the major issues solar power contractors must be concerned with are minor negligences that may be overlooked during the construction phase. Such negligences usually include improper cable and conduit assemblies, shortcuts in system grounding, lack of rigor in system tests, loose wires, and various NEC and UBC code infractions.

The following are photographs and descriptions of improper or inadequate construction practices.

EXPOSED AND UNPROTECTED SOLAR STRING DC WIRES

Figure 7.3 is a photograph of unprotected and exposed solar subarray cables that carry electrical currents with high voltage that range from 300 to 1000 V DC. Any physical damage to the cable may create a serious electrical shock to maintenance personnel and electrical equipment damage during wet or rainy conditions. Sparks generated between a damaged cable and an electrical grounded metallic object may result in an unpredictable consequence. Such negligence is usually a result of shortcuts to save manpower and material expenditure.

Under no circumstances should DC or AC wires in a solar power system be exposed or laid on the ground or a rooftop. All AC or DC current-carrying wires and cables must be must be protected by metallic or UV-protected PVC conduits.

Figures 7.4a and b are photographs of negligent and unacceptable measures used to cover unexposed cables. The photographs show cables covered by thin flexible tubing used for harnessing bundles of wires in enclosed electrical equipment cabinets. The tubing materials used are not approved for exterior use because they lack structural strength to withstand physical damage and do not have the physical characteristics to withstand solar UV radiation. Such shortcuts in solar power construction inevitably will be subject to severe citation during final acceptance, which must be avoided at all costs.

Figure 7.5 is another example of an unacceptable solar power construction practice that must be avoided. The photograph shows exposed and unprotected solar string

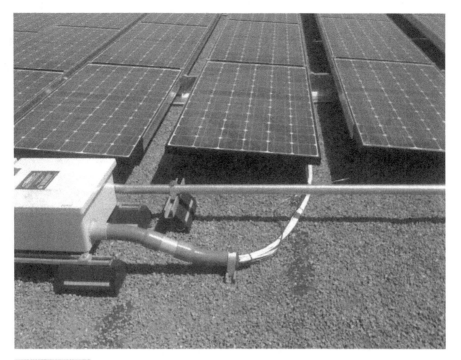

Figure 7.3 Unprotected exposed solar string cables.

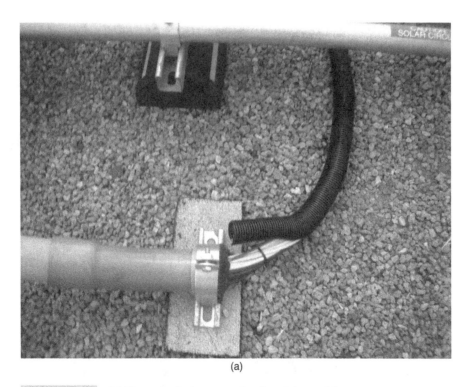

(a)

Figure 7.4 (*a*) Unprotected exposed solar string cables.

(b)

Figure 7.4 (*b*) Unprotected exposed solar string cables.

Figure 7.5 Unprotected exposed solar string grounding cables.

cables, as well as grounding wire. In addition to creating a trip-over hazard for maintenance personnel, small-gauge ground wires become subject to breakage and corrosion.
They are ultimately rendered useless to protect equipment against inadvertent electrical
shorts and lightning.

System Integration Tests and Commissioning

Unlike convention electrical systems, solar power systems require a well-organized document test and verification procedure. In view of hundreds of solar power PV strings, PV
arrays and subarrays, combiner and recombiner boxes, and feeder systems, solar power
systems require a continuous system testing of equipment during the integration. Since each
of the solar power PV strings are active miniature electrical power generators, each must
be tested and recoded individually after completing subarray construction. As such, prior
to final assembly of each combiner box, each PV string must be measured to verify that the
string will deliver the expected power output generation under a given insolation condition.
Such continuous testing also enables detection of anomalies that may be associated with
PV-module malfunction or cable interconnections. In addition, while interconnecting solar
power system DC and AC cables, all feeders must be tested for ground resistance.

It should be noted that system tests and measurements, as described, should not be
considered as a single test event that takes place at the end of the construction of the
entire solar power system. This is because an overall testing of an aggregated solar
array could not provide an indication of solar string performance characteristics or
detection of a string malfunction, which could otherwise be readily detected when
testing individual strings. In order to conduct a rigorous test, contractors must develop
a comprehensive standard system acceptance and test procedure.

Commissioning and Final Acceptance

In general, at the inception of initial commissioning there are no documented energy
output performance data, and all measurements taken must be based on real time,
instantaneous power output values of various solar power subsystems.

INSTANTANEOUS SOLAR POWER OUT PERFORMANCE MEASUREMENT

Note: All of the following performance characteristics have been covered in Chap. 2
of this book.

1 Measurement of peak DC power output. This value is the sum of all PV modules'
PSTC value (which is power measured under Standard Test Conditions measured at
25°C), as shown on manufacturer's module specification or the module nameplate.

2 Calculation of solar irradiance factor Ki. Solar irradiance is measured by a pyranometer. Irradiance readings are displayed in watts per square meter. When measuring solar irradiance, the pyranometer plane and inclination angle must be exactly the same as the PV module arrays, with the same azimuth and tilt angle.

Irradiance factor, Ki, is obtained by dividing measured irradiance by STC irradiance, which is 1000 W/m^2 at sea level and 25°C.

3 Calculation of PV module temperature factor K_T (see Chap. 4). PV cell temperature Tc is measured by an infrared thermometer (IRT), which is targeted at a module. To calculate K_T, the temperature coefficient C_T listed in PV module specification (generally −0.003/°C to −0.005/°C for mono-crystalline cells) is inserted in the following formula:

$$K_T = 1 + ((C_T \times (T_C - T_{STC}))$$

where C_T stands for PV module coefficient, T_C is ambient temperature, and T_{STC} is the temperature of Standard Test Condition, which is 25°C.

4 Calculation of solar power system derating factor, Ks. This factor is a product of all solar power system derating factors such as module nameplate tolerance, module mismatch inverter coefficient of efficiency module soiling, wiring losses, shading, system availability, sun-tracking efficiency, and aging efficiency loss factors. These are reflected in National Renewable Energy Laboratories' (NREL) Web-based PVWatts II calculation.

EVALUATION OF RANGES DEGRADATION FACTORS

The following are important considerations that must be taken into account when testing solar power system:

1 Module nameplate DC tolerance. Verify performance factor assigned to the PV module manufacturer's product quality, which involves electrical test performance, electrical output performance sorting, PV module batching, and quality assurance methodology.

2 Module mismatch is associated with ohmic resistive variations inherent to each PV module. Numerous parameters that result in varying electrical output of modules are inconsistencies in solar cell fabrication, cell crystalline structure, and intercell soldering.

3 Module soiling. Verify derating factors related to dust accumulation and settlement on the PV module. In general, for the first acceptance and commissioning period, because the PV modules are so clean, this factor could be assigned as 1. However, when recommissioning a solar power system, the factor must be assigned a lower value.

4 Wiring losses. Verify the derating factor associated with DC and AC cable losses.

5 Shading multiplier. Verify the derating factor that represents reduction in PV solar power output performance due to shading that could result from various buildings, objects, or trees that may cast shadow on the solar power platform. Verify overall equipment reliability and, most specifically, the methodology of the power system configuration, such as centralized or distributed DC-to-AC conversion system architecture, component failure repair, and system downtime mitigation.

6 System availability multiplier. This is an arbitrary figure that suggests proper system maintenance and is usually assigned a value of one.

7 Tracking efficiency. Verify the efficiency of varying types of solar power platform support structures.

It should be noted that irradiance variations result from overcast or hazy skies, as well as reflections from distant nimbus clouds. As such, looking at a clear sky should not be considered as an ideal environmental ambience.

GROUND INSULATION TESTS

All cable homeruns must be tested for ground resistance by a megger or megohmmeter. This practice will ensure that there are no damages to the cable and that there are no ground paths between current-carrying cables and metallic structures.

SOLAR OPEN VOLTAGE VOC TESTS

To perform a solar string open-circuit voltage Voc test at the DC combiner, remove fuses from the fuse holders and disengage the main DC disconnect.

After testing and recording the Voc of each PV string, insert all fuses into the fuse holders then measure the Voc of the combiner box. Verify discrepancies in paralleled string Voc measurement values of the combiner box and Voc measurements of a sample PV string, and then record deviations.

During the integration phase, all PV string Voc measurements must be logged bearing their specific tag designations (see Fig. 7.6).

PV STRING SHORT-CIRCUIT CURRENT (Isc) MEASUREMENT

Measure the log Isc and verify if all strings produce currents within the same margin under equal environmental conditions.

Note: Clamp the DC meter on the wire of the PV string while positive (+) and negative (−) leads are shorted at the combiner box. Under steady ambient temperature and climatic conditions, all Isc currents measured must fall between the same range, not exceeding +1 to 0.1 amperes.

Note any PV short circuit current discrepancies, also known as PV stings. If the discrepancy exceeds the PV specification value, check if it is a result of ground fault or a defective module.

POWER OUTPUT MEASUREMENT AND COMMISSIONING ACCEPTANCE TESTS

After completing Voc and Isc measurements of all PV strings, insert fuses into the combiner boxes and measure the DC voltage to verify if the value is within the expected nominal range of performance.

Note that the DC voltage at the input terminal of the inverter, because of voltage drop, may be slightly less than the Voc voltage values at the output of the combiner box. If so, note the discrepancy.

Measure the inverter DC input breaker, the current, and the voltage readings displayed. The readings should match the measurement values of the aggregate solar array system. Note any discrepancies.

INVERTER START-UP TEST PROCEDURE

Inverter testing during the commissioning phase must always be conducted according to the manufacturer's recommended test procedure, which involves the following:

- Visually inspect and verify all conduit and DC and AC cable connections.
- Check the output voltage at the AC disconnect breaker in open condition.
- Check the polarity of the incoming DC cables at the input disconnect in open condition.
- Close the inverter AC disconnect breaker.
- Close the DC disconnect breaker(s).
- Turn on the Inverter "ON" switch.
- Wait for the inverter to complete its start-up sequence.
- Wait for 20 to 30 minutes for the DC tracking to stabilize.
- Log the inverter AC output voltage and current measurement several times every 10 minutes.

FIELD TEST AND MEASUREMENT LOG SHEETS

Figures 7.6 through 7.12 are field test and measurement log sheets that could be used during construction and final acceptance and commissioning.

The Objectives of Solar Power System Commissioning

Commissioning of a solar power system is a form of formalized quality control. The process ensures that the installed system is safe and reliable and meets the following:

- Ensure short-term and long-term system performance
- Protects investment
- Maintains public confidence and goodwill toward the industry
- Satisfies installer and customers
- Minimizes callbacks
- Prevents safety hazard from fire and electrical shock

CABLE MEGGER TEST LOG SHEET

| PROJECT NAME – |
| JOB NO. – |
| ADDRESS – |
| DATE – |

VECTOR DELTA
DESIGN GROUP, INC.
1234 Olive Lane
La Canada, CA 91011
TEL. 818-864-6025

Project _____ Date ___/___/___   Page _____ Of _____

Technician _____

Cable ID Tag	Source	Destination	Cable Type	Insulation Class	Cable Gauge	Resistance OHMS	Conduit Type	Reference Drawing

Field Test Team

Name _____   Name _____

Employee Identification _____   Employee Identification _____

Test Date _____   Test Date _____

Page _____ of _____

Figure 7.6 Cable megger test log sheet.

LOW VOLTAGE FEEDER VERIFICATION CHECK LIST

| PROJECT NAME – |
| JOB NO. – |
| ADDRESS – |
| |
| DATE – |

VECTOR DELTA
DESIGN GROUP, INC.
1234 Olive Lane
La Canada, CA 91011
TEL. 818-864-6025

FEEDER INDENTIFICATION TAG _____

SOURCE ID _____   DESTINATION ID _____

BREAKER TYPE _____   BREAKER SIZE _____

CONECTED TO   _____

	Checked by Initial	Date	Comments
Feeder Size			
Conduit Size			
Feeder Cable Size			
Feeder Tag Verified			
Wire/Cable Type			
Conduit Strapping			
Ground Conductor			
NEMA Compliance			
Megger Value @ 1000 V Phase/Phase			
Megger Value Phase to GND			
Disconnect Switch Size			
Fuse Size			
Fuse Type			

Field Test Team

Name _____   Name _____

Employee Identification _____   Employee Identification _____

Test Date _____   Test Date _____

Page _____ of _____

Figure 7.7 **Low voltage feeder verification checklist.**

SOLAR POWER STRING TEST CHECKLIST

VECTOR DELTA
DESIGN GROUP, INC.
1234 Olive Lane
La Canada, CA 91011
TEL. 818-864-6025

PROJECT NAME –
JOB NO. –
ADDRESS –
DATE –

SPECIAL INSTRUCTION

- EACH STRING UNDER TEST MUST BE DISCONNECTED FROM COMBINER BOX
- VOLTAGE VARIATIONS OF GREATER THAN 6% BETWEEN THE SOLAR STRINGS MUST BE VERIFIED

SOLAR PLATFORM IDENTIFICATION   _____

SOLAR IRRADIANCE   _____

AMBIENT TEMPERTURE   _____

Combiner Box No ID	String Identification	Wire Tag Completed	Cable Type	Measured Voc Voltage	Isc Current	PV Module Temp. F

Field Test Team

Name   _____   Name   _____

Employee Identification   _____   Employee Identification   _____

Test Date _____   Test Date _____

Page _____ of _____

Figure 7.8   **PV string test log sheet.**

SOLAR POWER ACCEPTANCE TEST LOG

| PROJECT NAME – |
| JOB NO. – |
| ADDRESS – |
| |
| DATE – |

VECTOR DELTA
DESIGN GROUP, INC.
1234 Olive Lane
La Canada, CA 91011
TEL. 818-864-6025

Project _____ Date ___/___/___ Sky condition _____ Technician _____

Inverter ID	Time	Irrad .1	Watts 1	Irrad. 2	Watts 2	Irrad. 3	Watts 3	Ave. Irrad.	Ave. Watts

Field Test Team

Name   _____   Name   _____

Employee Identification _____   Employee Identification _____

Test Date _____   Test Date _____

Page _____ of _____

Figure 7.9   **Energy output measurement log sheet.**

Inverter Field Test Checklist

PROJECT NAME –	
JOB NO. –	
ADDRESS –	
DATE –	

VECTOR DELTA
DESIGN GROUP, INC.
1234 Olive Lane
La Canada, CA 91011
TEL. 818-864-6025

Inverter Identification Tag _____

Inverter Model:	KW Rating	
Grid Connected Yes _____	No _____	
		Check Status
Inverter Accessibility		
Name plate Rafting		
Ground Bus Connected to the Main Service or Building Ground		
Platform or Pedestal Anchoring		
Manufacture Startup Instructions Attached		
Cable Support System Check		
Spare Components		
Maintenance Test Documents		
Maintenance Tools		
Wire Lugs Bolts Torqued		
Bolt Torque Log Sheet Available		
Islanding Test		
Comments :		

Field Test Team

Name   _____   Name   _____

Employee Identification _____   Employee Identification _____

Test Date _____   Test Date _____

Page _____ of _____

Figure 7.10   Inverter test log sheet.

INVERTER STARTUP TEST

| PROJECT NAME – |
| JOB NO. – |
| ADDRESS – |
| |
| DATE – |

VECTOR DELTA
DESIGN GROUP, INC.
1234 Olive Lane
La Canada, CA 91011
TEL. 818-864-6025

Project_____ Date __/__/__   Page ___ Of___

Technician_____ Drawing Reference _____

Inverter ID	Make and Model No.	KW Rating	Source Combiner(s)	Input DC Voltages	Input DC Voltages

Field Test Team

Name   _____   Name   _____

Employee Identification _____   Employee Identification _____

Test Date _____   Test Date _____

Page _____ of _____

Figure 7.11   **Inverter startup test sheet.**

Electrical System Panel & Disconnect Check List

PROJECT NAME –	
JOB NO. –	
ADDRESS –	
DATE –	

VECTOR DELTA
DESIGN GROUP, INC.
1234 Olive Lane
La Canada, CA 91011
TEL. 818-864-6025

Panel Identification No.:	
Equipment Location	
Disconnect Switch Identification	
Incomint Feeder ID	
Incoming Feeder Size	
Main Disconnect Fuse Rating	
Panel Bus Rating A/C:	
Bus Type Copper _____ Aluminum _____	
Megger Test At 1000 Volt Phase-Phase	
Megger Test At 1000 Volts Phase to Ground	
Feeder Fuse Type	
Panel Directory Checked	☐
Nameplate Installed	☐
Ground Bus Installed	☐
Ground Bus Connected to Main Service Ground	☐
Door Interlocks Operational	☐
Front Guards Installed	☐
Unused Knockouts Closed	☐

Comments _____

Field Test Team

Name _____   Name _____

Employee Identification _____   Employee Identification _____

Test Date _____   Test Date _____

Page _____ of _____

Figure 7.12   Electrical system panel and disconnect checklist.

THE PROCESS

Design review and system test/commissioning are fundamentally undivided engineering and construction responsibility. Effectively, test and integration are considered validation of solar power system engineering. Therefore, commissioning can also be defined as a quality-oriented process which includes:

■ Verification of engineering design documents
■ System performance computations
■ Test and acceptance methodology specifications
■ Study and evaluation of defined design objectives

In essence, commissioning as a quality control task, is a process that must commence at project inception and continues through the life of the system (maintenance and system performance upkeep).

Commissioning Tasks   Commissioning process consists of tasks that must be executed during various phases of a project. The tasks include verification of the design, construction project management, construction engineering, construction supervision, test and acceptance commissioning methodology and customer training. In some instances it may also include extended maintenance planning. Tasks involved in commissioning include:

■ Verification of engineering documents
■ System installation and construction documents
■ Verification of test and commissioning specification and methodology
■ Verification of test and commissioning test and acceptance results
■ Verification of operations and maintenance documents
■ Verification of owner's manuals
■ Verification of customer training documents and training curriculum

FIELD INSTALLATION AND TEST

The significance of the task of field test and commissioning is to ensure that system engineering design and construction meet the owner's project's specification and system performance objectives.

Unfortunately, in most instances, solar power system commissioning occurs after installations are complete. It should be noted that quality assurance is a continuous task and it cannot be procrastinated to the end of the project.

Summary of field installation quality assurance include the following:

■ Ensure that system equipment installation is fully compliant with manufacturer's specifications and national and local construction codes.
■ Ensure that installation is safe. Solar power systems operate at 600–1000 volts DC; therefore, any negligence could result in serious life safety injury or fire hazard.

- Ensure that the installation is aesthetic
- Ensure structural integrity of the system
- Ensure that as-built documents are precise. Redline modifications on all engineering documents and drawings
- Perform precise, systematic power field performance measurements (as outlined in test and acceptance measurement methodology procedures)
- Verify system operation
- Establish performance benchmarks
- Complete test and acceptance documentation and report
- Ensure field participation of client's maintenance personnel

TEST AND COMMISSIONING TIMELINE

As mentioned above, test and commissioning is a dynamic process that must take place during the course of solar power system installation and integration. Therefore, methodology and procedures must be planned at the start of the design phase, and built in the system cost, and must be carried out up until project commissioning.

RE-COMMISSIONING AND RETRO-COMMISSIONING

In general, re-commissioning is a system re-testing procedure that must be performed if the initial commission was performed during less than optimal seasonal or environmental conditions, such as shading, extended cloudy or rainy weather. Ideal condition for testing a solar power system is summer. Re-commissioning essentially must be conducted to achieve a better result than the original commission.

Solar power systems must be re-commissioned if:

- Monitoring or data acquisition systems report excessive faults and alarms and low energy production
- Utility bills are not reduced as expected
- Some inverters within the solar power system appear to produce anticipated power better than others
- The total accumulated solar power output read over a period of time, is less than the anticipated power, beyond the so-called burn-in or system stabilization period

It should be noted that due to cleanliness of the PV modules and calculated oversizing of the system (for extended life cycle degradation), during first commissioning and startup, performance measurements made are usually high. On the other hand, re-commissioning after a burn-in period establishes true operational power output performance of a solar power system.

It should be noted that solar power commissioning must be scheduled during a favorable weather condition that is coincident with a suitable solar irradiance window. It is not productive to conduct field testing and commissioning when solar irradiance is less than 400 watts per meter square on a solar power array.

It is also important to note that solar power system test requires certain degree of rigor, sufficient time, and demands focus. Under cold or very hot weather conditions, test technicians would not be able to conduct thorough tests. Oversight or cursory testing of solar power system will only result in erroneous evaluation.

OBJECTIVITY IN TESTING AND COMMISSIONING

Unless it is a total turnkey design and installation (which includes, finance, installation, and maintenance contract), it is a known fact that due to conflicting priorities, installers or subcontractors may often be biased, and would perhaps be less concerned with the owner's best interests.

For large solar power systems, it is suggested that the owners (or the prime turnkey contractor) should consider employing the services of a qualified third-party specialist to oversee the commissioning process.

PERFORMANCE TESTS

The main objective of a solar power system performance test is to ensure that solar power output production is in concert with the expected performance results of the PV system design which meets the owner's requirements of energy production.

As a rule, based upon solar power system configuration and equipment used, engineering estimate of power production (within the United States) is most often calculated by use of a Web-based NREL's PV Watts estimating program. However, upon completion of a project, a series of field measurements are conducted to establish the actual system power output.

Parameters required for conducting a real-time solar power output production consist of simultaneous PV module surface temperature, solar power irradiance and inverter power output measurements that are inserted in specific solar power equations that result in derate multiplication factors, that when applied to manufacturer's Standard Test Condition (STC) maximum power output (Pmax), result in instantaneous power that forms the basis of solar array system power production.

Table 7.1 depicts fundamental test and measurements parameters, specification parameters, and computed results.

1 Column one denotes three measurements conducted for solar array, connected to inverter A. A1, A2, and A3 represent rows of three measurements and computations in different intervals that are a few minutes apart.
2 Column two represents the total PV module count within the solar array A.
3 Column three represents the PV module STC Pmax (as per manufacturer's specification).
4 Column four represents the total STC power of the solar array.
5 Column five represents the Ks multiplier.
6 Column six represents field solar irradiance measurement read from a pyranometer.
7 Column seven represents the ambient temperature measurement taken from the surface of a few PV modules by an IR thermometer.

TABLE 7.1 SOLAR POWER SYSTEM SITE TEST AND MEASUREMENT CALCULATION

MODULE	KYOCER	KD235GX-LPB

TEMP. MULTIPLIER $K_T = 1 + ((C_T \times (T_C - T_{STC}))$

STC WATTS	TEMP COEF. PER. CENTIGR.	Ks-DERATE MULTIPLYER	STC IRRADIANCE WATTS/MM	T_{STC} CENTIGRADE
235	-0.038	0.85	1000	25

INVERTER	PV MODULE COUNT	STC WATTS	TOTAL STC WATTS	SYSTEM Ks	IM MEASURED IRRADIANCE W/mm	MEASURED PV SURFACE TEMPERATURE	Ki IRRAD. FACT	Kt-TEMP MULTIPLIER	Wp PREDICTED WATTS	Wm INVERTER MEASURED WATTS	Wm/Wp % RATIO
TEST-A1	1360	235	319600	0.85	780	27	0.78	0.924	212836	20200	1.05%
TEST-A2	1360	235	319600	0.85	750	27	0.75	0.924	204650	20440	1.00%
TEST-A3	1360	235	319600	0.85	765	27	0.765	0.924	208743	20200	1.03%
AVERAGE-SITE A									208743	20280	1.0294
TEST-B1	1660	235	390100	0.85	768	27	0.768	0.924	255789	25507	1.00%
TEST-B2	1660	235	390100	0.85	774	27	0.774	0.924	257787	25540	1.01%
TEST-B3	1660	235	390100	0.85	774	27	0.774	0.924	257787	25230	1.02%
AVERAGE SITE B									257121	25426	1.0113
TEST-C1	2400	235	564000	0.85	770	26	0.77	0.962	401902	39500	1.02%
TEST-C2	2400	235	564000	0.85	773	26	0.773	0.962	403468	40040	1.01%
TEST-C3	2400	235	564000	0.85	775	26	0.775	0.962	404512	40230	1.01%
AVERAGE SITE C									403294	39923	101%

IRRADIANCE FACTOR Ki = Im/Istc

8 Column eight represents the computation result of the temperature derate factor.

9 Column nine represents the aggregated result of multiplication of all derates factors (PV Watts, temperature, and irradiance).

10 Column eleven represents the synchronized output measurements from the inverter, which is coincident with irradiance measurements.

11 Column twelve represents predicted versus inverter output watts percent ratio (the ratio should not exceed more than 5%).

12 Column ten represents calculated or predicted solar array power output based on DC output measurements.

SOLAR POWER STRING TEST

String Voc field measurements All solar power arrays while being assembled or integrated must be tested for PV module and string functionality. In some instances, an infant mortality or reverse diode failure could cause string power losses which may go unnoticed after integration. Since combiner boxes connect positive and negative output of incoming strings, Voc measured within the combiner boxes is a representation of the combined string voltages. In order to measure Voc of individual solar power strings, voltage measurements must be performed by removing the fuses from individual fuse holders.

Individually measured string Voc measurements, under proper working conditions, should display a variance of no more than 4–5 percent. Larger deviations are generally result of PV impedance mismatches, which must be corrected by verifying the Voc output of individual PV modules within the string. Even though the process appears to be somewhat time consuming, it is well worth the effort, since string mismatches in some instances may result in string malfunction sometime during the life cycle of the operation.

String Isc measurements Short-circuit current measurements is conducted by simple shorting of the positive and negative string leads. A DC clamp meter is used to record the Isc current. Deviation of Isc current measurements between various strings should not exceed 2–3%.

Inverter startup test sequence The following is a typical startup test procedure for solar power inverters. Testing shall always be in complete conformance with the particular inverter manufacturers' instructions:

- Verify all connections
- Verify if AC voltage is at the disconnect or breaker
- Verify the DC voltage ant polarity at the DC disconnect or breaker
- Close the AC breaker
- Verify the AC voltage at the inverter output terminals
- Close the DC breakers
- Verify the DC voltage and polarity at the inverter input terminals

- Switch the inverter ON and allow the inverter to complete its internal startup sequence
- Upon start-up of the inverter, the inverter must be allowed about 15 minutes for internal temperature stabilization

FIELD MEASUREMENT EQUIPMENT

Field test equipment required includes the following:

- To ensure PV module and string functionality Volt meter—0-1000 Vdc range, required for measuring the string Voc
- Clamp type current meter—0-100 for measuring string Isc
- Infrared (IR) thermometer—for measuring PV surface temperature
- Hand-held pyranometer—for measuring solar irradiance

SAFETY CONSIDERATIONS

Solar power measurement must only be delegated to electrical/solar power technicians with ample experience. At all instances, when conducting a solar power test, personnel must wear adequate and appropriate fire wear attire and insulating gloves.

Troubleshooting Procedures

All PV modules become active and produce electricity when illuminated in the presence of natural solar or high ambient lighting. Solar power equipment should be treated with the same caution and care as regular electric power service. Unlicensed electricians or inexperienced maintenance personnel should not be allowed to work with solar power systems.

In order to determine the functional integrity of a PV module, the output of one module must be compared with that of another under the same field operating conditions.

Note that the output of a PV module is a function of sunlight and prevailing temperature conditions. As such, electrical output can fluctuate from one extreme to another.

One of the best methods to check module output functionality is to compare the voltage of one module to that of another. A difference of greater than 20 percent or more will indicate a malfunctioning module.

When measuring electric current and voltage output values of a solar power module, Isc and Voc values must be compared with the manufacturer's product specifications.

To obtain the Isc value, a multimeter ampere meter must be placed between the positive and negative output leads shorting the module circuit. To obtain the Voc reading a multimeter voltmeter should simply be placed across the positive and negative leads of the PV module.

For larger current-carrying cables and wires, current measurements must be carried out with a clamping meter. Since current clamping meters do not require circuit opening or line disconnection, different points of the solar arrays could be measured at the same time. An excessive differential reading can indicate a malfunctioning array.

It should be noted that when a PV system operates at the startup and commissioning, problems seldom result from module malfunction or failure. Rather, most malfunctions result from improper connections or loose or corroded terminals.

In the event of a damaged connector or wiring, a trained or certified technician should be called upon to perform the repairs. Malfunctioned PV modules, which are usually guaranteed for an extended period of time, should be sent back to the manufacturer or installer for replacement.

Please be cautioned not to disconnect DC feed cables from the inverters unless the entire solar module is deactivated or covered with a canvas or a nontransparent material.

It is recommended that roof-mount solar power installations should have 3/4-in water hose bibs installed at appropriate distances to allow periodic washing and rinsing of solar modules.

All safety warning signs must be permanently secured to solar power system components.

CUSTOMER TRAINING CURRICULUM

After completing a solar power project, solar power contractors must provide adequate training materials and a training syllabus for clients' technical personnel. The curriculum must at its minimum cover the basic physics of PV phenomenon; an overview of solar power technologies with emphasis on the project-specific hardware and software elements; a review of operation and performance characteristics of all hardware systems; a review of the fundamentals of the solar power system configurations; a review of solar power system arrays, subarrays, PV elements, cables, conduits, and equipment tagging and identification; and a review of test and measurement methodologies for PV modules and PV strings, DC combiner boxes, and AC power output performance measurements. The instructor teaching the course must also discuss equipment and life safety issues and emergency mitigation measures.

As discussed earlier in this chapter, large-scale solar power system installations involve a considerable breadth of technologies. As such, maintenance and operation of each equipment and solar power system component requires technical knowledge and expertise. Customer training is particularly important for the client's technical personnel, as well as maintenance personnel who may have the ultimate maintenance responsibility for a solar project.

The customer training material and training syllabus provided in the following text have been specifically tailored to address all fundamental knowledge required to train the technical personnel who may have varying backgrounds and experiences.

Familiarity with solar power system technologies is of utmost importance for the client's technical and operation personnel. In view of the fact that solar power technologies involve orchestrating multiple disciplines, educational curriculum must include a comprehensive spectrum of courses that are specifically designed to inform and educate technical, operations, and maintenance personnel. The following is a suggested outline for solar power educational curriculum that encompasses comprehensive coverage of all aspects of solar power technologies.

1 Introduction to seminar
- Discussions on environmental pollution, global warming, and impact of energy use on ecology

2 Principle physics of PV technologies
- PV effect
- PV physics
- PV technologies

3 Solar power manufacturing overview
- Monocrystalline technology
- Polycrystalline technology
- Film technology
- Dye-sensitized solar nanotechnologies
- Multijunction technologies
- Concentrator technologies

4 Solar power systems system application
- Solar power system configuration classification such as roof-mount, ground-mount, or carport
- Stand-alone, grid-connected systems, hybrid configurations
- Materials and components
- PV modules
- Collector boxes
- Inverters
- PV support system
- Battery backup systems

5 Solar power systems applications
- Grid-connected single residential applications
- Grid-connected commercial systems
- Grid-connected industrial systems
- Stand-alone irrigation system
- Solar farms
- NEC code compliance
- Electrical service and solar power integration procedure and coordination
- System grounding considerations

6 Overview of solar power system engineering design and system integration
- Platform analysis and power generation potential
- Solar insolation physics
- Shading analysis
- Service switchgear and grid connectivity system requirements
- Service provider coordination
- Rebate system evaluation
- Feasibility study report

7 Review of specific case studies
- Residential
- Commercial
- Industrial
- Industrial application
- Solar power farms
- Agricultural solar power irrigation

8 Solar power system feasibility study
- Energy audit
- Service provider coordination and tariff studies
- Rebate and local tax exemption review
- Rebate application process
- Electrical power switchgear and metering system load burden analysis
- Geoclimatic system analysis
- LEED design
- Pollution footprint

9 Solar power costing and economic analysis
- Energy cost escalation
- Equipment and labor costing

10 Solar power design methodology
- Roof-mount platform capacity assessment
- Ground-mount platform capacity assessment
- Shading-analysis for existing and new projects
- Detail design solar power capacity evaluation procedure
- Commercial roof-mount support stanchions and hardware

11 Solar power design methodology continued
- Mega concentrators
- Sun-tracker systems
- A case study of platinum-rated building design

12 Student project design workshop
- Open forum discussion, question and answer

13 Student project design workshop continued
- Project design review and critique

Overview of Solar Power System Safety Hazards

Solar PV modules, when exposed to solar irradiance, produce DC electrical currents that cannot be turned off unless physically shorted across the output leads or covered by a canvas.

When connected in series strings, solar PV modules produce DC currents at high voltages that may range from 300 to 1000 V. Inadvertent exposure to high-voltage circuits may result in serious bodily injury. Disruption of high-voltage circuitry also results in electrical sparks that can spark fires under certain conditions. This may result in life safety and material damage.

SOLAR POWER SYSTEM HAZARDS

When designing solar power systems, the following are life safety issues that must be taken into consideration:

- In building fires where firefighters must penetrate the roofs, exposure to high-voltage DC under wet conditions presents a risk of an electrical shock hazard. This may result in severe burns, bodily injuries, and consequential litigations.
- Earthquakes may dislodge solar power support structures, which will result in high-voltage cable and conduit breakage. Feeder breakages may result in electrical sparks that under certain conditions could result in building fire and property damage.
- In high-DC voltage environments, maintenance technicians are always subject to inadvertent electrical shock hazards.
- During an earthquake, carport solar power systems could experience dislocation exposed and underground feeders, which may pose severe electrical shorts and shock hazards to maintenance personnel.
- Large-scale solar farms in particular are prone to serious fire and life safety hazards since environmental conditions, such as earthquakes, forest fires, flooding, lightning, and severe windstorms all could dislodge large segments of solar arrays. This could create large electrical sparks from ruptures of heavy currents carrying conductors (such as 1000 amps). This may result in significant life safety hazards, material and property loss; it can also cause ground fires or result in serious environmental damage.

Hazard mitigation   To mitigate the life safety risks for fire hazard issues (for each of the previously described scenarios), all of the solar power system configurations must incorporate an adjunct safety device on PV strings. This would allow systematic shutdown of each and every PV module that could be activated from a central data acquisition and control system.

Such a system should have an internally embedded hard contact relay that could be activated or deactivated by either reception of its unit address or by a global broadcast shutdown message initiated from the central data acquisition and control system.

Hazard signals that could activate a global shutdown of the solar power system include:

- Dry contact input from a central or a local (pull station) fire alarm system
- Dry contact or digital signal from an earthquake detection mechanism
- Dry contact from fire or CO_2 detection device
- Dry contact from water level float
- Dry contact from an emergency pushbutton

The following are special features of a circuitry named WISPR (Wireless Intelligent Solar Power Reader), patented by the author:

■ Individual data acquisition and control from each individual PV module within a solar system
■ Real-time monitoring of voltage, current, and unit temperature of each PV module
■ Frequency-hopping communication system
■ Adaptability of the module as digital and analog data sensor and actuator
■ The unit acquires a few microamperes of its operational power from the PV module during daylight
■ Each module has a longlasting lithium ion battery backup system intended to maintain module operation during extended cloudy periods when PV modules are dormant
■ WISPR operation is completely transparent to PV module functional performance
■ The module input and output connectivity are designed to match standard U.S. PV module positive and negative IN/OUT pigtails
■ Unit physical package is configured to allow retrofit installation on existing solar power systems
■ WISPR electronics could also be manufactured in a compact hybrid circuit form for sale as an OEM product that could be imbedded within the PV module junction boxes during manufacturing process

Solar Power System Loss Prevention

The following are some factors and conditions that affect output power performance of solar power systems, which could be monitored and diagnosed by WISPR:

1 Premature PV module failure or infant mortality
2 Accelerated degradation caused by improper lamination
3 PV module internal reverse diode failure
4 PV string connectivity increased resistance caused by oxidation and corrosion
5 Lightning damage to a solar array segment
6 PV module or string soiling
7 Shading resulted from adjacent vegetation or tree growth
8 Dislocation of a PV module support system, such as fixed-angle support structures or single-axis solar-tracking systems
9 Earth settlement of support structures caused by changes in soil settlement or erosion
10 DC combiner box fuse failure
11 AC combiner box breaker trip
12 Snow and ice accumulation
13 Tracker system motor failure
14 Inverter failure
15 Feeder cable failure
16 Transformer failure

8

CONCENTRATOR PHOTOVOLTAIC
SYSTEMS

Introduction

The highly efficient conversion of solar energy to electrical energy through the use of photovoltaic (PV) cells is one of the key elements for the future world energy supply. Researchers in the field of photovoltaics have always been faced with the problem as to how they could increase the efficiency of this PV conversion process while reducing the cost significantly. The solution for attaining this goal has been reached with concentrator photovoltaic (CPV) technologies. With CPV technologies, the cost reduction is achieved by replacing expensive PV cell material with lower-cost optical systems that enable a larger PV receiver aperture.

The development of CPV systems dates back to the late 1970s, when researchers at the National Sandia Laboratories designed CPV systems based on a combination of large acrylic Fresnel lenses and solar cells made out of crystalline silicon. With conventional crystalline silicon solar cells, CPV systems (because of the large cost of manufacturing the optical concentrator system) did not succeed in becoming a viable commercial product that could justify the higher-output power efficiencies. However, in the past few years CPV technology has undergone substantial improvements in power output production and cost reduction. Most recently, manufacturers of CPV systems have been deploying state-of-the-art expensive multijunction III-V solar cells, which have demonstrated efficiencies above 40 percent. In the recent past, the CPV manufacturers developed system technologies at low production costs and succeeded in penetrating the large-scale grid-connected solar power systems markets. Currently, solar-to-grid CPV system efficiencies of up to 25 percent have been discovered, and have opened the gate for the lowest-level cost of electricity.

Currently installed power generation capacity throughout the world is 2 terawatts (TW). Between the growth in energy consumption and replacement of aging power plants, it is estimated that as much as 6 TW of capacity will be required by 2030.

Relying primarily upon fossil fuels to meet this increased demand will severely impact the environment and natural resource reserves, so alternative sources must be implemented. Solar power is the most abundant renewable in the world with one hour of solar energy hitting the earth capable of meeting the current world demand for one year. In essence, we have at our disposal free fuel for life; it just has to be harvested. Of the various solar technologies available, it is CPV that is increasingly recognized as the technology that holds the greatest promise in meeting the energy challenges facing the world.

Concentrator Photovoltaic Technologies

The concentration of solar radiation for PV applications used in manufacturing CPV systems has been motivated by the potential of cost reduction and increased power output production. As mentioned previously, the cost reduction in CPV technologies is achieved by replacing expensive PV cell materials and through cost reduction of the optical systems, which provide a large receiver aperture. The increased efficiency is related to the increase of the open-circuit voltage, Voc. Most recent laboratory tests of CPV systems have demonstrated that a concentration of 56,000 times of the solar irradiance can be achieved.

Advantages of CPV Compared to Conventional PV Technologies

CPV technologies are best suited for hot climates. The very low temperature coefficient of the III to V multijunction concentrator solar cell means the performance of CPV systems is much less affected by temperature than any other PV technology (which compares to approximately one-third of comparable crystalline silicon PV modules). This is extremely significant for solar sites located in hot desert climates. In comparison to other PV technologies, at high temperatures, the efficiency and the electricity production of CPV systems are only slightly affected.

CPV systems using a high concentration are always manufactured as two-axes tracking-system platforms. A two-axes tracking system allows for a homogeneous electricity production profile over the day because the panels are always oriented perpendicularly to the incident irradiation from the sun. The most important effect is that the power production is at high levels when the power demand peaks in the afternoon. The superior daily production profile, together with the high efficiency, enables CPV systems to achieve highest energy production per used area and highest temperature corrected capacity factors of up to 34 percent in sites with a very good solar resource.

On average, the highest-conversion efficiency enables the concentrator technologies to achieve the lowest-level costs of electricity (LCOE). Essentially the main challenges for reducing the cost of production of conventional PV technologies include solar cells, lamination materials, glass, steel, and copper.

Concentrator Design for Passively Cooled Modules

Before discussing the different optical designs, it is important to clarify that there are two fundamentally different types of CPV module or system approaches. The first type of CPV module consists of many small- to medium-sized concentrator-cell assemblies, which are arranged in an array that is passively cooled. The second type of CPV modules have a single large optical concentrator and a central receiver module that has to be cooled actively.

Solar CPV concentrator systems are also classified as imaging and nonimaging concentrators. This classification is based on the optical principles of reflection, refraction, dispersion, diffraction, and fluorescence. As mentioned earlier, the conventional types of CPV systems operate on the principle use of the Fresnel lens and concentrator-type mirrors. In general, the use of the Fresnel lenses and the concentrator mirrors are not sufficient to fulfill essential requirements of solar concentrators. To be cost-effective, CPV technologies use high optical efficiency concentrators that absorb uniform illumination and are not sensitive to tracking errors or atmospheric variations of the incident direct solar radiation.

Today, Fresnel lenses are manufactured out of acrylics or silicon rubber on glass (SOG) and are used as primary concentrator optics. Their major advantage is that modules can be easily constructed by having a first concentrating plane and a second receiver plane separated by the focal distance. This allows for an easy electrical interconnection of the receivers on the second plane because there is no constraint by the optical aperture. The major differences amongst various CPV designs are the dimensions of the primary optics and the type of solar cells used. In CPV modules with small lenses and cells, the technologies generally deploy imaging designs, whereas in larger-cell technologies, they often use non-imaging designs, which improve the homogeneity of the irradiation of the cell.

Most often mirrors for primary optics are used in off-axis paraboloids and compact imaging designs. With off-axis paraboloids, it is quite difficult to achieve a slim module. Secondary optics most often includes domes, truncated cones, inverted pyramids, and compound parabolic concentrators.

Recent Advances in CPV Technologies

Although silicon cell-based CPV was developed and tested in the late 1970s, it has been only a few years since high-concentration CPV using multijunction solar cells was developed. Currently, about 50 companies around the world are developing or marketing CPV system technologies. Recently, CPV systems have achieved efficiencies of over 30 percent.

With this technology, reflectors are used to increase power output by either increasing the intensity of light on the module or by extending the length of time that sunlight falls on the modules. The main disadvantage of concentrators is their inability to focus scattered light, which limits their use to areas of concentrated light, such as deserts.

In this chapter, we will review one of the most efficient and innovative concentrated PV systems, which has been specifically designed for large-scale, grid-connected solar power systems mainly used by electrical power service providers. The reader must note that information presented is not intended to promote the product or products.

Amonix CPV Solar System

The Amonix 7700 (Fig. 8.1) is a 53-kilowatt high-concentration concentrating PV solar power system based on the MegaModule®, a unit pairing a durable Amonix Fresnel lens with very high-efficiency multijunction cells. Each unit is comprised of seven proprietary MegaModules® utilizing dual-axis tracking. It is the first PV system capable of converting one-fourth of the sun's energy into usable electricity.

The Amonix 7700 is designed specifically for utility-scale deployment. At 77– × 49-feet, it is also the world's largest pedestal-mounted solar array.

The Amonix 7700 was introduced to the large-scale solar marketplace in 2009. In October 2009, Amonix opened its manufacturing facility in Seal Beach, California, for production of the 7700. In November 2009, Amonix raised a substantial amount of new

Figure 8.1   **AMONIX 7700 high-efficiency CPV system.** *Photo courtesy of Amonix.*

capital and augmented its commercial team to further penetrate the large-scale solar market. Using a $5.9 million investment tax credit from the Recovery Act awarded in 2010, Amonix plans to establish a new manufacturing facility in the city of North Las Vegas area by the end of 2010, which when fully operational will result in annual production of 150 MW of solar capacity. Figure 8.1 is photograph of the AMONIX 7700 high-efficiency CPV system.

PRIMARY FUNCTION

The Amonix 7700 Solar Power System is a high-concentration, high-efficiency, warranted, PV solar system that produces considerably larger amounts of solar power energy per acre of land than comparable flat-panel PV technologies. Solar power production efficiency is achieved by using highly efficient solar cells, concentrator Fresnel lenses, and a dual-axis tracker system. According to the manufacturer, the system is the first terrestrial CPV system capable of converting one-fourth of the sun's energy into usable electricity. In view of power output performance compared to flat-panel PV solar power systems, the near-term market for CPV technologies is enormous. According to the CEC, California alone is planning to install 3000 MW of new solar electric capacity by 2017.

FIELD INSTALLATION AND PERFORMANCE

Amonix 7700 generates a megawatt of electricity for every five acres of plant site, less than half of the land required by other solar technologies. In addition to its high-performance efficiency, the CPV system has optimized features that enhance solar plant installation and operation. Each of the MegaModule seven 7.5 kW module assemblies forming a single integrated CPV system are factory assembled by an automated robotic system. The automated factory integration of the lens, mounting structure, and solar cell into a single unit results into considerable production cost savings. The MegaModule concept keeps most assembly in the factory, reducing field installation time from months to days.

In order to minimize the field installation and construction costs, the AMONIX CPV system has been designed and developed as a total-systems product, which is readily transportable and easily installed in the field and requires minimal manpower and installation time. The MegaModule systems are transported whole in four segments that can be stacked on a flatbed truck. Field installation time required to erect a single 53-kW CPV system requires as little as three days. The solar system is pedestal-mounted, which minimizes the need for major excavation, and its high-ground clearance allows native vegetation to flourish beneath the units so the impact on the local environment is minimal. The systems can be installed on rugged and uneven terrain because the installations require no grading or special site preparation.

Operation for large-scale electricity production includes the ability to direct the tracking system for normal, emergency, or maintenance needs. The 7700 uses the Amonix proprietary, closed-loop dual-axis tracker to keep the sun's rays coming directly onto the Fresnel concentration lenses and onto the high-efficiency solar cells. This allows the system to maximize energy production throughout the day.

The Amonix 7700 has an industry-leading sunlight-to-electricity conversion efficiency with a system efficiency of 25 percent, module efficiency of 31 percent, and a cell efficiency of 39 percent. Amonix CPV systems are recognized as the world's most powerful and efficient on the market.

Maintaining the 7700 solar power system is easy because a failure of one solar cell does not affect the functioning of the other cells, and each receiver plate of 30 cells can be field-repaired or replaced to keep production at a maximum. This ability to replace components also presents the option to upgrade or "repower" the system to extend the solar power system life to 50 years.

COLLABORATION WITH NATIONAL RENEWABLE ENERGY LABORATORIES (NREL)

Before developing the 7700, Amonix had an existing CPV approach using crystalline-silicon PV cells that was quite successful. About 13 MW of this earlier technology has been deployed, which accounts for about 73 percent of all CPV systems installed worldwide. By substituting an extremely high-efficiency multijunction PV cell (~41.6 percent record efficiency) for the silicon cell (~27.6 percent record efficiency) the CPV system achieved a remarkable reduction in power output cost reduction as well as land use, both of which were essential for solar electricity to achieve grid parity with fossil fuels. Note that multijunction solar cells used in the CPV system have been routinely used in space missions.

AMONIX-CONCENTRATED PV MEGA CONCENTRATOR DESIGN

CPV systems use lenses or mirrors to concentrate sunlight onto solar cells, which enhances the cells' efficiency. By minimizing the size of the cell and employing high concentration, a small amount of semiconductor material can produce large amounts of power. One MegaModule in the 7700 uses 1080 high-efficiency multijunction PV cells. Amonix developed the multijunction cell for this high-concentration application in collaboration with Spectralab, a wholly owned subsidiary of Boeing Corp., which has deployed this technology in many of today's communications and defense satellites. This collaboration resulted in the solar cells used in the Amonix 7700.

Each cell, approximately 1 cm^2 in area, is exposed to focused sunlight equivalent to 500 suns using the Amonix Fresnel lens, which is made of an inexpensive, durable acrylic material. The Amonix Fresnel lens uses refractive optics to concentrate the sun's irradiance onto a solar cell, as illustrated in Fig. 8.2. A square Fresnel lens, incorporating circular facets, is used to direct the sun's rays to a central focal point. A solar cell is mounted at this focal point and converts the sun power into electrical power. Thirty of these Fresnel lenses are manufactured as a single piece, or parquet. The solar cells are mounted on a plate at locations corresponding to the focus of each Fresnel lens. A steel C-channel structure maintains the aligned positions of the lenses and receiver plates.

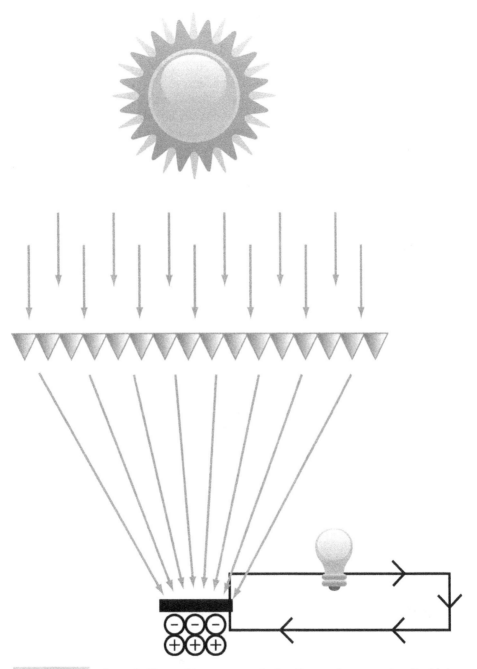

Figure 8.2 Amonix Fresnel lens concentrates the sun's rays onto ultrahigh-efficiency solar cells. *Graphics courtesy of Amonix.*

Each MegaModule is designed to produce about 7.5 kW of AC power at standard conditions of 850 W/m² [under direct normal irradiation (DNI) and 20°C ambient temperature]. Seven modules are mounted on a sun-tracking structure to obtain a 53-kW (AC) array. Each MegaModule contains 36 receiver plates. Each receiver plate has 30 cells and a matching sheet of 30 Fresnel lenses.

All CPV systems that use high concentration are equipped with trackers that follow the sun's position so that the sun's rays directly hit the lens or reflectors, which redirect them to the PV cell arrays. The 7700 has a superior tracking capability built into the system via the Amonix patented hydraulic-drive, dual-axis tracker.

The Amonix CPV system architecture minimizes individual solar cell failures. First, each cell is bypass diode-protected so that a failure of one solar cell does not affect the functioning of the other cells in the module or on the same string of cells. Second, the architecture is such that the solar cell can be field-repaired and replaced, providing further protection against solar cell failure. Third, this solar cell and receiver plate replacement capability opens up the option of repowering the Amonix CPV systems in the field. According to independent engineering firms that have evaluated the Amonix system, the 7700 delivers the following benefits:

■ The ability to upgrade and replace components is an advantage over flat-plate.
■ PV systems can extend the economic life of the systems.
■ Many of the Amonix system components, such as the steel housing and tracking support systems, have a physical life of 50 years.
■ Components with lives shorter than 50 years, such as cells and lenses, can be replaced, most likely with more efficient and less costly components. This "repowering" with new cells and lenses would effectively extend the physical life out to 50 years.

SYSTEM CONFIGURATION DETAIL

The Amonix 7700 is made up of the following major subsystems (illustrated in Fig. 8.2):

■ MegaModule. The MegaModules concentrate the sun's energy on a solar cell that converts it into electrical energy. It consists of Fresnel lenses, solar cells, and a mechanical structure. Each 7700 has seven MegaModules.
■ Drive subsystem. This rotates the MegaModules in azimuth and elevation to track the sun. The drive system consists of a foundation, pedestal, rotating bearing head, hydraulic actuators, and torque tube.
■ Hydraulic subsystem. This element applies hydraulic pressure to one side of the hydraulic actuators to move the torque tube and MegaModules in elevation and azimuth in order to keep the system pointing at the sun. The hydraulic system consists of hydraulic valves, an accumulator, a pump, a reservoir, and pressure sensors.
■ Tracking control subsystem. This monitors sensors on the system, calculates the required movement for the commanded operation, and applies signals to the hydraulic valves to move the system to the commanded position. The commanded position could be to track the sun, move to a night stow position, move to a wind stow position, or move to a maintenance position.

■ AC/DC control subsystem. This combines the DC power, converts it to AC power, and interfaces with the AC grid. It consists of DC fuses, circuit breakers, and an inverter.

TOTAL-SYSTEM MODULAR DESIGN

To produce the 7700, Amonix has instituted a total-systems approach that represents a "foundry-to-grid" way of thinking, with cost savings factored in at each step of the way. The process starts with assembling the ultrahigh-efficiency (~39%) III to V multi-junction cells into an integrated structure of the MegaModule at the Amonix factory. A capital-efficient manufacturing process enables distributed manufacturing. Integration of the lens, mounting structure, and solar cell into a single unit eliminates more than 75 percent of the parts and costs associated with other concentrators while providing special advantages in ease of manufacturing, installation, and maintenance. The MegaModule units have been designed to meet maximum-size limitation for surface transporting four MegaModules that could be stacked on a flatbed truck. Upon reaching the destination, the installation rate is rapid, twice as fast as that required by large thin-film modules. A significant feature of the MegaModule is that fabrication and field system integration time is reduced from one month to a single day.

Operating and maintaining the 7700 is streamlined based on the design modularity. The components of the 7700 have proven reliability, based on 15 years of field testing by Amonix.

POWER PRODUCTION

The Amonix 7700 is currently considered the largest CPV system in the market. At 53 kW (AC) per assembly, the Amonix 7700 provides twice as much as the largest competing CPV system that produces 30-kW per unit. This implies that the 7700 produces far "more power per tower" than any of its competitors. The physical size and deployment advantages of the Amonix CPV systems were recently acknowledged on the Greentech Media Web site (http://blogs.greentechmedia.com/articles/print/Solar-Roundup-Heliovolt-Hoku-Petra-Amonix-et-al/). Amonix system performance has a legacy of several years of field performance evaluation data. Field test of the Amonix Mega concentrators have a proven record that indicates an order of magnitude of power production per units compared to other CPV competitors. Figures 8.3 and 8.4 are photographs of the Amonix 7700 installation process.

BENEFITS OF AMONIX CPV SOLAR POWER SYSTEMS

Studies conducted by the U.S. Department of Energy (DOE) and the Electric Power Research Institute (EPRI) show that CPV systems can eventually achieve lower costs than flat-plate PV systems for the electric utility market. The lower cost results from the following:

■ Making a small amount of material go a long way. Because the semiconductor material for solar cells is a major cost element of all PV systems, reducing the required

Figure 8.3   **Amonix field installation and system integration.** *Photo courtesy of Amonix.*

cell area by concentrating a relatively large area of solar insolation onto a relatively small solar cell reduces the costs accordingly. Amonix uses low-cost, durable acrylic Fresnel lenses to focus the solar energy onto the cells, which reduces the required cell area/material by nearly 500 times. A 6-in. silicon wafer used in a flat-plate PV system will produce about 2.5 W. That same 6-in. wafer area will produce more than 1500 W in the Amonix 7700 system using multijunction cells.

■ Higher efficiency. Concentrating PV cells achieve higher efficiencies than do one-sun PV cells. Flat-plate silicon modules have DC efficiencies in the range of 15 to 20 percent, whereas Amonix MegaModules have DC efficiencies of ~31 percent more annual energy. Increased annual energy production is achieved via the use of tracking systems, which results in additional annual energy generation per installed kilowatt.

Currently, the manufacturer expects steady upward efficiency gains in the next 3 to 5 years with strong headroom remaining before CPV begins to approach cell-efficiency limits. Based on its highly modular design, the system offers a relatively inexpensive way to upgrade the solar power system by replacing the solar cells with higher-efficiency cells, without any impact on the system hardware and structure.

LIFECYCLE AND ENVIRONMENTAL FACTORS

In many important ways, CPV technologies using high concentration have the light-est "environmental footprint" among all solar technologies. Brookhaven National

Figure 8.4   **Amonix field installation and system integration.** *Photo courtesy of Amonix.*

Laboratory and Columbia University environmental scientists conducted a lifecycle analysis (LCA) of the Amonix 7700 system to estimate its "cradle-to-grave" energy requirements, greenhouse gas emissions, and toxic gas emissions.

The LCA conducted on the CPV system by the manufacturer has demonstrated that Amonix CPV systems operating in the southwestern United States will produce all of the energy required in their lifecycles in about 8 months of operation, and that their lifecycle GHG emissions are among the lowest in the PV industry. In another LCA analysis, the manufacturer has also established that the energy payback times of the Amonix 7700 operating in Las Vegas will be 0.70 years (8.4 months).

In general, optimal locations for utility-scale CPV systems are vast sunny areas in desert-like regions. For this reason, the technology often has to contend with the issue of water for cooling the CPV cells. However, design of the Amonix CPV system is such that it is cooled passively with air, and as a result requires no water for cooling during operation. This is a significant advantage, particularly in sunny and dry climates where water is increasingly scarce. A small amount of water is required occasionally for cleaning the modules.

On January 8, 2010, under the American Recovery and Reinvestment Act clean-energy manufacturing tax credit, Amonix was awarded a grant of $9.5 million for manufacturing work in Nevada and $3.6 million for work in Arizona. Currently, the company is making plans to begin construction of a new manufacturing facility in Nevada by the end of 2010.

COST AND MANUFACTURING

According to the manufacturer, special fabrication, patented technology, and speedy field installation gives the Amonix MegaModule concentrator systems significant cost advantages. The Amonix technology uses an inexpensive, but highly reliable acrylic Fresnel lens for concentration. Integration of the lens, mounting structure, and solar cell into a single unit eliminates more than 75 percent of the parts and costs associated with other concentrator designs while providing advantages in ease of manufacturing, installation, and maintenance. The manufacturing of the Amonix technology features a MegaModule assembly process that supports high-volume production. The automated assembly delivers multiple benefits, such as improved manufacturing by robotic assembly and quality control, which result in improved field performance.

OPERATIONS AND MAINTENANCE

The Amonix MegaModule's design modularity ensures that operation and maintenance of the system is simple and streamlined. The MegaModule can be readily repaired or replaced. In addition, cells can be upgraded without having to replace the entire system.

FIELD INSTALLATION PROCEDURE

The Amonix 7700 is deployable in increments of 53 kW (AC) and applicable to both distributed generation and centralized solar farms. The 7700 system is specifically designed for producing utility-scale PV power, making it an excellent fit for the solar market. Recent industry reports indicate that global solar market forecasts project a $100 billion industry by 2013, with utility installations expected to constitute a 75 percent market share.

According to a report by the Western and the Southwestern Governors' Association of United States, number of states will soon deploy very large distributed solar power virtual power generation centers that will harvest significant amounts of solar energy resources in California, Arizona, New Mexico, and Texas. California's Solar Power Initiative Program (CSI) potential market for solar energy technologies are expected to reach to $3.3 billion dollars.

California's immediate goal is to create 3000 MW of new solar electrical power production by 2017. The California Energy Commission (CEC) has also published a list of proposed energy projects, and Amonix has been certified by the CEC as qualifying for the California rebates.

Based on its large size, modularity, and ease of installation, the Amonix technology is of particular interest to utilities, many of which must adhere to Renewable Portfolio Standards (RPS) with solar set-asides. Currently, all of California's electric utility companies are required to use renewable energy to produce 20 percent of their power by 2010 and 33 percent by 2020, with a main source of that power being solar energy. The utility PG&E has announced plans to develop 500 MW of solar power in the next five years via a collection of midsized projects, from 1 to 20 MW each.

According to the Database of State Incentives for Renewables & Efficiency (DSIRE), 29 states and the District of Columbia have established an RPS. Wind, biomass, and hydropower will be the predominant resources of renewable energy sources that will be used to satisfy DSIRE obligations, however, growing numbers of states are incorporating solar set-asides, which stipulate that significant portions of the renewable energy percentage will most likely be derived from solar resources. Currently, 16 states and the District of Columbia have adopted broader distributed green-power generation set-asides that will be part of their RPS policies. In order to comply with federal- and state-mandated requirements, it is currently projected that electrical power generation from various alternative energy systems in the United States will reach 502 MW by 2011 and 8447 MW by 2025.

Concentrix FLATCON®
CPV Technology

The design of the FLATCON CPV module dates back to the late 1990s. The unique features of a cover and bottom plate made out of glass and a relatively small aperture of each of the primary lenses, which are assembled in an array, were developed by Concentrix. The cells are mounted on heat spreaders, which serve at the same time as contact pads for the internal electrical connection of the module. The major reasons for using two glass panes are high durability, low cost, and the low coefficient of thermal expansion (CTE), which insures that the foci remain on the cell at varying operating temperatures. As the CTE of glass is three times lower than that of aluminum, for example, it is possible to keep the focal position on the cell within 100 μm at all operating temperatures. It is important to note that the bottom plate need not be thermally conductive, as the heat spreading is already efficiently done by the heat spreader for which highly thermally conductive materials are used. Glass serves also as a scratch-resistant cover plate. The Fresnel lens array is replicated in one piece into a silicone rubber on glass (SOG), allowing for extremely UV-stable materials in a cost-effective mass fabrication. There are also good reasons for the relatively small lens aperture: thermal management and low module depth. In case of this primary lens, a simple heat spreader made out of a metal with a well-satisfying thermal conductance is sufficient for the thermal management. Cell temperature in a CPV module does not exceed 40 K above ambient temperature on average. Furthermore, a small lens allows for small cells, which are advantageous with respect to highest efficiency because of low resistance losses on the cell. The design is kept as simple as possible in order to provide highest robustness and lowest manufacturing cost. An identity element of a FLATCON module just consists of a primary lens and a solar cell plus bypass diode mounted on a small planar heat spreader. This planar design can be manufactured by using standard semiconductor assembly and printed circuit board machines. A FLATCON Gen II module CX-75 are shown in Fig. 8.5. On the top right-hand side of the photograph is the SOG lens array, which is manufactured in one piece; on the bottom is the solar cell assembly array, which is interconnected by wire bonds. The lens and bottom plate are mounted by using proven standard technologies from the architectural glazing industry. From the very beginning, the production methods were taken into account for the module design. The existing production line of Concentrix has a capacity of 50 modules per hour, equivalent to annual productions of 30 MW, and can be quickly scaled up on demand. Today, the FLATCON modules CX-75 have an average efficiency of 27 percent.

FIELD TEST AND TRIAL

The Concentrix FLATCON CPV system consists of the modules, the tracker, the inverter, auxiliaries, and the control and monitoring hardware and software. In 2008, trackers were installed with 120 Gen I modules, which had an aperture of 0.24 m² and 150 lenses resulting in a total module aperture of 28.8 m² per tracker. Depending on the evolution of the module efficiency, the nominal power of these

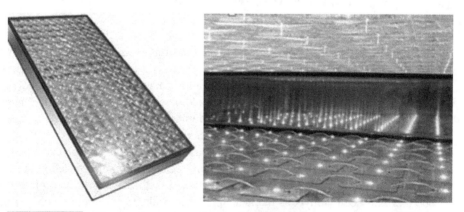

Figure 8.5   Photo of a module CX-75 on the left and view panel on right.
Photo courtesy of Concentrix.

trackers at a DNI of 850 W/m² is 5.4–5.6 kW. Installations were made at the ISFOC site at Puertollano, Spain, and at a site close to Seville, Spain. The tracker has accuracy determined to be better than 0.1° with a proprietary control system. Each of the trackers has its own inverter, which serves also as the control system of the mechanical tracking and as a communication port with own IP address. This very innovative inverter for CPV systems was developed by Fraunhofer ISE in collaboration with Concentrix, and currently has an efficiency of 96 percent. Figure 8.6 depicts the Concentrix concentrator geometry. Figure 8.7 depicts Concentrix system concentrator graphics.

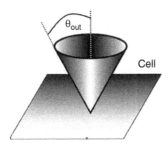

Figure 8.6   Concentrix concentrator geometry.

Figure 8.7   Concentrix system concentrator graphics.

The first of several 100-kW plants were installed and grid connected in 2008. Figure 8.8*a* and *b* depict the first system installation, which has been performing excellently and reliability in Puertollano, Spain.

In order to show the performance of a FLATCON CPV system, the field data from a demonstration system was also put into operation at a site close to Seville, Spain. This system has had an availability of 99 percent over period of several years of operation before writing this publication. Figure 8.9*a* and *b* display the daily DNI energy. The figures indicate that only in case of a low daily DNI energy does the system efficiency drop, mainly during wintertime. During summertime, when most of the electrical energy is gained, low-system efficiency data are rare. As a result, the system AC energy over the whole operation time is about 20 percent.

In 2009, the same tracker was installed, but with 90 modules CX-75 (i.e., same optical aperture on the tracker). Because of the very good and homogeneous quality

(a)

Figure 8.8   (*a*) FLATCON® CPV systems of Concentrix Solar installed in Puertollano at the ISFOC site in 2008.

(b)

Figure 8.8 (b) FLATCON® CPV systems of Concentrix Solar installed in Puertollano at the ISFOC site in 2008.

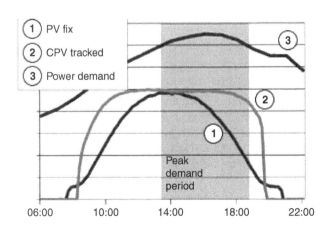

1. PV fix
2. CPV tracked
3. Power demand

Peak demand period

06:00 10:00 14:00 18:00 22:00

Figure 8.9 (a) Electricity production profiles of fixed and two-axes-tracked PV installations together with the power demand.

Graphic courtesy of Concentrix.

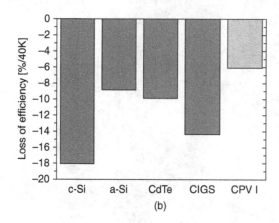

Figure 8.9 (*b*) **Efficiency loss for different PV technologies caused by a rise of temperature of 40 K.** *Graphic courtesy of Concentrix.*

of the module CX-75, which is the first CPV module produced on a fully automated production line, the measured system efficiency even outperformed the expectations. For this type of tracker, an AC system efficiency of 25 percent at a DNI of 850 W/m^2 was measured.

SolFocus Concentrator Photovoltaic Technology

The SolFocus concentrator photovoltaic (CPV) system is similar to concentrator technologies discussed previously, in that it converts light energy into electrical energy the same way that conventional PV technology does. The difference in the technologies lies in the addition of an optical system that focuses a large area of sunlight onto individual PV cells. The solar cells used in SolFocus CPV systems are different from silicon PV cells, as they are capable of converting very large amounts of sunlight into energy at high efficiency.

SolFocus concentrator optics systems are manufactured significantly less expensively than PV cells. The lower cell area used per unit concentrator reduces the overall manufacturing cost of the system. The SolFocus CPV system deploys a multijunction PV cell of 1 cm^2, which is illuminated by the sun magnified 650 times. This means that the sunlight covering 650 cm^2 is collected and redirected onto a single 1 cm^2 cell, which significantly reduces the cost per unit of energy as compared to conventional PV technologies.

In effect, the SolFocus CPV system is like a telescope; each unit requires an unobstructed view of the sun (direct radiation) and must actively track the sun in its progression across the sky. In general, the CPV systems do not respond well to indirect light or light-scatter caused by clouds or solar rays that are reflected off other objects. CPV systems are most effective when they are deployed in areas of clear weather and where there are a lot of sunshine hours, especially when combined with a precision sun-tracking mechanism. It should be noted that approximately 30 percent of the earth's surface conforms to the sunny conditions required for CPV-type systems. This represents over 40 percent of the world's population. In such climates, CPV can represent the best choice for harvesting lowest-cost energy and is the most effective method for generating clean and renewable electricity. Figure 8.10 shows a SolFocus solar power installation.

SolFocus CPV DESIGN

High-efficiency PV cells and precision optics are the essential building materials for all CPV technologies. To deliver the very best CPV solution, SolFocus marries the world's highest-performing PV cells with an efficient and durable optical design. The SolFocus design relies on precision glass components that are easy to produce in high volume and at low cost.

REFLECTIVE OPTICS

SolFocus utilizes reflective, non-imaging optics (Fig. 8.11*a* and *b*), which are precision-customized mirrors that collect and concentrate the sun's energy. The system uses a large mirror to collect direct sunlight and then focuses the reflected light onto a smaller secondary mirror. The secondary mirror then redirects the reflected light into a glass prism, channeling the sunlight onto the PV chip. The result is a compact and efficient CPV system.

Figure 8.10   **SolFocus solar power installation.** *Photo courtesy of SolFocus.*

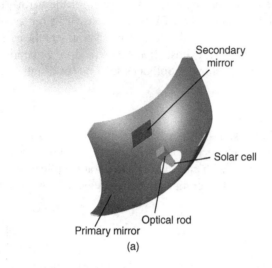

Secondary mirror

Solar cell

Optical rod

Primary mirror

(a)

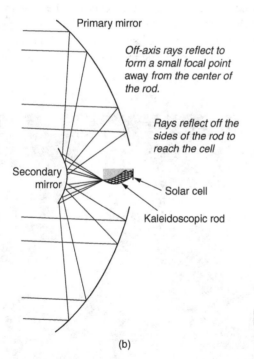

Primary mirror

Off-axis rays reflect to form a small focal point away from the center of the rod.

Rays reflect off the sides of the rod to reach the cell

Secondary mirror

Solar cell

Kaleidoscopic rod

(b)

Figure 8.11 (*a*) SolFocus reflective mirror system. (*b*) Diagrammatic presentation of SolFocus reflective mirror system. *Graphic courtesy of SolFocus.*

Research in the mechanical, electromechanical, and material aspects of the SolFocus design has resulted in a set of highly robust and efficient optical components. While some CPV solutions use refractive optics (usually plastic lenses through which the sun's light is focused), the SolFocus CPV system deploys reflective glass optics, which have superior performance and long-term durability against degradation. Reflective optics used enable higher efficiency and better concentration ratios. Furthermore, the reflective glass optics is known to maintain performance in high-temperature outdoor environments for decades.

The simple shape of the mirror's design is well suited for precision manufacturing using high-volume processes. The precision optics enables passive alignment in the assembly process. As a result, the assembled panels are inexpensive to manufacture and easily supported by high-volume automation.

HIGH-EFFICIENCY MULTIJUNCTION PHOTOVOLTAIC CELLS

The PV cells used in SolFocus CPV systems differ from crystalline silicon cells that make up traditional photovoltaic systems. SolFocus CPV cells, known as multijunction cells, provide energy-conversion efficiencies of approximately 38 percent, in contrast with the typical 12 to 17 percent of crystalline silicon. These cells are based on solar device technology used for space applications.

Note that current CPV cell technologies at present operate well below the theoretical efficiency limits of the device (at 60 percent). Currently, there are numerous industries that are researching novel advanced PV technologies that target efficiencies greater than 45 percent. Some of the research efforts for the development of advanced PV cells include the development of strain-balanced quantum wells, quantum dots, and intermediate band-gap designs. These research and product development efforts are believed to have the potential to yield PV cells that have a potential to surpass the 45 percent efficiency mark.

SolFocus CPV PANELS

Each SolFocus CPV panel, shown in Fig. 8.12, is made up of 20 individual optical assemblies or power units. The individual optical assemblies include a primary mirror, secondary mirror, nonimaging optical rod, and high-efficiency PV cell of approximately 1 cm^2. For the CPV panel, each set of optical assemblies are enclosed between an aluminum back pan and a thick protective sheet of glass. The resulting aluminum and glass enclosure provides long-term protection for all of the power-producing optical assemblies and facilitates easy cleaning.

The precision of components used and the design's simplicity result in CPV panels that are easily assembled using a fully automated process. No active alignment is needed for the enclosed power units to meet their designed efficiency. As a result, the manufacturing process is suited for rapid deployment and scalability, while maintaining both low cost and long-term reliability.

In order to ensure long-term performance, the design relies on proven durability of glass and aluminum to handle prolonged exposure to intense heat and sunlight as well as temperatures well below the freezing point.

TRACKING TECHNOLOGY

SolFocus CPV panels are mounted on a dual-axis tracker, which keeps them directly aligned with the sun. These SolFocus custom-designed trackers are engineered to

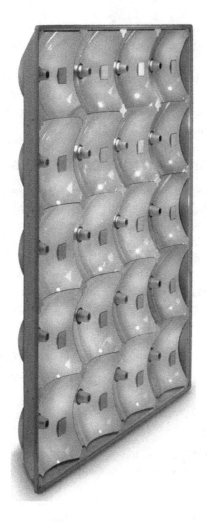

Figure 8.12   SolFocus panel. *Photo courtesy of SolFocus.*

provide sufficient pointing accuracy without sacrificing cost. The control system provides a 0.1° tracking precision through accurate gearing and a rigid panel-mounting frame. The control system uses the performance of the panels themselves to actively calibrate for any mechanical imprecision that occurs during system installation or operation. In addition, the control system is designed to collect and communicate telemetry data to a central server for analysis and predictive maintenance.

As mentioned above, SolFocus CPV panels use multijunction cells, which have an average 38 percent efficiency and an energy conversion efficiency of over 25 percent at the system level. It should be noted that CPV efficiency is the key factor that drives down the cost of energy produced by solar power systems. Figure 8.13 depicts a SolFocus reflective mirror.

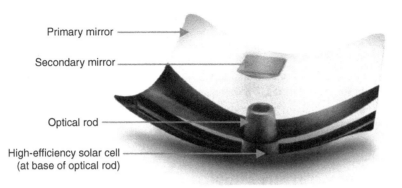

Primary mirror

Secondary mirror

Optical rod

High-efficiency solar cell
(at base of optical rod)

Figure 8.13   **SolFocus reflective mirror.** *Photo courtesy of SolFocus.*

HIGHEST ENERGY PERFORMANCE IN HOT CLIMATES

The specific engineering and product manufacturing methodology renders SolFocus technology considerably less susceptible to the effects of temperature degradation than silicon PV or thin film, so panels maintain high performance in the hottest climates. Figure 8.14 illustrates the power output from four panels in a location at 40°C ambient temperature. Although all panels are rated at 300 W, the PV and thin-film types show significantly less actual power than SolFocus panels. The manufacturer claims that SolFocus CPV arrays lose just 4 percent of their power output at an ambient temperature of 40°C/104°F (which are common in sunny regions of the world), whereas traditional silicon PV panels lose 22 percent.

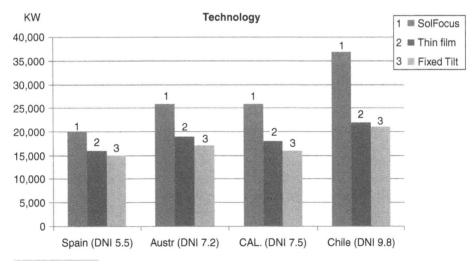

Figure 8.14   **Comparative performace chart of SolFocus versus thin-film and silicon PV systems tested in various countries.**

HIGHEST-ENERGY PRODUCTION PER MW INSTALLED

Capacity and power output are not the same for all technologies. SolFocus panels generate more energy per installed MW capacity in hot climates. This is because of the temperature effect, the precise tracking of the sun, and the typically higher percentage of direct light in hot locations. That means that a smaller-capacity power plant can be installed to meet a given energy production target using SolFocus CPV than with other technologies. Alternatively, SolFocus panels maximize energy production for a given site capacity.

HIGH-ENERGY OUTPUT PROFILE

SolFocus's high-efficiency tracked CPV solution provides higher daily energy output than fixed PV systems. Figure 8.14 shows how tracking CPV systems provide maximum power output and kWh production over the course of a typical day and not just in the midday period. By tracking the array platform to the sun, energy collection is maximized throughout the day from morning extending through to late afternoon when the electricity is of greatest value.

CPV Technology and Environmental Sustainability

A critical element in producing clean, green, renewable energy is assuring that the technology itself supports environmental sustainability.

Advantages and benefits of CPV technology could be summarized as follows:

- Scalability. CPV technologies are scalable, allowing the deployment of medium-scale to large-scale sites located close to the demand load. This reduces the need for power transmission, which is a source of significant power loss and disturbance to environmentally sensitive areas.
- Optimized land usage. The high-conversion efficiency of CPV technology yields high-energy output per area of land. As a result, less land is disrupted. Also, the technology can be deployed on irregularly shaped land, allowing use of already disrupted areas such as roadsides, power lines, and canal ways. By being mounted on single-pole trackers, the footprints of CPV systems are very small, requiring the use of only a small fraction of the land surface. In addition, dual use of the land is possible. By providing clearance under the arrays and occupying a small footprint, the land under an operating SolFocus power plant can be used for energy production and agricultural purposes at the same time.
- No permanent shadowing. CPV technology utilizes trackers that are high off of the ground and that move throughout the day, resulting in no permanent shadowing. Permanent shadowing is disruptive to both plant and animal life, and the natural ecosystem around the installation. In addition, natural revegetation can be allowed

to occur after construction is concluded, minimizing land maintenance. Many other technologies need to maintain gravel or other artificial surfaces around the plant, for maintenance and safety purposes.

■ Consumes no water. No water is consumed in the production of electricity utilizing CPV technology. Solar energy technologies, which require water (fresh or waste) to produce electricity, compete with water-starved desert ecosystems as well as industry and agriculture activity. The National Renewable Energy Laboratory estimates that solar electric generating systems that operate a wet-cooled, Rankin cycle system consume approximately 800 to 1000 gallons of water per MWH generated (CSP is an example).

■ Highly recyclable. CPV systems are manufactured using primarily glass, aluminum, and steel, all highly recyclable and nontoxic. With over 97 percent recyclability at end of life, the products can easily be transformed into future new uses.

■ Small carbon footprint and short energy payback. Compared to other solar technologies, CPV technologies provide a much smaller greenhouse gas (GHG) footprint and shorter energy payback. The energy payback for CPV technologies is quite superior compared to that of silicon PV systems.

9

SOLAR POWER SYSTEM
PROJECT MANAGEMENT

Introduction

Project management is a methodology and a discipline required for planning, organizing, and managing project manpower and resources. This is a prerequisite for successful completion of a project.

In general, all projects consist of certain tasks that have a beginning and an end, constrained by schedules and objectives that must be executed within the limits of date and budgets, and deliverables as outlined by the owner's specifications of specific requirements.

In order to meet the project objectives and goals, a technically qualified individual(s) designated as project manager is required to prepare a project execution plan that encompasses control methodology and procedures that could provide effective control and management of the project from start to the end.

In order to accomplish the above, the individual in charge of project management must be fully acquainted with all aspects of the project, which involve project planning, scheduling, engineering, system integration, and commissioning.

To be successful, the project manager must strive to meet project objectives and goals and ensure that the project meets its principle objective, while maintaining limitations of schedule and budgetary goals.

Project Development Phases

In general, all solar power system projects have five developmental stages which include the following:

1 Project initiation stage
2 Project planning or design stage

3 Project execution or production stage
4 Project monitoring and controlling systems
5 Project completion

In some instances, projects do not follow a structured planning and/or monitoring stage. For example, some projects will go through steps 2, 3, and 4 multiple times.

A typical construction project's development stages consist of preplanning, conceptual design, design development, schematic design, construction drawings, contract documents, and construction administration (all of which are valid for solar power-type projects).

Critical Chain Project Management

Critical Chain Project Management (CCPM) is a method of planning and managing projects that puts more emphasis on the resources (physical and human) needed in order to execute project tasks. A project management theory known as the Theory of Constraints (TOC) has been successfully used to establish goals that increase the rate of completion of projects. The TOC uses the first three of the five project developmental stages. In essence, all critical chains of events and constraints are given very high priority over all other activities. Furthermore, projects are planned and managed to ensure that the resources are ready when the critical chain tasks must start, subordinating all other resources to the critical chain.[1]

Regardless of project type, project planning emphasizes resource leveling, such that the longest sequence of resource-constrained tasks are identified as the critical chain. In multiproject environments such as large-scale solar power systems, resource leveling should be performed across the projects.

This chapter introduces specific project management methodologies and topics that specifically address issues unique to large-scale solar power projects.

Contractual Matters of Interest

In general, when planning for a large-scale solar power project (as mentioned previously), project managers must acquire familiarity with specific financial, economic, and technical aspects of the solar power project, which may include the following areas.

Technical Issues

- DC output size of the photovoltaic (PV) modules in kWh
- AC or PTC output of the photovoltaic modules in kWh
- Expected AC power output of the solar system in its first year of installation
- Expected lifecycle power output in kWh DC
- Expected lifecycle power output in kWh AC
- Guaranteed minimum annual power output performance in kWh AC

Financial Issues

- Terms of contractual agreement
- Penalty or compensation for performance failure
- Price structure at the end of the contract with client paying 0 percent of the cost
- Price structure at the end of the contract with client paying 50 percent of the cost
- Price structure at the end of the contract with client paying 100 percent of the cost
- Expected average yearly performance during lifecycle of the contract
- Expected mean yearly performance degradation during lifecycle of the contract
- Assumed PPA price per kWh of electrical energy
- Initial cost of power purchase agreement
- PPA yearly escalation cost as a percentage of the initial energy rate
- Net present value over 25 years
- Proposed cost reduction measures
- Net present value of reduction measures
- Annual inflation rate
- Projected annual electricity cost escalation
- First year avoided energy cost savings
- Total lifecycle energy saved in kWh
- Total lifecycle energy PPA payment
- Cost of PPA buyout at the end of lifecycle
- PPA expenses
- Total lifecycle pre-tax savings
- Total project completion time in months
- Customer training
- Insurance rating

All of the listed considerations are fundamental issues with which project managers must familiarize themselves at the outset of each program. Therefore, when starting a solar power project, it is imperative to engage a project manager throughout the entire lifecycle of a solar power project from inception to final project commissioning.

Technical Matters of Interest

In addition to project management skills, to achieve project objectives managers must fully understand the technical aspects of a solar power system, which may include the following components and technologies:

- PV module manufacturer and type
- PV module technology
- PV module efficiency rating

- PV module DC watts
- PV module PTC watts as listed under CEC equipment and product qualification listing
- Total PV module count
- Percent yearly solar power output degradation
- PV module warrantee in years after formal test acceptance and commissioning
- Inverter make and model as listed under CEC equipment and product qualification
- Inverter kilowatt rating
- Number of inverters used
- Inverter performance efficiency
- Inverter basic and extended warrantees
- Solar power tracking system (if used)
- Tracking system tilt angle in degrees east and west
- Number of solar power tracker assemblies
- Kilowatts of PV modules per tracker
- Ground or pedestal area requirement per 100 kW of tracker; for large solar power farms, tracker footprint must be accounted for in acres per megawatt of land required
- Tracker or support pedestal ground penetration requirements
- Tracker above-ground footing height
- Tracker below-ground footing height
- Wind shear withstand capability in miles per hour
- Environmental impact during and after system installation (if applicable)
- Lightning protection scheme
- Electrical power conversion and transformation scheme and equipment platform requirements
- Equipment mounting platforms
- Underground or aboveground DC or AC conduit installations
- PV module washing options, such as permanent water pressure bibs, automatic sprinklers, or mobile pressure washers
- Service options and maintenance during lifecycle of the PPA

Contractor Experience

In order to succeed in large-scale solar power system design and construction, solar power contractors must have the following engineering and project management skills:

- The contractor's staff must have substantial experience in PV and low- and medium-power engineering design.
- Contractors must preferably have their own installation crew. Subcontracted crews must be fully trained to ensure conformance with the contractor's construction methodology.

■ Contractors must have at least 3 to 5 years of construction experience in all types of large-scale solar power system installation.

■ Contractors must provide letters of recommendation and references of their previous constructions, which may include typical design documentation, shop drawings, test and integration procedures, customer training curriculums, and construction methodology.

Contractors must also be required to provide information about their permanent key personnel, personnel resumes, and their experience and roles in the project.

■ In the event of PPA-type contracts, the prime contractor must provide roster of at least three successful contracts and references.

■ Each contractor must disclose location of their principle management, engineering design team, and installation and maintenance depots.

■ Contractors must provide PV modules and solar power equipment PPAs that they have with major national and international manufacturers or suppliers.

■ Contractors must be required to disclose their financial standing and must be financially viable to obtain bonds, procure material, and maintain payroll.

Power Purchase Agreement (PPA) Contracts

Unlike conventional capital-intensive projects, PPA contracts completely bypass proven engineering design measures, which involve project feasibility studies, preliminary design and econometric analysis, design documentation, construction documentation, design specification, and procurement evaluations (which are based on job-specific criteria).

In order to ensure a measure of control and conformance to project needs, it is recommended that clients refer to an experienced consulting engineer or legal consultant who is familiar with PPA-type projects.

In order to avoid or minimize unexpected negative consequences associated with PPAs, clients are advised to incorporate the following documents, reports, and studies in their request for proposal (RFP):

■ For ground-mount installations, employ the services of qualified engineering consultants to prepare a negative environmental impact report, site grading, drainage report, and soil study.

■ Provide statistical power consumption and peak power analysis of present and future electrical demand loads.

■ Provide a set of electrical plans, including single-line diagrams, main service switchgear, and power demand calculations.

■ Conduct an energy audit.

■ Provide detailed data about site topology and present and future land use.

■ Provide data regarding local climatic conditions, such as wind, sand, or dust accumulation conditions and cyclic flooding conditions, if applicable.

■ For roof-mount systems, provide aerial photographs of roof plans which show mechanical equipment, air vents, and roof hatches. Drawings must also accompany architectural drawings that show parapet heights and objects that could cause shading.

■ Specifications should outline current electrical tariff agreements.

■ The document should also incorporate any and all special covenants, conditions, and restrictions associated with the leased property.

■ To ensure system hardware reliability at the posterity of the contract, the RFP must include a generic outline of hardware and data acquisition and monitoring software requirements.

■ The specifications must request provider to disclose all issues that may cause noncompliance.

■ Expected power output performance guarantees, as well projected annual power generation requirements must be delineated.

Clients are also advised to conduct preliminary renewable energy production studies that would enable them to evaluate energy production potential, as well as an economic analysis of possible alternatives.

POWER PURCHASE AGREEMENT CONTRACT STRUCTURE FOR SOLAR POWER SYSTEMS

In order to prepare a PPA request for a proposal document, the owner's legal counsel and management personnel must familiarize themselves with various elements of the contract agreement. Agreements involving third-party ownership consist of two parts: legal and technical. The following are some of the most significant points of PPA-type contracts that third-party purchase providers must respond to and evaluate accordingly.

Long-term financing agreements such as PPAs, as discussed in Chap. 4, are inherently complicated. As such, they demand extensive due diligence by the owner/client, legal counsel, and consulting engineers. In order to successfully execute a PPA, the owner must fully appreciate the importance of the collective effort of experts and the collaborative effort required in preparing specifications and requests for proposal documents.

Even though the initial investment in a solar power cogeneration system requires a large capital investment, the long-term financial and ecological advantages are so significant that their deployment in the existing project should be given special consideration. A solar power cogeneration system (if planned and executed as per recommendations) could be an excellent investment that could provide considerable energy savings over the lifecycle of the project and an excellent hedge against unavoidable energy cost escalation.

PROPOSAL EVALUATION

As mentioned, long-term financing agreements, such as PPAs are inherently complicated and demand extensive due diligence by the owner, legal counsel, and consulting engineers. In view of the above PPA, owners must fully appreciate the importance of close collaboration required in preparation of contract documents. Prior knowledge of specification and evaluation points (outlined above), when exercised properly, would prepare owners to evaluate comparative value advantages amongst PPA providers.

In view of the depletion of existing rebate funds, solar power systems designed in various states within the United States should review rebate programs at the outset of a project and apply for rebate funding at the earliest possible time. Furthermore, because of the design integration of the solar power system with the service grid, the decision to proceed with the program must be made at the commencement of the construction design document stage.

Project Completion Schedule

The following is a solar power project progress milestone and construction schedule for a typical 1 MW solar power installation.

1 Solar project site evaluation and feasibility study. Engineering and technical support time may require approximately 2 to 4 weeks. This engineering effort is mandatory for all types of solar power projects. The feasibility report essentially incorporates projected solar power generation potential and econometrics associated with the project. The study provides counts of all essential components such as PV modules, inverters, support structure counts, etc. In this phase, engineers assist the client to complete rebate application.

2 Detailed design documentation. Based on the results of the feasibility study information, solar power design engineering is initiated. At this stage, a detailed design documentation and construction specification is developed, which will enable solar power system integrators to provide costing based on a unified set of documents. The approximate time required to complete the engineering design, based on a 1-MW solar system, is about 14 to 16 weeks.

3 Project contract award and evaluation and negotiation. This process may require approximately 4 weeks.

4 Construction shop drawing preparation. This task may require 12 to 16 weeks.

5 Site preparation and grading. The process from commencement to the end may require approximately 8 to 12 weeks.

6 Solar power system foundation works. Time required to complete the work, depending on type of solar power system foundation and infrastructure works, may require approximately 12 to 20 weeks.

7 Solar power system installation period after site preparation. Assuming the contractor has experience in large-scale solar power system installations, this process may require approximately 12 to 16 weeks.

8 System test and commissioning. A system test must be an ongoing process that must be conducted during the period of solar power system integration. System testing is one of the most important responsibilities of any competent solar power integrator, and it must be carefully monitored by the project managers. The approximate time required is 12 to 16 weeks, which corresponds to system installation time.

9 Final test and acceptance. This is a task that must preferably be conducted at the end of solar power system integration by an independent expert. The final acceptance test must be conducted with the specifics as outlined in Chap. 7.

Solar Power System Expenditure Profile

The following expenditure profile is typical for solar power system projects and may apply to all types of solar power systems. In general, the order of expenditure progress shown below is correlated to the project development schedule:

1 Engineering design consultation payment schedule
- Feasibility study—Represents 10 to 15 percent of the total engineering fee.
- 50 percent design document completion—Represents 20 to 25 percent of the engineering fee.
- 100 percent design documentation completion—Represents 40 percent or the balance of the project.

2 Construction contract payment schedule
- Upon the construction's commencement, the contractors require 20 to 30 percent of the contract amount as a down payment for ordering long-term deliverables such as PV modules, inverters, and infrastructure support hardware.
- Upon the completion and acceptance of shop drawing by the engineer of record, the contractor requests 10 to 20 percent of the contract amount for procurement of electrical hardware material used in solar power construction.
- After delivery of long-term deliverables, the contractor will request 20 to 30 percent of the contract amount to pay off the suppliers and installation crews working on the project.
- After successfully completing the system test and acceptance, the contractor is paid the balance of the project amount, which is 20 percent of the contractual amount.

3 The cost of data acquisition and maintenance expenses are dispensed on a yearly basis and are generally paid by the owners/clients.

California Energy Commission and the Rebate Program

In view of the depletion of existing California rebate funds (CEC), it is recommended that applications for the rebate program be initiated at the earliest possible time. Furthermore, because of the design integration of the solar power system with the service grid, the decision to proceed with the program must be made at the commencement of the construction design document stage.

CALIFORNIA ASSEMBLY BILL 32

The following is a summary of the California assembly bill AB 32, a complete text of which can be accessed at http://www.environmentcalifornia.org/html/AB32-finalbill. pdf. The principle intent of the legislated act addresses health and safety codes, relating to air pollution. Under this act, also known as The Global Warming Solutions Act of 2006, the law mandates the State Air Resources Board, the State Energy Resources Conservation and Development Commission (Energy Commission), and the California Climate Action Registry to assume responsibilities with regard to the control of emissions of greenhouse gases. The Secretary for Environmental Protection is also mandated to coordinate emission reductions of greenhouse gases and climate change activity in state government.

To implement the bill, the state board is required to adopt regulations to require the reporting and verification of state-wide greenhouse gas emissions and to monitor and enforce compliance with this program, as specified. The bill requires the state board to adopt a state-wide greenhouse gas emissions limit equivalent to the state-wide greenhouse gas emissions levels in 1990 to be achieved by 2020. The bill would require the state board to adopt rules and regulations in an open public process to achieve the maximum technologically feasible and cost-effective greenhouse gas emission reductions. The bill also authorizes the state board to adopt market-based compliance mechanisms. Additionally, it requires the state board to monitor compliance with and enforce any rule, regulation, order, emission limitation, emissions reduction measure, or market-based compliance mechanism adopted by the state board, pursuant to specified provisions of existing law. The bill authorizes the state board to adopt a schedule of fees to be paid by regulated sources of greenhouse gas emissions, as specified.

The California Constitution requires the state to reimburse local agencies and school districts for certain costs mandated by the state. Statutory provisions also establish procedures for making that reimbursement.

Impact of California Assembly Bill 32   California currently produces 50 percent of its electrical energy mainly through coal, natural gas turbines, and nuclear power-generating stations. Hydroelectric and nuclear energy power represent a small percentage of state local electrical power. The state of California, therefore, imports a significant balance of its electrical energy from outside energy providers; this energy

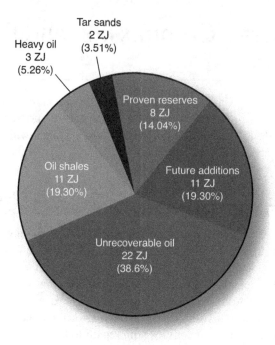

Figure 9.1   Worldwide crude oil reserves. *Graphic courtesy of DOE.*

is produced by coal and gas-fired turbines, hydroelectric power, and nuclear power–generating stations.

By mandating reduction of green gas production, California will within the near future abstain from the purchase and import of electrical energy from sources that use coal-based electrical turbines. Most significantly, the entire coal-based electrical power generating stations within California will be mandated to use less-polluting natural gas turbines.

With reference to Fig. 9.1, the cost of electric energy production historically has been going up in an accelerated rate. As indicated, the average annual cost of electric energy in California has escalated at an average of 4.18 percent. However, because of the enactment of AB 32 and other factors, it is expected that in the near future the rate of electrical energy cost will increase at higher rate.

ELECTRICAL ENERGY COST ESCALATION

Factors that will affect the electric energy cost escalation in California, soon to be followed by other states, are as follows:

■ In the past five years, the cost of natural gas has gone up by 13 percent. Natural gas production has been decreased over last decade.
■ No new natural gas refineries are being built in the United States.

■ Within three years, all electrical power-generating utility companies in the state of California using coal-fired turbines will be converted to natural gas.
■ The demand on natural gas within the near future will cause the prices to increase by 20 to 25 percent, which will inevitably be passed on to the consumers.
■ Currently, only 7 percent of electrical power produced in California is generated by use of natural gas; this percentage is expected to increase to 50 to 60 percent within the near future.
■ Since the cost of natural gas electrical energy production is higher than hydroelectric power, because of market forces, the cost of electrical power will most likely be level with the higher utility rates.

In addition to the listed risk factors (which may also have an effect on energy cost escalation), there is geopolitical unrest in the Middle East and Venezuela, international terrorism, and an accelerated demand for fossil fuels by India and China, which have experienced 8 to 14 percent gross domestic product (GDP) growth in the recent past. It is therefore not unreasonable to predict that for the foreseeable future annual electrical cost increases would range between 6 to 12 percent. In view of the above energy cost escalation, the initial capital investment required by solar power programs appears to be fully justified. Figure 9.1 is graphic of the worldwide crude oil reserves. Figure 9.2 is a breakdown of energy use by type. Figure 9.3 is a graphic representation of worldwide hydroelectric energy production in terrawatt hours, and Fig. 9.4 is the U.S. natural gas consumption. Figure 9.5 is the 2009 U.S. electricity generation source data. Figure 9.6 is graphic presentation of world energy resources, and Fig. 9.7 is a graphic of the world renewable energy use at the end of 2009.

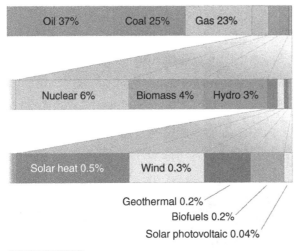

Figure 9.2   **Breakdown of energy use.**

Country ⋈	Annual Hydroelectric Energy Production (TWh)	Installed Capacity (GW)	Capacity Factor	Percent of All Electricity
China (2009)	652.05	196.79	0.37	22.25
Canada	369.5	88.974	0.59	61.12
Brazil	363.8	69.080	0.56	85.56
United States	250.6	79.511	0.42	5.74
Russia	167.0	45.000	0.42	17.64
Norway	140.5	27.528	0.49	98.25
India	115.6	33.600	0.43	15.80
Venezuela	86.8	-	-	67.17
Japan	69.2	27.229	0.37	7.21
Sweden	65.5	16.209	0.46	44.34
Paraguay (2006)	64.0	-	-	-
France	63.4	25.335	0.25	11.23

Figure 9.3   Hydroelectric energy production in terrawatt hours.

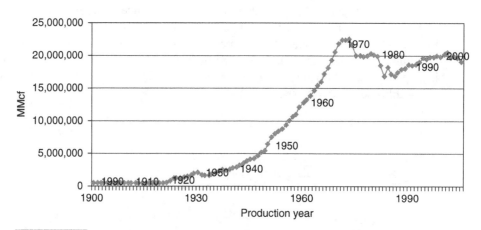

Figure 9.4   U.S. natural gas consumption. *Graphic courtesy of DOE.*

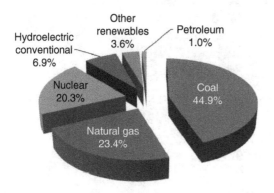

Figure 9.5   2009 U.S. electricity generation sources. *Graphic courtesy of DOE.*

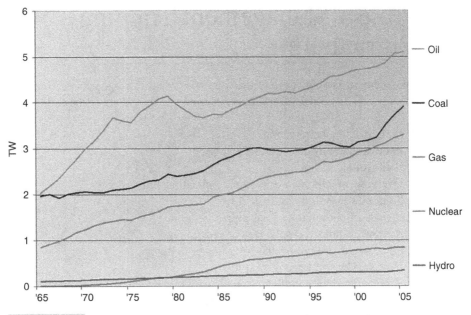

Figure 9.6   **World energy resources.** *Graphic courtesy of DOE.*

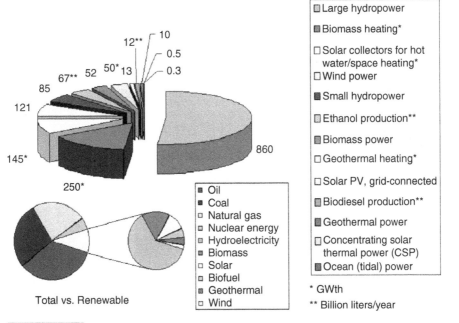

Figure 9.7   **World renewable energy use at the end of 2008.** *Graphic courtesy of DOE.*

Project Management Training Curriculum

The following is a condensed curriculum for project management personnel which could be staged in 2 to 3 days. The course curriculum should cover the following topics:

- Photovoltaic phenomenon
- Photovoltaic physics
- Photovoltaic technologies
- Overview of solar power system application and systems
- Solar power system construction
- Review of commercial solar power systems
- Review of industrial solar power systems
- Large-scale grid connected solar power systems
- Field survey and feasibility study
- Service provider coordination and tariff studies
- Rebate and local tax exemption review
- Rebate application process
- Electrical power switchgear and metering system load burden analysis
- Geoclimatic system analysis
- System installation platform analysis
- Solar power system tests and integration procedures
- Acceptance test and commissioning procedures and standards
- Solar power costing and economics overview
- Energy cost escalation
- Equipment and labor costing
- Construction coordination
- Power Purchase Agreement (PPA) financing

Reference

1. http://en.wikipedia.org/wiki/Project_management

10

SMART-GRID SYSTEMS

Introduction

In view of the accelerated proliferation of large-scale solar and wind power installations worldwide, existing electrical power transmission lines and grids can no longer sustain the extended burden of additional power transmission capacity. In addition, existing grid networks lack intelligence to regulate and manage dynamic supply-and-demand loads essential for solar and wind energy power generation systems' interconnection.

The principal objective of smart-grid systems is to deliver electricity from various sources of supplies, such as electrical power-generating stations, and geothermal, wind, and solar power farms, to consumers. These supplies use two-way digital technologies to control end-user loads such as appliances at consumers' homes to save energy, reduce cost, and increase reliability and transparency. In essence, smart-grid systems overlay electrical distribution grids with an information and net-metering system. Currently, such grid modernizations are being promoted worldwide as a means for addressing energy independence, global warming, and national security.

Smart-grid systems also include intelligent monitoring systems that keep track of all electricity flowing in the system. Additionally, smart-grid systems will incorporate use of innovative superconductive transmission lines that can conduct significantly larger currents with minimal power losses. In essence, smart grid systems are essential for integrating a wide network of future renewable electricity systems, such as solar and wind. Smart-grid end-user power consumption management is achieved through selective control of home appliances, such as washing machines, or factory processes that can run at low-peak energy-demand hours. This results in reduced energy consumption at peak energy hours. Figure 10.1 is diagram of an electrical power generation and distribution grid system.

In principle, the smart grid is an upgrade of twentieth-century power grids. These grids broadcast power from central power-generating stations to a large number of users. However, the new system will be capable of routing power in more optimal ways to respond to a wide range of conditions, and will also be capable of regulating the grid

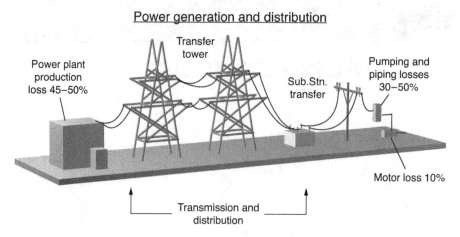

Figure 10.1   **Electrical power generation and distribution grid system.**

peak power demand through imposition of high-premium utility rates to customers that use energy at peak hours.

United States Unified Smart Grid

The Unified National Smart Grid is a proposal for a U.S.-wide area grid that is a national interconnected network relying on a high-capacity backbone of electric power transmission lines. These lines link all the nation's local electrical networks that have been upgraded to smart grids. Europe's analogous project is sometimes referred to as the SuperSmart Grid, a term that also appears in the literature describing the Unified Smart Grid. President Barack Obama asked the United States Congress to "act without delay" to pass legislation that included doubling renewable-energy production in the near future and building a new electricity "smart grid."[1]

The International Smart-Grid Systems

The concept of an intelligent grid system was explored in the early 1950s in Europe and was intended to unify and synchronize the European Continental Grid system serving 24 countries (spanning 13 time zones), which would unify the European grid with that of Russia, the Ukraine, and other countries of the former Soviet Union. In view of the immense scale of the project and its associated problems, such as network complexity, transmission congestion, and the need for rapid diagnostic, coordination, and control systems, the program has yet to materialize. However, advocates of the smart-grid schemes are convinced that such a major technological upgrade is essential for establishing a transcontinental mega-grid system.

Prospect of Smart-Grid Systems and Special Features

Proposed future smart transmission systems in the United States will most probably be configured as high-voltage capacity transmission, which will be capable of transmitting direct current at 800-KV potential. Such transmission networks would link all local electric utilities and power-generation facilities throughout the country.

Existing long-distance interconnections in the United States and North America include 1400 KM of transmission lines that intertie Los Angeles and the Pacific Northwest. The Pacific Intertie carries up to 3.1 GW on two 500-KV overhead lines. Another high-voltage DC transmission system spanning 1200 kilometers interconnects Québec, Canada, to New England, which has 2 GW of capacity.

The proposed United States Unified Smart Grid is not intended to be a collection of point-to-point interconnections between regional systems with some communications intelligence. Rather, the current conceptual topology incorporates many grid-node access points that would allow formation of virtual power generation clusters, which will consist of local electric-utility providers, solar and wind farms, or grid-energy storage facilities.

SuperSmart Grid

Currently, the SuperSmart grid system is a conceptual plan intended to be a grid system capable of delivering inexpensive, high-capacity, low-loss transmission system that would interconnect producers and consumers of electricity across vast distances. Smart-grid capabilities use the local grid's transmission and distribution network to coordinate distributed generation, grid storage, and consumption into a cluster that appears to the super grid as a virtual power plant.

The wide-area SuperSmart grid system involves two concepts: a centralized smart-grid control system and a small-scale, local and decentralized smart-grid system. These two approaches are often perceived as being mutually exclusive alternatives. The SuperSmart grid system, in essence, is intended to reconcile the two approaches and considers them complementary and necessary to realize a transition toward a fully decarbonized electricity system. It should be noted that a SuperSmart Grid system only refers to a network superimposed on local grid networks and should not be confused with a SuperGrid system, which refers to evolving technology in an electricity distribution system.

GRID POWER CONTROL AND SURPLUS POWER MANAGEMENT

The amount of data required to monitor and switch appliances off automatically is very small compared to residential or commercial voice, security, Internet, and TV system services. Most smart-grid bandwidth upgrades are already supported by consumer communication services. However, since government power and communications companies are generally separate commercial enterprises in North America and Europe, smart-grid system implementation will require considerable government aid and large vendors to facilitate and encourage various enterprises to coordinate specific communication methodologies. For instance, a large communication enterprise, such as Cisco, will be in position to provide specific smart-grid communication and control devices to consumers currently offered by Silver Spring Networks or Google (enterprises that, in principle are data integrators rather than vendors of equipment).

SCOPE OF SMARTGRID SYSTEM

As referenced earlier, smart-grid systems would be the intelligent interconnecting backbone of the electrical power distribution system that provides layers of coordination above the local grids. Regardless of development of specific terminologies, smart grid project objectives are to allow continental and national interconnections that prevent a local or regional grid failure to cause local smart grid shut-down. Therefore, all power distribution networks with the smart grid would have the capability to function independently and ration whatever power is available to critical needs.

Types of Electrical Power Grids

MUNICIPAL GRIDS

Municipalities, whether generating or purchasing electrical power, have the primary responsibility and a legal mandate to control power distribution during emergencies. They must also frequently ration power to ensure the distribution of power to critical clients such as hospitals, fire stations, and emergency shelters during power outages.

In fact, most large municipalities in the United States have actively taken a lead in enforcing integration standards for smart-grid metering. In recent years, a number of municipalities generating electricity have installed fiber-optic communication networks and power control transit exchange mechanisms that will all smooth integration with future smart-grid systems.

RESIDENTIAL NETWORKING SYSTEM

A residential or home-grid network consists of electrical and electronic hardware communication and control devices that use an electrical power distribution power line to establish communication with equipment and appliances within a house. Currently, smart-grid network communication is established through the use of radio frequency (RF) standards by a number of organizations such as Zigbee, INSTEON, Zwave, the Wi-Fi Alliance, and others. The most prevalent smart-grid communication standard developed by the National Institute of Standards and Technology (NIST) is promoting interoperability between the different standards. Another communication standard developed by OSHAN,[2] enables device interoperability at home.

In general, communication standards developed for smart-power grids and home-area networks support more bandwidth than is required for power control; as such they may impose a residual utility cost increase.

Smart-grid communication will use the existing 802.11 home networking system, which has wide-bandwidth multi-megabits that accommodate a wide variety of communication services used by security, fire, and medical and environmental alarm systems, closed circuit television (CCTV), local area networks (LAN), and cable TV networks.

Currently, consumer electronics devices consume over half the power in a typical U.S. home. Consequently, the ability to shut down or to enforce standby power-mode devices during peak power hours could result in substantial curtailment of energy use. However, the main issue of concern is that electric service providers could at their discretion decide to turn power off without prior notice.

Home devices that could assist utilities' efforts to shed load during times of peak demand include air conditioning units, electric water heaters, pool pumps, and other high-wattage devices. In the near future, smart-grid companies will represent one of the biggest and fastest-growing sectors in the clean technology markets. In the past, smart-grid technologies have received a substantial portion of the venture capital investments.

Principle Function and Architecture of Smart-Grid Systems

There are currently multiple networks, power generation companies, and power distribution operational centers employing varying levels of manual communication and control protocols. Smart grids, on the other hand, increase power transmission and distribution connectivity by means of automation. This promotes coordination among electrical power providers, consumers, and networks that perform local or long-distance power transmission tasks.

In general, power transmission networks, which operate from 345 kV to 800 kV over AC and DC lines transport bulk electricity over long distances. On the other hand, local networks move power in one direction, distributing the bulk power to consumers and businesses via transmission lines that operate at less than 132 kV.

In the past, residential and commercial solar power systems, as well as wind turbines, generated an energy surplus which they sold back to utilities. To modernize the existing power generation and distribution system, it is necessary to incorporate real-time power-flow management that would enable the bidirectional metering needed to compensate local producers of power. Even though transmission networks controlled in real time already exist, many U.S. states and European countries operate under antiquated standards, which would be incompatible to integration with smart-grid networking systems.

Smart-grid systems, as discussed, are modernizations of both the transmission and distribution grids. The main goal of grid modernization is to facilitate greater competition between providers, enabling greater use of variable energy sources by establishing automation and monitoring capabilities. These would allow bulk electrical power transmission across continental distances, which would force markets to enact energy conservation.

The smart grid will enable energy suppliers to charge variable electric rates and tariffs that will reflect differences in the cost of generating electricity during peak or off-peak periods. Such measures will allow load-control switches to make large energy-consuming devices, such as air conditioning systems and hot water heaters, operate at low-peak energy cycles.

PEAK POWER CONSUMPTION CONTROL

In order to reduce electrical power demand during the high-cost peak-usage periods, which mostly occur from 12:00 a.m. to 5:00 p.m. in summertime, smart-grid communications and metering technologies inform smart residential or industrial devices as to when energy demand is high and track how much electricity is used by the device. In order to motivate clients to reduce or cut back energy use during peak hours, also referred to as curtailment or peak power, prices of electricity will be considerably increased during high-demand periods and decreased during low-demand periods. As a result, consumers and businesses will be motivated to consume less during high-demand periods.

Impact of Smart Grid on Renewable Energy Production

As discussed in earlier chapters, renewable energy sources, such as solar- or wind-power generating systems (produced through natural environmental phenomenon) produce intermittent electrical energy that is not fully compatible with current electrical grid systems. Consequently, clients who plan to use intermittent renewable energy power produced by solar power systems must have power consumption schemes that can automatically control their electrical loads. They would do this by arming and disarming synchronously with power outputs of solar PV systems, output power which is constantly affected by phenomena. By setting lower electrical utility tariffs for peak solar power output periods and higher tariffs for conventional electrical power, consumers will be encouraged to schedule their power consumption. The main drawback of such tariff variations is that they may create unpredictable energy cost control that could be subject to climatic and environmental conditions.

Synchronized Grid Interconnections

Forthcoming plans to synchronize and interconnect the North American Grid, also referred to as the Wide Area Synchronous Grid, will integrate regional scale networks of electrical power transmission systems that will operate at a synchronized frequency.

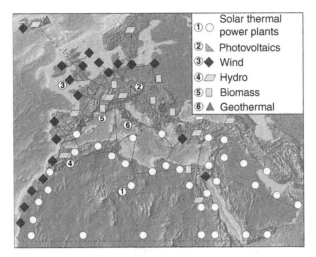

Solar thermal
① ○ power plants
② ◣ Photovoltaics
③ ◆ Wind
④ ▱ Hydro
⑤ ▢ Biomass
⑥ ▲ Geothermal

Figure 10.2   Smart-grid system, which is planned to interconnect Europe and African renewable electrical energy production centers.

Currently, such synchronized zones interconnect the Continental Europe (ENTSO-E) grid with 603 GW of electrical power generation. The widest segment of the synchronized grid, referred to as the IPS/UPS system, serves most countries of the former Soviet Union. In 2008, the ENTSO-E grid transacted over 350,000 MWh energy per day on the European Energy Exchange (EEX). Figure 10.2 depicts the European smart grid system (SuperSmart Grid), which is planned to interconnect European and African renewable electrical energy production centers.

Some of the interconnects in North America are synchronized at an average of 60 Hz, whereas those in Europe run at 50 Hz. Interconnections can either be tied to each other via high-voltage DC power transmission lines referred to as DC ties, or with variable frequency transformers (VFTs), which permit a controlled flow of energy while also functionally isolating independent AC frequencies of each side (60 or 50 Hz).

Significant benefits of synchronous zones include the collectivization or pooling of electrical power generation, which results in lower energy costs; the lowering of transmission line load burden; power distribution equalization; common provisioning of reserves; avoidance of load disturbances; and development of new energy markets.

Western American Interconnection

The Western American Interconnection is currently one of the two major AC power grids in North America. The other wide-area synchronous grid is the Eastern Interconnection. Currently, there are also three minor interconnections, the Hydro Québec (Canada) Interconnection, the Texas Interconnection, and the Alaska Interconnection.

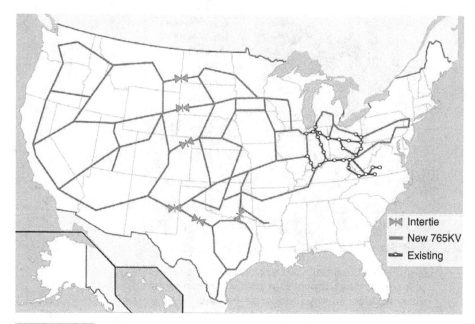

Figure 10.3   United States smart-grid network system.

The Western Interconnection currently extends from Western Canada through the Rockies, the Great Plains of the Western United States to Baja California in Mexico. All of the above-grid interconnections operate at a synchronized frequency of 60 Hz. The interconnections are tied to each either by high-voltage DC current, or with VFTs, which permit functional isolation of independent AC frequencies for each zone. The Western Interconnection is also coupled to the Eastern Interconnection with six DC ties.

All of the electric utilities in the Eastern Interconnection are electrically tied together during normal system conditions, and they operate at a synchronized frequency at an average of 60 Hz. The Eastern Interconnection reaches from Central Canada eastward to the Atlantic Coast, south to Florida, and back west to the foot of the Rockies (the intertie exclude both Québec and most of Texas). Figure 10.3 is the map of the United States smart-grid network system.

Smart-Grid Advanced Services and Devices

In order to allow customer load grid connection and control, industries involved in smart-grid technology will be required to develop a variety of communications network protocols, wireless communication actuators, advanced energy consumption and generation sensors, and distributed computing technology. This technology would have to provide efficient, reliable, and safe power delivery and use. In the near future, smart grid technologies will open up great prospects for new services and instrumentation, such as fire alarm monitoring and power control management systems that will allow power shut-down and load shedding. Figure 10.4 is a map of U.S. and Canada regional transmission organizations.

LOAD CONTROL SWITCHES

A load control switch is a remotely controlled relay that is placed adjacent to home appliances that consume large amounts of electricity, such as air conditioner units and electric water heaters. A load control switch consists of a communication module and a relay switch, which are used to turn on and off appliances that enable smart-grid systems

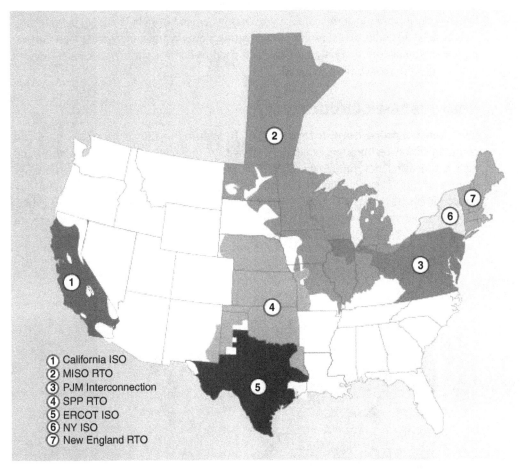

① California ISO
② MISO RTO
③ PJM Interconnection
④ SPP RTO
⑤ ERCOT ISO
⑥ NY ISO
⑦ New England RTO

Figure 10.4   **U.S. and Canada regional transmission organizations.** *Graphic courtesy of the Department of Energy.*

to improve energy consumption efficiency. Such a switch operates in a manner similar to telephone pagers. It allows the systems to receive command signals from power companies to turn off appliance power during peak electrical demand hours. In general, such devices also incorporate a timer that allows automatic resetting of the load control switch back to a normal state after a pre-set time.

Most load control switches are designed to have only one-way communication with a power company. However, some advanced load control switches also have the capability for two-way communication, which allows power companies to locate faulty load control switches. Most power companies in the United States offer special incentives and free load control switch installations to their customers. Load control switches are excellent devices that could prevent power brown-outs and blackouts, which would enable electrical power service companies to respond to emergencies and that would considerably minimize and avoid shutting off all power to their customers.

Another important feature of load control switch deployment is that it will permit power companies to reduce costs for electric tariffs by reducing the amount of expensive electricity that they must purchase during peak power energy demand from bulk energy providers.

Figure 10.5*a* and *b* are examples of residential energy management and smart meters. Figure 10.6 is a commercial control switch used to control various electrical devices that are used to reduce peak demand.

GRID SYSTEM CONDUCTORS

High-temperature conductors (HTC) In addition to the smart-grid control and communication technologies discussed above, to increase power transportation and distribution efficiency, modern transmission lines will be required to carry substantially larger currents. These currents will surpass the physical characteristics of present-day conventional transmission cables. To carry larger currents, transmission line cables must be replaced by conductors, which would be capable of withstanding very high temperatures. They will also have to be strong enough to span long crossings.

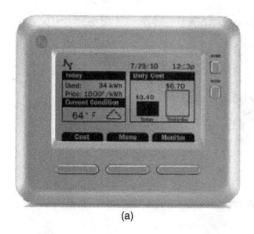

(a)

Figure 10.5 (*a*) A residential energy management system.

(b)

Figure 10.5   (*b*) **A smart meter system.** *Photo courtesy of Meterus.*

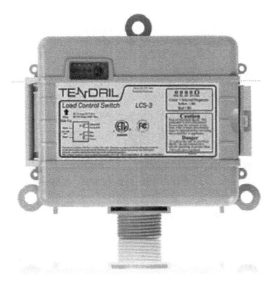

Figure 10.6   **A commercial load control switch.** *Photo courtesy of Tendril.*

Currently, three types of conductors are designed to withstand high temperatures: aluminum conductor composite reinforced (ACCR), aluminum conductor composite core (ACCC^{Tu}), and aluminum conductor steel supported (ACSS). The significant benefit of these conductors is that they can replace existing overhead power-line cables without the need for replacing existing transmission-line towers and structures. The temperature ratings of these conductors are quite high, so maintenance crews must take extreme care when repairing them.

Aluminum Conductor Composite Reinforced (ACCR)   ACCR conductors are composed of heat-resistant aluminum-zirconium alloy outer strands and aluminum oxide matrix core strands. The core of the ACCR is composed of a stranded-fiber reinforced metal matrix, constructed from an aluminum oxide fiber embedded in high-purity aluminum. The outer strands are either round or trapezoidal in shape. In addition to strength, the cables can withstand extremely high temperatures without softening or losing strength. Moreover, the cable has less thermal expansion than steel and retains its strength at high temperatures. ACCR conductors use similar stranding as conventional aluminum conductor steel-reinforced ACSR cables. The main features of ACCR cables are their lightweight core and their resistance to heat, which permits higher electrical conductivity and lower thermal expansion. This results in less sag, higher operating temperatures, and conductor ampacity. ACCR conductors and hardware are rated for 210°C continuous operating temperatures.

Aluminum Conductor Composite Core (ACCC)   Aluminum conductor composite core conductors are composed of trapezoidal aluminum stranded wires, which are wrapped around the composite core. The core of the ACCC conductor is a solid with no voids and is constructed from a carbon-glass-fiber polymer-matrix material core. This solid polymer-matrix core, in turn, is composed of carbon fibers surrounded by an outer shell of boron-free glass fibers that insulate the carbon from the aluminum conductor. The trapezoidal aluminum wires are fully annealed, which makes them softer compared to the hardened aluminum wires used in conventional transmission conductors. The aluminum strands, even though soft, are tempered by the composite core, which allows the conductor to convey considerably more current at high temperatures with minimal expansion. Similar to ACCR cables, ACCC conductors are rated for 180°C continuous operating temperatures. The main disadvantage of ACCC cables is that the softer temper of the aluminum wires makes them more susceptible to damage from improper installation and handling.

Aluminum Conductor Steel-Supported (ACSS)   Aluminum conductor steel-supported (ACSS) conductors are another type of high-temperature conductor; they are constructed with round or trapezoidal aluminum strands. ACSS conductors have similar operational characteristics to ACSR; however, the aluminum strands in ACSS are fully annealed and strengthened by steel outer wires that minimize sag characteristics. ACSS conductors are rated for 250°C continuous operating temperatures.

Similar to ACCC cables, because of the softer temper of the aluminum wires, the outer wire shells are more susceptible to damage from improper installation and handling.

Smart Grid and Mesh Networking

The principle backbone of all smart-grid systems is interwoven communication networks, referred to as mesh networks, which allow instantaneous global communication among all power generation, distribution, and energy consumption centers. Without mesh networking, it would be impossible to establish synchronized electrical power interconnections.

Mesh networking is a type of networking in which each node in the network, such as a power-generating center, may act as an independent communication router regardless of whether it is connected to another network or not. Such communication systems allow for continuous connections and reconfigurations around failed or congested paths by bypassing or hopping from node to node until information reaches the destination.

A mesh communication network in which all nodes are connected to one another is referred to as a fully connected network.

Mesh networks differ from one another since the component parts forming the mesh can connect various stationary nodes to each other via multiple hops. Mesh networks can also be considered an ad-hoc communication system network. A category of mesh networks referred to as Mobile Ad Hoc Networks (MANET) is closely interconnected by mobile communication nodes. Most often, mesh networks are self-healing, which allows them to operate when one or more nodes break down or become dysfunctional. In such a communication scheme, the network there is often more than one path between a source and a destination. MANET systems generally deploy wireless and hard-wired communication systems, which are entirely controlled by intricate software programs. Figure 10.7 is depiction of a smart-grid system's mesh network layout and nodal layout.

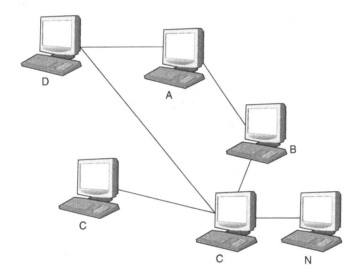

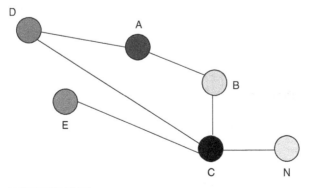

Figure 10.7   Smart grid system mesh network layout.

Meshed network architecture, using wireless communication systems, was originally developed for military applications. Today most nodal radio communication systems within each network are capable of supporting multiple functions, such as client access, backhaul services, and scanning. Moreover, considerable miniaturization of transmission and receivers have reduced mesh-networking system costs and have enabled mesh nodes to become more modular and versatile. This allows one node or device to contain multifunction communication circuitry that allows nodes to be customized to perform specific set communication frequency band control.

WIRELESS MESH NETWORKS IN THE UNITED STATES

In early 2007, an American communication company called Meraki undertook the installation of a small wireless mesh router in a rural location.[3] The wireless mesh network operates at speeds of up to 50 Mb/s. The radio within the Meraki Mini mesh network, which uses IEEE 802.11 communication standards, was later optimized for long-distance communication covering a span of 250 m. Such small communication systems using a simple single-radio mesh network will, in the near future, be used within communities that provide inexpensive multifunctional communication system infrastructures.

Recently, an MIT Media Lab project developed an inexpensive mesh-networking system on a laptop computer called the XO-1 laptop or OLPC, which is intended for underprivileged schools in developing nations. The OLPC mesh communication system also uses the IEEE 802.11's standard to create a robust and inexpensive communication infrastructure.[4] The feature of OLPS is that it establishes instantaneous connection via laptops that reduce the need for an external communication infrastructure such as the Internet, since the node can readily establish connection nodes nearby.

SMesh is another mesh communication system developed by Distributed Systems and Network Labs at Johns Hopkins University; it also uses the IEEE 802.11 communication standard and provides multihop wireless networking. A significant feature of the SMesh is that it provides a fast handoff scheme that allows mobile clients to roam within the network without interruption in connectivity, which is quite valuable for real-time applications.

Recent standards for wired communications incorporate innovative concepts that allow networking speeds of up to 1 Gb/s. These can readily be deployed in LAN using existing home wiring, power lines, phone lines, and coaxial cables.

Grid Home Network (G.hn) Communication Standard

Grid home network (G.hn) is a new standard for existing wired home networking, which is a complementary counterpart to the wireless Wi-Fi home-networking standard. G.hn standard is intended to provide extremely high data communication and operate at 1 Gb/s data rates, which will also accommodate data propagation on conventional communication media such as phone wires, coaxial cables, and carriers.[5] Additionally, devices such as televisions, set-top boxes, residential gateways, personal computers, and network-attached storage devices will also be communicating on the G.hn networks. G.hn will also facilitate integration with all home loads via a centralized appliance power control center. In the near future, G.hn systems will become a universal wired home-networking standard worldwide.

G.hn AS A UNIVERSAL SINGLE ELECTRONIC DEVICE

The main objective of the G.hn system will eventually be a single large-scale integrated semiconductor device that can be used for networking over any home-wiring system. Expected benefits of the final standard will be to lower equipment development and deployment costs for customers and service providers alike by allowing customers to self-install the G.hn system.

G.hn SECURITY

G.hn communication protocol uses an encryption algorithm with a 128-bit key length to ensure confidentiality and message integrity.

G.hn specifies point-to-point security inside a domain, which means that each transmitter and receiver pair uses a unique encryption key that is not shared by other devices in the same domain. For example, if node A transmits data to node B, node E within the same domain will not be able to eavesdrop on their communication.

G.hn also supports a concept referred to as relaying, in which one device can receive a message from one node and deliver it to another node farther away in the same domain. Relaying provides extended range for large networks. To ensure security in scenarios with relays, G.hn specifies end-to-end encryption. This means that node A can send data to node B using node N as an intermediate relay. The transmitted data is encrypted in such a way that node N cannot decrypt it or modify it. Another alternative security measure that may be used in the future could be hop-by-hop encryption, in which data can be sent from node A to N, decrypted by N. It will then be encrypted again by N, which will ultimately be transmitted to node B then decrypted by B. The drawback of a hop-by-hop scenario is that data will always be available in plain text while it's being relayed by node N, which will make the system susceptible to attack at the midway node.

HOME GRID FORUM

The Home Grid Forum is a global, nonprofit trade group that promotes the International Telecommunication Union's G.hn standardization efforts for next-generation home networking. The Home Grid Forum promotes adoption of technical standards, marketing, address certifications, and interoperability of compliant products.

Members of the Home Grid Forum include large international organizations, such as Intel, Lantiq, Panasonic, Best Buy, British Telecom, Texas Instruments, K-Micro,

Ikanos Communications, Aware, DS2, Gigle Networks, Sigma Designs, University of New Hampshire InterOperability Laboratory, IC Plus Corp., Korea Electrotechnology Research Institute (KERI), and Polaris Networks.

IP VENDORS

Currently, numerous silicon vendors, such as Infineon, Metanoia, and Intel are actively involved in the development of G.hn control and communication devices for all types of existing home-wiring systems discussed above. These will make markets easier to expand. Manufacturers such as Intel's home-networking products division and Sigma Designs are currently developing the next generation of products that will be available to consumers within the near future.

SERVICE PROVIDERS

In February 2009, AT&T (which makes extensive use of wire-line home networking systems) announced its support for a G.hn home networking standards developed by ITU-T. Most utility service providers, such as AT&T, will be major benefactors from G.hn system technologies because they could provide communication connection to any room regardless of type of wiring, which will allow customers to self-install the system.

EQUIPMENT VENDORS

In March 2009 Best Buy, the largest retailer of consumer electronics in the United States, joined the board of directors of the Home Grid Forum and expressed its support for G.hn technology as the single standard for wired home networks. They also announced that Best Buy will support the global adoption of the ITU-T's G.hn technology as the single wired standard for connecting devices together that operate with coax cable, power-line carriers, and phone lines.

Note that although Wi-Fi technology is the most popular choice for consumer home networks today, G.hn will be capable of using the technology as well. In essence, G.hn will in the near future be able to offer adequate solutions for consumers who are currently using wireless systems that communicate with stationary devices such, as television and other network equipment.

CONSUMER ELECTRONICS DEVICES

A recent trend in many types of consumer devices is connectivity. Today most consumer products include Internet connectivity, using such technologies such as Wi-Fi, Bluetooth, or Ethernet. Others, such as TVs and high-fidelity (Hi-Fi) products, which are not associated with computers, now have options that allow them to be connected to the Internet or a computer by use of home network hardware products. In the near future, G.hn will be able to provide high-speed connectivity to most consumer equipment components and products that have high-definition (HD) displays. Such consumer power appliance integration within the G.hn network will result in potential energy savings. Special consumer devices such as home electrical heaters and water boilers that use considerable amounts of electrical energy will represent significant savings to homeowners.

Smart Grid and Climate Change

In addition to intelligent power transmission and demand load control, it is estimated that smart grids could also reduce greenhouse gas emissions (GHG) by 5 to 9 percent from 2005 levels. They would also facilitate atmospheric pollution reduction to nearly one-quarter of the proposed Waxman-Markey GHG reduction targets set for 2030 through energy conservation and end-use efficiency improvements, by grid-efficiency optimization, the integration of large-scale renewable energy production distribution, and by providing vehicle electrification that would be powered by renewable energies. Figure 10.8 is graph of U.S. residential household energy use, and Figure 10.9 is bar graph of yearly electrical energy consumption by residential housing.

ENERGY CONSERVATION

The smart-grid system will enhance end-user energy conservation by providing real-time feedback on energy usage and communicating time-sensitive price information to customers. Studies have shown that reduction of 5 to 15 percent in electricity consumption can be achieved by implementing load management through feedback of energy usage. Just a 2 percent reduction in end-user energy consumption is expected to reduce about 100 million tons carbon dioxide CO_2 and GHG in the United States.

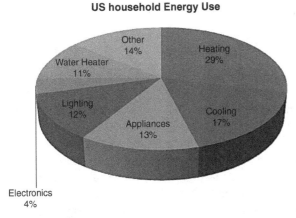

Figure 10.8   **U.S. household energy use.** *Graphic courtesy of DOE.*

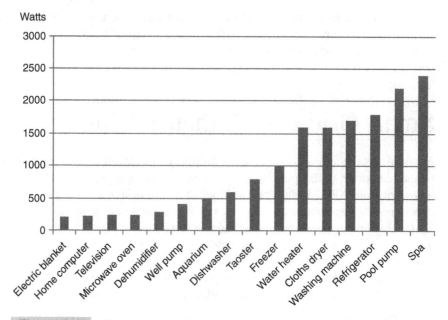

Figure 10.9   Yearly electrical energy consumption by residential housing.

GRID EFFICIENCY

Currently, up to 10 percent of the electricity generated at power plants is lost during electrical power transmission to the source. Smart-grid infrastructure can improve grid efficiency to reduce these line losses by networking distribution automation devices that minimize reactive power flow through adaptive voltage controls. A 1 percent reduction in grid losses from smart-grid networked infrastructure will translate into 30 million tons of CO_2 and GHG reductions in the United States.

RENEWABLE ENERGY INTEGRATION

Renewable energy produced by solar and wind power and other generators will substantially benefit the smart grid. As discussed previously, using smart grids will inevitably enable demand response control that will transform static demand loads into active loads that can offset intermittency associated with renewable generation. Furthermore, smart-grid networking is essential for utilities to identify and manage, and therefore minimize safety risks to line workers, which are associated with conventional decentralized power generation systems. In the long run, both utility-scale and distributed renewable energy power generation will benefit from smart-grid–networked energy storage, which consequently will render renewable energy power by storing off-peak generation for on-peak sales.

EFFECTS OF SMART GRID ON ELECTRICAL VEHICLES

Electric vehicles (EVs) present a powerful opportunity for the electric grid to reduce U.S. GHG emissions significantly by displacing internal combustion with electric power. Pacific Northwest National Laboratory estimates that EVs could reduce *total* U.S. carbon emissions by as much as 27 percent by utilizing off-peak electrical energy, which could significantly reduce imported petroleum.

In fact, smart-grid networking will be necessary for EVs to take full advantage of off-peak renewable power generation since electrical power generated by renewable energy systems would effectively eliminate carbon emissions generated by transportation. Just a 50 percent increase in electrical vehicle use is estimated to reduce U.S. CO_2 and GHG by about 100 million tons.

References

1. http://en.wikipedia.org/wiki/Unified_Smart_Grid.
2. http://en.wikipedia.org/wiki/Smart_grid - cite_note-7.
3. http://en.wikipedia.org/wiki/Mesh_network - cite_note-1.
4. http://en.wikipedia.org/wiki/Mesh_network - cite_note-5.
5. http://en.wikipedia.org/wiki/Power_line_communication.

11

SOLAR THERMAL POWER

Introduction

Greek legend claims that Archimedes used polished shields to concentrate sunlight on the invading Roman fleet and repel them from Syracuse. As early as 1866, French Augustin Mouchot used a parabolic trough to produce steam for the first solar steam engine (Fig. 11.1).

In this chapter, we review the basics principles of passive solar energy and applications. The term "passive" implies that solar power energy is harvested with direct exposure of fluids, such as water or a fluid media, which absorb the heat energy and subsequently convert the energy to steam or vapor. This steam or vapor, in turn, is used to drive turbines or provide evaporation energy in refrigerating and cooling equipment.

Solar power is the sun's energy, without which life as we know it on our planet will cease to exist. Solar energy has been known and used by mankind throughout ages. As we all know, concentrated solar rays by a magnifying glass can provide intense heat energy that can burn wood or heat water to a boiling point. Recent technological developments of this simple principle are currently being used to harness solar energy and provide an abundance of electrical power.

The simplest form of harvesting energy is accomplished by exposing fluid-filled pipes to the sun's rays. Modern technology passive solar panels used for heating water for use in pools and for general household use are constructed from a combination of magnifying glasses and fluid-filled pipes. In some instances, pipes carry special heat-absorbing fluids, such as bromide, which heat up quite rapidly. In other instances, water is heated and circulated by small pumps. In most instances, pipes are painted black and are laid on a silver-colored reflective base, which further concentrates the solar energy. Another purpose of silver backboards is to further conserve the heat energy.

Some of the concentrating solar technologies consist of parabolic troughs, concentrating linear Fresnel reflectors, Stirling dishes, and solar power towers. Each of these technologies uses various techniques to track the sun and focus light. In all of these

Figure 11.1 **Historical use of passive solar power in printing press.**

systems, a working fluid is heated by the concentrated sunlight and is then used for power generation or energy storage.

Terrestrial solar power is a predictably intermittent energy source; in other words, solar power is not available at all times, but we can predict with a good degree of accuracy when it will and will not be available. Some technologies, such as solar thermal concentrators, have an element of thermal storage, such as molten salts. These reservoirs store spare solar energy in the form of heat, which can be made available overnight or during periods when solar power is not available to produce electricity. It should be noted that orbital solar power collection systems deployed in solar power satellites being constantly exposed to solar rays are not subject to intermittent issues (because of increased intensity of sunlight above the atmosphere), and they provide higher power generation efficiency.[1] Figure 11.2 is graphic of a passive solar water-heating panel, and Fig. 11.3 shows a passive solar water heating system.

Concentrating solar energy technologies basically consist of apparatuses that capture solar power radiation to produce high-temperature heat, which is then converted into electricity. The three most advanced CSP technologies currently in use are parabolic troughs (PT), central receivers (CR), and dish engines (DE). CSPs are considered one

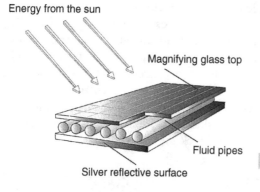

Energy from the sun

Magnifying glass top

Fluid pipes

Silver reflective surface

Figure 11.2 **Passive solar water-heating panel.**

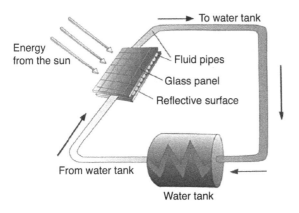

Figure 11.3   Passive solar water-heating system.

of today's most efficient power plants. Electrical power production capacity by concentrated solar energy is great, such that in the near future the technology could readily substitute electrical power production from fossil fuels. For instance, the solar resource for generating power from CSP systems is so high that it could provide sufficient electric power for the entire United States by covering only about 9 percent of state of Nevada, a plot of land 100 sq. miles with parabolic trough systems.[2]

The amount of power generated by CSP plants depends on the amount of direct sunlight. Similar to concentrating PV systems, solar thermal technologies use only the direct beam of the sunlight to concentrate the thermal energy of the sun. CSP systems can be sized from few kilowatts to large systems that could supply grid-connected power of up to 200 MW. Currently, a number of existing installations also use thermal storage during cloudy periods and are combined with natural gas, resulting in hybrid power plants that provide grid-connected dispatchable power.

Solar power-driven electric generator conversion efficiencies make concentrating technologies a viable renewable energy resource in the southwest United States. The U.S. Congress recently requested the Department of Energy to develop a plan for installing 1000 MW of CSP in the southwest over the next five years. CSP technologies are also considered an excellent source for providing thermal energy for commercial and industrial processes.

Benefits of Solar Thermal Concentrator Technologies

CSP technologies that are designed for use by heat-energy storage systems do not burn any fossil fuels, and therefore do not produce greenhouse gases, such as nitrogen oxide (NO_x) and sulfur oxide (SO_x) emissions. CSP technologies in the past couple of decades have also proven to be reliable and compatible for grid connectivity. In past decade, solar thermal plants have operated successfully in the southern California desert, providing enough power for 100,000 homes. Existing CSP plants in California now produce power for around 11¢/kWh, with projected costs dropping below 4¢/kWh within the next 20 years as technology refinements and economies of scale are implemented. Because CSP uses relatively conventional technologies and materials (glass, concrete, steel, and standard utility-scale turbines), capacity can rapidly be scaled up to several hundred MW/year.

Note that the emissions prevention benefits of solar thermal power system technologies depends on whether the power plants have their own storage capacity or are hybridized with other electricity or heat production technologies. CSP technologies with storage produce zero emissions; on the other hand, hybrid technologies can reduce emissions only by about 50 percent.

Trough Parabolic Heating System Technologies

In this technology, large fields of parabolic systems are secured on single-axis solar-tracking support systems and installed in modular parallel row configuration aligned in a north-south horizontal direction. Each of the solar parabolic collectors tracks the movement from east to west during daytime hours and focuses the sun's rays to a linear receiver tubing, which circulates a heat transfer fluid (HTF). The heated fluid, in turn, passes through a series of heat-exchanger chambers where the heat is transferred to superheated vapor that drives steam turbines. After propelling the turbine, the spent steam is condensed and returned to the heat exchanger via condensate pumps.

To harvest solar energy, parabolic heating systems make use of special parabolic reflectors that concentrate the solar energy rays into circular pipes that are located at the focal center of the parabola. Concentrated reflection of energy elevates the temperature of the circulating media, such as mineral liquid oil within the pipes, raising the temperature to such levels that allow considerable steam generation via special heat-exchangers that drive power turbines. Figure 11.4 is a graphic of a solar trough system.

A transparent glass tube placed in the focal line of the trough envelopes the receiver tube to reduce heat loss. Parabolic troughs usually employ single-axis or dual-axis tracking. In rare instances, they may be stationary.

CSP systems also use lenses or mirrors and tracking systems to focus a large area of sunlight into a small beam. The concentrated heat is then used as a heat source for a conventional power plant. Parabolic trough systems provide the best land-use factor of any solar technology.

Currently, the technology has been successfully applied in thermal electric power generation. For instance, a 354-MW solar power-generated electric plant installed in 1984 in California's Mojave Desert has been in operation with remarkable success.

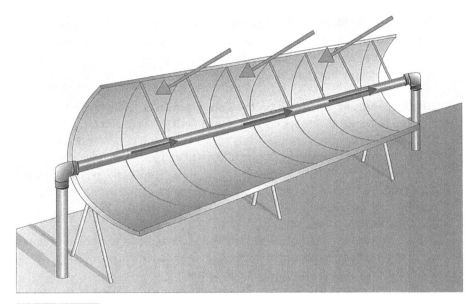

Figure 11.4   **Graphic representation of a solar trough system.**

In addition, completed in 2008, the 46-MW Moura PV power station in Portugal and the 40-MW Waldpolenz Solar Park in Germany appear to be characteristic of the trend toward larger PV power stations.

TYPES OF TROUGH COLLECTORS

Types of trough solar power systems include the Luz system, the EuroTrough, and the Solargenix.

Luz system collectors   Luz system collectors represent the standard by which all other collectors are measured. The collectors are constructed from galvanized steel, which makes them suitable for commercial power plant applications. Luz collector systems have proven to be highly reliable; so much so that they are deployed in most existing solar electric generation system power plants. There are two types of Luz system collectors: LS-2 and LS-3.

The LS-2 collector features a very accurate design. Their tubular structures are simple to erect and provide torsional stiffness. LS-2 has six torque-tube collector modules, three on either side of the drive. Each torque tube has two 4-m-long receivers. The main drawback of the system is that it uses a lot of steel and requires precise manufacturing to build.

In order to reduce manufacturing costs, Luz designers developed the LS-3 system, which has larger but lower manufacturing tolerances, and which requires less steel. To date, LS-3 has proved to be a very reliable design. The LS-3 uses a bridge truss structure in place of the torque-tube. Luz's LS-3 collector has truss assemblies on either side of the drive. Each LS-3 truss assembly has three 4-m-long receivers. Evidently, development of the LS-3 truss design did not lower manufacturing costs as much as expected. It also suffered from insufficient torsional stiffness, which led to lower-than-expected optical and thermal performance.

Linear receiver or heat collection elements   The parabolic trough linear receiver, also called a heat collection element (HCE), is one of the primary reasons for the high efficiency of the original Luz parabolic trough collector design.

The receiver is a 4-m-long, 70-mm diameter stainless steel tube with a special solar-selective absorber surface, surrounded by an antireflective evacuated 115-mm diameter glass tube. Located at the mirror focal line of the parabola, the receiver heats a special heat-transfer fluid as it circulates through the receiver tube. Figure 11.5 is a depiction of a solar thermal trough system.

The receiver has glass-to-metal seals and metal bellows to accommodate for differing thermal expansions between the steel tubing and the glass envelope. They also help achieve the necessary vacuum-tight enclosure.

The vacuum-tight enclosure primarily serves to significantly reduce heat losses at high-operating temperatures. It also protects the solar-selective absorber surface from oxidation.

The selective coating on the steel tube has good solar absorption and a low thermal emittance for reducing thermal radiation losses. The glass cylinder features an antireflective coating to maximize the solar transmittance. The original Luz receiver design suffered from poor reliability of the glass-to-metal seal. Subsequently, two companies (Solel Solar Systems and Schott Glass) have developed newer designs that have substantially improved the Luz system technology by improving the receiver reliability and the optical and thermal performance, while extending the receiver lifecycle.

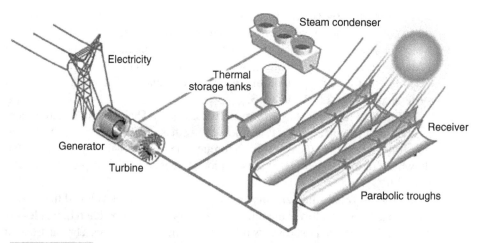

Figure 11.5   **Solar thermal trough system.** *Graphic courtesy of DOE.*

In addition to the parabolic collectors, the overall solar thermal system includes a computerized control and communications system and a number of other key components that make up the balance of system, including structural pylons, foundations, a drive mechanism, and collector interconnection linkages.

Pylons and Foundations   The pylons support the collector structure. They allow the collector to rotate and track the sun. The pylons are mounted on a concrete foundation that can support the weight and wind loading on the collector. Pylons also support the drive and controls at the center of the collector, and the bearings between each solar collector elements, such as truss or torque tubes that are located at the end of each collector.

Parabolic Reflector Drive System   Each solar collector assembly includes one drive. The drive positions the collector to track the sun during the day. The sun's beam radiation continuously reflects off the mirrors and onto the linear receiver.

The drive is located at the center of the collector. It can be either a standard motor and gear box configuration used in LS-2 or can use a hydraulic-drive system deployed in the LS-3, the EuroTrough, and Solargenix SGX-1 described below. The drive mechanisms of these systems are designed to accurately position the collector for tracking.

Computerized Controls   Each solar collector assembly has its own local controller that controls its operation. The local controller controls the tracking of the collector. It also monitors for any alarm conditions, such as a high- or low-fluid temperature in the receiver. A ball joint assembly connects the receivers on two adjacent collectors and allows them to track independently.

The local controllers also communicate with a supervisory computer in the power plant control building. The supervisory computer sends commands to the local controllers telling them when to start tracking the sun, or when to stop tracking at the end of the day.

Collector Interconnect   Each solar collector assembly operates independently from the adjacent collector. A set of insulated flexible hoses at the end of the collectors connects receivers to header pipings, which are located between two adjacent collectors. A flex hose allows the collectors to rotate independently.

EuroTrough collector   Recently, a European consortium, EuroTrough, initiated the development of a new collector design intended to build on the advantages of the Luz technology. The EuroTrough collector utilizes a torque-box design that integrates the torsional stiffness of a torque tube and the lower steel content of a truss design.

Solargenix collector   Under the U.S. Department of Energy's USA Trough Initiative, Solargenix Energy developed a new collector structure through a cost-shared, R&D contract with NREL. The Solargenix collector is made from extruded aluminum. It uses a unique organic hubbing structure, which Gossamer Spaceframes initially developed for buildings and bridges. The new design has the following specific design features:

- Weighs less than steel designs
- Requires very few fasteners
- Requires no welding or specialized manufacturing
- Assembles easily
- Requires no field alignment

Solargenix solar-trough technology uses parabolic-shaped reflectors or mirrors to concentrate the sun's rays to heat a mineral oil between 250 and 570 degrees Fahrenheit. The fluid then enters the Ormat engine, passing first through a heat exchanger to vaporize a secondary working fluid. The vapor is used to spin a turbine, making electricity. It is then condensed back into a liquid before being vaporized once again.

Historically, solar-trough technology has required tens of megawatts of plant installation to produce steam from water to turn generation turbines. The solar trough system combines the relatively low cost of parabolic solar trough thermal technology with the commercially available, smaller turbines usually associated with low-temperature geothermal generation plants.

Example of solar parabolic commercial installation   Located at the company's Saguaro Power Plant in Red Rock, about 30 miles north of Tucson, the Arizona Power System (APS) Saguaro Solar Trough Generating Station has a 1-MW generating capacity, enough to provide for the energy needs of approximately 200 average-size homes. The solar thermal station was constructed by Solargenix Energy, and the project was completed in April 2005. Solargenix partnered with Ormat Corporation, which provided the engine to convert the solar heat to electricity. Figure 11.6 is a photograph of a solar parabolic heater installation.

In addition to generating electricity for the Arizona power customers, the solar trough plant helps to meet the goals of the Arizona Corporation Commission's Environmental Portfolio Standard, which required APS to generate 1.1 percent of its energy through renewable sources by 2007. Figure 11.7 is photograph of a trough parabolic solar farm.

Solar Tower Technology

Another use of solar concentrator technology that generates electric power from the sun is constructed use of focusing concentrated solar radiation on a tower-mounted heat exchanger. The system basically is configured from thousands of sun-tracking mirrors commonly referred to as heliostats, which reflect the sun's rays onto the tower.

Figure 11.6   **Solar parabolic heater installation.** *Photo courtesy of Solargenix.*

Figure 11.7   Solar parabolic farm.

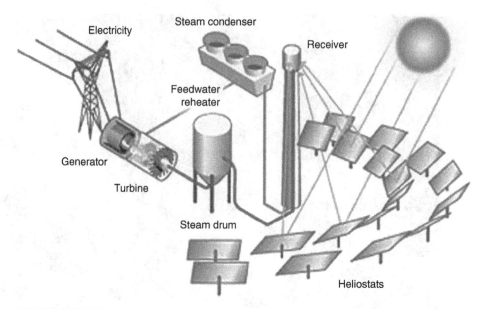

Figure 11.8   Graphic of solar thermal tower system. *Graphic courtesy of DOE.*

The receiver contains a fluid that once heated, similar to the trough parabolic system, transfers the absorbed heat in the heat exchanger to produce steam that then drives a turbine to produce electricity. Figure 11.8 depicts a solar thermal tower system.

These systems use a circular field array of heliostats, large individually tracking mirrors, to focus sunlight onto a central receiver mounted on top of a tower, which absorbs the heat energy that is then utilized in driving a turbine electric generator. A computer-controlled, dual-axis tracking system keeps the heliostats properly aligned, so that the reflected rays of the sun are always aimed at the receiver. Fluid circulating through the receiver transports heat to a thermal storage system, which can turn a turbine to generate electricity or provide heat directly for industrial applications. Temperatures achieved at the receiver range from 538 to 1482°C. Figure 11.9 is photograph of a solar thermal tower power plant.[3]

The first power tower "Solar One," built near Barstow in southern California, successfully demonstrated this technology for electricity generation. This facility operated in the mid-1980s and used a water/steam system to generate 10 MW of power. In 1992, U.S. utilities decided to retrofit Solar One to demonstrate a molten-salt receiver and thermal storage system. The addition of this thermal storage capability made power towers unique among solar technologies by promising dispatchable power at load factors of up to 65 percent. In this system, molten-salt is pumped from a "cold" tank at 288°C and is cycled through the receiver, where it is heated to 565°C and returned to a "hot" tank. The hot salt can then be used to generate electricity when needed. Current designs allow storage ranging from 3 to 13 hours.

Figure 11.9   Solar thermal tower power plant.

"Solar Two," a power tower electricity-generating plant in California, is a 10-MW prototype for large-scale commercial power plants. This facility first generated power in April 1996, and is scheduled to run for a three-year test, evaluation, and power production phase to prove the molten-salt technology. The reservoir which stores the sun's energy in molten salt at 550°C has allowed the plant to generate power, day and night, rain or shine. The successful completion of Solar Two has since prompted development and deployment of large scale multi-megawatt commercial solar power towers in the United States and Spain. Currently, power generated from this technology produces up to 400 MW of electricity.

Solar power towers are more cost effective, and offer higher efficiency and better energy storage capability than other CSP technologies. Projects such as Solar Two in Barstow, California, and the Planta Solar 10 in Sanlucar la Mayor, Spain, are excellent representatives of this technology, which generates multimegawatts of power.[3]

Solar Dish Technologies

Stirling solar dishes, or dish engine systems, consist of stand-alone parabolic reflectors that concentrate light onto a receiver positioned at the reflector's focal point. The reflectors track the sun along two axes. Parabolic dish systems give the highest efficiency among solar thermal concentrator-type technologies. A 500 m^2 solar dish system called ANU "Big Dish" in Canberra, Australia, is an example of this technology that combines a parabolic concentrating dish with a Stirling heat engine, which normally drives an electric generator. The advantages of Stirling solar over PV cells are higher efficiency of converting sunlight into electricity and longer lifetime. A solar power tower uses an array of tracking reflectors (heliostats) to concentrate light on a central receiver atop a tower. A solar bowl is a spherical dish mirror that is fixed in place. As opposed to a point focus with tracking parabolic mirrors, the Stirling engine receiver follows the line focus created by the dish. Figure 11.10 is a graphic of a solar thermal dish system.

Fluid in the Stirling engine receiver is heated up to 1000°C and is utilized directly to generate electricity in a small engine attached to the receiver. Engines currently under consideration include Stirling and Brayton cycle engines. Several prototype dish/engine systems, ranging in size from 7 to 25 kW have been deployed in various locations in the United States. High-optical efficiency and low start-up losses make dish/engine systems the most efficient of all solar technologies. Currently, a Stirling engine/parabolic dish system holds the world's record for converting sunlight into electricity. In 1984, a 27 percent net efficiency was measured at Rancho Mirage, California.

In addition, the modular design of dish/engine systems makes them a good match for both remote power needs in the kilowatt range as well as hybrid end-of-the-line grid-connected utility applications in the megawatt range. Figure 11.11 is a photograph of a solar thermal dish system.

EXAMPLE OF SOLAR DISH TECHNOLOGY DEPLOYMENT

Recently, this technology has been successfully demonstrated in a number of applications.

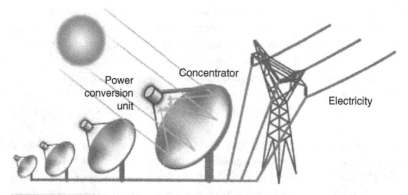

Figure 11.10   **A solar thermal dish system.** *Graphic courtesy of DOE.*

Figure 11.11   **Parabolic solar thermal dish system.** *Photo courtesy of Solar Power Technologies.*

One such project, named Solar Total Energy Project (STEP), operated between 1982 and 1989 in Shenandoah, Georgia, and consisted of 114 dishes, each 7 m in diameter. The system furnished high-pressure steam for electricity generation, medium-pressure steam for knitwear pressing, and low-pressure steam that ran the air conditioning system for a nearby knitwear factory. Failure of the main electrical power-generating turbine led to the system's decommissioning in October 1989.

At a later cooperative venture between Sandia National Lab and Cummins Power Generation, an attempt was made to commercialize a 7.5 kW dish/engine system. However, currently the lack of sufficient validation means the production has not reached a validation stage.

U.S. Department of Energy companies which presently manufacture solar thermal systems include Stirling Technology, Stirling Thermal Motors, and Detroit Diesel have made entry into the parabolic dish/Stirling technology and have teamed up with Science Applications International Corporation in a $36 million joint venture with the Department of Energy to develop a 25-kW membrane dish/Stirling system. Currently, the National Renewable Energy Laboratory (NREL) and the Cummins Engine Company are testing two new receivers for dish/engine solar thermal power systems: the pool-boiler receiver and the heat-pipe receiver. The pool-boiler receiver operates like a double boiler on a stove. It boils a liquid metal and transfers the heat energy to an engine on top. The heat-pipe receiver also uses a liquid metal, but instead of pooling the liquid, it uses a wick to transfer the molten liquid to a dome receiver.

STIRLING ENGINE SUNFLOWER™

The Stirling Engine Sunflower shown in Fig. 11.12 uses a radical concept because it does not use a stationary PV cell technology; rather, it is constructed from light-weight polished aluminum covered plastic-reflector petals, each being adjusted by a microprocessor-based motor controller, which enables the petals to track the sun in an independent fashion. This heat engine is essentially used to produce hot water by concentrating solar rays onto a low-profile water chamber.

Currently, the technology is being refined to produce higher-efficiency and more cost-effective production models. The company is working on larger-scale models for use in large-scale solar water-heating installations.

Figure 11.12   Stirling Engine Sunflower. *Photo courtesy of Idea Laboratories.*

Concentrating Linear Fresnel Reflectors

Concentrating Linear Fresnel reflectors are CSP plants that use many thin mirror strips instead of parabolic mirrors to concentrate sunlight onto two tubes with working fluids. These types of technologies have the advantage of flat mirrors, which can be used and are much cheaper than parabolic mirrors. In addition, more reflectors can be placed in the same amount of space, allowing more of the available sunlight to be used. Concentrating linear Fresnel reflectors can be used in either large or more compact plants. Figure 11.13 is a graphic of a linear Fresnel solar thermal system.

Experimental Solar Thermal Power Systems

A solar updraft tower, also known as a solar chimney, or solar tower, consists of a large greenhouse that funnels into a central tower. As sunlight shines on the greenhouse, the air inside is heated and expands. The expanding air flows toward the central tower, where a turbine converts the air flow into electricity. A 50-kW prototype was constructed in Ciudad Real, Spain, and operated for eight years before decommissioning in 1989.

Thermoelectric Solar Power System

A thermoelectric or "thermovoltaic" system is a mechanism or device that converts a temperature difference between dissimilar materials into an electric current. First proposed as a method to store solar energy, the system was proposed by the French solar pioneer Mouchout in the 1800s.

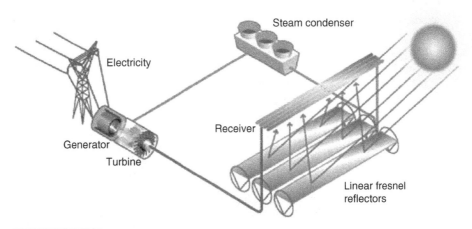

Figure 11.13   **Graphic of a linear Fresnel solar thermal system.** *Graphic courtesy of DOE.*

In 1930s, thermoelectric systems reemerged in the Soviet Union. Under the direction of Soviet scientist Abram Ioffe, a concentrating system was used to thermoelectrically generate power for a 1-hp engine. Thermogenerators were later used in the U.S. space program as an energy-conversion technology for powering deep-space missions, such as Cassini, Galileo, and Viking. Research in this area is focused on raising the efficiency of these devices from 7 to 8 percent to 15 to 20 percent.

Recently a new technology developed by the Idaho National Laboratory, uses nano-antennas to harvest solar power. Nanoantennas use the infrared radiation of the sun to convert energy. During the day, the earth's atmosphere lets some of the infrared radiation to pass through it and absorbs the rest. At night, the earth emits it.

Table 11.1 highlights the key features of the three solar technologies. Towers and troughs are best suited for large, grid-connected power projects in the 30 to 200 MW size, whereas dish/engine systems are modular and can be used in single-dish applications or

TABLE 11.1 A COMPARISON OF MAJOR SOLAR THERMAL TECHNOLOGIES

	PARABOLIC TROUGH	DISH/ENGINE	POWER TOWER
Size	30–320 MW	5–25 kW	10–200 MW
Operating Temperature (°C/°F)	390/734	750/1382	565/1049
Annual Capacity Factor	23–50%	25%	20–77%
Peak Efficiency	20%(d)	29.4%(d)	23%(p)
Net Annual Efficiency	11(d)-16%	12–25%(p)	7(d)-20%
Commercial Status	Commercially Scale-up Prototype	Demonstration	Available Demonstration
Technology Development Risk	Low	High	Medium
Storage Available	Limited	Battery	Yes
Hybrid Designs	Yes	Yes	Yes
Cost USD/W	$2.7–$ 4.0	$1.3–$12.6	$ 2.5–$4.4
Applications	Grid-connected electric plants; process heat for industrial use	Stand-alone small power systems; grid support	Grid-connected electric plants; process heat for industrial use
Advantages	Dispatchable peaking electricity; commercially available with 4,500 GWh operating experience; hybrid (solar/fossil) operation	Dispatchable electricity, high-conversion efficiencies; modularity; hybrid (solar/fossil) operation	Dispatchable base load electricity; high-conversion efficiencies; energy storage; hybrid (solar/fossil) operation

(p) = predicted; (d) = demonstrated
Source: Table is courtesy of DOE.

grouped in dish farms to create larger multimegawatt projects. Parabolic trough plants are the most mature solar power technology available today and the technology most likely to be used for near-term deployments. Power towers, with low cost and efficient thermal storage, promise to offer dispatchable, high-capacity factor, solar-only power plants in the near future. The modular nature of dishes will allow them to be used in smaller, high-value applications. Towers and dishes offer the opportunity to achieve higher solar-to-electric efficiencies and lower cost than parabolic trough plants, but uncertainty remains as to whether these technologies can achieve the necessary capital cost reductions and availability improvements. Parabolic troughs are currently a proven technology primarily waiting for an opportunity to be developed. Power towers require the operability and maintainability of the molten-salt technology to be demonstrated and the development of low-cost heliostats. Dish/engine systems require the development of at least one commercial engine and the development of a low-cost concentrator.

Solar Thermal Power Cost and Development Issues

Cost of electricity from solar thermal power systems depends on a multitude of factors. These factors include capital and operating and maintenance costs, and system performance. However, it is important to note that the technology's cost and the eventual cost of electricity generated is significantly influenced by factors "external" to the technology itself. As an example, for troughs and power towers, small stand-alone projects will be very expensive. In order to reduce the technology costs to compete with current fossil technologies, it will be necessary to scale-up projects to larger plant sizes and to develop solar power parks where multiple projects are built at the same site in time-phased succession. In addition, because these technologies, in essence replace conventional fuel with capital equipment, the cost of capital and taxation issues related to capital-intensive technologies will have a strong effect on their competitiveness.

COST VERSUS VALUE

Through the use of thermal storage and hybridization, solar thermal electric technologies can provide a firm and dispatchable source of power. "Firm" implies that the power source has a high reliability and will be able to produce power when the utility needs it. As a result, firm dispatchable power is of value to a utility because it offsets the utility's need to build and operate new power plants. "Dispatchability" implies that power production can be shifted to the period when it is needed. This means that even though a solar thermal plant might cost more, it can have a higher value.

BENEFITS OF SOLAR THERMAL SYSTEMS

Solar thermal power plants create two and one-half times as many skilled, high-paying jobs as do conventional power plants that use fossil fuels.

A California Energy Commission study shows that even with existing tax credits, a solar thermal electric plant pays about 1.7 times more in federal, state, and local taxes than an equivalent natural gas combined cycle plant. If the plants paid the same level of taxes, their cost of electricity would be roughly the same.

Utilizing only 1 percent of the earth's deserts to produce clean solar thermal electric energy would provide more electricity than is currently being produced on the entire planet by fossil fuels.

FUTURE OF SOLAR THERMAL POWER TECHNOLOGIES

It is estimated that solar thermal electric systems deployed by the year 2010 in the United States and internationally exceeded 5000 MW and currently serves the residential needs of 7 million people, which saves the energy equivalent of 46 million barrels of oil per year.

Summary

Solar thermal power technologies based on concentrating technologies are in different stages of development. Trough technology is commercially available today, with 354 MW currently operating in the Mojave Desert in California. Power towers are in the demonstration phase, with the 10-MW Solar Two pilot plant located in Barstow, California currently undergoing testing and power production. Several system designs are currently under engineering development in Golden, Colorado. Solar thermal power technologies have distinct features that make them attractive energy options in the expanding renewable energy market worldwide.

Solar thermal electricity-generating systems have come a long way over the past few decades. Increased research and development of solar thermal technology will make these systems more cost-competitive with fossil fuels, increase their reliability, and will become a serious alternative for meeting or supplying increased electricity demand.

References

1. http://www.quatrobioworld.com
2. http://www.concentratedsolarpower.com
3. http://energy.saving.nu/solarenergy/thermal.shtml

12

SOLAR POWER REBATE, FINANCING, AND FEED-IN TARIFF PROGRAMS

Introduction

This chapter explores financial topics that affect the viability of large-scale solar power system programs. Like any other capital investment projects, the proper appreciation of financial issues associated with large-scale solar power projects becomes a prerequisite for justifying the validity of a project. As such, the overall financial assessment becomes as important as designing the system. The rebate and tariff programs discussed in this chapter reflect specifics of California's programs, which have set the standards for the United States. Therefore, the readers are advised to inquire about the status of the programs that prevail in other states.

California Solar Initiative Rebate Program

This chapter highlights a summary of the California Solar Incentive (CSI) and Feed-In Tariff programs, which are essential to understand when dealing with large-scale solar power systems. It should be noted that the California Energy Commission (CEC) has developed all regulatory policy and solar power system equipment certification standards, which have been used throughout the United States and abroad. The CSI and feed-in tariffs discussed in this chapter are applicable to all states within the country.

On January 1, 2007, the state of California introduced solar rebate funding for the installation of PV power co-generation; this was authorized by the California Public Utilities Commission (CPUC) and the California senate. The bill referenced as SB1 has allotted a budget of $ 2.167 billion that will be used over a ten-year period.

The rebate funding program known as the California Solar Initiative (CSI) is a program that awards the incentive plans on the basis of performances rather than earlier programs that allotted rebates based on calculated projections of system energy output. The new rebate award system categorizes solar power installations into two incentive groups. An incentive program referred to as the performance-based incentive (PBI) addresses PV installation of 100 KW or larger and provides rebate dollars based on the solar power cogeneration's actual output over a five-year period. Another rebate referred to as expected performance-based buydown (EPBB) is a one-time lump-sum incentive payment program for solar power systems with performance capacities of less than 100 KW. This is a payment based on the system's expected future performance.

The distribution and administration of the CSI funds are delegated through three major utility providers, which service various state territories. The three main service providers that administer the program in California are:

- Pacific Gas and Electric (PG&E), which serves northern California
- Southern California Edison (SCE), which serves mid-California
- San Diego Regional Energy Office (SDREO)/San Diego Gas and Electric (SD&G), which serves San Diego and southern California

Note that municipal electric utility customers are not eligible to receive the CSI funds from the three administrator agencies.

Each of the service providers administering the CSI program has Web pages that enable clients to access online registration databases, and to download program handbooks, reservation forms, contract agreements, and all other forms required by the CSI program. All CSI application and reservation forms are available at www.csi.com.

The principle object of the CSI program is to ensure that 3000 MW of new solar energy facilities are installed throughout California by year 2017.

CSI FUND DISTRIBUTION

The CSI fund distribution administered by the three main agencies in California has a specific budget allotment, which is proportioned according to demographics of power demand and distribution. CSI budget allotment values are shown in Table 12.1.

The CSI budget shown in Table 12.1 is actually divided into two customer segments: residential and nonresidential. Table 12.2 shows relative allocations of CSI solar power generation by customer sector.

TABLE 12.1   CSI PROGRAM BUDGET BY ADMINISTRATOR

UTILITY	TOTAL BUDGET %
Pacific Gas & Electric	43.7
Southern California Edison	46.0
San Diego Gas & Electric	10.3

TABLE 12.2 CSI SOLAR POWER PRODUCTION TARGETS FOR RESIDENTIAL, COMMERCIAL, AND GOVERNMENT SECTORS						
		EPBB PAYMENTS (PER WATT)			PBI PAYMENTS (PER KWH)		
	MW	RES	COM	GOV'T NP	RES	COM	GOV'T/NP
1	50	n/a	n/a	n/a	n/a	n/a	n/a
2	70	$ 2.50	$ 2.50	$ 3.25	$ 0.39	$ 0.39	$ 0.50
3	100	$ 2.20	$ 2.20	$ 2.95	$ 0.34	$ 0.34	$ 0.46
4	130	$ 1.90	$ 1.90	$ 2.65	$ 0.26	$ 0.26	$ 0.37
5	160	$ 1.55	$ 1.55	$ 2.30	$ 0.22	$ 0.22	$ 0.32
6	190	$ 1.10	$ 1.10	$ 1.85	$ 0.15	$ 0.15	$ 0.26
7	215	$ 0.65	$ 0.65	$ 1.40	$ 0.09	$ 0.09	$ 0.19
8	250	$ 0.35	$ 0.35	$ 1.10	$ 0.05	$ 0.05	$ 0.15
9	285	$ 0.25	$ 0.25	$ 0.90	$ 0.03	$ 0.03	$ 0.12
10	350	$ 0.20	$ 0.20	$ 0.70	$ 0.03	$ 0.03	$ 0.10

CSI POWER GENERATION TARGETS

In order to offset the high installation costs associated with solar power installation and promote PV industry development, the CSI incentive program has devised a plan that encourages customer sectors to take immediate advantage of rebate initiatives, which are intended to last for a limited duration of 10 years. The incentive program is currently planned to be reduced automatically over the duration of the 3000 MW of solar power reservation, in 10 step-down trigger levels that gradually distribute the power generation over 10 allotted steps. CSI MW power production targets are proportioned amongst the administrative agencies by residential and nonresidential customer sectors (see Table 12.2).

In each of the 10 steps, CSI applications are limited to the trigger levels. Table 12.2 shows set trigger stages for SCE and PG&E client sectors. Once the trigger level allotments are complete, the reservation process is halted and restarted at the next trigger level. In the event of trigger-level power surplus, the excess of energy allotment is transferred forward to the next trigger level.

The CSI power production targets shown in Table 12.2 are based on the premise that solar power industry production output and client sector awareness will gradually be increased within the next decade, and that the incentive program will eventually promote a viable industry that will be capable to provide reliable source of renewable energy base in California.

INCENTIVE PAYMENT STRUCTURE

As mentioned, CSI offers PBI and EPBB incentive programs, both of which are based on verifiable PV system output performance. EPBB incentive output characteristics are basically determined by factors such as location of solar platforms, system size, shading conditions, tilt angle, and all of the factors discussed in previous chapters. On the other hand, PBI incentives are strictly based on predetermined flat rate per kilowatt hour output payments over a five-year period. Incentive payment levels have been devised to be reduced automatically over the duration of the program in ten steps, which are directly proportional to the MW volume reservation shown in Table 12.2.

As seen from the incentive distribution table, the rebate payments diminish as the targeted solar power program reaches its 3000-MW energy output. The main reasoning behind downscaling the incentive is based on the presumption that the solar power manufacturers will (within the next decade) be a in position to produce larger quantities of more efficient and less expensive PV modules. As a result of the economy of scale, the state will no longer be required to extend special incentives to promote the PV industry by using public funds.

EXPECTED PERFORMANCE-BASED BUYDOWN (EPBB)

As mentioned previously, EPBB is a one-time, up-front incentive based upon a PV power cogeneration's estimated or predicated future performance. This program is targeted to minimize program administration works for relatively small systems that do not exceed 100 KWh. Factors that affect the computations of the estimated power performance are relatively simple and consider such factors as panel count, PV module certified specifications, the location of the solar platform, insolation, PV panel orientation, tilt angle, and shading losses; these are all entered into a predetermined equation that results in a buydown incentive rate.

The EPBB program applies to all new projects other than systems that have customized building-integrated photovoltaics (BIPV). EPBB's one-time incentive payment calculation is based on the following formula:

$$\text{EPBB Incentive Payment} = \text{Incentive Rate} \times \text{System Rating (kW)} \times \text{Design Factor}$$

$$\text{System Rating (kW)} = \text{Number of PV modules} \times \text{CEC PTS value} \times \text{CEC Inverter Listed Efficiency}/(1000) \text{ for Kilowatt Conversion}$$

Special design requirements imposed are as follows:

- All PV modules must be oriented between 180 and 270°.
- Optimal tilt for each compass direction shall be in the range of 180° for optimized summer power output efficiency.
- The system must take into account derating factors associated with weather and shading analysis.
- The system must be on an optimal reference and location.
- The PV tilt must correspond to the local latitude.

Note that all residential solar power installations are also subject to EPBB incentive payment formulations.

PERFORMANCE-BASED INCENTIVE (PBI)

As of January 1, 2007, this incentive applies to solar power system installations equal to or exceeding 100 kW. As of January 1, 2008, the base power output reference was reduced to 50 kW and as of January 1, 2010 to 30 kW. Each of BPI payments is limited to duration of five years following completion of the system acceptance test. Also included in the plan are custom-made BIPV systems.

HOST CUSTOMERS

All beneficiaries of CSI programs are referred to as host customers, which includes not only the electric utility customers but also retail electric distribution organizations, such as PG&E, SCE, and SDG&E. Under the rules of the CSI program, all entities who apply for an incentive are referred to as an applicant, a host, or a system owner.

In general, host customers must have an outstanding account with a utility provider at the location of solar power cogeneration. In other words, the project in California must be located within the service territory of one of the three listed program administrators.

After approving the reservation, the host customer is considered the system owner and retains sole rights to the reservation. The reservation is a payment guarantee by CSI that cannot be transferred by the owner; however, the system installer can be designated to act on behalf of the owner.

To proceed with the solar power program, the applicant or the owner must receive a written confirmation letter from the administrating agency, and then apply for authorization for grid connectivity. In the event of project delays beyond the permitted period of fund reservation, the customer must reapply for another rebate to obtain authorization.

According to SCI regulations, several categories of customers do not qualify to receive the incentive. Customers exempted from the program are organizations that are in the business of power generation and distribution, publicly owned gas and electricity distribution utilities, and any entity that purchases electricity or natural gas for wholesale or retail purposes.

As a rule, the customer assumes full ownership upon receipt of the incentive payment and technically becomes responsible for the operation and maintenance of the overall solar power system.

It should be noted that a CSI applicant is recognized as the entity that completes and submits the reservation forms; that becomes the main contact person who must communicate with the program administrator throughout the duration of the project. However, the applicant may also designate an engineering organization or a system integrator, an equipment distributor, or even an equipment lessor to act as the designated applicant.

SOLAR POWER CONTRACTORS AND EQUIPMENT SELLERS

Contractors from California who specialize in solar power installation must hold an appropriate State of California contractor's license. In order to qualify as an installer by the program administrator, the solar power system integrator must provide the following information:

- Business name and address
- Principal's name or contact
- Business registration or license number
- Contractor's license number
- Contractor's bond (if applicable)and corporate limited-liability entities
- Reseller's license number (if applicable)

All equipment, such as PV modules, inverters, and meters sold by equipment sellers must be UL approved and certified by the California Energy Commision. All equipment provided must be new and have been tested for at least one year. The use of refurbished equipment is not permitted. It should be noted that experimental, field-demonstrated, or proof-of-concept operation-type equipment and materials are not approved and do not qualify for rebate incentive. All equipment used, therefore, must have UL certification and performance specifications that would allow program administrators to evaluate equipment performance.

According to CEC certification criteria, all grid-connected PV systems must carry a 10-year warranty and meet the following certification requirements:

- All PV modules must be certified to UL-1703 standards.
- All grid-connected solar watt-hour meters for systems under 10 kW must have an accuracy of (+) to (−) 5%. Watt-hour meters for systems over 10 kW must have measurement accuracy of (+) to (−) 2%.
- All inverters must be certified to UL-1741 standards.

PV SYSTEM SIZING REQUIREMENTS

Note that the primary objective of solar power cogeneration is to produce a certain amount of electricity to offset a certain portion of the electrical demand load. Therefore, power production of PV systems is set in a manner as not to exceed the actual energy consumption during the previous 12 months. The formula applied for establishing the maximum system capacity is:

$$\text{Maximum system power output (kW)} = 12 \text{ months of previous energy used (kWh)}/ (0.18 \times 8760 \text{ hours/year})$$

The factor of $0.18 \times 8760 = 1577$ hours/year can be translated into an average of 4.32 hours/day of solar power production, which essentially includes system performance and derating indexes applied in CEC PV system energy output calculations.

The maximum PV system under the current CSI incentive program is limited to 1000 kW or 1 MW. However, if the calculation limits permit, customers are allowed to install grid-connected systems of up to 5 MW, for which only 1 MW will be considered for receiving the incentive.

For new construction where the project has no history of previous energy consumption, an applicant must substantiate system power demand requirements by engineering system demand load calculations, which will include present and future load growth projections. All calculations must be substantiated by corresponding equipment specification, panel schedules, single-line diagrams, and building energy simulation programs such as eQUEST, EnergyPro, and DOE-2.

ENERGY-EFFICIENCY AUDIT

Rules enacted in January 2007 state that all existing residential and commercial customers applying for CSI rebates are obligated to provide a certified energy-efficiency audit for their existing building. The audit certification, along with the solar PV rebate application forms, must be provided to the program administrator for evaluation purposes.

Energy audits can be conducted either by calling an auditor or accessing a special Web page provided by each administrative entity. In some instances, energy audits could be waived if the applicants can provide a copy of an audit conducted in the past three years, or provide a proof of a California Title 24 Energy Certificate of Compliance, which is usually calculated by mechanical engineers. Projects that have a national LEED™ certification are also exempt from energy audit.

WARRANTEE AND PERFORMANCE PERMANENCY REQUIREMENTS

As mentioned previously, all major system components are required to have a minimum of 10 years of warranty by both manufacturers and installers. All equipment, including PV modules and inverters, in the event of malfunction are required to be replaced at no cost to the client. System power output performance calculation must also include an adequate percent of power output degradation as PV module specification. The power output degradation must reflect solar power system performance for the duration of the solar PV module lifecycle.

To be eligible for CSI, all solar power system installations must be permanently attached or secured to their platforms. PV modules supported by quick disconnect means and installed on wheeled platforms or trailers are not considered legitimate stationary installations.

During the course of project installation, the owner or its designated representative must maintain continuous communication with the program administrator and provide all required information regarding equipment specification, warranties, platform configuration, design revisions and system modifications, updated construction schedules, and the construction status on regular basis.

In the event the location of PV panels are changed and panels are removed or relocated within the same project perimeters or service territory, the owner must inform the CSI administrator and establish a revised PBI payment period.

INSURANCE

Currently, CSI requires the owner or host customer of all systems equal or above 30 kW receiving CSI to carry a minimum level of general liability insurance. Installers also must carry workman's compensation and business auto insurance coverage. Since U.S. government entities are self-insured, the program administrators will only require a proof of coverage.

GRID INTERCONNECTION AND METERING REQUIREMENTS

The main criterion for grid system integration is that the solar power cogeneration system must be permanently connected to the main electrical service network. As such, portable power generators are not considered eligible. In order to receive the incentive payment, the administrator must receive proof of grid interconnection record. In order to receive additional incentives, customers whose power demand coincides with California's peak electricity demand become eligible to apply for time-of-use (TOU) tariffs, which could increase their energy payback.

All meters installed must be easily physically located, to allow an administrator's authorized agents to have easy access to for tests or inspections.

INSPECTION

All systems rated from 30 to 100 kW that have not adopted a performance-based incentive (PBI) will be inspected by specially designated inspectors. In order to receive the incentive payment, the inspectors must verify system operational performance, installation conformance to applications, and the eligibility criteria and grid interconnection.

System owners who have opted for EPBB incentives must install the PV panels in proper orientation and produce power that is reflected in the incentive application.

In the event of inspection failure, the owner will be advised by the administrator about shortcomings regarding material or compliance, which must be mitigated within 60 days. Failure to correct the problem could result in cancellation of the application and a strike against the installer, applicant, seller, or any party deemed responsible.

Entities identified as responsible for mitigating the problem (if failed three times) will be disqualified from participating in CSI programs for a period of one year.

CSI INCENTIVE LIMITATIONS

The prerequisites for processing CSI program are based upon the premise that a project's total installed out-of-pocket expenses by the owner do not exceed the eligible costs. For this reason, the owner or the applicant must prepare a detailed project cost breakdown, which will highlight only relative embedded costs of the solar power system. A worksheet designed for this purpose is available on the California Solar Incentive Web page, www.csi.com.

It is important to note that clients are not permitted to receive incentives under other sources. In the event a project may be qualified to receive an additional incentive from another source for the same power cogenerating system, the first incentive amount will be discounted by the amount of the second incentive received. Nevertheless, the overall combined incentive amount must be assured not to exceed total eligibility costs.

At all times during the project construction, administrators reserve the right to conduct periodic spot checks and random audits to make certain that all payments received were made in accordance to CSI rules and regulations.

CSI RESERVATION STEPS

The following are summary steps for EPBB applications:

1 Reservation form must be completed and submitted with owner's or applicant's wet signature.
2 Proof of electric utility service or account number for the project site must be shown on the application. In case of a new project, the owner must procures a tentative service account number.
3 System description worksheet available on the CSI Web page must be completed.
4 Electrical system sizing documents computations must be attached to the application form.
5 If the project is subject to tax exemption, form AB1407 compliance for government and nonprofit organizations must be attached to the application.
6 For existing projects, energy-efficiency audit or Title 24 calculations must be submitted as well.
7 To calculate the EPBB, the CSI Web page calculator (www.csi-epbb.com) should be used.
8 A copy of the executed purchase agreement from solar system contractor or provider must be attached.
9 A copy of the executed contract agreement if the system ownership is given to another party must be attached.
10 A copy of grid interconnection agreement must be attached if available; otherwise the administrator must be informed about steps taken to secure the agreement.

To submit a payment claim, the following documents must be provided to the administrator.

1 Submit a wet-signed claim form, available on the CSI Web page
2 Proof of authorization for grid integration
3 Copy of the building permit and final inspection sign-off
4 Proof of warranty from the installer and equipment suppliers
5 Final project cost breakdown
6 Final project cost affidavit

For projects categorized as BPI or non-residential systems above 10 kW or larger, the owner must use the following process:

1 Reservation form must be completed and submitted with owner's or applicant's wet signature.

2 Proof of electric utility service or account number for the project site must be shown on the application. In case of a new project, the owner must procure a tentative service account number.

3 A system description worksheet, available on the CSI Web page, must be completed.

4 Electrical system sizing calculation must be attached to the application form.

5 An application fee (1 percent of the requested CSI incentive amount) must be attached.

6 If the project is subject to tax exemption, form AB1407 compliance for government and nonprofit organizations must be attached to the application.

7 For existing projects, an energy-efficiency audit or Title 24 calculations must be submitted as well.

8 The printout of the calculated PBI must be forwarded. Use the CSI Web page calculator.

9 A copy of the executed purchase agreement from the solar system contractor or provider must be attached.

10 A copy of executed contract agreement must be attached if the system ownership is given to another party.

11 A copy of grid interconnection agreement must be attached if available; otherwise the administrator must be informed about steps taken to secure the agreement.

12 A completed proof-of-project milestone must be attached.

13 A copy of host customer certificate of insurance must be attached.

14 A copy of system owner's certificate of insurance (if different from the host) must be attached.

15 A copy of the project cost breakdown worksheet must be attached.

16 A copy of an alternative system ownership such as a lease/buy agreement must be attached.

17 A copy of the RFP or solicitation document must be attached if customer is a government, a nonprofit organization, or a public entity.

To submit the claim for the incentive documents to the administrator, the owner or the contractor must provide the following:

1 Wet-signed claim form, available on the CSI Web page

2 Proof of authorization for grid integration

3 Copy of the building permit and final inspection sign-off

4 Proof of warranty from the installer and equipment suppliers

5 Final project cost breakdown

6 Final project cost affidavit

In the event of incomplete document submittal, the administrators will allow the applicant to provide missing documentation or information within 20 days. Information provided must be in a written form mailed using the U.S. postal system. Faxes or hand-delivered systems are not allowed.

All changes to the reservation must be undertaken by a formal letter that explains legitimate delay justification. Requests to extend the reservation expiration date are capped to a maximum of 180 calendar days. Written time-extension requests must explicitly highlight circumstances that were beyond the control of reservation holder, such as permitting process, manufacturing delays, and extended delivery of PV modules or critical equipment, acts of nature, and so forth. All correspondence associated with the delay must be transmitted by letter.

INCENTIVE PAYMENTS

After completing the final field acceptance and submission of the above-referenced documents (for an EPBB project), the program administrator will, within a period of 30 days, issue a complete payment. For PBI programs, the first incentive payment is commenced and issued within 30 days of the first scheduled performance reading from the wattmeter. All payments are made to the host customer or the designated agent.

In some instances, a host could request the administrator to assign the entire payments to a third party. For payment reassignment, the host must complete a special set of forms provided by the administrator.

The EPBB one-time lump sum payment calculation is based on the following formula:

EPBB Incentive Payment = Final CSI system size × Reserved EPBB Incentive rate

PBI payments for PV systems of 100 kW or greater are made on a monthly basis over a period of five years. The payments are based on the actual electric energy output of the PV system. If chosen by the owner, systems less than 100 kW could also be paid on the PBI incentive basis. The PBI payment calculation is based on the following formula:

Monthly PBI Incentive Payment = Reserved Incentive Rate × Measured kWh output

In the event of PV system size change, the original reservation request forms must be updated and incentive amount recalculated.

AN EXAMPLE OF THE PROCEDURE FOR CALCULATING CALIFORNIA SOLAR INCENTIVE (CSI) REBATE

The following example is provided to assist the reader with details of CSI reservation calculation requirements. Reservation forms in addition to completing host customer requirements and project site information, require supporting calculations that project solar power system output performance. Regardless of system size and classification as EPBB or PBI, supportive documentation for calculating the solar power system cogeneration remains identical.

In order to commence reservation calculations, the designer must undertake the following preliminary design measures:

- Outline the solar power cogeneration system net unobstructed platform area.
- Use a rule of thumb to determine watts per square feet of a particular PV module intended for use. For example, a PV module area of $2.5' \times 5' = 12.5$ ft^2 which produces 158 W PTC will have an approximate power output of about 14 W/ft^2.
- By dividing the available PV platform area by 12.5 ft^2 we can determine the number of panels required.

In order to complete the CSI reservation forms referenced above, the designer must also determine type, model, quantity, and efficiency of CEC-approved inverter. For example, let us assume that we are planning for a ground-mount solar farm with an output capacity of 1 MW. The area available for the project is 6 acres, which is adequate for installing a single-axis solar-tracking system. Our chosen solar power module and inverter chosen are from approved CEC-listed equipment.

Solar power system components used are:

- PV module: SolarWorks Powermax 175 p, unit DC watts = 175, PTC = 158.3 watts AC, number of panels required 6680 units
- Inverter: Xantrex Technology PV225S-480P, 225 kW, efficiency 94.5 percent

Prior to completing the CSI reservation form, the designer must use CSI EPBB calculator (available on the Web at www.csi-epbb.com) to determine the rebate for systems that are smaller than 30 KW. Even though BPI calculation is automatically determined by the CSI reservation form spreadsheet, EPBB calculation determines a design factor number required by the form.

EPBB calculation procedure   To conduct EPBB calculations, the designer must enter the following data in the blank field areas:

- Project area zip code, for example, 92596.
- Project address information.
- Customer type such as residential, commercial or government/nonprofit.
- PV module manufacture, model, associated module DC and PTC rating and unit count.
- Inverter manufacturer, model, output rating, and percent efficiency
- Shading measurement and analysis
- Array tilt, such as 30°. For maximum efficiency, the tilt angle should be close to latitude.
- Array azimuth in degrees, which determines orientation; for north facing south use 180°.

Upon entering the above datum, the CSI calculator will output the following results:

- Optimal tilt angle at proposed azimuth
- Annual kWh output at optimal tilt facing south
- Summer month output from May to October
- CEC-AC rating—a comment will indicate if the system is greater than 100 kW
- Design correction factor—required for calculating the CSI reservation form
- Geographic correction
- Design factor
- Incentive rate—dollars/watt
- Rebate incentive cash amount if system is qualified as EPBB

California Solar Initiative reservation form calculations Like EPBB referenced above, the CSI reservation is also a Web page spreadsheet that can be accessed at www.csi.com. The data required to complete the EPBB are the same as the ones used for PBI, except that EPBB design factors derived from the previous calculation must be inserted in the project incentive calculations.

The California Solar Initiative Program Reservation form consists of the following six major information input data fields:

1 Host Customer: Information required includes customer name, business class and company information, tax payer identification, contact person's name, title, mailing address, telephone, fax, and e-mail.

2 Applicant information if system procurer is not the host customer.

3 System owner information.

4 Project site information, same as EPBB. System platform information, such as available building or ground area must be specified. In this field the designer must also provide the electrical utility service account and meter numbers if available. For new projects, there should be a letter attached to the reservation form indicating the account procurement status.

5 PV and inverter hardware information identical to the ones used in calculating EPBB.

6 Project incentive calculation. In this field, the designer must enter the system rating in kW (CEC) and design factor data obtained from EPBB calculations.

Upon completing the above entries, the CSI reservation spreadsheet will automatically calculate the project system power output size in watts. In a field designated total eligible project cost, the designer must insert the projected cost of the system. CSI automatically produces a per-watt installed cost, CSI rebate amount, and system owner's out-of-pocket expenses.

The following CSI reservation request calculation is based on the same hardware information used in the above EPBB calculation. The data entry information and calculation steps are as follows:

- Platform—single-axis tracker
- Shading—none
- Insolation for zip code 92596, San Bernardino, California = 5.63 average hour/day
- PV module—SolarWorld model SW175 mono/P
- DC watts—175
- PTC—158.2
- PV count—6680
- Total power output—1057 kW AC
- Calculated CSI system size by the spread sheet—1032 kW AC
- Inverter—Xantrex Technology Model PV225S-480P
- Power output capacity—225 kW AC
- Efficiency—94.5%
- Resulting output = $1057 \times 94.5\% = 999$ kW

- CSI–EPBB Design Factor = 0.975
- CSI system size = 999 × 0.975 = 974 kWh
- PV system daily output = 974 × 5.63 (insolation) = 5484 kWh/day
- Annual system output = 5484 × 365 (days) = 2,001,660 kWh per year
- Assuming an incentive class for a government or a nonprofit organization for year 2007, allocated performance output per kWh = $0.513
- Total incentive over five years = 5 × 2,004,196 × $0.513 = $5,134,258.00
- Projected installed cost = $8,500,000.00 ($ 8.50/watt)
- System owner out-of-pocket cost = $3,365,742.00
- Application fee of 1% of incentive amount = $33,657.00

EQUIPMENT DISTRIBUTORS

Eligible ERP manufacturers and companies who sell system equipment must provide the CEC with the following information on the equipment seller information form (CEC-1038 R4). For all CEC rebate forms see Appendix C.

- Business name, address, phone, fax, and e-mail address
- Owner or principal contact
- Business license number
- Contractor license number (if applicable)
- Proof of good standing in the records of the California secretary of state, as required for corporate and limited-liability entities
- Reseller's license number

SPECIAL FUNDING FOR AFFORDABLE HOUSING PROJECTS

California Assembly Bill 58 mandates the CEC to establish an additional rebate for systems installed on affordable housing projects. These projects are entitled to qualify for an extra 25 percent rebate above the standard rebate level, provided that the total amount rebated does not exceed 75 percent of the system cost. The eligibility criteria for qualifying are as follows.

The affordable housing project must adhere to California health and safety codes. The property must expressly limit residency to extremely low-, very low-, lower-, or moderate-income persons and must be regulated by the California Department of Housing and Community Development.

Each residential unit (apartments, multifamily homes, etc.) must have individual electric utility meters. The housing project must be at least 10 percent more energy-efficient than current standards specified (see Chap. 5).

Conclusion

Even though the initial investment in a solar power cogeneration system requires a large capital investment, the long-term financial and ecological advantages are so significant that their deployment in specific project should be given special consideration.

A solar power cogeneration system, if applied as per the recommendations reviewed here, will provide considerable energy expenditure savings over the lifespan of the recreation facility and provide a hedge against unavoidable energy cost escalation.

Special note: In view of the depletion of existing CEC rebate funds, it is recommended that applications for the rebate program be initiated at the earliest possible time. Furthermore, because of the design integration of the solar power system with the service grid, the decision to proceed with the program must be made at the commencement of the construction design document stage.

California Feed-In Tariffs

CALIFORNIA ENERGY COMMISSION REPORT ON FEED-IN TARIFFS

The following feed-in tariff discussion is a summary of a report of a 2008 workshop report that was published by California Energy Commission. California has a Renewables Portfolio Standard (RPS) that requires the state's investor-owned utilities, energy service providers, and community choice aggregators to serve 20 percent of retail sales with renewable resources by 2010; publicly owned utilities are required to develop RPS programs as well. As indicated in the *2007 Integrated Energy Policy Report* (IEPR), California set a 20 percent green energy target which was to be met by 2010, (which has not been realized to date). California has also set a renewable energy target of 33 percent by 2020 and is expected to need new policy tools to meet this aggressive target.

The RPS referenced above explores the potential approaches to expanding the use of feed-in tariffs as a mechanism to aid in making California's renewable generation objectives a reality. There are a great variety of potential feed-in tariff policy design options and policy paths. The report examines feed-in tariff options for design issues, such as appropriate tariff structure, eligibility, and pricing. The report considers policy goals and objectives, stakeholder comments on materials presented in the Energy Commission's feed-in tariff design issues and options workshop held in June 30, 2008, as well as data accumulated from feed-in tariff experience in Spain and Germany. The report identifies six representative policies for consideration. The pros and cons of each of the six policies are explored and analyzed in detail. In addition, the report explores the potential interaction of the policy paths, examines the interaction of feed-in tariff policies with other related policies, and discusses issues related to potential for establishing a specific program for the California feed-in tariff program.

The six policy paths that are examined in the report span a range of directions, as well as timing and scope.

Policy Path 1—This policy is designed to be similar to the feed-in tariff system currently in place in Germany, but is conditional in that it will be triggered only if California's 20 percent renewable energy goal is not met by 2010. Under this option, tariffs would become available in the 2012–2013 timeframe to ensure that the 33 percent renewables target would be met by 2020. The policy does not mandate any restrictions on generator size, and all contracts are fixed price and long term. The tariffs intended would be differentiated by technology and project size. Fundamentally the program is intended to be cost based, and the preliminary price settings would be set competitively and not administratively. The use of emerging resources would be capped, so as to limit ratepayer impacts. In addition, the use of long-term contract and technology differentiation would provide a degree of price stability to investors while promoting a diversity of renewable resources.

Policy Path 2—The second policy is essentially a pilot program that would involve a single utility that could generate over 20 megawatts (MWs) of electrical energy. It would go into effect immediately without any sort of trigger mechanism. Long-term fixed-price contracts would be available for projects coming on-line within a three-year window, after which the policy would be reevaluated. There would be no limit to the quantity of generation eligible to use this tariff, as the limited duration would serve to constrain its overall use. Tariff payments under this option would be value based, with payments differentiated only by production profiles. These include times of production, contribution to peak, etc. (which also takes into account environmental issues). The value-based payments are intended to alleviate some ratepayer concerns relative to the cost-based alternatives. However, this path may not promote the resource diversity that Policy Path 1 discussed above, does.

Policy Path 3—This policy would be triggered by the establishment of a Competitive Renewable Energy Zone (CREZ) designated for feed-in tariff procurement in the 2010/2011 timeframe, allowing generation within the CREZ to proceed aggressively with development once transmission expansion is committed (without being constrained by the timing and risk of a RPS competitive solicitation). The policy is essentially a cost-based system; however, tariff prices would be set administratively rather than through use of competitive benchmarks. This option would be limited geographically by the CREZ footprint, and the quantity eligible to take the feed-in tariff price would be capped at the CREZ transmission limit. This option would also target generators over 1.5 MW. Based on the renewable resource potential and available/planned transmission in the CREZ, this option would help alleviate worries of undersubscription of new transmission lines and support a diverse mix of renewable resources.

Policy Path 4—This policy is supposed to be a solar-only pilot feed-in tariff. It will include elements of Policy Paths 1 and 3 in that are cost based, which will use competitive benchmark rates. Rather than being limited to a specific window of time, however, the pilot scale for the tariff would be accomplished by limiting long-term contract availability to a single utility territory. Eligibility would be limited to solar installations larger than the net metering limit of 1 MW. This policy would also have a capacity cap option. Although this option could provide incentives for larger systems, since solar energy is above market, it may not provide enough renewable energy and diversity for the state to meet its goals. This option could be established independently or in concert with another policy path.

Policy Path 5—This policy is limited to sustainable biomass. Tariffs would be cost based and differentiated by size and by biomass fuel feedstock. Unlike the solar-only option, the biomass path would be available in every market, rather than on a pilot scale in a single utility, and would not be capped. Finally, unlike all of the other policy paths that would incorporate long-term contracts or price guarantees, the contract term would be either short- or medium- term in acknowledgement of the fuel price risk that longer term contracts would place on biomass developers and investors. As discussed below, this option could be established independently or in concert with another policy path.

Policy Path 6—This policy is intended to establish feed-in tariffs that will be available statewide to generators up to 20 MW in size without any conditions, which would help to address a perceived gap in the current RPS solicitation process. This policy would offer cost-based, long-term prices differentiated by size and technology. Unlike Policy Path 1 however, prices would not be based on a competitive benchmark, and the tariff quantity would be uncapped. It is not limited to one technology, and therefore might be helpful in enabling the state to meet its diversity goals. Nevertheless, the California feed-in tariff policy report is intended to stimulate various stakeholder inputs on which feed-in tariffs options could best help California meet its renewable energy objectives.

As discussed above, a feed-in tariff is an offering of a fixed-price contract over a specified term with specified operating conditions to eligible renewable energy generators. Feed-in tariffs can be either an all-inclusive rate or a fixed premium payment on top of the prevailing spot market price for power. The tariff price paid will represent estimates of either the cost or value of renewable generation. In the future, feed-in tariffs will be offered by the interconnecting utilities, which will set standing prices for each category of eligible renewable generators; the prices will be available to all eligible generators. Tariffs will essentially be based upon technology type, resource quality, or project size, and may decline on a set schedule over time.

BENEFITS AND LIMITATIONS OF FEED-IN TARIFFS

As with other policies, feed-in tariffs provide benefits and limitations, a number of which depend upon the design of the tariff system. From the generator's perspective, the benefits of a feed-in tariff include the availability of a guaranteed price, buyer, and long-term revenue stream without the cost of solicitation. Market access is enhanced by feed-in tariffs, as project timing is not constrained by periodic scheduled solicitations. In addition, completion dates may not be constrained by contractual requirements, quantities are often uncapped, and interconnection is typically guaranteed. Together, these characteristics can help to reduce or alleviate generator revenue uncertainty, project risk, and associated financing concerns. Feed-in tariffs reduce transaction costs for both buyer and seller and are more transparent to administer than the current system. Because responding to standing tariffs is likely to be less costly and less complex than competitive solicitations, feed-in tariffs may also increase the ability of smaller projects or developers to help the state meet its Renewables Portfolio Standard (RPS) and greenhouse gas emission reduction goals. Policy makers can target feed-in tariffs to encourage specific types of projects and technologies if so desired. However, there are limitations to how a feed-in tariff might function in California. Total feed-in tariff costs cannot be predicted accurately because, despite the predetermined payments, the quantity of generation responding to a feed-in tariff is not typically predetermined.

One key concern is how the tariff fits in a deregulated market structure, including questions of who pays, how payments are distributed, what portion of rates would be used to recover tariff costs, and how to integrate electric production purchased through feed-in tariffs into utility power supplies. Another question specific to California is whether feed-in tariffs would work in concert with California's existing RPS law or would require changes in that law.

Getting the price right can be challenging. If the price is set too high, the tariff introduces the risk of overpaying and overstimulating the market. This risk may be exacerbated when the tariff is open to large projects in regions with ample resource potential. On the other hand, if the tariff is set too low to provide adequate returns to eligible projects, it may have little effect on stimulating development of new renewable energy generation. A range of approaches for setting the price are discussed in the six options considered in this report.

UNIT CONVERSION AND DESIGN

REFERENCE TABLES

Renewable Energy Tables and Important Solar Power Facts

1 Recent analysis by the Department of Energy (DOE) shows that by year 2025, one-half of the new U.S. electricity generation could come from the sun.

2 The United States generated only 4 GW (1 GW is 1000 MW) of solar power. By the year 2030, it is estimated to be 200 GW.

3 A typical nuclear power plant generates about 1 GW of electric power, which is equal to 5 GW of solar power (daily power generation is limited to an average of 5 to 6 hours per day).

4 Global sales of solar power systems have been growing at a rate of 45 percent in the past few years.

5 It is projected that by the year 2020, the United States will be producing about 7.2 GW of solar power.

6 Shipment of U.S. solar power systems has fallen by 10 percent annually, but has increased by 45 percent throughout Europe.

7 Annual sales growth of solar power photovoltaic technologies globally has been 35 percent.

8 Current cost of solar power modules on the average is \$2.33/W. By 2030 it should be about \$0.38/W.

9 World production of solar power is 1 GW/year.

10 Germany has a \$0.50/W grid-feed incentive that will be valid for the next 20 years. The incentive is to be decreased by 5 percent per year.

11 In the past few years, Germany installed 130 MW of solar power per year.

12 Japan has a 50 percent subsidy for solar power installations of 3- to 4-kW systems and has about 800 MW of grid-connected solar power systems. Solar power in Japan has been in effect since 1994.

13 California, in 1996, set aside $540 million for renewable energy, which has provided a $4.50/W to $3.00/W buyback as a rebate.

14 In the years 2015 through 2024, it is estimated that California could produce an estimated $40 billion of solar power sales.

15 In the United States, 20 states have a solar-rebate program. Nevada and Arizona have set aside a state budget for solar programs.

16 Total U.S. production of photovoltaic modules has been just about 18 percent of global production.

17 For each megawatt of solar power produced, the United States employs 32 people.

18 A solar power collector, sized 100 miles × 100 miles, in the southwest United States could produce sufficient electric power to satisfy the country's yearly energy needs.

19 For every kilowatt of power produced by nuclear or fossil fuel plants, $1/2$ gal of water is used for scrubbing, cleaning, and cooling. Solar power practically does not require any water usage.

20 Significant impact of solar power cogeneration:
   a. Boosts economic development.
   b. Lowers cost of peak power.
   c. Provides greater grid stability.
   d. Lowers air pollution.
   e. Lowers greenhouse gas emissions.
   f. Lowers water consumption and contamination.

22 A mere 6.7 mi/gal efficiency increase in cars driven in the United States could offset our share of imported Saudi oil.

23 Current state of solar power technology:
   - Crystalline
   - Polycrystalline
   - Amorphous
   - Thin- and thick-film technologies

24 State of solar power technology in the future:
   - Plastic solar cells
   - Nano-structured materials
   - Dye-synthesized cells

Energy Conversion Table

ENERGY UNITS
1 J (joule) = 1 W · s = 4.1868 cal
1 GJ (gigajoule) = 10 E9 J
1 TJ (terajoule) = 10 E12 J
1 PJ (petajoule) = 10 E15 J
1 kWh (kilowatt-hour) = 3,600,000 J
1 toe (ton oil equivalent) = 7.4 barrels of crude oil in primary energy
$\qquad$ = 7.8 barrels in total final consumption
$\qquad$ = 1270 m³ of natural gas
$\qquad$ = 2.3 metric tons of coal
Mt (million ton oil equivalent) = 41.868 PJ

POWER
Electric power is usually measured in watts (W), kilowatts (kW), megawatts (MW), and so forth. Power is energy transfer per unit of time.
1 kW = 1000 W
1 MW = 1,000,000
1 GW = 1000 MW
1 TW = 1,000,000 MW
Power (e.g., in W) may be measured at any point in time, whereas energy (e.g., in kWh) has to be measured during a certain period; for example, a second, an hour, or a year.

UNIT ABBREVIATIONS
m = meter = 3.28 feet (ft)
s = second
h = hour
W = watt
hp = horsepower
J = joule
cal = calorie
toe = tons of oil equivalent

(Continued)

UNIT ABBREVIATIONS
Hz = hertz (cycles per second)
10 E–12 = pico (p) = 1/1000,000,000,000
10 E–9 = nano (n) = 1/1,000,000,000
10 E–6 = micro (µ) = 1/1000,000
10 E–3 = milli (m) = 1/1000
10 E–3 = kilo (k) = 1000 = thousands
10 E–6 = mega (M) = 1,000,000 = millions
10 E–9 = giga (G) = 1,000,000,000
10 E–12 = tera (T) = 1,000,000,000,000
10 E–15 = peta (P) = 1,000,000,000,000,000
WIND SPEEDS
1 m/s = 3.6 km/h = 2.187 mi/h = 1.944 knots
1 knot = 1 nautical mile per hour = 0.5144 m/s = 1.852 km/h = 1.125 mi/h

Voltage Drop Formulas and DC Cable Charts

Single-phase VD = A L × 2K/C.M.
Three-phase VD = A L × 2K × 0.866/C.M.
Three-phase VD = A L × 2K × 0.866 × 1.5/C.M.(for two-pole systems)
where A = amperes
L = distance from source of supply to load
C.M. = cross-sectional area of conductor in circular mills:
K = 12 for copper more than 50 percent loading
K = 11 for copper less than 50 percent loading
K = 18 for aluminum

VOLTAGE DROP CALCULATION FOR COPPER WIRES				
WIRE	**THHN AMPACITY**	**THWN AMPACITY**	**MCM**	**CONDUIT DIAMETER (IN)**
2,000			2,016,252	
1,750			1,738,503	
1,500			1,490,944	
1,250			1,245,699	
1,000	615	545	999,424	
900	595	520	907,924	
800	565	490	792,756	
750	535	475	751,581	
700	520	460	698,389	4
600	475	420	597,861	4
500	430	380	497,872	4
400	380	335	400,192	4
350	350	310	348,133	3
300	320	285	299,700	3
250	290	255	248,788	3
4/0	260	230	211,600	$2\frac{1}{2}$
3/0	225	200	167,000	2
2/0	195	175	133,100	2
1/0	170	150	105,600	2
1	150	130	83,690	$1\frac{1}{2}$
2	130	115	66,360	$1\frac{1}{4}$
3	110	100	52,620	$1\frac{1}{2}$
4	95	85	41,740	$1\frac{1}{2}$
6	75	65	26,240	1
8	55	50	15,510	$\frac{3}{4}$
10	30	30	10,380	$\frac{1}{2}$
12	20	20	6,530	$\frac{1}{2}$

240 NEC ALLOWED CABLE DISTANCES FOR 240 VOLT AC OR OR DC CABLE – CHART FOR VOLTAGE DROP OF 2% VOLT AC OR DC CABLE CHART = VOLTAGE DROP OF 2% NEC CODE ALLOWED CABLE DISTANCES

AMPS	WATTS	AWG #14	AWG #12	AWG #10	AWG #8	AWG #6	AWG #4	AWG #2	AWG #1/0	AWG #2/0	AWG #3/0
2	480	338	525								
4	960	150	262	413							
6	1,440	113	180	262	450						
8	1,920	82	180	218	338	427					
10	2,400	67	105	173	270	285					
15	3,600	45	67	105	180	266	450				
20	4,800		52	82	144	250	338	540			
25	6,000			67	105	218	270	434			
30	7,200			53	90	173	225	360	578		
40	9,600				67	142	173	270	434	540	
50	12,000				54	82	137	218	345	434	547

AMPS	WATTS	AWG #14	AWG #12	AWG #10	AWG #8	AWG #6	AWG #4	AWG #2	AWG #1/0	AWG #2/0	AWG #3/0
2	240	169	262								
4	480	75	131	206							
6	720	56	90	131	225						
8	960	41	90	109	169	266					
10	1,200	34	52	86	135	214					
15	1,800	22	34	52	90	142	225				
20	2,400		26	41	72	109	169	270			
25	3,000			34	52	86	135	217			
30	3,600			26	45	71	112	180	289		
40	4,800				34	125	86	135	217	270	
50	6,000				27	41	68	109	172	217	274

NEC ALLOWED CABLE DISTANCES FOR 48 VOLT DC—CHART FOR VOLTAGE DROP OF 2% VOLT DC CABLE
CHART = VOLTAGE DROP OF 2% NEC CODE ALLOWED CABLE DISTANCES

AMPS	WATTS	AWG #14	AWG #12	AWG #10	AWG #8	AWG #6	AWG #4	AWG #2	AWG #1/0	AWG #2/0	AWG #3/0
1	48	135	210	330	540						
2	96	67	105	166	270	426					
4	192	30	53	82	135	214					
6	288	22	36	53	90	142	226				
8	384	17	26	43	67	106	173				
10	480	14	21	34	54	86	135	216			
15	720	9	14	21	36	57	90	144	231		
20	960		10	17	30	43	67	108	174	216	274
25	1,200			14	21	34	54	86	138	174	219
30	1,440			10	18	29	45	72	115	138	182
40	1,920				14	21	34	54	86	115	137
50	2,400				9	17	27	43	69	86	138

24 NEC ALLOWED CABLE DISTANCES FOR 24 VOLT DC – CHART FOR VOLTAGE DROP OF 2% VOLT DC CABLE CHART = VOLTAGE DROP OF 2% NEC CODE ALLOWED CABLE DISTANCES

AMPS	WATTS	AWG #14	AWG #12	AWG #10	AWG #8	AWG #6	AWG #4	AWG #2	AWG #1/0	AWG #2/0	AWG #3/0
1	24	68	105	165	270						
2	48	34	52	83	135	213					
4	96	15	26	41	68	107					
6	144	11	18	26	45	71	113				
8	192	8	13	22	34	53	86				
10	240	7	10	17	27	43	68	108			
15	360	4	7	10	18	28	45	72	116		
20	480		5	8	15	22	34	54	87	108	137
25	600			7	10	17	27	43	69	87	110
30	720			5	9	14	22	36	58	69	91
40	960				7	10	17	27	43	58	68
50	1,200				4	8	14	22	34	43	89

12 NEC ALLOWED CABLE DISTANCES FOR 12 VOLT DC — CHART FOR VOLTAGE DROP OF 2% VOLT DC CABLE CHART = VOLTAGE DROP OF 2% NEC CODE ALLOWED CABLE DISTANCES

AMPS	WATTS	AWG #14	AWG #12	AWG #10	AWG #8	AWG #6	AWG #4	AWG #2	AWG #1/0	AWG #2/0	AWG #3/0
1	12	84	131	206	337	532					
2	24	42	66	103	168	266	432	675			
4	48	18	33	52	84	133	216	337	543	672	
6	72	14	22	33	56	89	141	225	360	450	570
8	96	10	16	27	42	66	108	168	272	338	427
10	120	9	13	22	33	53	84	135	218	270	342
15	180	6	9	13	22	35	56	90	144	180	228
20	240		7	10	16	27	42	67	108	135	171
25	300			8	13	22	33	54	86	108	137
30	360			7	11	18	28	45	72	90	114
50	480				8	13	21	33	54	67	85

CROSS REFERENCE OF AMERICAN WIRE GAUGE (AWG) AND METRIC SYSTEM (mm)			
AWG	mm^2	AWG	mm^2
30	0.05	6	16
28	0.08	4	25
26	0.14	2	35
24	0.25	1	50
22	0.34	1/0	55
21	0.38	2/0	70
20	0.5	3/0	95
18	0.75	4/0	120
17	1	300 MCM	150
16	1.5	350 MCM	185
14	2.5	500 MCM	240
12	4	600 MCM	300
10	6	750 MCM	400
8	10	1,000 MCM	500

Solar Photovoltaic Module Tilt Angle Correction Table

SOLAR PANEL ORIENTATION TILT CORRECTION FACTOR						
	COLLECTOR TILT ANGLE FROM HORIZONTAL (DEGREES)					
	0	15	30	45	60	90
South	0.89	0.97	1.00	0.97	0.88	0.56
SSE or SSW	0.89	0.97	0.99	0.96	0.87	0.57
SE or SW	0.89	0.95	0.96	0.93	0.85	0.59
ESE or WSW	0.89	0.92	0.91	0.87	0.79	0.57
East or West	0.89	0.88	0.84	0.78	0.7	0.51

TILT ANGLE EFFICIENCY MULTIPLIER TABLE

	COLLECTOR TILT ANGLE FROM HORIZONTAL (DEGREES)					
	0	15	30	45	60	90
FRESNO						
South	0.90	0.98	1.00	0.96	0.87	0.55
SSE, SSW	0.90	0.97	0.99	0.96	0.87	0.56
SE, SW	0.90	0.95	0.96	0.92	0.84	0.68
ESE, WSW	0.90	0.92	0.91	0.87	0.79	0.57
E, W	0.90	0.88	0.86	0.78	0.70	0.51
DAGGETT						
South	0.88	0.97	1.00	0.97	0.88	0.56
SSE, SSW	0.88	0.96	0.99	0.96	0.87	0.58
SE, SW	0.88	0.94	0.96	0.93	0.85	0.59
ESE, WSW	0.88	0.91	0.91	0.86	0.78	0.57
E, W	0.88	0.87	0.83	0.77	0.69	0.51
SANTA MARIA						
South	0.89	0.97	1.00	0.97	0.88	0.57
SSE, SSW	0.89	0.97	0.99	0.96	0.87	0.58
SE, SW	0.89	0.95	0.96	0.93	0.86	0.59
ESE, WSW	0.89	0.92	0.91	0.87	0.79	0.67
E, W	0.89	0.88	0.84	0.78	0.70	0.52
LOS ANGELES						
South	0.89	0.97	1.00	0.97	0.88	0.57
SSE, SSW	0.89	0.97	0.99	0.96	0.87	0.58
SE, SW	0.89	0.95	0.96	0.93	0.85	0.69
ESE, WSW	0.89	0.92	0.91	0.87	0.79	0.57
E, W	0.89	0.88	0.85	0.78	0.70	0.51
SAN DIEGO						
South	0.89	0.98	1.00	0.97	0.88	0.57
SSE, SSW	0.89	0.97	0.99	0.96	0.87	0.58
SE, SW	0.89	0.95	0.96	0.92	0.54	0.59
ESE, WSW	0.89	0.92	0.91	0.87	0.79	0.57
E, W	0.89	0.88	0.85	0.78	0.70	0.51

Solar Insolation Table for Major Cities in the United States*

STATE	CITY	HIGH	LOW	AVG.	STATE	CITY	HIGH	LOW	AVG.
AK	Fairbanks	5.87	2.12	3.99	GA	Griffin	5.41	4.26	4.99
AK	Matanuska	5.24	1.74	3.55	HI	Honolulu	6.71	5.59	6.02
AL	Montgomery	4.69	3.37	4.23	IA	Ames	4.80	3.73	4.40
AR	Bethel	6.29	2.37	3.81	IL	Boise	5.83	3.33	4.92
AR	Little Rock	5.29	3.88	4.69	IL	Twin Falls	5.42	3.42	4.70
AZ	Tucson	7.42	6.01	6.57	IL	Chicago	4.08	1.47	3.14
AZ	Page	7.30	5.65	6.36	IN	Indianapolis	5.02	2.55	4.21
AZ	Phoenix	7.13	5.78	6.58	KS	Manhattan	5.08	3.62	4.57
CA	Santa Maria	6.52	5.42	5.94	KS	Dodge City	4.14	5.28	5.79
CA	Riverside	6.35	5.35	5.87	KY	Lexington	5.97	3.60	4.94
CA	Davis	6.09	3.31	5.10	LA	Lake Charles	5.73	4.29	4.93
CA	Fresno	6.19	3.42	5.38	LA	New Orleans	5.71	3.63	4.92
CA	Los Angeles	6.14	5.03	5.62	LA	Shreveport	4.99	3.87	4.63
CA	Soda Springs	6.47	4.40	5.60	MA	E. Wareham	4.48	3.06	3.99
CA	La Jolla	5.24	4.29	4.77	MA	Boston	4.27	2.99	3.84
CA	Inyokern	8.70	6.87	7.66	MA	Blue Hill	4.38	3.33	4.05
CO	Grandbaby	7.47	5.15	5.69	MA	Natick	4.62	3.09	4.10
CO	Grand Lake	5.86	3.56	5.08	MA	Lynn	4.60	2.33	3.79
CO	Grand Junction	6.34	5.23	5.85	MD	Silver Hill	4.71	3.84	4.47
CO	Boulder	5.72	4.44	4.87	ME	Caribou	5.62	2.57	4.19
DC	Washington	4.69	3.37	4.23	ME	Portland	5.23	3.56	4.51
FL	Apalachicola	5.98	4.92	5.49	MI	Sault Ste. Marie	4.83	2.33	4.20
FL	Belle Isle	5.31	4.58	4.99	MI	E. Lansing	4.71	2.70	4.00
FL	Miami	6.26	5.05	5.62	MN	St. Cloud	5.43	3.53	4.53
FL	Gainesville	5.81	4.71	5.27	MO	Columbia	5.50	3.97	4.73
FL	Tampa	6.16	5.26	5.67	MO	St. Louis	4.87	3.24	4.38
GA	Atlanta	5.16	4.09	4.74	MS	Meridian	4.86	3.64	4.43

(Continued)

STATE	CITY	HIGH	LOW	AVG.	STATE	CITY	HIGH	LOW	AVG.
MT	Glasgow	5.97	4.09	5.15	PA	Pittsburgh	4.19	1.45	3.28
MT	Great Falls	5.70	3.66	4.93	PA	State College	4.44	2.79	3.91
MT	Summit	5.17	2.36	3.99	RI	Newport	4.69	3.58	4.23
NM	Albuquerque	7.16	6.21	6.77	SC	Charleston	5.72	4.23	5.06
NB	Lincoln	5.40	4.38	4.79	SD	Rapid City	5.91	4.56	5.23
NB	N. Omaha	5.28	4.26	4.90	TN	Nashville	5.2	3.14	4.45
NC	Cape Hatteras	5.81	4.69	5.31	TN	Oak Ridge	5.06	3.22	4.37
NC	Greensboro	5.05	4.00	4.71	TX	San Antonio	5.88	4.65	5.3
ND	Bismarck	5.48	3.97	5.01	TX	Brownsville	5.49	4.42	4.92
NJ	Sea Brook	4.76	3.20	4.21	TX	El Paso	7.42	5.87	6.72
NV	Las Vegas	7.13	5.84	6.41	TX	Midland	6.33	5.23	5.83
NV	Ely	6.48	5.49	5.98	TX	Fort Worth	6.00	4.80	5.43
NY	Binghamton	3.93	1.62	3.16	UT	Salt Lake City	6.09	3.78	5.26
NY	Ithaca	4.57	2.29	3.79	UT	Flaming Gorge	6.63	5.48	5.83
NY	Schenectady	3.92	2.53	3.55	VA	Richmond	4.50	3.37	4.13
NY	Rochester	4.22	1.58	3.31	WA	Seattle	4.83	1.60	3.57
NY	New York City	4.97	3.03	4.08	WA	Richland	6.13	2.01	4.44
OH	Columbus	5.26	2.66	4.15	WA	Pullman	6.07	2.90	4.73
OH	Cleveland	4.79	2.69	3.94	WA	Spokane	5.53	1.16	4.48
OK	Stillwater	5.52	4.22	4.99	WA	Prosser	6.21	3.06	5.03
OK	Oklahoma City	6.26	4.98	5.59	WI	Madison	4.85	3.28	4.29
OR	Astoria	4.76	1.99	3.72	WV	Charleston	4.12	2.47	3.65
OR	Corvallis	5.71	1.90	4.03	WY	Lander	6.81	5.50	6.06
OR	Medford	5.84	2.02	4.51					

*Values are given in kilowatt-hours per square meter per day.

Longitude and Latitude Tables

	LONGITUDE	LATITUDE
ALABAMA		
Alexander City	32° 57'N	85° 57'W
Anniston AP	33° 35'N	85° 51'W
Auburn	32° 36'N	85° 30'W
Birmingham AP	33° 34'N	86° 45'W
Decatur	34° 37'N	86° 59'W
Dothan AP	31° 19'N	85° 27'W
Florence AP	34° 48'N	87° 40'W
Gadsden	34° 1'N	86° 0'W
Huntsville AP	34° 42'N	86° 35'W
Mobile AP	30° 41'N	88° 15'W
Mobile Co	30° 40'N	88° 15'W
Montgomery AP	32° 23'N	86° 22'W
Selma-Craig AFB	32° 20'N	87° 59'W
Talladega	33° 27'N	86° 6'W
Tuscaloosa AP	33° 13'N	87° 37'W
ALASKA		
Anchorage AP	61° 10'N	150° 1'W
Barrow	71° 18'N	156° 47'W
Fairbanks AP	64° 49'N	147° 52'W
Juneau AP	58° 22'N	134° 35'W
Kodiak	57° 45'N	152° 29'W
Nome AP	64° 30'N	165° 26'W
ARIZONA		

	LONGITUDE	LATITUDE
Blythe AP	33° 37'N	114° 43'W
Burbank AP	34° 12'N	118° 21'W
Chico	39° 48'N	121° 51'W
Concord	37° 58'N	121° 59'W
Covina	34° 5'N	117° 52'W
Crescent City AP	41° 46'N	124° 12'W
Downey	33° 56'N	118° 8'W
El Cajon	32° 49'N	116° 58'W
El Cerrito AP	32° 49'N	115° 40'W
Escondido	33° 7'N	117° 5'W
Eureka/Arcata AP	40° 59'N	124° 6'W
Fairfield-Travis AFB	38° 16'N	121° 56'W
Fresno AP	36° 46'N	119° 43'W
Hamilton AFB	38° 4'N	122° 30'W
Laguna Beach	33° 33'N	117° 47'W
Livermore	37° 42'N	121° 57'W
Lompoc, Vandenberg AFB	34° 43'N	120° 34'W
Long Beach AP	33° 49'N	118° 9'W
Los Angeles AP	33° 56'N	118° 24'W
Los Angeles Co	34° 3'N	118° 14'W
Merced-Castle AFB	37° 23'N	120° 34'W
Modesto	37° 39'N	121° 0'W
Monterey	36° 36'N	121° 54'W
Napa	38° 13'N	122° 17'W
Needles AP	34° 36'N	114° 37'W

Location	Lat.	Long.	Location	Lat.	Long.
Douglas AP	31° 27'N	109° 36'W	Oakland AP	37° 49'N	122° 19'W
Flagstaff AP	35° 8'N	111° 40'W	Oceanside	33° 14'N	117° 25'W
Fort Huachuca AP	31° 35'N	110° 20'W	Ontario	34° 3'N	117° 36'W
Kingman AP	35° 12'N	114° 1'W	Oxnard	34° 12'N	119° 11'W
Nogales	31° 21'N	110° 55'W	Palmdale AP	34° 38'N	118° 6'W
Phoenix AP	33° 26'N	112° 1'W	Palm Springs	33° 49'N	116° 32'W
Prescott AP	34° 39'N	112° 26'W	Pasadena	34° 9'N	118° 9'W
Tucson AP	32° 7'N	110° 56'W	Petaluma	38° 14'N	122° 38'W
Winslow AP	35° 1'N	110° 44'W	Pomona Co	34° 3'N	117° 45'W
Yuma AP	32° 39'N	114° 37'W	Redding AP	40° 31'N	122° 18'W
ARKANSAS			Redlands	34° 3'N	117° 11'W
Blytheville AFB	35° 57'N	89° 57'W	Richmond	37° 56'N	122° 21'W
Camden	33° 36'N	92° 49'W	Riverside-March AFB	33° 54'N	117° 15'W
El Dorado AP	33° 13'N	92° 49'W	Sacramento AP	38° 31'N	121° 30'W
Fayetteville AP	36° 0'N	94° 10'W	Salinas AP	36° 40'N	121° 36'W
Fort Smith AP	35° 20'N	94° 22'W	San Bernardino, Norton AFB	34° 8'N	117° 16'W
Hot Springs	34° 29'N	93° 6'W	San Diego AP	32° 44'N	117° 10'W
Jonesboro	35° 50'N	90° 42' W	San Fernando	34° 17'N	118° 28'W
Little Rock AP	34° 44'N	92° 14'W	San Francisco AP	37° 37'N	122° 23'W
Pine Bluff AP	34° 18'N	92° 5'W	San Francisco Co	37° 46'N	122° 26'W
Texarkana AP	33° 27'N	93° 59' W	San Jose AP	37° 22'N	121° 56'W
CALIFORNIA			San Luis Obispo	35° 20'N	120° 43'W
Bakersfield AP	35° 25'N	119° 3'W	Santa Ana AP	33° 45'N	117° 52'W
Barstow AP	34° 51'N	116° 47'W			

(Continued)

CALIFORNIA (Continued)	LONGITUDE	LATITUDE
Santa Barbara MAP	34° 26'N	119° 50'W
Santa Cruz	36° 59'N	122° 1'W
Santa Maria AP	34° 54'N	120° 27'W
Santa Monica CIC	34° 1'N	118° 29'W
Santa Paula	34° 21'N	119° 5'W
Santa Rosa	38° 31'N	122° 49'W
Stockton AP	37° 54'N	121° 15'W
Ukiah	39° 9'N	123° 12'W
Visalia	36° 20'N	119° 18'W
Yreka	41° 43'N	122° 38'W
Yuba City	39° 8'N	121° 36'W
COLORADO		
Alamosa AP	37° 27'N	105° 52'W
Boulder	40° 0'N	105° 16'W
Colorado Springs AP	38° 49'N	104° 43'W
Denver AP	39° 45'N	104° 52'W
Durango	37° 17'N	107° 53'W
Fort Collins	40° 45'N	105° 5'W
Grand Junction AP	39° 7'N	108° 32'W
Greeley	40° 26'N	104° 38'W
La Junta AP	38° 3'N	103° 30'W
Leadville	39° 15'N	106° 18'W

	LONGITUDE	LATITUDE
Gainesville AP	29° 41'N	82° 16'W
Jacksonville AP	30° 30'N	81° 42'W
Key West AP	24° 33'N	81° 45'W
Lakeland Co	28° 2'N	81° 57'W
Miami AP	25° 48'N	80° 16'W
Miami Beach Co	25° 47'N	80° 17'W
Ocala	29° 11'N	82° 8'W
Orlando AP	28° 33'N	81° 23'W
Panama City, Tyndall AFB	30° 4'N	85° 35'W
Pensacola Co	30° 25'N	87° 13'W
St. Augustine	29° 58'N	81° 20'W
St. Petersburg	27° 46'N	82° 80'W
Sarasota	27° 23'N	82° 33'W
Stanford	28° 46'N	81° 17'W
Tallahassee AP	30° 23'N	84° 22'W
Tampa AP	27° 58'N	82° 32'W
West Palm Beach AP	26° 41'N	80° 6'W
GEORGIA		
Albany, Turner AFB	31° 36'N	84° 5'W
Americus	32° 3'N	84° 14'W
Athens	33° 57'N	83° 19'W
Atlanta AP	33° 39'N	84° 26'W
Augusta AP	33° 22'N	81° 58'W

Location	Latitude	Longitude
Pueblo AP	38° 18'N	104° 29'W
Sterling	40° 37'N	103° 12'W
Trinidad	37° 15'N	104° 20'W
CONNECTICUT		
Bridgeport AP	41° 11'N	73° 11'W
Hartford, Brainard Field	41° 44'N	72° 39'W
New Haven AP	41° 19'N	73° 55'W
New London	41° 21'N	72° 6'W
Norwalk	41° 7'N	73° 25'W
Norwich	41° 32'N	72° 4'W
Waterbury	41° 35'N	73° 4'W
Windsor Locks, Bradley Fld	41° 56'N	72° 41'W
DELAWARE		
Dover AFB	39° 8'N	75° 28'W
Wilmington AP	39° 40'N	75° 36'W
DISTRICT OF COLUMBIA		
Andrews AFB	38° 5'N	76° 5'W
Washington, National AP	38° 51'N	77° 2'W
FLORIDA		
Belle Glade	26° 39'N	80° 39'W
Cape Kennedy AP	28° 29'N	80° 34'W
Daytona Beach AP	29° 11'N	81° 3'W
E Fort Lauderdale	26° 4'N	80° 9'W
Fort Myers AP	26° 35'N	81° 52'W
Fort Pierce	27° 28'N	80° 21'W
Brunswick	31° 15'N	81° 29'W
Columbus, Lawson AFB	32° 31'N	84° 56'W
Dalton	34° 34'N	84° 57'W
Dublin	32° 20'N	82° 54'W
Gainesville	34° 11'N	83° 41'W
Griffin	33° 13'N	84° 16'W
LaGrange	33° 1'N	85° 4'W
Macon AP	32° 42'N	83° 39'W
Marietta, Dobbins AFB	33° 55'N	84° 31'W
Savannah	32° 8'N	81° 12'W
Valdosta-Moody AFB	30° 58'N	83° 12'W
Waycross	31° 15'N	82° 24'W
HAWAII		
Hilo AP	19° 43'N	155° 5'W
Honolulu AP	21° 20'N	157° 55'W
Kaneohe Bay MCAS	21° 27'N	157° 46'W
Wahiawa	21° 3'N	158° 2'W
IDAHO		
Boise AP	43° 34'N	116° 13'W
Burley	42° 32'N	113° 46'W
Coeur D'Alene AP	47° 46'N	116° 49'W
Idaho Falls AP	43° 31'N	112° 4'W
Lewiston AP	46° 23'N	117° 1'W
Moscow	46° 44'N	116° 58'W
Mountain Home AFB	43° 2'N	115° 54'W

(Continued)

	LONGITUDE	LATITUDE
Pocatello AP	42° 55'N	112° 36'W
Twin Falls AP	42° 29'N	114° 29'W
ILLINOIS		
Aurora	41° 45'N	88° 20'W
Belleville, Scott AFB	38° 33'N	89° 51'W
Bloomington	40° 29'N	88° 57'W
Carbondale	37° 47'N	89° 15'W
Champaign/Urbana	40° 2'N	88° 17'W
Chicago, Midway AP	41° 47'N	87° 45'W
Chicago, O'Hare AP	41° 59'N	87° 54'W
Chicago Co	41° 53'N	87° 38'W
Danville	40° 12'N	87° 36'W
Decatur	39° 50'N	88° 52'W
Dixon	41° 50'N	89° 29'W
Elgin	42° 2'N	88° 16'W
Freeport	42° 18'N	89° 37'W
Galesburg	40° 56'N	90° 26'W
Greenville	38° 53'N	89° 24'W
Joliet	41° 31'N	88° 10'W
Kankakee	41° 5'N	87° 55'W
La Salle/Peru	41° 19'N	89° 6'W
Macomb	40° 28'N	90° 40'W
Moline AP	41° 27'N	90° 31'W

	LONGITUDE	LATITUDE
Muncie	40° 11'N	85° 21'W
Peru, Grissom AFB	40° 39'N	86° 9'W
Richmond AP	39° 46'N	84° 50'W
Shelbyville	39° 31'N	85° 47'W
South Bend AP	41° 42'N	86° 19'W
Terre Haute AP	39° 27'N	87° 18'W
Valparaiso	41° 31'N	87° 2'W
Vincennes	38° 41'N	87° 32'W
IOWA		
Ames	42° 2'N	93° 48'W
Burlington AP	40° 47'N	91° 7'W
Cedar Rapids AP	41° 53'N	91° 42'W
Clinton	41° 50'N	90° 13'W
Council Bluffs	41° 20'N	95° 49'W
Des Moines AP	41° 32'N	93° 39'W
Dubuque	42° 24'N	90° 42'W
Fort Dodge	42° 33'N	94° 11'W
Iowa City	41° 38'N	91° 33'W
Keokuk	40° 24'N	91° 24'W
Marshalltown	42° 4'N	92° 56'W
Mason City AP	43° 9'N	93° 20'W
Newton	41° 41'N	93° 2'W
Ottumwa AP	41° 6'N	92° 27'W

Station	Lat	Long		Station	Lat	Long
Mt Vernon	38° 19'N	88° 52'W		Sioux City AP	42° 24'N	96° 23'W
Peoria AP	40° 40'N	89° 41'W		Waterloo	42° 33'N	92° 24'W
Quincy AP	39° 57'N	91° 12'W		**KANSAS**		
Rantoul, Chanute AFB	40° 18'N	88° 8'W		Atchison	39° 34'N	95° 7'W
Rockford	42° 21'N	89° 3'W		Chanute AP	37° 40'N	95° 29'W
Springfield AP	39° 50'N	89° 40'W		Dodge City AP	37° 46'N	99° 58'W
Waukegan	42° 21'N	87° 53'W		El Dorado	37° 49'N	96° 50'W
INDIANA				Emporia	38° 20'N	96° 12'W
Anderson	40° 6'N	85° 37'W		Garden City AP	37° 56'N	100° 44'W
Bedford	38° 51'N	86° 30'W		Goodland AP	39° 22'N	101° 42'W
Bloomington	39° 8'N	86° 37'W		Great Bend	38° 21'N	98° 52'W
Columbus, Bakalar AFB	39° 16'N	85° 54'W		Hutchinson AP	38° 4'N	97° 52'W
Crawfordsville	40° 3'N	86° 54'W		Liberal	37° 3'N	100° 58'W
Evansville AP	38° 3'N	87° 32'W		Manhattan, Ft Riley	39° 3'N	96° 46'W
Fort Wayne AP	41° 0'N	85° 12'W		Parsons	37° 20'N	95° 31'W
Goshen AP	41° 32'N	85° 48'W		Russell AP	38° 52'N	98° 49'W
Hobart	41° 32'N	87° 15'W		Salina	38° 48'N	97° 39'W
Huntington	40° 53'N	85° 30'W		Topeka AP	39° 4'N	95° 38'W
Indianapolis AP	39° 44'N	86° 17'W		Wichita AP	37° 39'N	97° 25'W
Jeffersonville	38° 17'N	85° 45'W		**KENTUCKY**		
Kokomo	40° 25'N	86° 3'W		Ashland	38° 33'N	82° 44'W
Lafayette	40° 2'N	86° 5'W		Bowling Green AP	35° 58'N	86° 28'W
La Porte	41° 36'N	86° 43'W		Corbin AP	36° 57'N	84° 6'W
Marion	40° 29'N	85° 41'W		Covington AP	39° 3'N	84° 40'W

(Continued)

	LONGITUDE	LATITUDE
KENTUCKY (Continued)		
Hopkinsville, Ft Campbell	36° 40'N	87° 29'W
Lexington AP	38° 2'N	84° 36'W
Louisville AP	38° 11'N	85° 44'W
Madisonville	37° 19'N	87° 29'W
Owensboro	37° 45'N	87° 10'W
Paducah AP	37° 4'N	88° 46'W
LOUISIANA		
Alexandria AP	31° 24'N	92° 18'W
Baton Rouge AP	30° 32'N	91° 9'W
Bogalusa	30° 47'N	89° 52'W
Houma	29° 31'N	90° 40'W
Lafayette AP	30° 12'N	92° 0'W
Lake Charles AP	30° 7'N	93° 13'W
Minden	32° 36'N	93° 18'W
Monroe AP	32° 31'N	92° 2'W
Natchitoches	31° 46'N	93° 5'W
New Orleans AP	29° 59'N	90° 15'W
Shreveport AP	32° 28'N	93° 49'W
MAINE		
Augusta AP	44° 19'N	69° 48'W
Bangor, Dow AFB	44° 48'N	68° 50'W
Caribou AP	46° 52'N	68° 1'W
Lewiston	44° 2'N	70° 15'W

	LONGITUDE	LATITUDE
MICHIGAN		
Adrian	41° 55'N	84° 1'W
Alpena AP	45° 4'N	83° 26'W
Battle Creek AP	42° 19'N	85° 15'W
Benton Harbor AP	42° 8'N	86° 26'W
Detroit	42° 25'N	83° 1'W
Escanaba	45° 44'N	87° 5'W
Flint AP	42° 58'N	83° 44'W
Grand Rapids AP	42° 53'N	85° 31'W
Holland	42° 42'N	86° 6'W
Jackson AP	42° 16'N	84° 28'W
Kalamazoo	42° 17'N	85° 36'W
Lansing AP	42° 47'N	84° 36'W
Marquette Co	46° 34'N	87° 24'W
Mt Pleasant	43° 35'N	84° 46'W
Muskegon AP	43° 10'N	86° 14'W
Pontiac	42° 40'N	83° 25'W
Port Huron	42° 59'N	82° 25'W
Saginaw AP	43° 32'N	84° 5'W
Sault Ste. Marie AP	46° 28'N	84° 22'W
Traverse City AP	44° 45'N	85° 35'W
Ypsilanti	42° 14'N	83° 32'W
MINNESOTA		
Albert Lea	43° 39'N	93° 21'W
Alexandria AP	45° 52'N	95° 23'W

Location	Latitude	Longitude
Millinocket AP	45° 39'N	68° 42'W
Portland	43° 39'N	70° 19'W
Waterville	44° 32'N	69° 40'W
MARYLAND		
Baltimore AP	39° 11'N	76° 40'W
Baltimore Co	39° 20'N	76° 25'W
Cumberland	39° 37'N	78° 46'W
Frederick AP	39° 27'N	77° 25'W
Hagerstown	39° 42'N	77° 44'W
Salisbury	38° 20'N	75° 30'W
MASSACHUSETTS		
Boston AP	42° 22'N	71° 2'W
Clinton	42° 24'N	71° 41'W
Fall River	41° 43'N	71° 8'W
Framingham	42° 17'N	71° 25'W
Gloucester	42° 35'N	70° 41'W
Greenfield	42° 3'N	72° 4'W
Lawrence	42° 42'N	71° 10'W
Lowell	42° 39'N	71° 19'W
New Bedford	41° 41'N	70° 58'W
Pittsfield AP	42° 26'N	73° 18'W
Springfield, Westover AFB	42° 12'N	72° 32'W
Taunton	41° 54'N	71° 4'W
Worcester AP	42° 16'N	71° 52'W
Bemidji AP	47° 31'N	94° 56'W
Brainerd	46° 24'N	94° 8'W
Duluth AP	46° 50'N	92° 11'W
Faribault	44° 18'N	93° 16'W
Fergus Falls	46° 16'N	96° 4'W
International Falls AP	48° 34'N	93° 23'W
Mankato	44° 9'N	93° 59'W
Minneapolis/St. Paul AP	44° 53'N	93° 13'W
Rochester AP	43° 55'N	92° 30'W
St. Cloud AP	45° 35'N	94° 11'W
Virginia	47° 30'N	92° 33'W
Willmar	45° 7'N	95° 5'W
Winona	44° 3'N	91° 38'W
MISSISSIPPI		
Biloxi, Keesler AFB	30° 25'N	88° 55'W
Clarksdale	34° 12'N	90° 34'W
Columbus AFB	33° 39'N	88° 27'W
Greenville AFB	33° 29'N	90° 59'W
Greenwood	33° 30'N	90° 5'W
Hattiesburg	31° 16'N	89° 15'W
Jackson AP	32° 19'N	90° 5'W
Laurel	31° 40'N	89° 10'W
McComb AP	31° 15'N	90° 28'W
Meridian AP	32° 20'N	88° 45'W
Natchez	31° 33'N	91° 23'W

(Continued)

	LONGITUDE	LATITUDE		LONGITUDE	LATITUDE
Tupelo	34° 16'N	88° 46'W	Omaha AP	41° 18'N	95° 54'W
Vicksburg Co	32° 24'N	90° 47'W	Scottsbluff AP	41° 52'N	103° 36'W
MISSOURI			Sidney AP	41° 13'N	103° 6'W
Cape Girardeau	37° 14'N	89° 35'W	**NEVADA**		
Columbia AP	38° 58'N	92° 22'W	Carson City	39° 10'N	119° 46'W
Farmington AP	37° 46'N	90° 24'W	Elko AP	40° 50'N	115° 47'W
Hannibal	39° 42'N	91° 21'W	Ely AP	39° 17'N	114° 51'W
Jefferson City	38° 34'N	92° 11'W	Las Vegas AP	36° 5'N	115° 10'W
Joplin AP	37° 9'N	94° 30'W	Lovelock AP	40° 4'N	118° 33'W
Kansas City AP	39° 7'N	94° 35'W	Reno AP	39° 30'N	119° 47'W
Kirksville AP	40° 6'N	92° 33'W	Reno Co	39° 30'N	119° 47'W
Mexico	39° 11'N	91° 54'W	Tonopah AP	38° 4'N	117° 5'W
Moberly	39° 24'N	92° 26'W	Winnemucca AP	40° 54'N	117° 48'W
Poplar Bluff	36° 46'N	90° 25'W	**NEW HAMPSHIRE**		
Rolla	37° 59'N	91° 43'W	Berlin	44° 3'N	71° 1'W
St. Joseph AP	39° 46'N	94° 55'W	Claremont	43° 2'N	72° 2'W
St. Louis AP	38° 45'N	90° 23'W	Concord AP	43° 12'N	71° 30'W
St. Louis Co	38° 39'N	90° 38'W	Keene	42° 55'N	72° 17'W
Sedalia, Whiteman AFB	38° 43'N	93° 33'W	Laconia	43° 3'N	71° 3'W
Sikeston	36° 53'N	89° 36'W	Manchester, Grenier AFB	42° 56'N	71° 26'W
Springfield AP	37° 14'N	93° 23'W	Portsmouth, Pease AFB	43° 4'N	70° 49'W
MONTANA			**NEW JERSEY**		
Billings AP	45° 48'N	108° 32'W	Atlantic City Co	39° 23'N	74° 26'W
Bozeman	45° 47'N	111° 9'W			

	Lat.	Long.		Lat.	Long.
Butte AP	45° 57'N	112° 30'W	Long Branch	40° 19'N	74° 1'W
Cut Bank AP	48° 37'N	112° 22'W	Newark AP	40° 42'N	74° 10'W
Glasgow AP	48° 25'N	106° 32'W	New Brunswick	40° 29'N	74° 26'W
Glendive	47° 8'N	104° 48'W	Paterson	40° 54'N	74° 9'W
Great Falls AP	47° 29'N	111° 22'W	Phillipsburg	40° 41'N	75° 11'W
Havre	48° 34'N	109° 40'W	Trenton Co	40° 13'N	74° 46'W
Helena AP	46° 36'N	112° 0'W	Vineland	39° 29'N	75° 0'W
Kalispell AP	48° 18'N	114° 16'W	**NEW MEXICO**		
Lewiston AP	47° 4'N	109° 27'W	Alamogordo, Holloman AFB	32° 51'N	106° 6'W
Livingstown AP	45° 42'N	110° 26'W	Albuquerque AP	35° 3'N	106° 37'W
Miles City AP	46° 26'N	105° 52'W	Artesia	32° 46'N	104° 23'W
Missoula AP	46° 55'N	114° 5'W	Carlsbad AP	32° 20'N	104° 16'W
NEBRASKA			Clovis AP	34° 23'N	103° 19'W
Beatrice	40° 16'N	96° 45'W	Farmington AP	36° 44'N	108° 14'W
Chadron AP	42° 50'N	103° 5'W	Gallup	35° 31'N	108° 47'W
Columbus	41° 28'N	97° 20'W	Grants	35° 10'N	107° 54'W
Fremont	41° 26'N	96° 29'W	Hobbs AP	32° 45'N	103° 13'W
Grand Island AP	40° 59'N	98° 19'W	Las Cruces	32° 18'N	106° 55'W
Hastings	40° 36'N	98° 26'W	Los Alamos	35° 52'N	106° 19'W
Kearney	40° 44'N	99° 1'W	Raton AP	36° 45'N	104° 30'W
Lincoln Co	40° 51'N	96° 45'W	Roswell, Walker AFB	33° 18'N	104° 32'W
McCook	40° 12'N	100° 38'W	Santa Fe Co	35° 37'N	106° 5'W
Norfolk	41° 59'N	97° 26'W	Silver City AP	32° 38'N	108° 10'W
North Platte AP	41° 8'N	100° 41'W			

(Continued)

	LONGITUDE	LATITUDE		LONGITUDE	LATITUDE
Socorro AP	34° 3'N	106° 53'W	Henderson	36° 22'N	78° 25'W
Tucumcari AP	35° 11'N	103° 36'W	Hickory	35° 45'N	81° 23'W
NEW YORK			Jacksonville	34° 50'N	77° 37'W
Albany AP	42° 45'N	73° 48'W	Lumberton	34° 37'N	79° 4'W
Albany Co	42° 39'N	73° 45'W	New Bern AP	35° 5'N	77° 3'W
Auburn	42° 54'N	76° 32'W	Raleigh/Durham AP	35° 52'N	78° 47'W
Batavia	43° 0'N	78° 11'W	Rocky Mount	35° 58'N	77° 48'W
Binghamton AP	42° 13'N	75° 59'W	Wilmington AP	34° 16'N	77° 55'W
Buffalo AP	42° 56'N	78° 44'W	Winston-Salem AP	36° 8'N	80° 13'W
Cortland	42° 36'N	76° 11'W	**NORTH DAKOTA**		
Dunkirk	42° 29'N	79° 16'W	Bismarck AP	46° 46'N	100° 45'W
Elmira AP	42° 10'N	76° 54'W	Devils Lake	48° 7'N	98° 54'W
Geneva	42° 45'N	76° 54'W	Dickinson AP	46° 48'N	102° 48'W
Glens Falls	43° 20'N	73° 37'W	Fargo AP	46° 54'N	96° 48'W
Gloversville	43° 2'N	74° 21'W	Grand Forks AP	47° 57'N	97° 24'W
Hornell	42° 21'N	77° 42'W	Jamestown AP	46° 55'N	98° 41'W
Ithaca	42° 27'N	76° 29'W	Minot AP	48° 25'N	101° 21'W
Jamestown	42° 7'N	79° 14'W	Williston	48° 9'N	103° 35'W
Kingston	41° 56'N	74° 0'W	**OHIO**		
Lockport	43° 9'N	79° 15'W	Akron-Canton AP	40° 55'N	81° 26'W
Massena AP	44° 56'N	74° 51'W	Ashtabula	41° 51'N	80° 48'W
Newburgh, Stewart AFB	41° 30'N	74° 6'W	Athens	39° 20'N	82° 6'W
NYC-Central Park	40° 47'N	73° 58'W	Bowling Green	41° 23'N	83° 38'W

City	Latitude	Longitude
NYC-Kennedy AP	40° 39'N	73° 47'W
NYC-La Guardia AP	40° 46'N	73° 54'W
Niagara Falls AP	43° 6'N	79° 57'W
Olean	42° 14'N	78° 22'W
Oneonta	42° 31'N	75° 4'W
Oswego Co	43° 28'N	76° 33'W
Plattsburg AFB	44° 39'N	73° 28'W
Poughkeepsie	41° 38'N	73° 55'W
Rochester AP	43° 7'N	77° 40'W
Rome, Griffiss AFB	43° 14'N	75° 25'W
Schenectady	42° 51'N	73° 57'W
Suffolk County AFB	40° 51'N	72° 38'W
Syracuse AP	43° 7'N	76° 7'W
Utica	43° 9'N	75° 23'W
Watertown	43° 59'N	76° 1'W
NORTH CAROLINA		
Asheville AP	35° 26'N	82° 32'W
Charlotte AP	35° 13'N	80° 56'W
Durham	35° 52'N	78° 47'W
Elizabeth City AP	36° 16'N	76° 11'W
Fayetteville, Pope AFB	35° 10'N	79° 1'W
Goldsboro, Seymour-Johnson	35° 20'N	77° 58'W
Greensboro AP	36° 5'N	79° 57'W
Greenville	35° 37'N	77° 25'W

City	Latitude	Longitude
Cambridge	40° 4'N	81° 35'W
Chillicothe	39° 21'N	83° 0'W
Cincinnati Co	39° 9'N	84° 31'W
Cleveland AP	41° 24'N	81° 51'W
Columbus AP	40° 0'N	82° 53'W
Dayton AP	39° 54'N	84° 13'W
Defiance	41° 17'N	84° 23'W
Findlay AP	41° 1'N	83° 40'W
Fremont	41° 20'N	83° 7'W
Hamilton	39° 24'N	84° 35'W
Lancaster	39° 44'N	82° 38'W
Lima	40° 42'N	84° 2'W
Mansfield AP	40° 49'N	82° 31'W
Marion	40° 36'N	83° 10'W
Middletown	39° 31'N	84° 25'W
Newark	40° 1'N	82° 28'W
Norwalk	41° 16'N	82° 37'W
Portsmouth	38° 45'N	82° 55'W
Sandusky Co	41° 27'N	82° 43'W
Springfield	39° 50'N	83° 50'W
Steubenville	40° 23'N	80° 38'W
Toledo AP	41° 36'N	83° 48'W
Warren	41° 20'N	80° 51'W
Wooster	40° 47'N	81° 55'W

(Continued)

	LONGITUDE	LATITUDE
Youngstown AP	41° 16'N	80° 40'W
Zanesville AP	39° 57'N	81° 54'W
OKLAHOMA		
Ada	34° 47'N	96° 41'W
Altus AFB	34° 39'N	99° 16'W
Ardmore	34° 18'N	97° 1'W
Bartlesville	36° 45'N	96° 0'W
Chickasha	35° 3'N	97° 55'W
Enid, Vance AFB	36° 21'N	97° 55'W
Lawton AP	34° 34'N	98° 25'W
McAlester	34° 50'N	95° 55'W
Muskogee AP	35° 40'N	95° 22'W
Norman	35° 15'N	97° 29'W
Oklahoma City AP	35° 24'N	97° 36'W
Ponca City	36° 44'N	97° 6'W
Seminole	35° 14'N	96° 40'W
Stillwater	36° 10'N	97° 5'W
Tulsa AP	36° 12'N	95° 54'W
Woodward	36° 36'N	99° 31'W
OREGON		
Albany	44° 38'N	123° 7'W
Astoria AP	46° 9'N	123° 53'W
Baker AP	44° 50'N	117° 49'W
Bend	44° 4'N	121° 19'W

	LONGITUDE	LATITUDE
Pittsburgh Co	40° 27'N	80° 0'W
Reading Co	40° 20'N	75° 38'W
Scranton/Wilkes-Barre	41° 20'N	75° 44'W
State College	40° 48'N	77° 52'W
Sunbury	40° 53'N	76° 46'W
Uniontown	39° 55'N	79° 43'W
Warren	41° 51'N	79° 8'W
West Chester	39° 58'N	75° 38'W
Williamsport AP	41° 15'N	76° 55'W
York	39° 55'N	76° 45'W
RHODE ISLAND		
Newport	41° 30'N	71° 20'W
Providence AP	41° 44'N	71° 26'W
SOUTH CAROLINA		
Anderson	34° 30'N	82° 43'W
Charleston AFB	32° 54'N	80° 2'W
Charleston Co	32° 54'N	79° 58'W
Columbia AP	33° 57'N	81° 7'W
Florence AP	34° 11'N	79° 43'W
Georgetown	33° 23'N	79° 17'W
Greenville AP	34° 54'N	82° 13'W
Greenwood	34° 10'N	82° 7'W
Orangeburg	33° 30'N	80° 52'W
Rock Hill	34° 59'N	80° 58'W

Location	Latitude	Longitude
Corvallis	44° 30'N	123° 17'W
Eugene AP	44° 7'N	123° 13'W
Grants Pass	42° 26'N	123° 19'W
Klamath Falls AP	42° 9'N	121° 44'W
Medford AP	42° 22'N	122° 52'W
Pendleton AP	45° 41'N	118° 51'W
Portland AP	45° 36'N	122° 36'W
Portland Co	45° 32'N	122° 40'W
Roseburg AP	43° 14'N	123° 22'W
Salem AP	44° 55'N	123° 1'W
The Dalles	45° 36'N	121° 12'W
PENNSYLVANIA		
Allentown AP	40° 39'N	75° 26'W
Altoona Co	40° 18'N	78° 19'W
Butler	40° 52'N	79° 54'W
Chambersburg	39° 56'N	77° 38'W
Erie AP	42° 5'N	80° 11'W
Harrisburg AP	40° 12'N	76° 46'W
Johnstown	40° 19'N	78° 50'W
Lancaster	40° 7'N	76° 18'W
Meadville	41° 38'N	80° 10'W
New Castle	41° 1'N	80° 22'W
Philadelphia AP	39° 53'N	75° 15'W
Pittsburgh AP	40° 30'N	80° 13'W
Spartanburg AP	34° 58'N	82° 0'W
Sumter, Shaw AFB	33° 54'N	80° 22'W
SOUTH DAKOTA		
Aberdeen AP	45° 27'N	98° 26'W
Brookings	44° 18'N	96° 48'W
Huron AP	44° 23'N	98° 13'W
Mitchell	43° 41'N	98° 1'W
Pierre AP	44° 23'N	100° 17'W
Rapid City AP	44° 3'N	103° 4'W
Sioux Falls AP	43° 34'N	96° 44'W
Watertown AP	44° 55'N	97° 9'W
Yankton	42° 55'N	97° 23'W
TENNESSEE		
Athens	35° 26'N	84° 35'W
Bristol-Tri City AP	36° 29'N	82° 24'W
Chattanooga AP	35° 2'N	85° 12'W
Clarksville	36° 33'N	87° 22'W
Columbia	35° 38'N	87° 2'W
Dyersburg	36° 1'N	89° 24'W
Greenville	36° 4'N	82° 50'W
Jackson AP	35° 36'N	88° 55'W
Knoxville AP	35° 49'N	83° 59'W
Memphis AP	35° 3'N	90° 0'W
Murfreesboro	34° 55'N	86° 28'W

(Continued)

	LONGITUDE	LATITUDE		LONGITUDE	LATITUDE
Nashville AP	36° 7'N	86° 41'W	Vernon	34° 10'N	99° 18'W
Tullahoma	35° 23'N	86° 5'W	Victoria AP	28° 51'N	96° 55'W
TEXAS			Waco AP	31° 37'N	97° 13'W
Abilene AP	32° 25'N	99° 41'W	Wichita Falls AP	33° 58'N	98° 29'W
Alice AP	27° 44'N	98° 2'W	**UTAH**		
Amarillo AP	35° 14'N	100° 42'W	Cedar City AP	37° 42'N	113° 6'W
Austin AP	30° 18'N	97° 42'W	Logan	41° 45'N	111° 49'W
Bay City	29° 0'N	95° 58'W	Moab	38° 36'N	109° 36'W
Beaumont	29° 57'N	94° 1'W	Ogden AP	41° 12'N	112° 1'W
Beeville	28° 22'N	97° 40'W	Price	39° 37'N	110° 50'W
Big Spring AP	32° 18'N	101° 27'W	Provo	40° 13'N	111° 43'W
Brownsville AP	25° 54'N	97° 26'W	Richfield	38° 46'N	112° 5'W
Brownwood	31° 48'N	98° 57'W	St George Co	37° 2'N	113° 31'W
Bryan AP	30° 40'N	96° 33'W	Salt Lake City AP	40° 46'N	111° 58'W
Corpus Christi AP	27° 46'N	97° 30'W	Vernal AP	40° 27'N	109° 31'W
Corsicana	32° 5'N	96° 28'W	**VERMONT**		
Dallas AP	32° 51'N	96° 51'W	Barre	44° 12'N	72° 31'W
Del Rio, Laughlin AFB	29° 22'N	100° 47'W	Burlington AP	44° 28'N	73° 9'W
Denton	33° 12'N	97° 6'W	Rutland	43° 36'N	72° 58'W
Eagle Pass	28° 52'N	100° 32'W	**VIRGINIA**		
El Paso AP	31° 48'N	106° 24'W	Charlottesville	38° 2'N	78° 31'W
Fort Worth AP	32° 50'N	97° 3'W	Danville AP	36° 34'N	79° 20'W
Galveston AP	29° 18'N	94° 48'W	Fredericksburg	38° 18'N	77° 28'W
Greenville	33° 4'N	96° 3'W	Harrisonburg	38° 27'N	78° 54'W

Location	Latitude	Longitude	Location	Latitude	Longitude
Harlingen	26° 14'N	97° 39'W	Lynchburg AP	37° 20'N	79° 12'W
Houston AP	29° 58'N	95° 21'W	Norfolk AP	36° 54'N	76° 12'W
Houston Co	29° 59'N	95° 22'W	Petersburg	37° 11'N	77° 31'W
Huntsville	30° 43'N	95° 33'W	Richmond AP	37° 30'N	77° 20'W
Killeen, Robert Gray AAF	31° 5'N	97° 41'W	Roanoke AP	37° 19'N	79° 58'W
Lamesa	32° 42'N	101° 56'W	Staunton	38° 16'N	78° 54'W
Laredo AFB	27° 32'N	99° 27'W	Winchester	39° 12'N	78° 10'W
Longview	32° 28'N	94° 44'W	**WASHINGTON**		
Lubbock AP	33° 39'N	101° 49'W	Aberdeen	46° 59'N	123° 49'W
Lufkin AP	31° 25'N	94° 48'W	Bellingham AP	48° 48'N	122° 32'W
McAllen	26° 12'N	98° 13'W	Bremerton	47° 34'N	122° 40'W
Midland AP	31° 57'N	102° 11'W	Ellensburg AP	47° 2'N	120° 31'W
Mineral Wells AP	32° 47'N	98° 4'W	Everett, Paine AFB	47° 55'N	122° 17'W
Palestine Co	31° 47'N	95° 38'W	Kennewick	46° 13'N	119° 8'W
Pampa	35° 32'N	100° 59'W	Longview	46° 10'N	122° 56'W
Pecos	31° 25'N	103° 30'W	Moses Lake, Larson AFB	47° 12'N	119° 19'W
Plainview	34° 11'N	101° 42'W	Olympia AP	46° 58'N	122° 54'W
Port Arthur AP	29° 57'N	94° 1'W	Port Angeles	48° 7'N	123° 26'W
San Angelo,Goodfellow AFB	31° 26'N	100° 24'W	Seattle-Boeing Field	47° 32'N	122° 18'W
San Antonio AP	29° 32'N	98° 28'W	Seattle Co	47° 39'N	122° 18'W
Sherman, Perrin AFB	33° 43'N	96° 40'W	Seattle-Tacoma AP	47° 27'N	122° 18'W
Snyder	32° 43'N	100° 55'W	Spokane AP	47° 38'N	117° 31'W
Temple	31° 6'N	97° 21'W	Tacoma, McChord AFB	47° 15'N	122° 30'W
Tyler AP	32° 21'N	95° 16'W	Walla Walla AP	46° 6'N	118° 17'W

(Continued)

	LONGITUDE	LATITUDE		LONGITUDE	LATITUDE
Wenatchee	47° 25'N	120° 19'W	Madison AP	43° 8'N	89° 20'W
Yakima AP	46° 34'N	120° 32'W	Manitowoc	44° 6'N	87° 41'W
WEST VIRGINIA			Marinette	45° 6'N	87° 38'W
Beckley	37° 47'N	81° 7'W	Milwaukee AP	42° 57'N	87° 54'W
Bluefield AP	37° 18'N	81° 13'W	Racine	42° 43'N	87° 51'W
Charleston AP	38° 22'N	81° 36'W	Sheboygan	43° 45'N	87° 43'W
Clarksburg	39° 16'N	80° 21'W	Stevens Point	44° 30'N	89° 34'W
Elkins AP	38° 53'N	79° 51'W	Waukesha	43° 1'N	88° 14'W
Huntington Co	38° 25'N	82° 30'W	Wausau AP	44° 55'N	89° 37'W
Martinsburg AP	39° 24'N	77° 59'W	**WYOMING**		
Morgantown AP	39° 39'N	79° 55'W	Casper AP	42° 55'N	106° 28'W
Parkersburg Co	39° 16'N	81° 34'W	Cheyenne	41° 9'N	104° 49'W
Wheeling	40° 7'N	80° 42'W	Cody AP	44° 33'N	109° 4'W
WISCONSIN			Evanston	41° 16'N	110° 57'W
Appleton	44° 15'N	88° 23'W	Lander AP	42° 49'N	108° 44'W
Ashland	46° 34'N	90° 58'W	Laramie AP	41° 19'N	105° 41'W
Beloit	42° 30'N	89° 2'W	Newcastle	43° 51'N	104° 13'W
Eau Claire AP	44° 52'N	91° 29'W	Rawlins	41° 48'N	107° 12'W
Fond Du Lac	43° 48'N	88° 27'W	Rock Springs AP	41° 36'N	109° 0'W
Green Bay AP	44° 29'N	88° 8'W	Sheridan AP	44° 46'N	106° 58'W
La Crosse AP	43° 52'N	91° 15'W	Torrington	42° 5'N	104° 13'W

AP = airport, AFB = air force base.

CANADA LONGITUDES AND LATITUDES

	LONGITUDE	LATITUDE		LONGITUDE	LATITUDE
ALBERTA			Trail	49° 8~ N	117° 44~ W
Calgary AP	51° 6~ N	114° 1~ W	Vancouver AP	49° 11~ N	123° 10~ W
Edmonton AP	53° 34~ N	113° 31~ W	Victoria Co	48° 25~ N	123° 19~ W
Grande Prairie AP	55° 11~ N	118° 53~ W	**MANITOBA**		
Jasper	52° 53~ N	118° 4~ W	Brandon	49° 52~ N	99° 59~ W
Lethbridge AP	49° 38~ N	112° 48~ W	Churchill AP	58° 45~ N	94° 4~ W
McMurray AP	56° 39~ N	111° 13~ W	Dauphin AP	51° 6~ N	100° 3~ W
Medicine Hat AP	50° 1~ N	110° 43~ W	Flin Flon	54° 46~ N	101° 51~ W
Red Deer AP	52° 11~ N	113° 54~ W	Portage La Prairie AP	49° 54~ N	98° 16~ W
BRITISH COLUMBIA			The Pas AP	53° 58~ N	101° 6~ W
Dawson Creek	55° 44~ N	120° 11~ W	Winnipeg AP	49° 54~ N	97° 14~ W
Fort Nelson AP	58° 50~ N	122° 35~ W	**NEW BRUNSWICK**		
Kamloops Co	50° 43~ N	120° 25~ W	Campbellton Co	48° 0~ N	66° 40~ W
Nanaimo	49° 11~ N	123° 58~ W	Chatham AP	47° 1~ N	65° 27~ W
New Westminster	49° 13~ N	122° 54~ W	Edmundston Co	47° 22~ N	68° 20~ W
Penticton AP	49° 28~ N	119° 36~ W	Fredericton AP	45° 52~ N	66° 32~ W
Prince George AP	53° 53~ N	122° 41~ W	Moncton AP	46° 7~ N	64° 41~ W
Prince Rupert Co	54° 17~ N	130° 23~ W	Saint John AP	45° 19~ N	65° 53~ W

(Continued)

CANADA LONGITUDES AND LATITUDES (Continued)

	LONGITUDE	LATITUDE		LONGITUDE	LATITUDE
NEWFOUNDLAND			Sudbury AP	46° 37~ N	80° 48~ W
Corner Brook	48° 58~ N	57° 57~ W	Thunder Bay AP	48° 22~ N	89° 19~ W
Gander AP	48° 57~ N	54° 34~ W	Timmins AP	48° 34~ N	81° 22~ W
Goose Bay AP	53° 19~ N	60° 25~ W	Toronto AP	43° 41~ N	79° 38~ W
St John's AP	47° 37~ N	52° 45~ W	Windsor AP	42° 16~ N	82° 58~ W
Stephenville AP	48° 32~ N	58° 33~ W	**PRINCE EDWARD ISLAND**		
NORTHWEST TERRITORIES			Charlottetown AP	46° 17~ N	63° 8~ W
Fort Smith AP	60° 1~ N	111° 58~ W	Summerside AP	46° 26~ N	63° 50~ W
Frobisher AP	63° 45~ N	68° 33~ W	**QUEBEC**		
Inuvik	68° 18~ N	133° 29~ W	Bagotville AP	48° 20~ N	71° 0~ W
Resolute AP	74° 43~ N	94° 59~ W	Chicoutimi	48° 25~ N	71° 5~ W
Yellowknife AP	62° 28~ N	114° 27~ W	Drummondville	45° 53~ N	72° 29~ W
NOVA SCOTIA			Granby	45° 23~ N	72° 42~ W
Amherst	45° 49~ N	64° 13~ W	Hull	45° 26~ N	75° 44~ W
Halifax AP	44° 39~ N	63° 34~ W	Megantic AP	45° 35~ N	70° 52~ W
Kentville	45° 3~ N	64° 36~ W	Montreal AP	45° 28~ N	73° 45~ W
New Glasgow	45° 37~ N	62° 37~ W	Quebec AP	46° 48~ N	71° 23~ W
Sydney AP	46° 10~ N	60° 3~ W	Rimouski	48° 27~ N	68° 32~ W
Truro Co	45° 22~ N	63° 16~ W	St Jean	45° 18~ N	73° 16~ W
Yarmouth AP	43° 50~ N	66° 5~ W	St Jerome	45° 48~ N	74° 1~ W

ONTARIO

Location	Latitude	Longitude
Belleville	44° 9~ N	77° 24~ W
Chatham	42° 24~ N	82° 12~ W
Cornwall	45° 1~ N	74° 45~ W
Hamilton	43° 16~ N	79° 54~ W
Kapuskasing AP	49° 25~ N	82° 28~ W
Kenora AP	49° 48~ N	94° 22~ W
Kingston	44° 16~ N	76° 30~ W
Kitchener	43° 26~ N	80° 30~ W
London AP	43° 2~ N	81° 9~ W
North Bay AP	46° 22~ N	79° 25~ W
Oshawa	43° 54~ N	78° 52~ W
Ottawa AP	45° 19~ N	75° 40~ W
Owen Sound	44° 34~ N	80° 55~ W
Peterborough	44° 17~ N	78° 19~ W
St Catharines	43° 11~ N	79° 14~ W
Sarnia	42° 58~ N	82° 22~ W
Sault Ste Marie AP	46° 32~ N	84° 30~ W

Location	Latitude	Longitude
Sept. Iles AP	50° 13~ N	66° 16~ W
Shawinigan	46° 34~ N	72° 43~ W
Sherbrooke Co	45° 24~ N	71° 54~ W
Thetford Mines	46° 4~ N	71° 19~ W
Trois Rivieres	46° 21~ N	72° 35~ W
Val D'or AP	48° 3~ N	77° 47~ W
Valleyfield	45° 16~ N	74° 6~ W

SASKATCHEWAN

Location	Latitude	Longitude
Estevan AP	49° 4~ N	103° 0~ W
Moose Jaw AP	50° 20~ N	105° 33~ W
North Battleford AP	52° 46~ N	108° 15~ W
Prince Albert AP	53° 13~ N	105° 41~ W
Regina AP	50° 26~ N	104° 40~ W
Saskatoon AP	52° 10~ N	106° 41~ W
Swift Current AP	50° 17~ N	107° 41~ W
Yorkton AP	51° 16~ N	102° 28~ W

YUKON TERRITORY

Location	Latitude	Longitude
Whitehorse AP	60° 43~ N	135° 4~ W

	LONGITUDE	LATITUDE
AFGHANISTAN		
Kabul	34° 35~ N	69° 12~ E
ALGERIA		
Algiers	36° 46~ N	30° 3~ E
ARGENTINA		
Buenos Aires	34° 35~ S	58° 29~ W
Cordoba	31° 22~ S	64° 15~ W
Tucuman	26° 50~ S	65° 10~ W
AUSTRALIA		
Adelaide	34° 56~ S	138° 35~ E
Alice Springs	23° 48~ S	133° 53~ E
Brisbane	27° 28~ S	153° 2~ E
Darwin	12° 28~ S	130° 51~ E
Melbourne	37° 49~ S	144° 58~ E
Perth	31° 57~ S	115° 51~ E
Sydney	33° 52~ S	151° 12~ E
AUSTRIA		
Vienna	48° 15~ N	16° 22~ E
AZORES		
Lajes (Terceira)	38° 45~ N	27° 5~ W
BAHAMAS		
Nassau	25° 5~ N	77° 21~ W

	LONGITUDE	LATITUDE
BURMA		
Mandalay	21° 59~ N	96° 6~ E
Rangoon	16° 47~ N	96° 9~ E
CAMBODIA		
Phnom Penh	11° 33~ N	104° 51~ E
CHILE		
Punta Arenas	53° 10~ S	70° 54~ W
Santiago	33° 27~ S	70° 42~ W
Valparaiso	33° 1~ S	71° 38~ W
CHINA		
Chongquing	29° 33~ N	106° 33~ E
Shanghai	31° 12~ N	121° 26~ E
COLOMBIA		
Baranquilla	10° 59~ N	74° 48~ W
Bogota	4° 36~ N	74° 5~ W
Cali	3° 25~ N	76° 30~ W
Medellin	6° 13~ N	75° 36~ W
CONGO		
Brazzaville	4° 15~ S	15° 15~ E
CUBA		
Guantanamo Bay	19° 54~ N	75° 9~ W
Havana	23° 8~ N	82° 21~ W

	Latitude	Longitude
BANGLADESH		
Chittagong	22° 21~ N	91° 50~ E
BELGIUM		
Brussels	50° 48~ N	4°21~ E
BELIZE		
Belize	17° 31~ N	88° 11~ W
BERMUDA		
Kindley AFB	33° 22~ N	64° 41~ W
BOLIVIA		
La Paz	16° 30~ S	68° 9~ W
BRAZIL		
Belem	1° 27~ S	48° 29~ W
Belo Horizonte	19° 56~ S	43° 57~ W
Brasilia	15° 52~ S	47° 55~ W
Curitiba	25° 25~ S	49° 17~ W
Fortaleza	3° 46~ S	38° 33~ W
Porto Alegre	30° 2~ S	51° 13~ W
Recife	8° 4~ S	34° 53~ W
Rio de Janeiro	22° 55~ S	43° 12~ W
Salvador	13° 0~ S	38° 30~ W
Sao Paulo	23° 33~ S	46° 38~ W
BULGARIA		
Sofia	42° 42~ N	23° 20~ E
Strasbourg	48° 35~ N	7° 46~ E
CZECHOSLOVAKIA		
Prague	50° 5~ N	14° 25~ E
DENMARK		
Copenhagen	55° 41~ N	12° 33~ E
DOMINICAN REPUBLIC		
Santo Domingo	18° 29~ N	69° 54~ W
EGYPT		
Cairo	29° 52~ N	31° 20~ E
EL SALVADOR		
San Salvador	13° 42~ N	89° 13~ W
EQUADOR		
Guayaquil	2° 0~ S	79° 53~ W
Quito	0° 13~ S	78° 32~ W
ETHIOPIA		
Addis Ababa	90° 2~ N	38° 45~ E
Asmara	15° 17~ N	38° 55~ E
FINLAND		
Helsinki	60° 10~ N	24° 57~ E
FRANCE		
Lyon	45° 42~ N	4° 47~ E
Marseilles	43° 18~ N	5° 23~ E
Nantes	47° 15~ N	1° 34~ W
Nice	43° 42~ N	7° 16~ E
Paris	48° 49~ N	2° 29~ E

(Continued)

	LONGITUDE	LATITUDE
FRENCH GUIANA		
Cayenne	4° 56~ N	52° 27~ W
GERMANY		
Berlin (West)	52° 27~ N	13° 18~ E
Hamburg	53° 33~ N	9° 58~ E
Hannover	52° 24~ N	9° 40~ E
Mannheim	49° 34~ N	8° 28~ E
Munich	48° 9~ N	11° 34~ E
GHANA		
Accra	5° 33~ N	0° 12~ W
GIBRALTAR		
Gibraltar	36° 9~ N	5° 22~ W
GREECE		
Athens	37° 58~ N	23° 43~ E
Thessaloniki	40° 37~ N	22° 57~ E
GREENLAND		
Narsarssuaq	61° 11~ N	45° 25~ W
GUATEMALA		
Guatemala City	14° 37~ N	90° 31~ W
GUYANA		
Georgetown	6° 50~ N	58° 12~ W

	LONGITUDE	LATITUDE
IRAN		
Abadan	30° 21~ N	48° 16~ E
Meshed	36° 17~ N	59° 36~ E
Tehran	35° 41~ N	51° 25~ E
IRAQ		
Baghdad	33° 20~ N	44° 24~ E
Mosul	36° 19~ N	43° 9~ E
IRELAND		
Dublin	53° 22~ N	6° 21~ W
Shannon	52° 41~ N	8° 55~ W
IRIAN BARAT		
Manokwari	0° 52~ S	134° 5~ E
ISRAEL		
Jerusalem	31° 47~ N	35° 13~ E
Tel Aviv	32° 6~ N	34° 47~ E
ITALY		
Milan	45° 27~ N	9° 17~ E
Naples	40° 53~ N	14° 18~ E
Rome	41° 48~ N	12° 36~ E
IVORY COAST		
Abidjan	5° 19~ N	4° 1~ W

Place	Latitude	Longitude
HAITI		
Port au Prince	18° 33~ N	72° 20~ W
HONDURAS		
Tegucigalpa	14° 6~ N	87° 13~ W
HONG KONG		
Hong Kong	22° 18~ N	114° 10~ E
HUNGARY		
Budapest	47° 31~ N	19° 2~ E
ICELAND		
Reykjavik	64° 8~ N	21° 56~ E
INDIA		
Ahmenabad	23° 2~ N	72° 35~ E
Bangalore	12° 57~ N	77° 37~ E
Bombay	18° 54~ N	72° 49~ E
Calcutta	22° 32~ N	88° 20~ E
Madras	13° 4~ N	80° 15~ E
Nagpur	21° 9~ N	79° 7~ E
New Delhi	28° 35~ N	77° 12~ E
INDONESIA		
Djakarta	6° 11~ S	106° 50~ E
Kupang	10° 10~ S	123° 34~ E
Makassar	5° 8~ S	119° 28~ E
Medan	3° 35~ N	98° 41~ E
Palembang	3° 0~ S	104° 46~ E
Surabaya	7° 13~ S	112° 43~ E
JAPAN		
Fukuoka	33° 35~ N	130° 27~ E
Sapporo	43° 4~ N	141° 21~ E
Tokyo	35° 41~ N	139° 46~ E
JORDAN		
Amman	31° 57~ N	35° 57~ E
KENYA		
Nairobi	1° 16~ S	36° 48~ E
KOREA		
Pyongyang	39° 2~ N	125° 41~ E
Seoul	37° 34~ N	126° 58~ E
LEBANON		
Beirut	33° 54~ N	35° 28~ E
LIBERIA		
Monrovia	6° 18~ N	10° 48~ W
LIBYA		
Benghazi	32° 6~ N	20° 4~ E
MADAGASCAR		
Tananarive	18° 55~ S	47° 33~ E
MALAYSIA		
Kuala Lumpur	3° 7~ N	101° 42~ E
Penang	5° 25~ N	100° 19~ E
MARTINIQUE		
Fort de France	14° 37~ N	61° 5~ W

(Continued)

	LONGITUDE	LATITUDE
MEXICO		
Guadalajara	20° 41~ N	103° 20~ W
Merida	20° 58~ N	89° 38~ W
Mexico City	19° 24~ N	99° 12~ W
Monterrey	25° 40~ N	100° 18~ W
Vera Cruz	19° 12~ N	96° 8~ W
MOROCCO		
Casablanca	33° 35~ N	7° 39~ W
NEPAL		
Katmandu	27° 42~ N	85° 12~ E
NETHERLANDS		
Amsterdam	52° 23~ N	4° 55~ E
NEW ZEALAND		
Auckland	36° 51~ S	174° 46~ E
Christchurch	43° 32~ S	172° 37~ E
Wellington	41° 17~ S	174° 46~ E
NICARAGUA		
Managua	12° 10~ N	86° 15~ W
NIGERIA		
Lagos	6° 27~ N	3° 24~ E
NORWAY		
Bergen	60° 24~ N	5° 19~ E

	LONGITUDE	LATITUDE
RUSSIA		
Alma Ata	43° 14~ N	76° 53~ E
Archangel	64° 33~ N	40° 32~ E
Kaliningrad	54° 43~ N	20° 30~ E
Krasnoyarsk	56° 1~ N	92° 57~ E
Kiev	50° 27~ N	30° 30~ E
Kharkov	50° 0~ N	36° 14~ E
Kuibyshev	53° 11~ N	50° 6~ E
Leningrad	59° 56~ N	30° 16~ E
Minsk	53° 54~ N	27° 33~ E
Moscow	55° 46~ N	37° 40~ E
Odessa	46° 29~ N	30° 44~ E
Petropavlovsk	52° 53~ N	158° 42~ E
Rostov on Don	47° 13~ N	39° 43~ E
Sverdlovsk	56° 49~ N	60° 38~ E
Tashkent	41° 20~ N	69° 18~ E
Tbilisi	41° 43~ N	44° 48~ E
Vladivostok	43° 7~ N	131° 55~ E
Volgograd	48° 42~ N	44° 31~ E
SAUDI ARABIA		
Dhahran	26° 17~ N	50° 9~ E
Jedda	21° 28~ N	39° 10~ E

City	Latitude	Longitude
Oslo	59° 56~ N	10° 44~ E
PAKISTAN		
Karachi	24° 48~ N	66° 59~ E
Lahore	31° 35~ N	74° 20~ E
Peshwar	34° 1~ N	71° 35~ E
PANAMA		
Panama City	8° 58~ N	79° 33~ W
PAPUA NEW GUINEA		
Port Moresby	9° 29~ S	147° 9~ E
PARAGUAY		
Asuncion	25° 17~ S	57° 30~ W
PERU		
Lima	12° 5~ S	77° 3~ W
PHILIPPINES		
Manila	14° 35~ N	120° 59~ E
POLAND		
Krakow	50° 4~ N	19° 57~ E
Warsaw	52° 13~ N	21° 2~ E
PORTUGAL		
Lisbon	38° 43~ N	9° 8~ W
PUERTO RICO		
San Juan	18° 29~ N	66° 7~ W
RUMANIA		
Bucharest	44° 25~ N	26° 6~ E
Riyadh	24° 39~ N	46° 42~ E
SENEGAL		
Dakar	14° 42~ N	17° 29~ W
SINGAPORE		
Singapore	1° 18~ N	103° 50~ E
SOMALIA		
Mogadiscio	2° 2~ N	49° 19~ E
SOUTH AFRICA		
Cape Town	33° 56~ S	18° 29~ E
Johannesburg	26° 11~ S	28° 3~ E
Pretoria	25° 45~ S	28° 14~ E
SOUTH YEMEN		
Aden	12° 50~ N	45° 2~ E
SPAIN		
Barcelona	41° 24~ N	2° 9~ E
Madrid	40° 25~ N	3° 41~ W
Valencia	39° 28~ N	0° 23~ W
SRI LANKA		
Colombo	6° 54~ N	79° 52~ E
SUDAN		
Khartoum	15° 37~ N	32° 33~ E
SURINAM		
Paramaribo	5° 49~ N	55° 9~ W
SWEDEN		
Stockholm	59° 21~ N	18° 4~ E

(Continued)

	LONGITUDE	LATITUDE
SWITZERLAND		
Zurich	47° 23~ N	8° 33~ E
SYRIA		
Damascus	33° 30~ N	36° 20~ E
TAIWAN		
Tainan	22° 57~ N	120° 12~ E
Taipei	25° 2~ N	121° 31~ E
TANZANIA		
Dar es Salaam	6° 50~ S	39° 18~ E
THAILAND		
Bangkok	13° 44~ N	100° 30~ E
TRINIDAD		
Port of Spain	10° 40~ N	61° 31~ W
TUNISIA		
Tunis	36° 47~ N	10° 12~ E
TURKEY		
Adana	36° 59~ N	35° 18~ E
Ankara	39° 57~ N	32° 53~ E
Istanbul	40° 58~ N	28° 50~ E
Izmir	38° 26~ N	27° 10~ E
UNITED KINGDOM		
Belfast	54° 36~ N	5° 55~ W

	LONGITUDE	LATITUDE
Birmingham	52° 29~ N	1° 56~ W
Cardiff	51° 28~ N	3° 10~ W
Edinburgh	55° 55~ N	3° 11~ W
Glasgow	55° 52~ N	4° 17~ W
London	51° 29~ N	0° 0~ W
URUGUAY		
Montevideo	34° 51~ S	56° 13~ W
VENEZUELA		
Caracas	10° 30~ N	66° 56~ W
Maracaibo	10° 39~ N	71° 36~ W
VIETNAM		
Da Nang	16° 4~ N	108° 13~ E
Hanoi	21° 2~ N	105° 52~ E
Ho Chi Minh City (Saigon)	10° 47~ N	106° 42~ E
YUGOSLAVIA		
Belgrade	44° 48~ N	20° 28~ E
ZAIRE		
Kinshasa (Leopoldville)	4° 20~ S	15° 18~ E
Kisangani (Stanleyville)	0° 26~ S	15° 14~ E

ENERGY SYSTEMS

The following is a summarized narrative discussed in the Wikipedia Web encyclopedia.[1] In order to differentiate forms of alternative energy sources, it is important to understand the various definitions of energy. Energy in physics, chemistry, and nature occurs in numerous forms, all of which imply similar connotations as the ability to perform work. In physics and other sciences, energy is a scalar quantity that is a property of objects and systems and is conserved by nature.

Several different forms of energy, including kinetic energy, potential energy, thermal energy, gravitational energy, electromagnetic radiation energy, chemical energy, and nuclear energy have been defined to explain all known natural phenomena.

Conservation of Energy

Energy is transformation from one form to another, but it is never created or destroyed. This principle, the law of conservation of energy, was first postulated in the early nineteenth century, and applies to any isolated system. The total energy of a system does not change over time, but its value may depend on the frame of reference. For example, a seated passenger in a moving vehicle has zero kinetic energy relative to the vehicle, but does indeed have kinetic energy relative to earth.

The Concept of Energy in Various Scientific Fields

- In chemistry, the energy differences between chemical substances determine whether, and to what extent, they can be converted into or react with other substances.
- In biology, chemical bonding is often broken and made during metabolism. Energy is often stored by the body in the form of carbohydrates and lipids, both of which release energy when reacting with oxygen.

■ In earth sciences, continental drift, volcanic activity, and earthquake are phenomena that can be explained in terms of energy transformation in the earth's interior. Meteorological phenomena like wind, rain, hail, snow, lightning, tornado, and hurricanes are all a result of energy transformations brought about by solar energy.

Energy transformations in the universe are characterized by various kinds of potential energy that have been available since the Big Bang, later "released" to be transformed into more active types of energy.

NUCLEAR DECAY

Examples of such processes include those in which energy that was originally "stored" in heavy isotopes such as uranium and thorium and are released by nucleosynthesis. In this process, gravitational potential energy, released from the gravitational collapse of supernovae, is used to store energy in the creation of these heavy elements before their incorporation into the solar system and earth. This energy is triggered and released in nuclear fission bombs.

FUSION

In a similar chain of transformations at the dawn of the universe, the nuclear fusion of hydrogen in the sun released another store of potential energy that was created at the time of the Big Bang. Space expanded, and the universe cooled too rapidly for hydrogen to completely fuse into heavier elements. Hydrogen thus represents a store of potential energy that can be released by nuclease fusion.

SUNLIGHT ENERGY STORAGE

Light from our sun may again be stored as gravitational potential energy after it strikes the earth. After being released at a hydroelectric dam, this water can be used to drive turbines and generators to produce electricity. Sunlight also drives all weather phenomena, including events such as hurricanes, in which large unstable areas of warm ocean, heated over months, suddenly give up some of their thermal energy to power intense air movement.

KINETIC VERSUS POTENTIAL ENERGY

An important distinction should be made between kinetic and potential energy. Potential energy is the energy of matter attributed to its position or arrangement. This stored energy can be found in any lifted objects, which have the force of gravity bringing them down to their original positions. Kinetic energy is the energy that an object possesses because of its motion. A great example of this is seen with a ball that falls under the influence of gravity. As it accelerates downward, its potential energy is converted into kinetic energy. When it hits the ground and deforms, the kinetic energy converts into elastic potential energy. Upon bouncing back up, this potential energy once again becomes kinetic energy.

The two forms, though seemingly very different, play important roles in complementing each other.

GRAVITATIONAL POTENTIAL ENERGY

The gravitational force near the earth's surface is equal to the mass, m, multiplied by the gravitational acceleration, $g = 9.81$ m/s².

Temperature

On the macroscopic scale, temperature is the unique physical property that determines the direction of heat flow between two objects placed in thermal contact.

If no heat flow occurs, the two objects have the same temperature, as heat flows from the hotter object to the colder object. These two basic principles are stated in the Zeroth Law of thermodynamics and the Second Law of thermodynamics, respectively. For a solid, these microscopic motions are principally the vibrations of its atoms about their sites in the solid.

In most of the world (except for the United States, Jamaica, and a few other countries), the degree-Celsius scale is used for most temperature-measuring purposes. The global scientific community, with the United States included, measures temperature using the Celsius scale and thermodynamic temperature using the Kelvin scale, in which 0 K$= -273.15°$C, or absolute zero.

The United States is the last major country in which the degree-Fahrenheit scale is popularly used in everyday life. In the field of engineering in the United States, the Centigrade degree is relied upon when working in thermodynamic-related subjects.

SPECIFIC HEAT CAPACITY

Specific heat capacity, also known as specific heat, is a measure of the energy that is needed to raise the temperature of a quantity of a substance by a certain temperature.

Chemical Energy

Chemical energy is defined as the work done by electric forces during the rearrangement of electric charges, electrons, and protons in the process of aggregation.

If the chemical energy of a system decreases during a chemical reaction, it is transferred to the surroundings in some form of energy (often heat). On the other hand, if the chemical energy of a system increases as a result of a chemical reaction, it is from the conversion of another form of energy from its surroundings.

Moles are the typical units used to describe change in chemical energy, and values can range from tens to hundreds of kJ/mol.

Radiant Energy

Radiant energy is the energy of an electromagnetic wave, or sometimes of other forms of radiation. Like all forms of energy, its unit is the joule. The term is used especially when radiation is emitted by a source into the surrounding environment.

As electromagnetic (EM) radiation can be conceptualized as a stream of photons, radiant energy can be seen as the energy carried by these photons. EM radiation can also be seen as an electromagnetic wave that carries energy in its oscillating electric and magnetic fields. Quantum field theory reconciles these two views.

EM radiation can have a range of frequencies. From the viewpoint of photons, the energy carried by each photon is proportional to its frequency. From the wave viewpoint, the energy of a monochromatic wave is proportional to its intensity. Thus, it can be implied that if two EM waves have the same intensity but different frequencies, the wave with the higher frequency contains fewer photons.

When EM waves are absorbed by an object, their energy is typically converted to heat. This is an everyday phenomenon, seen, for example, when sunlight warms the surfaces it irradiates. This is often associated with infrared radiation, but any kind of EM radiation will warm the object that absorbs it. EM waves can also be reflected or scattered, causing their energy to be redirected or redistributed.

Energy can enter or leave an open system in the form of radiant energy. Such a system can be man-made, as with a solar energy collector, or natural, as with the earth's atmosphere. Greenhouse gasses trap the sun's radiant energy at certain wavelengths, allowing it to penetrate deep into the atmosphere or all the way to the earth's surface, where it is reemitted as longer wavelengths. Radiant energy is produced in the sun because of the phenomenon of nuclear fusion.

Reference

1. http://en.wikipedia.org/wiki/Energy_systems

LOS ANGELES CITY FIRE

DEPARTMENT REQUIREMENT

Solar Photovoltaic System

The following are the Los Angeles City Fire Department's minimum requirements for solar photovoltaic system installations.

References

A. State Fire Marshal
  1 Solar Photovoltaic Installation Guidelines
B. International Fire Code (IFC)—2006
  1 1003.3.3—Horizontal projections
  2 1003.6—Means of egress continuity
  3 1014.3—Common path of egress travel
C. Los Angeles Municipal Code
  1 57.12.03—Storage on roofs
  2 57.12.04—Passageways on roofs
  3 57.138.04—Access aisles and operating clearances

SCOPE

This requirement regulates the installation of solar photovoltaic systems and their ancillary devices. Included are requirements regulating access, fire protection, and other measures and general precautions relating to solar photovoltaic systems.

DEFINITIONS

The following words and phrases whenever used in this requirement shall be construed as defined in this section.

Array—An uninterrupted section of solar photovoltaic panels or a group of interconnected subarrays.

Grid—The electrical system that is on the service side of the meter.

Inverter—A device used to convert direct current (DC) electricity from the solar system to alternating current (AC) electricity for use in the building's electrical system or the grid.

Required access pathway—Required walking pathway that is designed to provide emergency access that meets the requirements of Sections 57.46.06 and 57.46.09 of Fire Code.

Solar photovoltaic system—A system of solar power components and parts that receives sunlight and converts it to electricity.

Subarray—Uninterrupted sections of solar photovoltaic panels interconnected into an array.

Travel distance—The walking distance between two points.

Venting cutout—Section(s) in an array that are designed to accommodate emergency ventilation procedures.

Plan Review

All solar installations on buildings shall be approved by the Construction Services Unit prior to installation. At a minimum, the following information shall be presented for approval:

Site plan (to scale) of the structure, on which the photovoltaic arrays are to be installed showing the following:

- Footprint of the building and north reference point
- Location of all structures on site
- Street address of building
- Access from street to building
- Location of arrays
- Location of disconnects
- Location of required signage
- Location of required access pathways

- Plan and elevation views of building clearly showing the following:
  - Array placement
  - Roof ridgelines
  - Eave lines
  - Equipment on roof
  - Other objects that may be present on the roof, such as vent lines, skylights, and roof hatches
- Location and verbiage of all markings, labels, and warning signs
- Building photographs that may be useful in the evaluation of the array placement

Markings, Labels, and Warning Signs

Purpose: Provides emergency responders with appropriate warning and guidance with respect to isolating the solar electrical system. This can facilitate identifying energized electrical lines that connect the solar panels to the inverter, as these should not be cut when venting for smoke removal.

MAIN SERVICE DISCONNECT

Residential buildings—The marking may be placed within the main service disconnect. The marking shall be placed on the outside cover if the main service disconnect is operable with the service panel closed.

Commercial buildings—The marking shall be placed adjacent to the main service disconnect clearly visible from the location where the lever is operated.

Markings: Verbiage, Format, and Type of Material
Verbiage: CAUTION: SOLAR ELECTRIC SYSTEM CONNECTED
Format: White lettering on a red background with minimum 3/8 inches letter height; all letters shall be capitalized Arial or similar font, non-bold.
Material: Reflective, weather resistant material suitable for the environment (use UL—969 as standard for weather rating). Durable adhesive materials meet this requirement.

Marking Requirements on DC Conduit, Raceways, Enclosures, Cable Assemblies, DC Combiners, and Junction Boxes:

Markings: Placement, Verbiage, Format, and Type of Material.

Placement: Markings shall be placed every 10 feet on all interior and exterior DC conduits, raceways, enclosures, and cable assemblies, at turns, above and for below penetrations, at all DC combiners, and at junction boxes.
Verbiage: CAUTION: SOLAR CIRCUIT
Inverters—Are not required to have caution markings.

ACCESS PATHWAYS AND SMOKE VENTILATION

Solar Photovoltaic Systems for One- and Two-Family Dwelling Units:
All plans are required to be reviewed by the Fire Department.
Access:
Buildings with a hip roof layout:

- Panels shall be located in a manner that provides one 3-foot-wide clear-access pathway from the eave to the ridge on each roof slope where panels are located.
- The access pathway shall be located at a structurally strong location on the building (i.e., bearing wall).

Buildings with a single ridge:

- Panels shall be located in a manner that provides two 3-foot-wide access pathways from the eave to the ridge on each slope where panels are located.
- Access pathway clear width shall not include any eave overhang.

Hips and valleys:

- Panels shall be located no closer than $1^1/_2$ feet to a hip or valley if placed on both sides of the hip or valley.
- If the panels are to be located on only one side of a hip or valley (of equal length), then the panels may be placed directly adjacent to the hip or valley.

Dead ends:

- Where there are two or more access pathways, the clear pathways shall be arranged so there are no dead ends greater than 25 feet in length.
- If any access pathway leading to a dead end is greater than 25 feet in distance it shall continue on to the next access pathway.
- At no time shall any access pathway cause a person's travel distance to exceed 150 feet before arriving at another required access pathway.

Ventilation:

- An uninterrupted section of photovoltaic panels (array) shall not exceed 150 feet by 150 feet in dimension in either axis.
- Panels shall be located no higher than 3 feet below the ridge.

EXCEPTION: The panels may be located 2 feet below the ridge if the Department has determined that an approved product or method will provide an equal or greater opportunity for ventilation.

Commercial buildings and residential housing comprised of three or more units or single-family attached buildings, one- and two-family residential unit access and ventilation requirements.

Access:

■ A minimum 6-foot wide clear perimeter is required around the edges of the roof.

EXCEPTION: If either axis of the building is 250 feet or less, there shall be a minimum 4-foot-wide clear perimeter around the edges of the roof.

■ Pathways: Shall be established in the design of the solar installation and meet the following requirements:
  ■ Located over structurally supported members.
  ■ Centerline axis pathways shall be provided in both axes of the roof. Centerline axis pathways shall run on structurally supported members or over the next closest structurally supported member nearest to the centerlines of the roof.
  ■ A minimum of 4-feet clearance straightline pathway shall be provided from the access path to skylights and/or ventilation hatches.
  ■ A minimum of 4 feet clear straightline pathway shall be provided from the access to roof standpipes.
  ■ Not less than a 4-feet clearance around roof access hatches with a minimum of one pathway that is straight and not less than 4 feet clear to the parapet or roof edge.
■ Ventilation: Arrays shall be no greater than 150 feet by 150 feet in dimension in either axis. Ventilation options between array sections shall be one of the following:
  ■ An access pathway 8 feet or greater in width.
  ■ The access pathways shall be 4 feet or greater in width and bordering on the existing roof skylights or ventilation hatches.
  ■ The access pathways shall be 4 feet or greater in width with bordering 4-foot by 8-foot "venting cutouts" every 20 feet on alternating sides of the pathway.

DIRECT CURRENT (DC) CONDUCTOR LOCATIONS

Conduits, wiring systems, and raceways:

■ Located as close as possible to the ridge, hip, or valley and from the hip or valley as directly as possible to an outside wall.
■ Conduit runs between subarrays and DC combiner boxes shall:
■ Use design guidelines that minimize the total amount of conduit used on the roof by taking the shortest path from the array to the DC combiner box.
■ The DC combiner boxes are to be located such a way that conduit runs could be minimized in pathways between arrays.

DC wiring:

■ DC wiring shall be run in metallic conduit or raceways when located within enclosed spaces in a building.
■ When possible DC wiring shall run along the bottom of load-bearing members.

GROUND-MOUNTED PHOTOVOLTAIC ARRAYS

Setback requirements:

- Does not apply to ground-mounted free-standing photovoltaic arrays.
- A 10-foot minimum clearance is required around ground-mounted PV systems.
- Shall not obstruct Fire Department access.

OVERHEAD ARRAYS ON ROOF TOPS (E.G., TRELLIS SYSTEMS)

Minimum requirements:

- Overhead arrays shall comply with the same marking, labeling, and warning signs as required of roof-mounted systems.
- There shall be an unobstructed clearance of 7 feet or more between the roof deck surface and the underside of the overhead array.
- The regulations in 57.12.03 and 57.138.04 of the Los Angeles Fire Code shall be complied with.
- An uninterrupted section of solar photovoltaic panels shall not exceed 150 feet by 150 feet in dimension in either axis.
- The overhead clear width between arrays or subarrays shall be 4 feet or greater extending from the edge of the array(s) to the roof deck surface, thereby maintaining an unobstructed access pathway, and providing for emergency ventilation procedures.

The use of the area below arrays is prohibited.

PHOTO GALLERY

Figure D.1 Southern California Metropolitan Water District Museum of Water & Life. *Photo courtesy of Vector Delta Design & Lehrer Gangi Architects.*

Figure D.2   Southern California Metropolitan Water District Museum of Water & Life. *Photo courtesy of Vector Delta Design & Lehrer Gangi Architects.*

Figure D.3 **Amonix megaconcentrator solar farm.** *Photo courtesy of Amonix.*

Figure D.4 **Broad view of Amonix megaconcentrator solar farm.** *Photo courtesy of Amonix.*

Figure D.5   **SolFocus CPV system installation.** *Photo courtesy of SolFocus.*

Figure D.6   **SolFocus CPV system installation.** *Photo courtesy of SolFocus.*

Figure D.7   **Skinner Lake water treatment plant.** *Photo courtesy of Vector Delta Design Group, Inc.*

Figure D.8   **Concentrix CPV system installation.** *Photo courtesy of Concentrix.*

Figure D.9   **Concentrix CPV system installation.** *Photo courtesy of Concentrix.*

Figure D.10   **Solar carport parking structure.** *Photo courtesy of ProtekPark.*

Figure D.11   Solar carport parking structure. *Photo courtesy of ProtekPark.*

Figure D.12   Solar carport parking structure. *Photo courtesy of ProtekPark.*

Figure D.13   Solar carport parking structure. *Photo courtesy of ProtekPark.*

Figure D.14   Solar carport parking structure. *Photo courtesy of ProtekPark.*

Figure D.15   Solar carport parking structure. *Photo courtesy of ProtekPark.*

Figure D.16 **Protek solar carport parking structure.** *Photo courtesy of EATON Corporation.*

Figure D.17 **Solar power carport canopy installation.** *Photo courtesy of EATON Corporation.*

Figure D.18 **Solar carport parking structure.** *Photo courtesy of ProtekPark.*

Figure D.19   Solar power system monument, southern Spain.

Figure D.20   Solar power system monument, southern Spain.

STATE-WIDE SOLAR

INITIATIVE PROGRAMS

Arizona

INCENTIVES/POLICIES FOR RENEWABLES & EFFICIENCY

Solar and Wind Equipment Sales Tax Exemption

Last DSIRE Review: 05/13/2010

State:	Arizona
Incentive Type:	Sales Tax Incentive
Eligible Renewable/Other Technologies:	Passive Solar Space Heat, Solar Water Heat, Solar Space Heat, Solar Thermal Electric, Photovoltaics, Wind, Solar Pool Heating, Daylighting
Applicable Sectors:	Commercial, Residential, General Public/Consumer
Amount:	100% of sales tax on eligible equipment
Maximum Incentive:	No maximum
Start Date:	1/1/1997
Expiration Date:	12/31/2016
Web Site:	http://www.azsolarcenter.com/economics/taxbreaks.html

Authority 1:
A.R.S. § 42-5061 (N)
Date Effective:
1/1/1997
Expiration Date
12/31/2016

Authority 2:
A.R.S. § 42-5075 (14)
Date Effective:
1/1/1997
Expiration Date
12/31/2016

Authority 3:
HB 2700
Date Enacted:
5/10/2010
Date Effective:
5/10/2010
Expiration Date
12/31/2016

Arizona provides a sales tax exemption* for the retail sale of solar energy devices and for installation of solar energy devices by contractors. The statutory definition of "solar energy device" includes wind electric generators and wind-powered water pumps in addition to daylighting, passive solar heating, active solar space heating, solar water heating, and photovoltaics. The sales tax exemption does not apply to batteries, controls, etc., that are not part of the system. (Note that HB 2429, enacted in June 2006, eliminated the $5,000 limit per device.)

To take advantage of these exemptions from tax, a solar energy retailer or a solar energy contractor must register with the Arizona Department of Revenue prior to selling or installing solar energy devices. (Arizona Form 6015, Solar Energy Devices—Application for Registration)

The Arizona Department of Commerce Energy Office has compiled a guide to the solar energy devices that qualify for exemption under the statutory definition. It is possible to petition the Arizona Department of Commerce to add additional items if they qualify per the statutory definition.

According to the Arizona Solar Center's Web site, another provision of Arizona sales tax exemption may apply without value limit to the basic power generating part of the system (consisting of at least PV modules, structure, array wiring, and controls; the limits have not been clearly defined). This further exemption requires the filling out of form ADOR 5000 titled "Transaction Privilege Tax Exemption Certificate" and checking reason #15, "Machinery, equipment or transmission lines used directly in producing or transmitting electrical power, but not including distribution."

Most cities have a 0.5 to 2% city privilege ("sales") tax that is applicable to sales or installations of solar energy devices, unless a city specifically exempts such sales under its city tax code. Solar energy retailers should check with the city in which the retail business is located to find out whether city privilege tax is applicable. Solar energy contractors should check with the city in which the installation will be performed to find out whether city privilege tax is applicable.

Technically, the law allows retailers to deduct the amount received from the sale of solar energy devices from their transaction privilege tax base, and similarly, it allows prime contractors to deduct proceeds from a contract to provide and install a device from their transaction privilege tax base.

Non-Residential Solar and Wind Tax Credit (Personal)

Last DSIRE Review: 05/13/2010

State:	Arizona
Incentive Type:	Personal Tax Credit
Eligible Renewable/Other Technologies:	Passive Solar Space Heat, Solar Water Heat, Solar Space Heat, Solar Thermal Electric, Solar Thermal Process Heat, Photovoltaics, Wind, Solar Cooling, Solar Pool Heating, Daylighting
Applicable Sectors:	Commercial, Industrial, Nonprofit, Schools, Local Government, State Government, Tribal Government, Fed. Government, Agricultural, Institutional
Amount:	10% of installed cost
Maximum Incentive:	$25,000 for any one building in the same year and $50,000 per business in total credits in any year
Eligible System Size:	No size restrictions specified
Carryover Provisions:	Unused credits may be carried forward for not more than five consecutive taxable years
Program Administrator:	Arizona Department of Revenue
Program Budget:	$1 million annually
Start Date:	1/1/2006
Expiration Date:	12/31/2018
Web Site:	http://www.azcommerce.com/BusAsst/Incentives/
Authority 1: A.R.S. §43-1085 Date Effective: 1/1/2006 Expiration Date 12/31/2018	
Authority 2: A.R.S. §43-1164 Date Effective: 1/1/2006 Expiration Date 12/31/2018	

Authority 3:
A.R.S. §41-1510.01

Authority 4:
HB 2700
Date Enacted:
5/10/2010
Date Effective:
5/10/2010
Expiration Date
12/31/2018

Authority 5:
Tax Credit Guidelines

Arizona's tax credit for solar and wind installations in commercial and industrial applications was established in June 2006 (HB 2429). In May 2007, the credit was revised by HB 2491 to extend the credit to all non-residential entities, including those that are tax-exempt. Third parties who install or manufacture systems for non-residential applications are now eligible to claim the credit—not only those that finance a system as allowed in the original legislation. These provisions are retroactive to January 1, 2006.

The tax credit, which may be applied against corporate or personal taxes, is equal to 10% of the installed cost of qualified "solar energy devices" and applies to taxable years beginning January 1, 2006 and extending through December 31, 2018.

A solar energy device is defined as "a system or series of mechanisms designed primarily to provide heating, to provide cooling, to produce electrical power, to produce mechanical power, to provide solar daylighting or to provide any combination of the foregoing by means of collecting and transferring solar generated energy into such uses either by active or passive means, including wind generator systems that produce electricity. Solar energy systems may also have the capability of storing solar energy for future use. Passive systems shall clearly be designed as a solar energy device, such as a trombe wall, and not merely as a part of a normal structure, such as a window."
The maximum credit per taxpayer is $25,000 for any one building in the same year and $50,000 in total credits in any year. If the allowable credit exceeds the taxpayer's income tax liability, the amount of the claim not used to offset taxes may be carried forward for not more than five consecutive taxable years as a credit against subsequent years' income tax liability.

To qualify for the tax credits, a business must submit an application to the Arizona Department of Commerce (DOC). The DOC will review applications, provide an initial certification to qualifying installations, and issue a credit certificate to the business once the installation is complete and approved. The Arizona Department of Revenue will also receive a copy of the credit certificate. The DOC may certify tax credits up to a total of $1 million each calendar year. The DOC and the Department of Revenue will collaborate in adopting rules to implement the tax credit program.

TEP-Renewable Energy Credit Purchase Program

Last DSIRE Review: 05/07/2010

State:	Arizona
Incentive Type:	Utility Rebate Program
Eligible Renewable/Other Technologies:	Solar Water Heat, Solar Space Heat, Photovoltaics, Landfill Gas, Wind, Biomass, Geothermal Electric, Solar Pool Heating, Daylighting, Anaerobic Digestion, Small Hydroelectric
Applicable Sectors:	Commercial, Residential
Amount:	Up-front incentives for PV may be de-rated based on expected performance Residential grid-tied PV: $3.00/W-DC Residential off-grid PV: $2.00/W-DC Non-Residential grid-tied PV (100 kW or less): $2.50/W-DC or production-based incentive Non-Residential grid-tied PV (more than 100 kW): production-based incentive Non-Residential off-grid PV: $2.00/W-DC Wind (grid-tied): $2.25/W-AC Wind (off-grid): $1.80/W-AC or a performance-based incentive Daylighting (non-residential only): $0.18 per kWh savings for 5 years Residential Solar Water Heater: $0.25/kWh-equivalent, plus $750 up to a maximum amount of $1,750 Non-Residential Solar Water Heater: $0.50/kWh-equivalent, plus $750 Non-Residential Solar Water Heater: performance base incentive; all other eligible technologies can receive performance based incentives which vary depending on the technology and the length of the contract. Contract options for performance-based incentives are 10, 15 and 20 years
Maximum Incentive:	Residential PV systems greater than 20 kW AC will receive rebates on the first 20 kW AC only TEP incentive for PV can not exceed 60% of the total project cost TEP incentive for PV may be combined with other state and federal incentives, but combined they can not pay for more than 85% of the total project cost
Eligible System Size:	PV: Minimum size is 1.2 kW.

Equipment Requirements:	Photovoltaic modules must be covered by a manufacturer's warranty of at least 20 years to qualify for the up-front incentive
	SWH systems must be certified to SRCC OG-300 standards
	Eligible small wind systems must be certified and nameplate rated by the CEC
	Wind systems must have at least a 10-year manufacturer's warranty to qualify for the up-front incentive; wind systems must have at least a 5-year manufacturer's warranty to qualify for the performance-based incentive
Ownership of Renewable Energy Credits:	TEP
Program Administrator:	Tucson Electric Power
Web Site:	http://www.tep.com/Green/

Tucson Electric Power (TEP) created the SunShare Program in 2001 to encourage residential and business customers to install new photovoltaic (PV) equipment. TEP transitioned away from their original incentive structure for PV and added incentives for a variety of other technologies in May 2008 under the new Renewable Energy Credit Purchase Program (RECPP). The technologies now eligible for funding through the RECPP all qualify under Arizona's renewable energy standard (RES). TEP offers these incentives in exchange for the renewable energy certificates they generate. Incentives for the 2010 program year are as follows:

- **Residential PV (grid-tied):** $3.00/W up front for qualified systems. Systems with less than a 20-year warranty on the module or a 10-year warranty on the inverter, or with a building-integrated PV (BIPV) system over 5kW must receive the 10, 15, or 20 year performance-based incentive (PBI), which is based on the actual metered electricity output.
- **Residential PV (off-grid):** $2.00/W.
- **Non-Residential PV (grid-tied):** $2.50/W for systems 100kW or less. Systems greater than 100kW must take the PBI.
- **Non-Residential PV (off-grid):** $2.00/W.
- **Solar Domestic Water Heating and Solar Space Heating:** $0.25/kWh equivalent, plus $750 up to a maximum incentive of $1,750.
- **Non-Residential Solar Water Heating and Solar Space Heating:** $0.50/kWh-equivalent, plus $750.
- **Daylighting (Non-residential only):** $0.18/kWh equivalent for 5 years.
- **Wind (grid-tied, up to 1 MW):** $2.25/W-AC.
- **Wind (off-grid, up to 1 MW):** $1.80/W-AC.
- **On-grid small hydro, biomass-biogas systems, pool heating (non-residential only), space cooling, and geothermal (electric, cooling, and heating systems)** are all eligible to receive PBIs. See program Web site for full details including contract options, equipment requirements, and PBI amounts.

California Solar Initiative-PV Incentives

Last DSIRE Review: 01/04/2010

State:	California
Incentive Type:	State Rebate Program
Eligible Renewable/ Other Technologies:	Solar Space Heat, Solar Thermal Electric, Solar Thermal Process Heat, Photovoltaics
Applicable Sectors:	Commercial, Industrial, Residential, Nonprofit, Schools, Local Government, State Government, Fed. Government, Multi-Family Residential, Low-Income Residential, Agricultural, Institutional (All customers of PG&E, SDG&E, SCE; Bear Valley eligible only for NSHP)
Amount:	Varies by sector and system size (see below)
Equipment Requirements:	System components must be on the CEC's list of eligible equipment Systems must be grid-connected Inverters and modules must each carry a 10-year warranty PV modules must be UL 1703-certified Inverters must be UL 1741-certified, and tested by the Energy Commission
Installation Requirements:	Systems must be installed by appropriately licensed California solar contractors or self-installed by the system owner Installer certification by NABCEP is encouraged
Ownership of Renewable Energy Credits:	Remains with customer-generator
Program Administrator:	SCE, CCSE, PG&E
Program Budget:	$3.2 billion over 10 years
Start Date:	7/1/2009
Web Site:	http://www.cpuc.ca.gov/PUC/energy/solar

Authority 1:
SB 1
Date Enacted:
8/21/2006

Authority 2:
CSI Handbook (2009)
Date Effective:
7/1/2009

Authority 3:
CPUC decision 06-01-024

Authority 4:
CPUC Proceeding R0803008

In January 2006, the California Public Utilities Commission (CPUC) adopted a program—the California Solar Initiative (CSI)—to provide more than $3 billion in incentives for solar-energy projects with the objective of providing 3,000 megawatts (MW) of solar capacity by 2016. The CPUC manages the solar program for non-residential projects and projects on existing homes ($2+ billion), while the CEC oversees the *New Solar Homes Partnership*, targeting the residential new construction market (~$400 million). Together, these two programs comprise the effort to expand the presence of photovoltaics (PV) throughout the state, Go Solar California.

Originally limited to customers of the state's investor-owned utilities, the CSI was expanded in August 2006, as a result of Senate Bill 1, to encompass municipal utility territories as well. Municipal utilities were required to offer incentives beginning in 2008 (nearly $800 million); many already offer PV rebates.

CSI Incentives for Non-Residential Buildings and Existing Homes

The CSI includes a transition to performance-based and expected performance-based incentives (as opposed to capacity-based buydowns), with the aim of promoting effective system design and installation. CSI incentive levels will automatically be reduced over the duration of the program in 10 steps based on the aggregate capacity of solar installed. In this way, incentive reductions are linked to levels of solar demand rather than an arbitrary timetable.

Expected Performance-Based Buydowns for systems under 30 kW began in 2007 at $2.50/W AC for residential and commercial systems (adjusted based on expected performance) and $3.25/W AC for government entities and nonprofits (adjusted based on expected performance). The incentive levels decline as the aggregate capacity of PV installations increases. Incentives will be awarded as a one-time, up-front payment based on expected performance, which is calculated using equipment ratings and installation factors such as geographic location, tilt, orientation, and shading. Systems under 30 kW also have the option of opting for a performance-based incentive rather than the incentive based on expected performance.

Performance-Based Incentives (PBI) for systems 30 kW and larger began in 2007 at $0.39/kWh for the first five years for taxable entities, and $$0.50/kWh for the first five years for government entities and nonprofits. The incentive levels decline as the aggregate capacity of PV installations increases. PBI will be paid monthly based on the actual amount of energy produced for a period of five years. Residential and small commercial projects under the 30 kW threshold can also choose to opt in to the PBI rather than the upfront Expected Performance-Based Buydown approach. However, all installations of 30 kW or larger must take the PBI.

The program is managed by the Pacific Gas and Electric Company (PG&E), Southern California Edison (SCE), and the California Center for Sustainable Energy.

Low-Income Programs

Ten percent of the CSI Program budget ($216 million) has been allocated to two low-income solar incentive programs. As of March 2009, the single-family low income program is still being developed, but SCE, PG&E and CCSE are accepting applications for Track 1 of the multi-family affordable solar housing (MASH) program. Rebates are available through Track 1 in the amount of $3.30/W for PV systems offsetting common area loads, and $4.00/W for systems offsetting tenant loads. As required by the CPUC, the utilities are developing *virtual* net energy metering (VNEM) tariffs which will allow MASH participants to allocate the kWh credits from a single solar system across several electric accounts at the same building complex.

Incentives for Other Solar Electric Generating Technologies

The *CSI Handbook* released in January 2008 clarified the eligibility of other solar electric generating technologies which either produce electricity or displace electricity. Incentives for other solar electric generating technologies are available for CSI incentives effective October 1, 2008. The CPUC specifically recognizes electric generating solar thermal as including dish stirling, solar trough, and concentrating solar technologies, while technologies that displace electricity include solar forced air heating and solar cooling or air conditioning. The budget for electric displacing technologies is capped at $100.8 million. While solar water heaters can also displace electricity, the CPUC excludes them from the CSI because they plan to offer incentives for solar water heaters through a separate program based on the pilot program currently in operation within the service territory of San Diego Gas and Electric. Future CSI rulemaking activities will address energy-efficiency requirements, additional affordable housing incentives, and other program elements.

CPUC Program Administrators:

Pacific Gas and Electric (PG&E)

Web Site: www.pge.com/solar
E-mail Address: solar@pge.com
Contact Person: Program Manager, California Solar Initiative Program
Telephone: 877-743-4112
Fax: 415-973-2510
Mailing Address:
PG&E Integrated Processing Center
P.O. Box 7265
San Francisco, CA 94120-7265

California Center for Sustainable Energy (CCSE)

Web Site: www.energycenter.org
E-mail Address: csi@energycenter.org
Contact Person: John Supp, Program Manager

Telephone: 858-244-1177/(866)-sdenergy
Fax: 858-244-1178
Mailing Address:
California Center for Sustainable Energy
Attn: SELFGEN Program Manager
8690 Balboa Avenue Suite 100
San Diego, CA 92123

Southern California Edison (SCE)
Web Site: www.sce.com/rebatesandsavings/CaliforniaSolarInitiative
E-mail Address: greenh@sce.com
Contact Person: Program Manager, California Solar Initiative Program
Telephone: 1-800-799-4177
Fax: 626-302-6253
Mailing Address:
Southern California Edison
6042A Irwindale Avenue
Irwindale, CA 91702

California State Energy Code

Last DSIRE Review: 01/14/2010

State:	California
Incentive Type:	Building Energy Code
Eligible Efficiency Technologies:	Comprehensive Measures/Whole Building
Eligible Renewable/Other Technologies:	Passive Solar Space Heat, Solar Water Heat, Photovoltaics
Applicable Sectors:	Commercial, Residential
Residential Code:	State developed code, Title 24, Part 6, exceeds 2006 IECC, and is mandatory statewide
Commercial Code:	State developed code, Title 24, Part 6, meets or exceeds ASHRAE/IESNA 90.1-2004, and is mandatory statewide
Code Change Cycle:	Three-year code change cycle. The 2005 California Energy Efficiency Standards became effective October 1, 2005. The Commission adopted the new 2008 Standards on April 23, 2008. They will be effective January 1, 2010.
Web Site:	http://bcap-ocean.org/state-country/california

Much of the information presented in this summary is drawn from the U.S. Department of Energy's (DOE) Building Energy Codes Program and the Building Codes Assistance Project (BCAP). For more detailed information about building energy codes, visit the DOE and BCAP Web sites.

The California Building Standards Commission (BSC) is responsible for administering California's building standards adoption, publication, and implementation. Since 1989, the BSC has published triennial editions of the code, commonly referred to as Title 24, in its entirety every three years. On July 17, 2008 the BSC unanimously approved the nation's first statewide voluntary green building code. In January 2010, the BSC adopted a final version of the new building code, CALGreen, which became mandatory on January 1, 2011. The new code includes provisions to ensure the reduction water use by 20%, require separate water meters for nonresidential building's indoor and outdoor water use, improve indoor air quality, require the diversion of 50% of construction waste from landfills, and a number of other green building principles.

Title 24 applies to all buildings that are heated and/or mechanically cooled and are defined under the Uniform Building Code as A, B, E, H, N, R, or S occupancies, except registered historical buildings. Additions and renovations are also covered by the code. Institutional buildings, including hospitals and prisons are not covered.

For residential low-rise buildings the current code provision include compliance credits for high-performance ducts and building envelope features. The size of credit depends on the action taken. For example, simply designing ducts to Air Conditioner Contractor's Association guidelines or properly sealing duct joints provided lower levels of credit than having the HVAC system tested for duct leaks. To take credit for these measures the installer and inspector must be trained and certified.

Local governmental agencies can modify the state energy code to be more stringent when documentation is provided to the California Energy Commission.

AB 1103, passed in October 2007, requires annual energy-use reporting for all of California's nonresidential buildings effective January 2009. Beginning in 2010, owners of commercial buildings must disclose their energy usage and Energy Star rating to potential buyers, leasers, and financiers. The state Department of General Services has been working closely with utilities to streamline the reporting process.

Federal

INCENTIVES/POLICIES FOR RENEWABLES & EFFICIENCY

Business Energy Investment Tax Credit (ITC)

Last DSIRE Review: 06/10/2009

State:	Federal
Incentive Type:	Corporate Tax Credit
Eligible Renewable/ Other Technologies:	Solar Water Heat, Solar Space Heat, Solar Thermal Electric, Solar Thermal Process Heat, Photovoltaics, Wind, Biomass, Geothermal Electric, Fuel Cells, Geothermal Heat Pumps, CHP/Cogeneration, Solar Hybrid Lighting, Microturbines, Geothermal Direct-Use
Applicable Sectors:	Commercial, Industrial, Utility
Amount:	30% for solar, fuel cells and small wind** 10%** for geothermal, microturbines and CHP
Maximum Incentive:	Fuel cells: $1,500 per 0.5 kW Microturbines: $200 per kW Small wind turbines placed in service 10/4/08-12/31/08: $4,000 Small wind turbines placed in service after 12/31/08: no limit All other eligible technologies: no limit
Eligible System Size:	Small wind turbines: 100 kW or less** Fuel cells: 0.5 kW or greater Microturbines: 2 MW or less CHP: 50 MW or less**
Equipment Requirements:	Fuel cells, microturbines and CHP systems must meet specific energy-efficiency criteria
Program Administrator:	U.S. Internal Revenue Service
Authority 1: 26 USC § 48	
Authority 2: Instructions for IRS Form 3468	
Authority 3: IRS Form 3468	

Note: The American Recovery and Reinvestment Act of 2009 (H.R. 1) allows taxpayers eligible for the federal renewable electricity production tax credit (PTC)** to take the federal business energy investment tax credit (ITC) or to receive a grant from the U.S. Treasury Department instead of taking the PTC for new installations. The new law also allows taxpayers eligible for the business ITC to receive a grant from the U.S. Treasury Department instead of taking the business ITC for new installations. The Treasury Department issued Notice 2009-52 in June 2009, giving limited guidance *on how to take the federal business energy investment tax credit instead of the federal renewable electricity production tax credit. The Treasury Department will issue more extensive guidance at a later time.*

The federal business energy investment tax credit available under 26 USC § 48 was expanded significantly by the *Energy Improvement and Extension Act of 2008* (H.R. 1424), enacted in October 2008. This law extended the duration—by eight years—of the existing credits for solar energy, fuel cells, and microturbines; increased the credit amount for fuel cells; established new credits for small wind-energy systems, geothermal heat pumps, and combined heat and power (CHP) systems; extended eligibility for the credits to utilities; and allowed taxpayers to take the credit against the alternative minimum tax (AMT), subject to certain limitations. The credit was further expanded by *The American Recovery and Reinvestment Act of 2009*, enacted in February 2009.

In general, credits are available for eligible systems placed in service on or before December 31, 2016:

- **Solar.** The credit is equal to 30% of expenditures, with no maximum credit. Eligible solar energy property includes equipment that uses solar energy to generate electricity, to heat or cool (or provide hot water for use in) a structure, or to provide solar process heat. Hybrid solar lighting systems, which use solar energy to illuminate the inside of a structure using fiber-optic distributed sunlight, are eligible. Passive solar systems and solar pool-heating systems are *not* eligible. (Note that the Solar Energy Industries Association has published a three-page document that provides answers to frequently asked questions regarding the federal tax credits for solar energy.)
- **Fuel Cells.** The credit is equal to 30% of expenditures, with no maximum credit. However, the credit for fuel cells is capped at $1,500 per 0.5 kilowatt (kW) of capacity. Eligible property includes fuel cells with a minimum capacity of 0.5 kW that have an electricity-only generation efficiency of 30% or higher. (Note that the credit for property placed in service before October 4, 2008, is capped at $500 per 0.5 kW.)
- **Small Wind Turbines.** The credit is equal to 30% of expenditures, with no maximum credit for small wind turbines placed in service after December 31, 2008. Eligible small wind property includes wind turbines up to 100 kW in capacity. (In general, the maximum credit is $4,000 for eligible property placed in service after October 3, 2008, and before January 1, 2009. *The American Recovery and Reinvestment Act of 2009* removed the $4,000 maximum credit limit for small wind turbines.)
- **Geothermal Systems.** The credit is equal to 10% of expenditures, with no maximum credit limit stated. Eligible geothermal energy property includes geothermal heat pumps and equipment used to produce, distribute or use energy derived from a geothermal deposit. For electricity produced by geothermal power, equipment qualifies only up to, but not including, the electric transmission stage. For geothermal heat pumps, this credit applies to eligible property placed in service after October 3, 2008.
- **Microturbines.** The credit is equal to 10% of expenditures, with no maximum credit limit stated (explicitly). The credit for microturbines is capped at $200 per kW of

capacity. Eligible property includes microturbines up to two megawatts (MW) in capacity that have an electricity-only generation efficiency of 26% or higher.

■ **Combined Heat and Power (CHP).** The credit is equal to 10% of expenditures, with no maximum limit stated. Eligible CHP property generally includes systems up to 50 MW in capacity that exceed 60% energy efficiency, subject to certain limitations and reductions for large systems. The efficiency requirement does not apply to CHP systems that use biomass for at least 90% of the system's energy source, but the credit may be reduced for less-efficient systems. This credit applies to eligible property placed in service after October 3, 2008.

In general, the original use of the equipment must begin with the taxpayer, or the system must be constructed by the taxpayer. The equipment must also meet any performance and quality standards in effect at the time the equipment is acquired. The energy property must be operational in the year in which the credit is first taken.

Significantly, *The American Recovery and Reinvestment Act of 2009* repealed a previous limitation on the use of the credit for eligible projects also supported by "subsidized energy financing." For projects placed in service after December 31, 2008, this limitation no longer applies. Businesses that receive other incentives are advised to consult with a tax professional regarding how to calculate this federal tax credit.

U.S. Department of Treasury—Renewable Energy Grants

Last DSIRE Review: 03/31/2010

State:	Federal
Incentive Type:	Federal Grant Program
Eligible Renewable/ Other Technologies:	Solar Water Heat, Solar Space Heat, Solar Thermal Electric, Solar Thermal Process Heat, Photovoltaics, Landfill Gas, Wind, Biomass, Hydroelectric, Geothermal Electric, Fuel Cells, Geothermal Heat Pumps, Municipal Solid Waste, CHP/Cogeneration, Solar Hybrid Lighting, Hydrokinetic, Anaerobic Digestion, Tidal Energy, Wave Energy, Ocean Thermal, Microturbines
Applicable Sectors:	Commercial, Industrial, Agricultural
Amount:	30% of property that is part of a qualified facility, qualified fuel cell property, solar property, or qualified small wind property 10% of all other property
Maximum Incentive:	$1,500 per 0.5 kW for qualified fuel cell property $200 per kW for qualified microturbine property 50 MW for CHP property, with limitations for large systems
Program Administrator:	U.S. Department of Treasury

Funding Source:	The American Recovery and Reinvestment Act (ARRA)
Start Date:	1/1/2009
Expiration Date:	12/31/2010 (construction must begin by this date)
Web Site:	http://www.treas.gov/recovery/1603.shtml

Authority 1:
H.R. 1: Div. B, Sec. 1104 & 1603 (The American Recovery and Reinvestment Act of 2009)
Date Enacted:
2/17/2009
Date Effective:
1/1/2009

Authority 2:
U.S. Department of Treasury: Grant Program Guidance
Date Enacted:
07/09/2009, subsequently amended

Note: The American Recovery and Reinvestment Act of 2009 (H.R. 1) allows taxpayers eligible for the federal business energy investment tax credit (ITC) to take this credit or to receive a grant from the U.S. Treasury Department instead of taking the business ITC for new installations. The new law also allows taxpayers eligible for the renewable electricity production tax credit (PTC) to receive a grant from the U.S. Treasury Department instead of taking the PTC for new installations. (It does not allow taxpayers eligible for the residential renewable energy tax credit to receive a grant instead of taking this credit.) Taxpayers may not use more than one of these incentives. Tax credits allowed under the ITC with respect to progress expenditures on eligible energy property will be recaptured if the project receives a grant. The grant is not included in the gross income of the taxpayer.

The *American Recovery and Reinvestment Act of 2009* (H.R. 1), enacted in February 2009, created a renewable energy grant program that will be administered by the U.S. Department of Treasury. This cash grant may be taken in lieu of the federal business energy investment tax credit (ITC). In July 2009 the Department of Treasury issued documents detailing guidelines for the grants, terms and conditions and a sample application. There is an online application process, and applications are currently being accepted. See the U.S. Department of Treasury program Web site for more information, including answers to frequently asked questions.

Grants are available to eligible property placed in service in 2009 or 2010, or placed in service by the specified credit termination date, if construction began in 2009 or 2010. The guidelines include a "safe harbor" provision that sets the beginning of construction at the point where the applicant has incurred or paid at least 5% of the total cost of the property, excluding land and certain preliminary planning activities. Generally, construction begins when "physical work of a significant nature" begins.

Below is a list of important program details as they apply to each different eligible technology.

- **Solar.** The grant is equal to 30% of the basis of the property for solar energy. Eligible solar-energy property includes equipment that uses solar energy to generate electricity, to heat or cool (or provide hot water for use in) a structure, or to provide solar process heat. Passive solar systems and solar pool-heating systems are *not* eligible. Hybrid solar-lighting systems, which use solar energy to illuminate the inside of a structure using fiber-optic distributed sunlight, are eligible.
- **Fuel Cells.** The grant is equal to 30% of the basis of the property for fuel cells. The grant for fuel cells is capped at $1,500 per 0.5 kilowatt (kW) in capacity. Eligible property includes fuel cells with a minimum capacity of 0.5 kW that have an electricity-only generation efficiency of 30% or higher.
- **Small Wind Turbines.** The grant is equal to 30% of the basis of the property for small wind turbines. Eligible small wind property includes wind turbines up to 100 kW in capacity.
- **Qualified Facilities.** The grant is equal to 30% of the basis of the property for qualified facilities that produce electricity. Qualified facilities include wind energy facilities, closed-loop biomass facilities, open-loop biomass facilities, geothermal energy facilities, landfill gas facilities, trash facilities, qualified hydropower facilities, and marine and hydrokinetic renewable energy facilities.
- **Geothermal Heat Pumps.** The grant is equal to 10% of the basis of the property for geothermal heat pumps.
- **Microturbines.** The grant is equal to 10% of the basis of the property for microturbines. The grant for microturbines is capped at $200 per kW of capacity. Eligible property includes microturbines up to two megawatts (MW) in capacity that have an electricity-only generation efficiency of 26% or higher.
- **Combined Heat and Power (CHP).** The grant is equal to 10% of the basis of the property for CHP. Eligible CHP property generally includes systems up to 50 MW in capacity that exceed 60% energy efficiency, subject to certain limitations and reductions for large systems. The efficiency requirement does not apply to CHP systems that use biomass for at least 90% of the system's energy source, but the grant may be reduced for less-efficient systems.

It is important to note that only tax-paying entities are eligible for this grant. Federal, state, and local government bodies, non-profits, qualified energy tax credit bond lenders, and cooperative electric companies are not eligible to receive this grant. Partnerships or pass-thru entities for the organizations described above are also not eligible to receive this grant, except in cases where the ineligible party only owns an indirect interest in the applicant through a taxable C corporation. Grant applications must be submitted by October 1, 2011. The U.S. Treasury Department will make payment of the grant within 60 days of the grant application date or the date the property is placed in service, whichever is later.

U.S. Department of Energy—Loan Guarantee Program

Last DSIRE Review: 12/14/2009

State:	Federal
Incentive Type:	Federal Loan Program
Eligible Efficiency Technologies:	Yes; specific technologies not identified
Eligible Renewable/Other Technologies:	Solar Thermal Electric, Solar Thermal Process Heat, Photovoltaics, Wind, Hydroelectric, Geothermal Electric, Fuel Cells, Daylighting, Tidal Energy, Wave Energy, Ocean Thermal, Biodiesel
Applicable Sectors:	Commercial, Industrial, Nonprofit, Schools, Local Government, State Government, Agricultural, Institutional, Any Non-Federal Entity, Manufacturing Facilities
Amount:	Varies; program focuses on projects with total project costs over $25 million
Maximum Incentive:	Not specified
Terms:	Full repayment is required over a period not to exceed the lesser of 30 years or 90% of the projected useful life of the physical asset to be financed
Program Administrator:	U.S. Department of Energy
Web Site:	http://www.lgprogram.energy.gov
Authority 1: 42 USC § 16511 et seq.	
Authority 2: 10 CFR 609	

Innovative Technology Loan Guarantee Program

Title XVII of the federal *Energy Policy Act of 2005* (EPAct 2005) authorized the U.S. Department of Energy (DOE) to issue loan guarantees for projects that "avoid, reduce or sequester air pollutants or anthropogenic emissions of greenhouse gases; and employ new or significantly improved technologies as compared to commercial technologies in service in the United States at the time the guarantee is issued." The loan guarantee program has been authorized to offer more than $10 billion in loan guarantees for energy efficiency, renewable energy, and advanced transmission and distribution projects.

DOE actively promotes projects in three categories: (1) manufacturing projects, (2) stand-alone projects, and (3) large-scale integration projects that may combine multiple eligible renewable energy, energy efficiency, and transmission technologies in accordance with a staged development scheme. Under the original authorization, loan guarantees were intended to encourage early commercial use of new or significantly improved technologies in energy projects. The loan guarantee program generally does not support research and development projects.

In July 2009, the U.S. DOE issued a new solicitation for projects that employ innovative energy efficiency, renewable energy, and advanced transmission and distribution technologies. Proposed projects must fit within the criteria for "New or Significantly Improved Technologies" as defined in 10 CFR 609. The solicitation provides for a total of $8.5 billion in funding and is to remain open until that amount is fully obligated. The initial due date for applicants was September 16, 2009.

Temporary Loan Guarantee Program

The American Recovery and Reinvestment Act of 2009 (ARRA) (H.R. 1), enacted in February 2009, extended the authority of the DOE to issue loan guarantees and appropriated $6 billion for this program. Under this act, the DOE may enter into guarantees until September 30, 2011. The act amended EPAct 2005 by adding a new section defining eligible technologies for new loan guarantees. Eligible projects include renewable energy projects that generate electricity or thermal energy and facilities that manufacture related components, electric power transmission systems, and innovative biofuels projects. Funding for biofuels projects is limited to $500 million. Davis-Bacon wage requirements apply to any project receiving a loan guarantee.

In July 2009, the U.S. DOE issued a solicitation for innovative energy efficiency, renewable energy, transmission, and distribution technologies. The solicitation is expected to support as much as $8.5 billion in lending to eligible projects.

Modified Accelerated Cost-Recovery System (MACRS) + Bonus Depreciation (2008–2009)

Last DSIRE Review: 01/20/2010

State:	Federal
Incentive Type:	Corporate Depreciation
Eligible Renewable/Other Technologies:	Solar Water Heat, Solar Space Heat, Solar Thermal Electric, Solar Thermal Process Heat, Photovoltaics, Landfill Gas, Wind, Biomass, Geothermal Electric, Fuel Cells, Geothermal Heat Pumps, Municipal Solid Waste, CHP/Cogeneration, Solar Hybrid Lighting, Anaerobic Digestion, Microturbines, Geothermal Direct-Use
Applicable Sectors:	Commercial, Industrial
Program Administrator:	U.S. Internal Revenue Service
Start Date:	1986
Authority 1: 26 USC § 168 Date Effective: 1986	
Authority 2: 26 USC § 48	

Note: While the general Modified Accelerated Cost Recovery System (MACRS) remains in effect, the provision authorizing additional first-year bonus depreciation of 50% of eligible costs expired December 31, 2009. Although it is possible that bonus depreciation could be renewed for projects placed in service in 2010, as of this writing no such renewal had been enacted.

Under the federal Modified Accelerated Cost-Recovery System (MACRS), businesses may recover investments in certain property through depreciation deductions. The MACRS establishes a set of class lives for various types of property, ranging from 3 to 50 years, over which the property may be depreciated. A number of renewable energy technologies are classified as five-year property (26 USC § 168(e)(3)(B)(vi)) under the MACRS, which refers to 26 USC § 48(a)(3)(A), often known as the energy investment tax credit or ITC to define eligible property. Such property currently includes:

- A variety of solar electric and solar thermal technologies
- Fuel cells and microturbines
- Geothermal electric
- Direct-use geothermal and geothermal heat pumps
- Small wind (100 kW or less)
- Combined heat and power (CHP)
- The provision that defines ITC technologies as eligible also adds the general term "wind" as an eligible technology, extending the five-year schedule to large wind facilities as well.

In addition, for certain other biomass property, the MACRS property class life is seven years. Eligible biomass property generally includes assets used in the conversion of biomass to heat or to a solid, liquid, or gaseous fuel, and to equipment and structures used to receive, handle, collect, and process biomass in a waterwall, combustion system, or refuse-derived fuel system to create hot water, gas, steam, and electricity.

The five-year schedule for most types of solar, geothermal, and wind property has been in place since 1986. The federal *Energy Policy Act of 2005* (EPAct 2005) classified fuel cells, microturbines and solar hybrid lighting technologies as five-year property as well by adding them to § 48(a)(3)(A). This section was further expanded in October 2008 by the addition of geothermal heat pumps, combined heat and power, and small wind under *The Energy Improvement and Extension Act of 2008.*

The federal *Economic Stimulus Act of 2008,* enacted in February 2008, included a 50% first-year bonus depreciation (26 USC § 168(k)) provision for eligible renewable-energy systems acquired and placed in service in 2008. This provision was extended (retroactively to the entire 2009 tax year) under the same terms by *The American Recovery and Reinvestment Act of 2009*, enacted in February 2009. To qualify for bonus depreciation, a project must satisfy these criteria:

- The property must have a recovery period of 20 years or less under normal federal tax depreciation rules
- The original use of the property must commence with the taxpayer claiming the deduction
- The property generally must have been acquired during 2008 or 2009
- The property must have been placed in service during 2008 or 2009

If property meets these requirements, the owner is entitled to deduct 50% of the adjusted basis of the property in 2008 and 2009. The remaining 50% of the adjusted basis of the property is depreciated over the ordinary depreciation schedule. The bonus depreciation rules do not override the depreciation limit applicable to projects qualifying for the federal business energy tax credit. Before calculating depreciation for such a project, including any bonus depreciation, the adjusted basis of the project must be reduced by one-half of the amount of the energy credit for which the project qualifies.

For more information on the federal MACRS, see *IRS Publication 946, IRS Form 4562: Depreciation and Amortization*, and *Instructions for Form 4562*. The IRS Web site provides a search mechanism for forms and publications. Enter the relevant form, publication name or number, and click "GO" to receive the requested form or publication.

Florida

INCENTIVES/POLICIES FOR RENEWABLES & EFFICIENCY

Renewable Energy Production Tax Credit

Last DSIRE Review: 06/17/2009

State:	Florida
Incentive Type:	Corporate Tax Credit
Eligible Renewable/Other Technologies:	Solar Thermal Electric, Photovoltaics, Wind, Biomass, Hydroelectric, Geothermal Electric, CHP/Cogeneration, Hydrogen, Tidal Energy, Wave Energy, Ocean Thermal
Applicable Sectors:	Commercial
Amount:	$0.01/kWh for electricity produced from 1/1/2007 through 6/30/2010
Maximum Incentive:	No maximum specified for individual projects Maximum of $5 million per state fiscal year for all credits under this program
Carryover Provisions:	Unused credit may be carried forward for up to 5 years
Program Administrator:	Florida Department of Revenue
Start Date:	7/1/2006
Expiration Date:	6/30/2010
Web Site:	http://www.myfloridaclimate.com/climate_quick _links/florida_energ

Authority 1:
Fla. Stat. § 220.193
Date Enacted:
6/19/2006
Date Effective:
7/1/2006
Expiration Date
6/30/2010

In June 2006, SB 888 established a renewable energy production tax credit to encourage the development and expansion of renewable energy facilities in Florida. This annual corporate tax credit is equal to $0.01/kWh of electricity produced and sold by the taxpayer to an unrelated party during a given tax year. For new facilities (placed in service after May 1, 2006) the credit is based on the sale of the facility's entire electrical production. For an expanded facility, the credit is based on the increases in the facility's electrical production that are achieved after May 1, 2006.

For the purposes of this credit, renewable energy is defined as "electrical, mechanical, or thermal energy produced from a method that uses one or more of the following fuels or energy sources: hydrogen, biomass, solar energy, geothermal energy, wind energy, ocean energy, waste heat, or hydroelectric power."

The credit may be claimed for electricity produced and sold on or after January 1, 2007 through June 30, 2010. Beginning in 2008 and continuing until 2011, each taxpayer claiming a credit under this section must first apply to the Department of Revenue (DOR) by February 1 of each year for an allocation of available credit. If the credit granted is not fully used in one year because of insufficient tax liability, the unused amount may be carried forward for up to five years.

The combined total amount of tax credits which may be granted for all taxpayers under this program is limited to $5 million per state fiscal year. If the amount of credits applied for each year exceeds $5 million, the DOR will award a prorated amount based on each applicant's increased production and sales.

A taxpayer cannot claim both this production tax credit and Florida's Renewable Energy Technologies Investment Tax Credit. In June 2008, Florida enacted HB 7135, which specified that a taxpayer's use of the credit does not reduce the amount of the Florida alternative minimum tax available to the taxpayer.

Renewable Energy Technologies Investment Tax Credit

Last DSIRE Review: 09/15/2009

State:	Florida
Incentive Type:	Corporate Tax Credit
Eligible Renewable/Other Technologies:	Fuel Cells, Hydrogen, Ethanol, Biodiesel
Applicable Sectors:	Commercial
Amount:	75% of all capital costs, operation and maintenance costs, and research and development costs
Maximum Incentive:	Varies by application
Carryover Provisions:	Unused amount may be carried forward and used in tax years beginning 1/1/2007 and ending 12/31/2012
Program Administrator:	Executive Office of the Governor

Start Date:	7/1/2006
Expiration Date:	6/30/2010 (tax credit provision)
Web Site:	http://myfloridaclimate.com/climate_quick_links/florida_energy
Authority 1: Fla. Stat. § 220.192 Date Enacted: 6/19/2006 Date Effective: 7/1/2006 Expiration Date 6/30/2010 (tax credit provision)	

In June 2006, a corporate tax credit was established in Florida (SB 888) to promote investment in (1) hydrogen-powered vehicles and hydrogen vehicle fueling stations; (2) commercial stationary hydrogen fuel cells; and (3) production, storage, and distribution of biodiesel and ethanol.

For tax years beginning on or after January 1, 2007 through and ending December 31, 2010, the tax credit amount for each technology is as follows:

Hydrogen-Powered Vehicles and Hydrogen Vehicle Fueling Station: 75% of all capital costs, operation and maintenance costs, and research and development costs incurred between July 1, 2006, and June 30, 2010, up to a limit of $3 million per state fiscal year for all taxpayers, in connection with an investment in hydrogen-powered vehicles and hydrogen vehicle fueling stations in the state, including, but not limited to, the costs of constructing, installing, and equipping such technologies in the state.

Commercial Stationary Hydrogen Fuel Cells: 75% of all capital costs, operation and maintenance costs, and research and development costs incurred between July 1, 2006, and June 30, 2010, up to a limit of $ 1.5 million per state fiscal year for all taxpayers, and limited to a maximum of $ 12,000 per fuel cell, in connection with an investment in commercial stationary hydrogen fuel cells in the state, including, but not limited to the costs of constructing, installing, and equipping such technologies in the state.

Biodiesel and Ethanol Production, Storage, and Distribution: 75% of all capital costs, operation and maintenance costs, and research and development costs incurred between July 1, 2006, and June 30, 2010, up to a limit of $ 6.5 million per state fiscal year for all taxpayers, in connection with an investment in the production, storage, and distribution of biodiesel (B10-B100) and ethanol (E10-E100) in the state, including the costs of constructing, installing, and equipping such technologies in the state. Gasoline fueling station pump retrofits for ethanol (E10-E100) distribution also qualify as an eligible cost.

If the credit is not fully used in any one tax year because of insufficient tax liability on the part of the corporate entity, the unused amount may be carried forward and used in tax years beginning January 1, 2007, and ending December 31, 2012, after which the credit carryover expires and may not be used. Beginning January 1, 2009, this credit is also transferrable.

To be eligible for the credit, corporations must submit a tax credit application to the Florida Energy and Climate Commission and attach their certification (if approved) to the tax return on which the credit is claimed.

The Florida Energy and Climate Commission will determine and publish on a regular basis the amount of available tax credits remaining in each fiscal year. If a taxpayer does not receive a tax credit allocation due to the exhaustion of the annual tax credit authorizations, the taxpayer may reapply in the following year for those eligible costs and will have priority over other applicants for the allocation of credits.

This legislation also created a sales tax refund for products relating to hydrogen-powered vehicles, commercial stationary hydrogen fuel cells, and materials used in distributing biodiesel and ethanol.

Investment tax credit applications and draft rules are available on the program Web site.

Renewable Energy Equipment Sales Tax Exemption

Last DSIRE Review: 01/14/2010

State:	Florida
Incentive Type:	Sales Tax Incentive
Eligible Renewable/Other Technologies:	Fuel Cells, Ethanol, Biodiesel
Applicable Sectors:	Commercial, Residential, General Public/ Consumer
Amount:	All
Maximum Incentive:	None
Program Administrator:	Florida Department of Revenue
Start Date:	07/01/2006
Expiration Date:	07/01/2010

Web Site:	http://myfloridaclimate.com/climate_quick_links/ florida_energy
Authority 1: Fla. Stat. § 212.08 Date Effective: 07/01/2006 Expiration Date 07/01/2010	

In June 2006, Senate Bill 888 created a sales and use tax refund for "equipment, machinery and other materials for renewable energy technologies." The renewable energy technologies include hydrogen-powered vehicles; hydrogen-fueling stations (up to $2M total), commercial stationary hydrogen fuel cells (up to $1M total), and materials used in the distribution of biodiesel (B10-B100) and ethanol (E10-E100), including fueling infrastructure, transportation, and storage (up to $1M total).

The refund is available to a purchaser only on previously paid Florida sales tax. Applications for Florida Renewable Energy Technologies Sales Tax Program will be reviewed by the Florida Energy & Climate Commission. The Florida Energy & Climate Commission will not issue the sales tax refund. The actual sales tax refund must be applied for with the Florida Department of Revenue using appropriate forms, which are available at the program Web site listed above.

Michigan

INCENTIVES/POLICIES FOR RENEWABLES & EFFICIENCY

Consumers Energy—Photovoltaic Purchase Tariff

Last DSIRE Review: 10/26/2009

State:	Michigan
Incentive Type:	Production Incentive
Eligible Renewable/Other Technologies:	Photovoltaics
Applicable Sectors:	Commercial, Industrial, Residential, Nonprofit, Schools, Local Government, State Government, Fed. Government, Multi-Family Residential, Institutional
Amount:	Residential: $0.65/kWh or $0.525/kWh (see summary for details) Non-Residential: $0.45/kWh or $0.375/kWh (see summary for details)
Maximum Incentive:	None specified
Terms:	Fixed rate contract for up to 12 years; participants pay system access charge ($6–$50 per month) to cover additional metering costs
Program Administrator:	Consumers Energy
Start Date:	08/27/2009
Expiration Date:	12/31/2010
Web Site:	http://www.consumersenergy.com/welcome.htm?/products/index.asp?

Authority 1:
Experimental Advanced Renewables Program Tariff (multiple sheets)
Date Effective:
08/27/2009
Expiration Date
12/31/2010

Authority 2:
EARP Application Instructions

Authority 3:
EARP Program Application

Note: While the overall program limit is set at 2 megawatts (MW), Consumer's Energy has already received applications for more than 6.3 MW of capacity. New applications continue to be accepted, but they will be placed in a queue for funding should it become available due to project failures or application withdrawals.

In addition, in response to concerns from installers and developers, the in-service deadline for residential and non-residential projects seeking the higher incentive amount has been extended to May 1, 2010. The higher incentive level is intended apply to the first 250 kW (residential) or 750 kW (non-residential) of capacity that comes on-line by the May deadline. Projects that miss this deadline will be eligible for the lower incentive. If a portion of a project's capacity will exceed the 250 kW or 750 kW limit, that project will be offered the rate that corresponds to the largest proportion of the project capacity above or below the limit.

Beginning in August 2009, Consumers Energy of Michigan is offering its residential and non-residential customers an experimental buy-back tariff—termed the Experimental Advanced Renewables Program (EARP)—for electricity produced by solar photovoltaic (PV) systems. Owners of residential systems from 1–20 kilowatts (kW) and non-residential systems from 20–150 kW are eligible to participate in the program. The minimum system size is 1 kW. Residential customers must receive electric service on tariff rate RS or RT in order to be eligible for the program. Non-residential customers on tariff rates RS, RT, GS, GSD, GP, and GPD are eligible for the program. The overall program is limited to 2 megawatts (MW) with 500 kW reserved for residential sites.

It is important to note that this is not a net metering program and program participants are not eligible for net metering. Under the program, Consumers Energy will purchase all of the electricity produced by the system through a fixed-rate contract of 1 to 12 years. Electricity production is metered separately from the customer's existing electricity source (i.e., the grid). Participants are assessed a monthly System Access Charge equivalent to the existing distribution account used to qualify for the program to cover metering costs. Systems with battery backup or any other type of energy storage capability are not eligible to participate in this program. Purchase rates are as follows:

- Residential: $0.65/kilowatt-hour (kWh) for systems available by May 1, 2010 (up to roughly 250 kW of capacity); $0.525/kWh for systems that do not qualify for the higher level.
- Non-residential: $0.45 kWh for systems available by May 1, 2010 (up to roughly 750 kW of aggregate capacity); $0.375/kWh for systems available in 2010 that do not qualify for the higher incentive level.

Solar systems that receive the residential tariff rate may not be located on property that is used for commercial purposes, such as rental properties, warehouses, workshops, office buildings, etc. Tax exempt entities are not eligible to participate in the program under the residential rates, although they are eligible under the non-residential rates. Third-party ownership structures are not eligible for the program. The applicant must be a Consumers Energy customer on one of the qualifying electricity rates and must own the generating system. If the generation system is located on property that is not owned by the applicant, the applicant must have a lease or other instrument that permits him to construct, own, and operate the system throughout the term of the contract. Systems installed on newly constructed buildings are eligible for this program as long as the applicant will receive electric service from the utility at that site, or an adjacent site.

In order to be eligible for the program, solar equipment must be manufactured in the state of Michigan or constructed by a Michigan workforce. The manufacturing requirement can be met if 50% or more of the equipment and material costs associated with the system are attributable to components manufactured or assembled in Michigan. In order to qualify as a system constructed by a Michigan workforce, at least 60% of the total labor hours associated with installing the system must be performed by Michigan residents. All systems must meet the requirements of UL 1741 and IEEE 1547.1 and be installed in compliance with all current local and state electric and construction code requirements. The utility owns all renewable energy credits (RECs) associated with electricity purchased under this program, including all Michigan RECs, Michigan Incentive RECs, and Federal RECs.

The program began accepting formal applications on August 3, 2009 and is scheduled to run through December 31, 2010, subject to the limitations on overall enrollment described above. For further information on this program, please see the EARP tariff and application documents above, and contact program personnel.

Nevada

INCENTIVES/POLICIES FOR RENEWABLES & EFFICIENCY

NV Energy—RenewableGenerations Rebate Program

Last DSIRE Review: 04/27/2010

State:	Nevada
Incentive Type:	State Rebate Program
Eligible Renewable/Other Technologies:	Photovoltaics, Wind, Small Hydroelectric
Applicable Sectors:	Commercial, Residential, Nonprofit, Schools, Local Government, State Government, Agricultural, Other Public Buildings
Amount:	**Solar (Step 1, 2010–2011 program year):** Schools and public and other property, including non-profits and churches: $5.00 per watt AC Residential and small business property: $2.30 per watt AC **Wind (Step 1, 2010–2011 program year):** Residential, small business, agriculture: $3.00 per watt Schools and Public Buildings: $4.00 per watt **Small Hydro (Step 1, 2010–2011 program year):** Non-net metered systems: $2.80/W Net metered systems: $2.50/W
Maximum Incentive:	**Solar (Step 1, 2010–2011 program year):** Public and other property, including non-profits and churches: $500,000 Schools: $250,000, or up to $500,000 if given permission by the utilities commission Residential: $23,000 Small business property: $115,000 **Wind (Step 1, 2010–2011 program year):** Residential: $180,000 Small Business: $750,000 Schools: $1.0 million Agriculture: $1.5 million Public Buildings: $2.0 million **Small Hydro (Step 1, 2010–2011 program year):** Non-net metered systems: $560,000 Net metered systems: $500,000
Eligible System Size:	Maximum of 1 MW

Equipment Requirements:	**Solar:** Systems must be in compliance with all applicable standards; must carry a minimum 7-year warranty on inverters, 20-year warranty on panels, and 2-year warranty on labor; modules and inverters must be on the California Energy Commission (CEC) approved equipment list. **Wind:** Systems must be in compliance with all applicable standards; generator must be listed or certified by at least one of the following organizations: American Wind Energy Association (AWEA), British Wind Energy Association (BWEA), California Energy Commission (CEC), New York State Energy and Research Development Authority (NYSERDA), Small Wind Certification Council (SWCC). **Hydro:** Systems must be in compliance with all applicable standards
Installation Requirements:	Installations must comply with all federal, state, and local codes and meet detailed siting criteria specified in program guidelines. Systems must be grid-connected and net metered. Solar systems must be installed by a Nevada-licensed electrical C-2 or C-2g electrical contractor. Wind and microhydro systems must be installed by a Nevada-licensed C-2 electrical contractor
Ownership of Renewable Energy Credits:	NV Energy
Program Administrator:	NV Energy
Web Site:	http://www.Nvenergy.com/renewablegenerations
Authority 1: NRS § 701B.010 et. seq.	
Authority 2: Senate Bill 358 Date Enacted: 5/28/2009	
Authority 3: LCB File R175-07	

Note: In January 2010, the Public Utilities Commission of Nevada (PUCN) approved new regulations for the RenewableGenerations programs that included changes to the application process and a new step system for reservations. The solar incentive program began accepting applications on April 21, 2010 for program year 2010–2011. SolarGenerations had 13.4 megawatts (MW) of solar capacity available for this program year, and had received 34.8 MW worth of applications within the first six hours.

The SolarGenerations program is now closed for this program year and NV Energy is no longer accepting applications. Applications are currently being accepted for the wind and hydro incentive programs. Only customers of NV Energy are eligible to participate in the program.

NV Energy (formerly Sierra Pacific Power and Nevada Power) administers the RenewableGenerations Rebate Program for photovoltaic (PV) systems and small wind and hydroelectric systems on behalf of the Nevada Task Force on Energy Conservation and Renewable Energy. With rebates originally available only for PV, the SolarGenerations Rebate Program was established in 2003 as a result of AB 431 ("the Solar Energy Systems Demonstration Program") and began in August 2004. Rebates are now available for grid-connected PV installations on residences, small businesses, public buildings, non-profits, and schools; small wind systems on residences, small businesses, agricultural sites, schools and public buildings; and small hydroelectric systems installed at grid-connected agricultural sites. Participants must be current Nevada customers of NV Energy to participate.

SB 358 of 2009 made adjustments to the administration of the RenewableGenerations program. After the utility approves the applicant, the utility will have 30 days to notify them in writing. Further, applicants will have 12 months to complete a project following their initial approval. If projects that have been approved miss the 12-month target date, they can become eligible again after the project is complete, but will receive an incentive at the current rate, rather than the rate when they received initial authorization.

Including three years as a demonstration program, SolarGenerations is now in its sixth program year. In June 2007 the program was made permanent (the planned end date had been June 2010 for a total of six years of demonstration program funding). As demonstrated above, incentive levels vary by technology type, customer class, and program year, with incentive levels stepping down with each program year. Each program year has a designated amount of installed capacity set aside for each customer class. Applications received after one step is fully subscribed for that customer class may be reserved for the next incentive step. However, applications reserved for a future program year will have 12 months to be installed following the date of approval, but will not receive rebate until that program year commences. NV Energy's Web site will be frequently updated to indicate current subscription levels.

There are no size restrictions for participating systems, aside from the net metering limit of 1 megawatt, but rebates will be limited to certain system sizes corresponding to the customer class and the technology.

NV Energy takes ownership of the renewable energy credits (RECs) associated with the electricity produced by a customer's PV, wind, or small hydro system. The RECs count toward the utility's' goals under Nevada's renewable portfolio standards (RPS).

Renewable Energy Sales and Use Tax Abatement

Last DSIRE Review: 07/07/2009

State:	Nevada
Incentive Type:	Sales Tax Incentive
Eligible Renewable/ Other Technologies:	Solar Thermal Electric, Solar Thermal Process Heat, Photovoltaics, Landfill Gas, Wind, Biomass, Hydroelectric, Geothermal Electric, Fuel Cells, Municipal Solid Waste, Facilities for the transmission of electricity produced from renewable energy or geothermal resources located in Nevada, Anaerobic digestion, Fuel Cells using Renewable Fuels
Applicable Sectors:	Commercial, Industrial, Utility, Agricultural, (Renewable Energy Power Producers)
Amount:	Purchaser is only required to pay sales and use taxes imposed in Nevada at the rate of 2.6% (effective through June 30, 2011) and at the rate of 2.25% (effective July 01, 2011–June 30, 2049)
Equipment Requirements:	Systems must have a generating capacity of at least 10 megawatts
Program Administrator:	Nevada State Office of Energy
Start Date:	7/1/2009
Expiration Date:	6/30/2049
Web Site:	http://renewableenergy.state.nv.us/TaxAbatement.htm
Authority 1: AB 522 Date Enacted: 5/30/2009 Date Effective: 7/1/2009 Expiration Date 6/30/2049	

New or expanded businesses in Nevada may apply to the Director of the State Office of Energy for a sales and use tax abatement for qualifying renewable energy technologies. Purchaser is only required to pay sales and use taxes imposed in Nevada at the rate of 2.6% (effective through June 30, 2011) and at the rate of 2.25% (effective July 01, 2011–June 30, 2049). The start date begins when the first piece of equipment is delivered to the designated facility or taxes are paid on the equipment.

The abatement applies to property used to generate electricity from renewable energy resources including solar, wind, biomass, fuel cells, geothermal or hydro. Generation facilities must have a capacity of at least 10 megawatts (MW). Facilities that use solar energy to generate at least 25,840,000 British thermal units of process heat per hour can also qualify for an abatement.

There are several job creation and job quality requirements that must be met in order for a project to receive an abatement. Depending on the population of the county or city where the project will be located, the project owners must:

- Employ a certain number of full-time employees during construction, a percentage of whom must be Nevada residents
- Ensure that the hourly wage paid to the facility's employees and construction workers is a certain percentage higher than the average statewide hourly wage
- Make a capital investment of a specified amount in the state of Nevada
- Provide the construction workers with health insurance, which includes coverage for the worker's dependents

Note that this exemption does not apply to residential property. A facility that is owned, operated, leased, or controlled by a governmental entity is also ineligible for this abatement. Note that this exemption does not apply to residential property or property that is owned, operated, leased, or controlled by a governmental entity.

History

This abatement went through significant revisions with AB 522, signed in May 2009. Notably, AB 522 raised the capacity minimum for eligible projects from 10 kilowatts (kW) to 10 MW. It also changed the abatement such that the purchaser is only required to pay sales and use taxes imposed in Nevada at the rate of 2.6% (effective through June 30, 2011) and at the rate of 2.25% (effective July 01, 2011–June 30, 2049), extended it to additional technologies, and increased the qualification requirements to ensure that incentivized projects result in more high-quality jobs. These changes took effect on July 1, 2009. AB 522 also created a property tax abatement for renewable energy producers.

New Mexico

INCENTIVES/POLICIES FOR RENEWABLES & EFFICIENCY

Renewable Energy Production Tax Credit (Corporate)

Last DSIRE Review: 05/12/2010

State:	New Mexico
Incentive Type:	Corporate Tax Credit
Eligible Renewable/Other Technologies:	Solar Thermal Electric, Photovoltaics, Landfill Gas, Wind, Biomass, Municipal Solid Waste, Anaerobic Digestion
Applicable Sectors:	Commercial, Industrial
Amount:	$0.01/kWh for wind and biomass $0.027/kWh (average) for solar (see below)
Maximum Incentive:	Wind and biomass: First 400,000 MWh annually for 10 years (i.e., $4,000,000/year) Solar electric: First 200,000 MWh annually for 10 years (annual amount varies) Statewide cap: 2,000,000 MWh plus an additional 500,000 MWh for solar electric
Eligible System Size:	Minimum of 1 MW capacity per facility
Equipment Requirements:	System must be in compliance with all applicable performance and safety standards; generators must be certified by the New Mexico Energy, Minerals, and Natural Resources Department (EMNRD).
Carryover Provisions:	Prior to 10/1/2007: Excess credit may be carried forward five years After 10/1/2007: Excess credit is refunded to the taxpayer
Program Administrator:	Taxation and Revenue Department
Start Date:	7/1/2002
Expiration Date:	1/1/2018
Web Site:	http://www.cleanenergynm.org

Authority 1:
N.M. Stat. § 7-2A-19
Date Enacted:
3/4/2002, amended 2003, 2007
Date Effective:
7/1/2002
Expiration Date
1/1/2018

Enacted in 2002, the New Mexico Renewable Energy Production Tax Credit provides a tax credit against the corporate income tax of one cent per kilowatt-hour for companies that generate electricity from wind or biomass. Companies that generate electricity from solar energy receive a tax incentive that varies annually according to the following scale:

- Year 1: 1.5¢/kWh
- Year 2: 2¢/kWh
- Year 3: 2.5¢/kWh
- Year 4: 3¢/kWh
- Year 5: 3.5¢/kWh
- Year 6: 4¢/kWh
- Year 7: 3.5¢/kWh
- Year 8: 3¢/kWh
- Year 9: 2.5¢/kWh
- Year 10: 2¢/kWh

According to the EMNRD, this incentive averages 2.7¢/kWh annually.

For wind and biomass generators, the credit is applicable only to the first 400,000 megawatt-hours (MWh) of electricity in each of 10 consecutive taxable years. For solar, the credit is applicable only to the first 200,000 MWh of electricity in each taxable year. To qualify, an energy generator must have a capacity of at least 1 megawatt and be installed before January 2018.

Total generation from both the corporate and personal tax credit programs combined must not exceed two million megawatt-hours of production annually, plus an additional 500,000 MWh produced by solar energy. Taxpayers cannot claim both the corporate and the personal tax credit for the same renewable energy system.

For electricity generated prior to October 1, 2007, excess credit may be carried forward for up to five consecutive taxable years. For electricity generated on or after October 1, 2007, excess credit shall be refunded to the taxpayer in order to allow project owners with limited tax liability to fully utilize the credit.

PNM-Performance-Based Customer Solar PV Program

Last DSIRE Review: 04/05/2010

State:	New Mexico
Incentive Type:	Production Incentive
Eligible Renewable/Other Technologies:	Photovoltaics

Applicable Sectors:	Commercial, Residential
Amount:	Systems up to 10 kW: $0.13/kWh for RECs Systems greater than 10 kW up to 1 MW: $0.15/kWh for RECs
Maximum Incentive:	None specified
Terms:	Systems up to 10 kW: 12-year contract Systems greater than 10 kW up to 1 MW: 20-year contract (System must be net-metered to be eligible)
Program Administrator:	PNM
Start Date:	3/1/2006
Web Site:	http://www.pnm.com/customers/pv/program.htm
Date Effective: 3/1/2006	

In March 2006, PNM initiated a renewable energy credit (REC) purchase program as part of its plan to comply with New Mexico's renewable portfolio standard (RPS). PNM will purchase RECs from customers who install photovoltaic (PV) systems up to one megawatt (MW). PNM will then be able to apply these RECs towards their obligations under the state's RPS, which requires 4% of the total generation capacity to come from solar electricity by 2020, and 0.6% from distributed generation in 2020.

REC payments are based on the system's total output. PNM will purchase RECs from each participant as part of the regular monthly billing process. Participants will receive a monthly bill documenting the number of kilowatt-hours (kWh) produced by the PV system, the number of RECs purchased by PNM, the purchase price per REC and the total price of RECs purchased that billing period. REC purchase payments will be applied as a credit to the participant's electric bill on a monthly basis.

Systems up to 10 kW
PNM will purchase RECs generated by small PV systems at a rate of $0.13/kWh for 12 years of the system's operation. If the amount paid for the RECs is greater than the total of the customer's monthly electric service plus kWh charges, the balance of the REC payment will be carried forward as a credit for the following month's bill if $20 or less. If the REC payment balance is greater than $20 after credits to the customer's electric bill have been made, the entire REC payment balance will be paid directly to the customer. Program participants must pay an application fee of $100 for residential customers, which includes the cost of installing a second meter to monitor system output. Customers also must pay a net-metering application fee of $50 to establish an approved interconnection with PNM.

Systems greater than 10 kW up to 1 MW
PNM will purchase RECs associated with the electricity generated by large PV systems and used on-site at a rate of $0.15/kWh for 20 years of the system's operation. If the

amount paid for the RECs is greater than the total of the customer's monthly electric service plus kWh charges, the balance of the REC payment will be carried forward as a credit for the following month's bill if $200 or less. If the REC payment balance is greater than $200 after credits to the customer's electric bill have been made, the entire REC payment balance will be paid directly to the customer. PNM does not pay for RECs associated with net excess generation. Program participants must pay an application fee of $350 for commercial customers, which includes the cost of installing a second meter to monitor system output. Customers also must pay an interconnection application fee of $100.00 up to 100 kW for interconnection + $1.00 for every kW above 100 kW up to 1 MW to establish an approved interconnection with PNM.

Oregon

INCENTIVES/POLICIES FOR RENEWABLES & EFFICIENCY

EWEB-Solar Electric Program (Production Incentive)

Last DSIRE Review: 09/30/2009

State:	Oregon
Incentive Type:	Production Incentive
Eligible Renewable/Other Technologies:	Photovoltaics
Applicable Sectors:	Commercial, Industrial, Residential, Nonprofit, Schools, Local Government, State Government, Agricultural, Institutional
Amount:	$0.076–$0.12/kWh for 10 years (subject to annual review), actual rate depends on level of monthly generation and season
Maximum Incentive:	Available to systems sized 10 kW to 1 MW
Terms:	System must be greater than 10 kW in capacity. System owners must execute an EWEB interconnection agreement and program agreement. A building permit is required. Systems must be inspected by city or county building officials, and by EWEB. All system equipment must be UL-listed. All PV modules and inverters must be listed and rated by the CEC
Program Administrator:	Eugene Water & Electric Board
Start Date:	1/25/2008
Web Site:	http://www.eweb.org/content.aspx/ee5003fe-cb03-484c-86e0-2bb16f5d
Authority 1: EWEB Solar Electric Program Information and Requirements Date Effective: 1/25/2008	

The Eugene Water & Electric Board's (EWEB) Solar Electric Program offers financial incentives for residential and commercial customers who generate electricity using solar photovoltaic (PV) systems. Rebates are available to customers who choose to net meter, and a production incentive is available to customers with systems greater than 10 kilowatts (kW) in capacity who choose *not* to net meter. Under the latter arrangement, all electricity generated is fed into the grid.

The rebate for residential customers who choose to net meter is $2.00 per watt-AC, with a maximum incentive of $10,000. The rebate for commercial customers who choose to net meter is $1.00 per watt-AC, with a maximum incentive of $25,000. Rebate amounts are based on the electrical output of the system after equipment and site losses are calculated. Under the rebate program, customers retain ownership of all renewable-energy credits (RECs) associated with customer generation.

PV systems sized 10 kW to 1 megawatt (MW) in capacity that are designed to generate and feed electricity directly into the grid—an arrangement under which the customer uses none of the electricity generated by the PV system—are eligible for a production payment of $0.076-$0.12 per kilowatt-hour (kWh) generated, payable for 10 years (but subject to annual review). The level of the incentive varies, depending on the season and level of monthly kWh generation. These "direct generation" systems require a separate EWEB service and electric meter to measure the amount of kWh generated. Under this program, EWEB assumes ownership of all RECs associated with customer generation.

All system owners must execute an EWEB interconnection agreement and program agreement. A building permit is required, and all systems must be inspected first by city or county building officials and then by EWEB. All system equipment must be UL-listed. All PV modules and inverters must be listed and rated in the California Energy Commission's Emerging Renewables Program. This list is available on the California Solar Initiative's Eligible Solar Equipment Web site.

Energy Trust-Industrial Production Efficiency Program

Last DSIRE Review: 06/12/2009

State:	Oregon
Incentive Type:	State Rebate Program
Eligible Efficiency Technologies:	Lighting, Lighting Controls/Sensors, Heat Pumps, Compressed Air, Motors, Motor-ASDs/VSDs, Agricultural Equipment, Custom/Others pending approval
Eligible Renewable/ Other Technologies:	Geothermal Heat Pumps
Applicable Sectors:	Industrial, Agricultural, Manufacturing, Water/Wastewater Treatment
Amount:	Varies depending on technology; incentives awarded per kilowatt-hour saved by project
Maximum Incentive:	Non-lighting projects: $0.25/kWh, up to 60% of cost until end of 2009 and up to 50% of cost after 2009 Lighting projects: $0.17/kWh, up to 50% of project cost Custom lighting incentives are 35% of project cost NEMA Premium efficiency motors: $10 per horsepower, up to 200 horsepower Municipal/service district project: $0.32/kWh, up to 50% of cost

Equipment Requirements:	Minimum efficiency levels for all equipment is available on program Web site
Program Administrator:	Energy Trust of Oregon
Web Site:	http://energytrust.org/Business/incentives/industrial/ production

Energy Trust of Oregon offers the Industrial Production Efficiency Program to industrial customers of Portland General Electric, Pacific Power, NW Natural, or Cascade Natural Gas. In order to qualify for these rebates, customers must be contributing to the Public Purpose Charge. Energy Trust offers technical assistance and cash incentives for industrial processes of all kinds—including large industrial, manufacturing, agriculture, and water/wastewater treatment. Standard prescriptive incentives include lighting, premium motors, heat pumps, variable speed drives, and premium HVAC equipment. Other rebates that are designed to fit the needs of specific industrial processes also exist. For example, there are irrigation system rebates for agricultural customers and compressed air rebates for small manufacturing customers. Customers interested in participating in the program should contact a production service representative to find out which rebates best fit their particular facility.

Steps for reserving incentive funds for projects vary depending on the type and magnitude of the project.

Tax Credit for Renewable Energy Equipment Manufacturers

Last DSIRE Review: 03/23/2010

State:	Oregon
Incentive Type:	Industry Recruitment/Support
Eligible Renewable/ Other Technologies:	Solar Water Heat, Solar Space Heat, Photovoltaics, Wind, Biomass, Geothermal Heat Pumps, Solar Pool Heating, Small Hydroelectric, Tidal Energy, Wave Energy
Applicable Sectors:	Commercial, Industrial
Amount:	50% of eligible costs (10% per year for 5 years)
Maximum Incentive:	$20 million
Program Administrator:	Oregon Department of Energy
Start Date:	6/20/2008
Web Site:	http://egov.oregon.gov/ENERGY/CONS/BUS/BETC.shtml
Authority 1: OAR 330-090-0105 to 330-090-0150 Date Effective: 6/20/2008	

Oregon's Business Energy Tax Credit (BETC) is for investments in energy conservation, recycling, renewable energy resources, sustainable buildings, and less-polluting transportation fuels. The Tax Credit for Renewable Energy Resource Equipment Manufacturing Facilities was enacted as a part of BETC in July 2007, with the passage of HB 3201. The tax credit equals 50% of the construction costs of a facility which will manufacture renewable energy systems, and includes the costs of the building, excavation, machinery and equipment which is used primarily to manufacture renewable energy systems. The credit may also be applied to the costs of improving an existing facility which will be used to manufacture renewable energy systems. The 50% credit is taken over the course of five years, at 10% each year. The original maximum credit of $10 million was expanded to $20 million (50% of a $40 million facility) upon the enactment of HB 3619 in March 2008. This legislation clarified the manufacturing credit and separated the revenue stream from the rest of BETC.

The credit applies to companies that manufacture systems that harness energy from wood waste or other wastes from farm and forest lands, non-petroleum plant or animal based biomass, the sun, wind, water, or geothermal resources. Prior to construction, a business must apply to the Oregon Department of Energy for preliminary certification. In addition to this preliminary certification, the manufacturing facility must apply for final certification. Another review required for manufacturing facilities is a financial feasibility review. The Oregon Department of Energy may establish other rules to govern the type of equipment, machinery or other manufactured products eligible for this credit, as well as minimum performance and efficiency standards for those manufactured products. The passage of HB 3680 in March 2010 set a sunset date for the tax credit. Renewable energy equipment manufacturing facilities must receive preliminary certification before January 1, 2014 in order to use the tax credit.

Texas

SOLAR AND WIND ENERGY BUSINESS FRANCHISE TAX EXEMPTION

Last DSIRE Review: 11/13/2009

State:	Texas
Incentive Type:	Industry Recruitment/Support
Eligible Renewable/Other Technologies:	Solar Water Heat, Solar Space Heat, Solar Thermal Electric, Solar Thermal Process Heat, Photovoltaics, Wind
Applicable Sectors:	Commercial, Industrial
Amount:	All
Maximum Incentive:	None
Terms:	N/A
Program Administrator:	Comptroller of Public Accounts
Start Date:	1982
Web Site:	http://www.seco.cpa.state.tx.us/re_incentives-taxcode-statutes.ht...

Authority 1:
Texas Tax Code § 171.056
Date Enacted:
1981
Date Effective:
1982

Companies in Texas engaged solely in the business of manufacturing, selling, or installing solar energy devices are exempted from the franchise tax. The franchise tax is Texas's equivalent to a corporate tax; their primary elements are the same. There is no ceiling on this exemption, so it is a substantial incentive for solar manufacturers.

For the purposes of this exemption, a solar energy device means "a system or series of mechanisms designed primarily to provide heating or cooling or to produce electrical or mechanical power by collecting and transferring solar-generated energy. The term includes a mechanical or chemical device that has the ability to store solar-generated energy for use in heating or cooling or in the production of power." Under this definition wind energy is also listed as an eligible technology.

Texas also offers a franchise tax deduction for solar energy devices which also includes wind energy as an eligible technology.

INCENTIVES/POLICIES FOR RENEWABLES & EFFICIENCY

Solar and Wind Energy Device Franchise Tax Deduction

Last DSIRE Review: 11/13/2009

State:	Texas
Incentive Type:	Corporate Deduction
Eligible Renewable/Other Technologies:	Solar Water Heat, Solar Space Heat, Solar Thermal Electric, Solar Thermal Process Heat, Photovoltaics, Wind
Applicable Sectors:	Commercial, Industrial
Amount:	10% of amortized cost
Maximum Incentive:	None
Program Administrator:	Comptroller of Public Accounts
Start Date:	1982
Web Site:	http://www.seco.cpa.state.tx.us/re_incentives-taxcode-statutes

Authority 1:
Texas Tax Code § 171.107
Date Enacted:
1981 (subsequently amended)
Date Effective:
1982

Texas allows a corporation or other entity subject the state franchise tax to deduct the cost of a solar energy device from the franchise tax. Entities are permitted to deduct 10% of the amortized cost of the system from their apportioned margin. This treatment is effective January 1, 2008 and replaces prior tax law that allowed a company to deduct (1) the total cost of the system from the company's taxable capital; or, (2) 10% of the system's cost from the company's earned surplus (i.e., income). The franchise tax is Texas's equivalent to a corporate tax.

For the purposes of this deduction, a solar energy device means "a system or series of mechanisms designed primarily to provide heating or cooling or to produce electrical or mechanical power by collecting and transferring solar-generated energy. The term includes a mechanical or chemical device that has the ability to store solar-generated energy for use in heating or cooling or in the production of power." Under this definition wind energy is also included as an eligible technology.

Texas also offers a franchise tax exemption for manufacturers, seller, or installers of solar energy systems which also includes wind energy as an eligible technology.

LoanSTAR Revolving Loan Program

Last DSIRE Review: 04/21/2010

State:	Texas
Incentive Type:	State Loan Program
Eligible Efficiency Technologies:	Lighting, Lighting Controls/Sensors, Chillers, Furnaces, Boilers, Heat Pumps, Central Air Conditioners, Heat Recovery, Programmable Thermostats, Energy Mgmt. Systems/Building Controls, Building Insulation, Motors, Motor-ASDs/VSDs, Custom/Others Pending Approval, LED Exit Signs
Eligible Renewable/ Other Technologies:	Passive Solar Space Heat, Solar Water Heat, Solar Space Heat, Photovoltaics, Wind, Geothermal Heat Pumps
Applicable Sectors:	Schools, Local Government, State Government, Hospitals
Amount:	Varies
Maximum Incentive:	$5 million
Terms:	Current interest rates are 3% APR. Loans are repaid through energy cost savings. Projects must have an average payback of 10 years or less
Program Administrator:	Comptroller of Public Accounts State Energy Conservation Office (SECO)
Funding Source:	Petroleum Violation Escrow Funds
Program Budget:	$98.6 million (revolving loan)
Start Date:	1989
Web Site:	http://seco.cpa.state.tx.us/ls/

Through the State Energy Conservation Office, the LoanSTAR Program offers low-interest loans to all public entities, including state, public school, colleges, university, and non-profit hospital facilities for Energy Cost Reduction Measures (ECRMs). Such measures include, but are not limited to HVAC, lighting, and insulation. Funds can be used for retrofitting existing equipment or, in the case of new construction, to finance the difference between standard and high efficiency equipment. The evaluation of on-site renewable energy options (e.g., solar water heating, photovoltaic panels, small wind turbines) is encouraged in the analysis of potential projects.

The LoanSTAR Program funds "Design, Bid, Built" or "Design, Built" projects. All projects are approved based on the Detailed Energy Assessment Report, which must be prepared according to LoanSTAR Technical Guidelines or the Performance

Contracting Guidelines. SECO performs design specification review and on-site construction monitoring at the very minimum when the project is 100% complete. Repayment of the loans does not begin until after construction is 100% completed.

As of November 2007, LoanSTAR had funded a total of 191 loans totaling over $240 million dollars and resulting in approximately $212 million in energy savings. The National Association of State Energy Officials (NASEO) reports that the LoanSTAR program helped state agencies save more than $20 million in energy costs during 2008 and that the program had a waiting list of $28 million in proposed projects as of winter 2009. Applications are available on the program Web site. The technical guidelines for the LoanSTAR program can be found on the program Web site.

LIST OF CALIFORNIA ENERGY
COMMISION CERTIFIED
SOLAR POWER EQUIPMENT

Photovoltaic Modules

Beginning July 1, 2009, only modules that are on the SB1 Guidelines-compliant module list will be eligible for incentive in California.

Please note that PTC values on this module list are calculated using laboratory-tested parameter values. After July 1, 2009, if a California solar electric incentive program uses PTC values, the SB1 Guidelines compliant PTC values must be used for new reservation applications.

MANUFACTURER NAME	MODULE MODEL NUMBER	DESCRIPTION	BIPV*	PTC**	NOTES
1Soltech	1STH-235-WH	235W Monocrystalline Module	N	205.1	
1Soltech	1STH-240-WH	240W Monocrystalline Module	N	209.6	
1Soltech	1STH-245-WH	245W Monocrystalline Module	N	214.1	
1Soltech	1STH-250-WH	250W Monocrystalline Module	N	218.6	
A10Green Technology	A10J-S72-175	175W Monocrystalline Module	N	151.2	
A10Green Technology	A10J-S72-180	180W Monocrystalline Module	N	155.7	
A10Green Technology	A10J-S72-185	185W Monocrystalline Module	N	160.2	
A10Green Technology	A10J-M60-220	220W Polycrystalline Module	N	189.1	
A10Green Technology	A10J-M60-225	225W Polycrystalline Module	N	193.5	
A10Green Technology	A10J-M60-230	230W Polycrystalline Module	N	203.5	
A10Green Technology	A10J-M60-235	235W Polycrystalline Module	N	208.1	
A10Green Technology	A10J-M60-240	240W Polycrystalline Module	N	212.7	
Aavid Thermalloy	ASMP-175M	175W Monocrystalline Module	N	154.1	
Aavid Thermalloy	ASMP-180M	180W Monocrystalline Module	N	158.6	
AblyTek	5MN6C175-A0	175W Monocrystalline Module	N	151.2	
AblyTek	5MN6C180-A0	180W Monocrystalline Module	N	155.7	
AblyTek	5MN6C185-A0	185W Monocrystalline Module	N	160.2	
AblyTek	6PN6A220-A0	220W Polycrystalline Module	N	189.1	
AblyTek	6PN6A225-A0	225W Polycrystalline Module	N	193.5	
AblyTek	6PN6A230-A0	230W Polycrystalline Module	N	203.5	
AblyTek	6PN6A235-A0	235W Polycrystalline Module	N	208.1	
AblyTek	6PN6A240-A0	240W Polycrystalline Module	N	212.7	
Abound Solar	AB1-55-A	55W Thin Film CdTe Module	N	48.9	
Abound Solar	AB1-57-A	57.5W Thin Film CdTe Module	N	52.3	
Abound Solar	AB1-57-B	57.5W Thin Film CdTe Module	N	52.3	

Abound Solar	AB1-60-A	60W Thin Film CdTe Module	N	54.6
Abound Solar	AB1-60-B	60W Thin Film CdTe Module	N	54.6
Abound Solar	AB1-62-A	62.5W Thin Film CdTe Module	N	56.9
Abound Solar	AB1-62-B	62.5W Thin Film CdTe Module	N	56.9
Abound Solar	AB1-65-A	65W Thin Film CdTe Module	N	59.2
Abound Solar	AB1-65-B	65W Thin Film CdTe Module	N	59.2
AccuSolar Power	ASP610-B230	230W Monocrystalline Module	N	201.2
Advanced Renewable Energy	AREi-210W-M6-G	210W Polycrystalline Module	N	187.4
Advanced Renewable Energy	AREi-215W-M6-G	215W Polycrystalline Module	N	192.0
Advanced Renewable Energy	AREi-220W-M6-G	220W Polycrystalline Module	N	196.6
Advanced Renewable Energy	AREi-225W-M6-G	225W Polycrystalline Module	N	201.1
Advanced Renewable Energy	AREi-230W-M6-G	230W Polycrystalline Module	N	205.7
Aleo Solar	S16.165	165W Polycrystalline Module	N	146.1
Aleo Solar	S16.170	170W Polycrystalline Module	N	150.6
Aleo Solar	S16.175	175W Polycrystalline Module	N	155.2
Aleo Solar	S16.180	180W Polycrystalline Module	N	159.7
Aleo Solar	S16.185	185W Polycrystalline Module	N	164.3
Aleo Solar	S18.210	210W Polycrystalline Module	N	186.2
Aleo Solar	S18.215	215W Polycrystalline Module	N	190.8
Aleo Solar	S18.220	220W Polycrystalline Module	N	195.3
Aleo Solar	S18.225	225W Polycrystalline Module	N	199.9
Aleo Solar	S18.230	230W Polycrystalline Module	N	204.5

(Continued)

MANUFACTURER NAME	MODULE MODEL NUMBER	DESCRIPTION	BIPV*	PTC**	NOTES
Andalay Solar	ST165-1	165W Monocrystalline Module	N	142.8	
Andalay Solar	ST170-1	170W Monocrystalline Module	N	152.9	
Andalay Solar	ST175-1	175W Monocrystalline Module	N	157.5	
Andalay Solar	KC180-1	180W Polycrystalline Module	N	162.5	
Andalay Solar	ST180-1	180W Monocrystalline Module	N	162.1	
Andalay Solar	KC205-1	205W Polycrystalline Module	N	185.2	
Andalay Solar	KC210-1	210W Polycrystalline Module	N	189.8	
Anji Dasol Solar Energy Science & Technology	DS-A4-210	210W Polycrystalline Module	N	182.7	
Anji Dasol Solar Energy Science & Technology	DS-A4-215	215W Polycrystalline Module	N	187.2	
Anji Dasol Solar Energy Science & Technology	DS-A4-220	220W Polycrystalline Module	N	191.7	
Anji Dasol Solar Energy Science & Technology	DS-A4-225	225W Polycrystalline Module	N	196.2	
Anji Dasol Solar Energy Science & Technology	DS-A4-230	230W Polycrystalline Module	N	200.7	
Anji Technology	AJP-S572-175	175W Monocrystalline Module	N	151.2	
Anji Technology	AJP-S572-180	180W Monocrystalline Module	N	155.7	
Anji Technology	AJP-S572-185	185W Monocrystalline Module	N	160.2	
Anji Technology	AJP-M660-220	220W Polycrystalline Module	N	189.1	
Anji Technology	AJP-M660-225	225W Polycrystalline Module	N	193.5	
Anji Technology	AJP-M660-230	230W Polycrystalline Module	N	203.5	
Anji Technology	AJP-M660-235	235W Polycrystalline Module	N	208.1	
Anji Technology	AJP-M660-240	240W Polycrystalline Module	N	212.7	
Anji Technology	AJP-M660-245	245W Polycrystalline Module	N	217.2	
Anji Technology	AJP-M660-250	250W Polycrystalline Module	N	221.8	

Apollo Solar Energy	ASEC-175G6M	175W Polycrystalline Module	N	155.9
Apollo Solar Energy	ASEC-175G6S	175W Monocrystalline Module	N	155.2
Apollo Solar Energy	ASEC-180G6M	180W Polycrystalline Module	N	160.5
Apollo Solar Energy	ASEC-180G6S	180W Monocrystalline Module	N	159.7
Apollo Solar Energy	ASEC-185G6M	185W Polycrystalline Module	N	165.1
Apollo Solar Energy	ASEC-185G6S	185W Monocrystalline Module	N	164.3
Apollo Solar Energy	ASEC-190G6M	190W Polycrystalline Module	N	167.8
Apollo Solar Energy	ASEC-190G6S	190W Monocrystalline Module	N	169.8
Apollo Solar Energy	ASEC-195G6M	195W Polycrystalline Module	N	172.4
Apollo Solar Energy	ASEC-195G6S	195W Monocrystalline Module	N	174.4
Apollo Solar Energy	ASEC-200G6M	200W Polycrystalline Module	N	176.9
Apollo Solar Energy	ASEC-200G6S	200W Monocrystalline Module	N	179.0
Apollo Solar Energy	ASEC-205G6M	205W Polycrystalline Module	N	181.5
Apollo Solar Energy	ASEC-205G6S	205W Monocrystalline Module	N	183.6
Apollo Solar Energy	ASEC-210G6M	210W Polycrystalline Module	N	186.1
Apollo Solar Energy	ASEC-210G6S	210W Monocrystalline Module	N	188.2
APOS Energy	AP130	130W Polycrystalline Module	N	115.4
APOS Energy	AP135	135W Polycrystalline Module	N	119.9
APOS Energy	AS135	135W Monocrystalline Module	N	120.9
APOS Energy	AP140	140W Polycrystalline Module	N	124.5
APOS Energy	AS140	140W Monocrystalline Module	N	125.4
APOS Energy	AS145	145W Monocrystalline Module	N	130.0
APOS Energy	AP175	175W Polycrystalline Module	N	155.4
APOS Energy	AP180	180W Polycrystalline Module	N	160.0
APOS Energy	AS180	180W Monocrystalline Module	N	161.2

(Continued)

MANUFACTURER NAME	MODULE MODEL NUMBER	DESCRIPTION	BIPV*	PTC***	NOTES
APOS Energy	AP185	185W Polycrystalline Module	N	164.5	
APOS Energy	AS185	185W Monocrystalline Module	N	165.8	
APOS Energy	AP190	190W Polycrystalline Module	N	168.6	
APOS Energy	AS190	190W Monocrystalline Module	N	170.4	
APOS Energy	AP195	195W Polycrystalline Module	N	173.1	
APOS Energy	AS195	195W Monocrystalline Module	N	174.5	
APOS Energy	AP200	200W Polycrystalline Module	N	177.7	
APOS Energy	AS200	200W Monocrystalline Module	N	179.1	
APOS Energy	AP205	205W Polycrystalline Module	N	182.3	
APOS Energy	AS205	205W Monocrystalline Module	N	183.7	
APOS Energy	AP210	210W Polycrystalline Module	N	186.3	
APOS Energy	AS210	210W Monocrystalline Module	N	188.3	
APOS Energy	AP215	215W Polycrystalline Module	N	190.9	
APOS Energy	AS215	215W Monocrystalline Module	N	192.9	
APOS Energy	AP220	220W Polycrystalline Module	N	195.4	
APOS Energy	AS220	220W Monocrystalline Module	N	197.0	
APOS Energy	AP225	225W Polycrystalline Module	N	200.0	
APOS Energy	AS225	225W Monocrystalline Module	N	201.6	
APOS Energy	AP230	230W Polycrystalline Module	N	204.6	
APOS Energy	AS230	230W Monocrystalline Module	N	206.1	
APOS Energy	AS235	235W Monocrystalline Module	N	210.7	
APOS Energy	AS240	240W Monocrystalline Module	N	215.4	
Applied Materials	1/4 Size Single Junction Version 1.0	86W Thin Film Single Junction Amorphous Silicon Module	N	80.8	
Applied Materials	1/4 Size Tandem Junction	114W Thin Film Tandem Junction Module	N	104.6	

Applied Materials	1/2-L Size Single Junction Version 1.0	172W Thin Film Single Junction Amorphous Silicon Module	N	161.6
Applied Materials	1/2-P Size Single Junction Version 1.0	172W Thin Film Single Junction Amorphous Silicon Module	N	161.6
Applied Materials	1/2-L Size Tandem Junction	229W Thin Film Tandem Junction Module	N	210.1
Applied Materials	1/2-P Size Tandem Junction	229W Thin Film Tandem Junction Module	N	210.1
Applied Materials	Full Size Single Junction Version 1.0	343W Thin Film Single Junction Amorphous Silicon Module	N	322.3
Applied Materials	Full Size Tandem Junction	458W Thin Film Tandem Junction Module	N	420.2
Applied Solar	OE-34	34W Polycrystalline Waterproof Built-in PV Roof Tile	Y	28.2
Applied Solar	OE-48	48W Polycrystalline Waterproof Built-in PV Roof Tile	Y	38.6
Applied Solar	OE-50	50W Polycrystalline Waterproof Built-in PV Roof Tile	Y	44.0
ASUN Energy	ASM190PCA0G101	190W Polycrystalline Module, Framed	N	159.1
ASUN Energy	ASM195PCA0G101	195W Polycrystalline Module, Framed	N	163.5
ASUN Energy	ASM200PCA0G101	200W Polycrystalline Module, Framed	N	167.8

(Continued)

MANUFACTURER NAME	MODULE MODEL NUMBER	DESCRIPTION	BIPV*	PTC**	NOTES
ASUN Energy	ASM205PCA0G101	205W Polycrystalline Module, Framed	N	172.1	
ASUN Energy	ASM210PCA0G101	210W Polycrystalline Module, Framed	N	176.5	
Atlantis Energy	AES-SS-100-C	14W Monocrystalline SunSlate	Y	11.5	
Atlantis Energy	AES-SS-100-W	14W Monocrystalline SunSlate	Y	11.5	
AU Optronics	PM220P01.0_220	220W Polycrystalline Module	N	189.1	
AU Optronics	PM220P01.0_225	225W Polycrystalline Module	N	193.5	
AU Optronics	PM220P01.0_230	230W Polycrystalline Module	N	203.5	
AU Optronics	PM220P01.0_235	235W Polycrystalline Module	N	208.1	
AU Optronics	PM220P01.0_240	240W Polycrystalline Module	N	212.7	
Auria Solar	M115000	115W Thin Film Module	N	107.5	
Auria Solar	M120000	120W Thin Film Module	N	112.2	
Auria Solar	M125000	125W Thin Film Module	N	116.9	
Auxin Solar	AXN-P6T160	160W Polycrystalline Module	N	137.7	
Auxin Solar	AXN-M5T165	165W Monocrystalline Module	N	144.7	
Auxin Solar	AXN-P6T165	165W Polycrystalline Module	N	142.1	
Auxin Solar	AXN-M5T170	170W Monocrystalline Module	N	149.2	
Auxin Solar	AXN-P6T170	170W Polycrystalline Module	N	146.5	
Auxin Solar	AXN-M5T175	175W Monocrystalline Module	N	153.7	
Auxin Solar	AXN-P6T175	175W Polycrystalline Module	N	151.0	
Auxin Solar	AXN-M5T180	180W Monocyrystalline Module	N	158.3	
Auxin Solar	AXN-P6T180	180W Polycrystalline Module	N	155.4	
Avancis	PowerMax 100 FB	100W Thin Film CIS Module	N	86.1	
Baoding Tianwei Solarfilms	TWSF-aSi-90W-1	90W Thin Film a-Si Module	N	84.7	

Manufacturer	Model	Description		
Baoding Tianwei Solarfilms	TWSF-aSi-95W-1	95W Thin Film a-Si Module	N	89.4
Bosch Solar Thin Film	μm-Si plus 105	105W Thin Film Tandem Junction Module	N	97.3
Bosch Solar Thin Film	μm-Si plus 110	110W Thin Film Tandem Junction Module	N	102.0
Bosch Solar Thin Film	μm-Si plus 115	115W Thin Film Tandem Junction Module	N	106.6
BP Solar	BP365TS	65W Polycrystalline Module, MC Connectors, 6V, Roof Tile Frame	Y	51.7
BP Solar	BP170B	170W Polycrystalline Module, MC3, Bronze Frame	N	152.6
BP Solar	BP3170B	170W Polycrystalline Module, MC4, U Frame	N	153.3
BP Solar	BP3170N	170W Polycrystalline Module, MC3, U Frame	N	152.6
BP Solar	BP3170N_Q	170W Polycrystalline Module, MC3, U Frame	N	152.6
BP Solar	BP3170T	170W Polycrystalline Module, MC4, U Frame	N	152.6
BP Solar	BP175B	175W Polycrystalline Module, MC4, Bronze Frame	N	157.3
BP Solar	BP175I	175W Polycrystalline Module, Multicontact MC3, Integra Frame	N	157.3
BP Solar	BP3175B	175W Polycrystalline Module, MC4, U Frame, Black/Black	N	157.2

(Continued)

MANUFACTURER NAME	MODULE MODEL NUMBER	DESCRIPTION	BIPV*	PTC**	NOTES
BP Solar	BP3175N	175W Polycrystalline Module, MC3, U Frame	N	157.2	
BP Solar	BP3175N_Q	175W Polycrystalline Module, MC3, U Frame	N	157.2	
BP Solar	BP3175T	175W Polycrystalline Module, MC4, U Frame	N	157.2	
BP Solar	BP4175B	175W Monocrystalline Module, MC4, U Frame, Black/Black	N	156.2	
BP Solar	BP4175T	175W Monocrystalline Module, MC4, U Frame	N	156.2	
BP Solar	BP3180N	180W Polycrystalline Module, MC3, U Frame	N	161.8	
BP Solar	BP3180N_Q	180W Polycrystalline Module, MC3, U Frame	N	161.8	
BP Solar	BP4180T	180W Monocrystalline Module, MC4, U Frame	N	160.7	
BP Solar	SX3190B	190W Polycrystalline Module, MC3, U Frame, Black/Black	N	166.0	
BP Solar	SX3190N	190W Polycrystalline Module, MC3, U Frame	N	166.0	
BP Solar	SX3190W	190W Polycrystalline Module, MC3, U Frame, White/Black	N	166.0	
BP Solar	SX3195B	195W Polycrystalline Module, MC3, U Frame, Black/Black	N	170.5	
BP Solar	SX3195N	195W Polycrystalline Module, MC3, U Frame	N	170.5	
BP Solar	SX3195W	195W Polycrystalline Module, MC3, U Frame, White/Black	N	170.5	
BP Solar	SX3200B	200W Polycrystalline Module, MC3, U Frame, Black/Black	N	175.0	

Manufacturer	Model	Description		Value
BP Solar	SX3200N	200W Polycrystalline Module, MC3, U Frame	N	175.0
BP Solar	SX3200W	200W Polycrystalline Module, MC3, U Frame, White/Black	N	175.0
BP Solar	BP3210N	210W Polycrystalline Module, MC3, P Frame	N	176.2
BP Solar	BP3210T	210W Polycrystalline Module, MC4, U Frame	N	184.1
BP Solar	BP3215B	215W Polycrystalline Module, MC4, U Frame, Black/Black	N	193.6
BP Solar	BP3220N	220W Polycrystalline Module, MC3, P Frame	N	193.1
BP Solar	BP3220T	220W Polycrystalline Module, MC4, U Frame	N	193.1
BP Solar	BP3225N	225W Polycrystalline Module, MC4, U Frame	N	197.7
BP Solar	BP3225T	225W Polycrystalline Module, MC4, U Frame	N	197.7
BP Solar	BP3230N	230W Polycrystalline Module, MC3, P Frame	N	202.2
BP Solar	BP3230T	230W Polycrystalline Module, MC4, U Frame	N	202.2
BP Solar	BP3237T	237W Polycrystalline Module, MC4, U Frame	N	208.5
Brightwatts	BWI-72-P175	175W Monocrystalline Module	N	159.4
Brightwatts	BWI-60-M190	190W Polycrystalline Module	N	168.1
Brightwatts	BI-156-200W	200W Polycrystalline Module	N	180.2
Brightwatts	BWI-60-M200	200W Polycrystalline Module	N	177.1

(Continued)

MANUFACTURER NAME	MODULE MODEL NUMBER	DESCRIPTION	BIPV*	PTC**	NOTES
Brightwatts	BWI-60-M220	220W Polycrystalline Module	N	201.6	
Brightwatts	BWI-96-M220	220W Monocrystalline Module	N	200.1	
Brightwatts	BWI-60-M225	225W Polycrystalline Module	N	206.3	
Brightwatts	BWI-96-M225	225W Monocrystalline Module	N	204.8	
Brightwatts	BWI-60-M230	230W Polycrystalline Module	N	211.0	
Brightwatts	BWI-96-M230	230W Monocrystalline Module	N	209.4	
CA Solar	MS-150M	150W Monocrystalline Module	N	131.7	
CA Solar	MS-155M	155W Monocrystalline Module	N	136.2	
CA Solar	MS-160M	160W Monocrystalline Module	N	140.7	
CA Solar	MS-165M	165W Monocrystalline Module	N	145.2	
CA Solar	MS-170M	170W Monocrystalline Module	N	149.7	
CA Solar	MS-175M	175W Monocrystalline Module	N	154.2	
CA Solar	MS-180M	180W Monocrystalline Module	N	158.8	
Canadian Solar	CS5C-80M	80W Monocrystalline Module	N	72.7	
Canadian Solar	CS5C-90M	90W Monocrystalline Module	N	82.0	
Canadian Solar	CS5A-150M	150W Monocrystalline Module	N	136.2	
Canadian Solar	CS6A-150P	150W Polycrystalline Module	N	130.7	
Canadian Solar	CS6A-150PE	150W Polycrystalline Module	N	132.6	
Canadian Solar	CS6A-155P	155W Polycrystalline Module	N	135.1	
Canadian Solar	CS6A-155PE	155W Polycrystalline Module	N	137.1	
Canadian Solar	CS5A-160M	160W Monocrystalline Module	N	145.4	
Canadian Solar	CS5A-160MX	160W Monocrystalline Module with ZEP	N	144.5	
Canadian Solar	CS6A-160P	160W Polycrystalline Module	N	139.6	
Canadian Solar	CS6A-160PE	160W Polycrystalline Module	N	141.7	

Canadian Solar	CS6P-160PE	160W Polycrystalline Module	N	141.0
Canadian Solar	CS5A-165M	165W Monocrystalline Module	N	150.1
Canadian Solar	CS5A-165MX	165W Monocrystalline Module with ZEP	N	149.1
Canadian Solar	CS6A-165P	165W Polycrystalline Module	N	144.1
Canadian Solar	CS6P-165PE	165W Polycrystalline Module	N	145.5
Canadian Solar	CS5A-170M	170W Monocrystalline Module	N	154.7
Canadian Solar	CS5A-170MX	170W Monocrystalline Module with ZEP	N	153.8
Canadian Solar	CS6A-170P	170W Polycrystalline Module	N	148.6
Canadian Solar	CS6P-170PE	170W Polycrystalline Module	N	150.0
Canadian Solar	CS5A-175M	175W Monocrystalline Module	N	159.4
Canadian Solar	CS5A-175MX	175W Monocrystalline Module with ZEP	N	158.4
Canadian Solar	CS6A-175P	175W Polycrystalline Module	N	153.1
Canadian Solar	CS6P-175PE	175W Polycrystalline Module	N	154.5
Canadian Solar	CS5A-180M	180W Monocrystalline Module	N	164.0
Canadian Solar	CS5A-180MX	180W Monocrystalline Module with ZEP	N	163.0
Canadian Solar	CS6A-180P	180W Polycrystalline Module	N	157.6
Canadian Solar	CS6P-180P	180W Polycrystalline Module	N	156.7
Canadian Solar	CS6P-180PE	180W Polycrystalline Module	N	159.0
Canadian Solar	CS5A-185M	185W Monocrystalline Module	N	167.7
Canadian Solar	CS5A-185MX	185W Monocrystalline Module with ZEP	N	167.7
Canadian Solar	CS6A-185P	185W Polycrystalline Module	N	162.1

(Continued)

MANUFACTURER NAME	MODULE MODEL NUMBER	DESCRIPTION	BIPV*	PTC**	NOTES
Canadian Solar	CS6P-185P	185W Polycrystalline Module	N	159.5	
Canadian Solar	CS6P-185PE	185W Polycrystalline Module	N	163.5	
Canadian Solar	CS5A-190M	190W Monocrystalline Module	N	172.3	
Canadian Solar	CS5A-190MX	190W Monocrystalline Module with ZEP	N	172.3	
Canadian Solar	CS6A-190P	190W Polycrystalline Module	N	166.6	
Canadian Solar	CS6P-190P	190W Polycrystalline Module	N	163.9	
Canadian Solar	CS6P-190PE	190W Polycrystalline Module	N	168.1	
Canadian Solar	CS5A-195M	195W Monocrystalline Module	N	177.0	
Canadian Solar	CS5A-195MX	195W Monocrystalline Module with ZEP	N	177.0	
Canadian Solar	CS5P-195M	195W Monocrystalline Module	N	177.0	
Canadian Solar	CS6P-195P	195W Polycrystalline Module	N	168.4	
Canadian Solar	CS6P-195PE	195W Polycrystalline Module	N	172.6	
Canadian Solar	CS5P-200M	200W Monocrystalline Module	N	181.6	
Canadian Solar	CS6P-200P	200W Polycrystalline Module	N	183.0	
Canadian Solar	CS6P-200PE	200W Polycrystalline Module	N	177.1	
Canadian Solar	CS6P-200PX	200W Polycrystalline Module with ZEP	N	183.0	
Canadian Solar	CS5P-205M	205W Monocrystalline Module	N	186.2	
Canadian Solar	CS6P-205P	205W Polycrystalline Module	N	187.6	
Canadian Solar	CS6P-205PX	205W Polycrystalline Module with ZEP	N	187.6	
Canadian Solar	CS5P-210M	210W Monocrystalline Module	N	190.8	
Canadian Solar	CS6P-210P	210W Polycrystalline Module	N	192.3	
Canadian Solar	CS6P-210PX	210W Polycrystalline Module with ZEP	N	192.3	

Canadian Solar	CS5P-215M	215W Monocrystalline Module	N	195.5
Canadian Solar	CS6P-215P	215W Polycrystalline Module	N	197.0
Canadian Solar	CS6P-215PX	215W Polycrystalline Module with ZEP	N	197.0
Canadian Solar	CS5P-220M	220W Monocrystalline Module	N	200.1
Canadian Solar	CS6P-220P	220W Polycrystalline Module	N	201.6
Canadian Solar	CS6P-220PX	220W Polycrystalline Module with ZEP	N	201.6
Canadian Solar	CS5P-225M	225W Monocrystalline Module	N	204.8
Canadian Solar	CS6P-225P	225W Polycrystalline Module	N	206.3
Canadian Solar	CS6P-225PX	225W Polycrystalline Module with ZEP	N	206.3
Canadian Solar	CS5P-230M	230W Monocrystalline Module	N	209.4
Canadian Solar	CS6P-230P	230W Polycrystalline Module	N	211.0
Canadian Solar	CS6P-230PX	230W Polycrystalline Module with ZEP	N	211.0
Canadian Solar	CS5P-235M	235W Monocrystalline Module	N	214.1
Canadian Solar	CS6P-235PX	235W Polycrystalline Module with ZEP	N	215.7
Canadian Solar	CS5P-240M	240W Monocrystalline Module	N	218.7
CEEG (Shanghai) Solar Science and Technology	SST160-72M	160W Monocrystalline Module	N	136.4
CEEG (Shanghai) Solar Science and Technology	SST160-72M-Roof	160W Monocrystalline Module	N	136.4
CEEG (Shanghai) Solar Science and Technology	SST165-72M	165W Monocrystalline Module	N	140.8

(Continued)

MANUFACTURER NAME	MODULE MODEL NUMBER	DESCRIPTION	BIPV*	PTC**	NOTES
CEEG (Shanghai) Solar Science and Technology	SST165-72M-Roof	165W Monocrystalline Module	N	140.8	
CEEG (Shanghai) Solar Science and Technology	SST170-72M	170W Monocrystalline Module	N	145.3	
CEEG (Shanghai) Solar Science and Technology	SST170-72M-Roof	170W Monocrystalline Module	N	145.3	
CEEG (Shanghai) Solar Science and Technology	SST175-72M	175W Monocrystalline Module	N	156.5	
CEEG (Shanghai) Solar Science and Technology	SST175-72M-Roof	175W Monocrystalline Module	N	156.5	
CEEG (Shanghai) Solar Science and Technology	SST180-72M	180W Monocrystalline Module	N	161.0	
CEEG (Shanghai) Solar Science and Technology	SST180-72M-Roof	180W Monocrystalline Module	N	161.0	
CEEG (Shanghai) Solar Science and Technology	SST185-72M	185W Monocrystalline Module	N	165.6	
CEEG (Shanghai) Solar Science and Technology	SST185-72M-Roof	185W Monocrystalline Module	N	165.6	
CEEG (Shanghai) Solar Science and Technology	SST190-72M	190W Monocrystalline Module	N	167.4	

CEEG (Shanghai) Solar Science and Technology	SST190-72M-Roof	190W Monocrystalline Module	N	167.4
CEEG (Shanghai) Solar Science and Technology	SST220-60M	220W Monocrystalline Module	N	198.8
CEEG (Shanghai) Solar Science and Technology	SST225-60M	225W Monocrystalline Module	N	203.4
CEEG (Shanghai) Solar Science and Technology	SST230-60M	230W Monocrystalline Module	N	208.0
CEEG (Shanghai) Solar Science and Technology	SST235-60M	235W Monocrystalline Module	N	212.6
CEEG (Shanghai) Solar Science and Technology	SST240-60M	240W Monocrystalline Module	N	217.3
CEEG (Shanghai) Solar Science and Technology	SST245-60M	245W Monocrystalline Module	N	221.9
CEEG (Shanghai) Solar Science and Technology	SST250-60M	250W Monocrystalline Module	N	226.6
CEEG (Shanghai) Solar Science and Technology	SST265-72M	265W Monocrystalline Module	N	239.5
CEEG (Shanghai) Solar Science and Technology	SST270-72M	270W Monocrystalline Module	N	244.1

(Continued)

MANUFACTURER NAME	MODULE MODEL NUMBER	DESCRIPTION	BIPV*	PTC**	NOTES
CEEG (Shanghai) Solar Science and Technology	SST275-72M	275W Monocrystalline Module	N	248.7	
CEEG (Shanghai) Solar Science and Technology	SST275-72P	275W Polycrystalline Module	N	241.6	
CEEG (Shanghai) Solar Science and Technology	SST280-72M	280W Monocrystalline Module	N	253.4	
CEEG (Shanghai) Solar Science and Technology	SST280-72P	280W Polycrystalline Module	N	246.1	
CEEG (Shanghai) Solar Science and Technology	SST285-72M	285W Monocrystalline Module	N	258.0	
CEEG (Shanghai) Solar Science and Technology	SST285-72P	285W Polycrystalline Module	N	250.6	
CEEG (Shanghai) Solar Science and Technology	SST290-72M	290W Monocrystalline Module	N	253.5	
CEEG (Shanghai) Solar Science and Technology	SST290-72P	290W Polycrystalline Module	N	255.2	
CEEG (Shanghai) Solar Science and Technology	SST295-72M	295W Monocrystalline Module	N	258.1	
CEEG (Shanghai) Solar Science and Technology	SST300-72M	300W Monocrystalline Module	N	262.6	

Centennial Solar	CS200	200W Polycrystalline Module	N	180.2
Centrosolar America	E165	165W Monocrystalline Module	N	146.5
Centrosolar America	E170	170W Monocrystalline Module	N	151.1
Centrosolar America	E175	175W Monocrystalline Module	N	155.7
Centrosolar America	E180	180W Monocrystalline Module	N	157.7
Centrosolar America	E185	185W Monocrystalline Module	N	162.2
Centrosolar America	E195	195W Polycrystalline Module	N	172.4
Centrosolar America	D200	200W Polycrystalline Module	N	183.0
Centrosolar America	E200	200W Polycrystalline Module	N	176.9
Centrosolar America	E205	205W Polycrystalline Module	N	181.5
Centrosolar America	D210	210W Polycrystalline Module	N	192.3
Centrosolar America	E210	210W Polycrystalline Module	N	186.1
Centrosolar America	D220	220W Polycrystalline Module	N	201.6
Centrosolar America	E220	220W Polycrystalline Module	N	198.2
Centrosolar America	E220B	220W Polycrystalline Module, Black Frame	N	198.2
Centrosolar America	E225	225W Polycrystalline Module	N	202.8
Centrosolar America	E225B	225W Polycrystalline Module, Black Frame	N	202.8
Centrosolar America	D230	230W Polycrystalline Module	N	211.0
Centrosolar America	E230	230W Polycrystalline Module	N	207.5
Centrosolar America	E230B	230W Polycrystalline Module, Black Frame	N	207.5
Centrosolar America	E260	260W Polycrystalline Module	N	231.2
Centrosolar America	E265	265W Polycrystalline Module	N	235.8
Centrosolar America	E265B	265W Polycrystalline Module, Black Frame	N	235.8

(Continued)

MANUFACTURER NAME	MODULE MODEL NUMBER	DESCRIPTION	BIPV*	PTC**	NOTES
Centrosolar America	E270	270W Polycrystalline Module	N	237.2	
Centrosolar America	E270B	270W Polycrystalline Module, Black Frame	N	237.2	
Centrosolar America	E275	275W Polycrystalline Module	N	241.7	
Centrosolar America	E275B	275W Polycrystalline Module, Black Frame	N	241.7	
Changzhou Eging Photovoltaic Technology	EGM-150	150W Monocrystalline Module	N	134.6	
Changzhou Eging Photovoltaic Technology	EGM-155	155W Monocrystalline Module	N	139.2	
Changzhou Eging Photovoltaic Technology	EGM-160	160W Monocrystalline Module	N	143.8	
Changzhou Eging Photovoltaic Technology	EGM-165	165W Monocrystalline Module	N	148.4	
Changzhou Eging Photovoltaic Technology	EGM-170	170W Monocrystalline Module	N	153.0	
Changzhou Eging Photovoltaic Technology	EGM-175	175W Monocrystalline Module	N	157.6	
Changzhou Eging Photovoltaic Technology	EGM-180	180W Monocrystalline Module	N	162.2	
Changzhou Nesl Solartech	DJ-165D	165W Monocrystalline Module	N	147.3	

Changzhou Nesl Solartech	DJ-170D	170W Monocrystalline Module	N	151.8
Changzhou Nesl Solartech	DJ-175D	175W Monocrystalline Module	N	156.4
Changzhou Nesl Solartech	DJ-180D	180W Monocrystalline Module	N	161.0
Changzhou Nesl Solartech	DJ-200P	200W Polycrystalline Module	N	175.9
Changzhou Nesl Solartech	DJ-210P	210W Polycrystalline Module	N	184.9
Changzhou Nesl Solartech	DJ-220P	220W Polycrystalline Module	N	194.0
Changzhou Nesl Solartech	DJ-230P	230W Polycrystalline Module	N	203.1
Changzhou Nesl Solartech	DJ-240P	240W Polycrystalline Module	N	211.7
Changzhou Nesl Solartech	DJ-250P	250W Polycrystalline Module	N	220.7
Changzhou Nesl Solartech	DJ-260P	260W Polycrystalline Module	N	229.8
Changzhou Nesl Solartech	DJ-270P	270W Polycrystalline Module	N	238.9
Changzhou Nesl Solartech	DJ-280P	280W Polycrystalline Module	N	248.0
Chi Mei Energy	CSSU-100A	100W Thin Film a-Si Module	N	93.7
Chi Mei Energy	CSSU-100B	100W Thin Film a-Si Module	N	93.7
China Sunergy	CSUN 290M-160	160W Monocrystalline Module	N	136.4
China Sunergy	CSUN160D-24/D	160W Monocrystalline Module	N	136.4

(Continued)

MANUFACTURER NAME	MODULE MODEL NUMBER	DESCRIPTION	BIPV*	PTC**	NOTES
China Sunergy	CSUN 290M-165	165W Monocrystalline Module	N	140.8	
China Sunergy	CSUN165D-24/D	165W Monocrystalline Module	N	140.8	
China Sunergy	CSUN 290M-170	170W Monocrystalline Module	N	145.3	
China Sunergy	CSUN170D-24/D	170W Monocrystalline Module	N	145.3	
China Sunergy	CSUN 290M-175	175W Monocrystalline Module	N	156.5	
China Sunergy	CSUN175D-24/D	175W Monocrystalline Module	N	156.5	
China Sunergy	CSUN 290M-180	180W Monocrystalline Module	N	161.0	
China Sunergy	CSUN180D-24/D	180W Monocrystalline Module	N	161.0	
China Sunergy	CSUN 290M-185	185W Monocrystalline Module	N	165.6	
China Sunergy	CSUN185D-24/D	185W Monocrystalline Module	N	165.6	
China Sunergy	CSUN 240M-220	220W Monocrystalline Module	N	198.8	
China Sunergy	CSUN 240M-225	225W Monocrystalline Module	N	203.4	
China Sunergy	CSUN 240M-230	230W Monocrystalline Module	N	208.0	
China Sunergy	CSUN 240M-235	235W Monocrystalline Module	N	212.6	
China Sunergy	CSUN 240M-240	240W Monocrystalline Module	N	217.3	
China Sunergy	CSUN 240M-245	245W Monocrystalline Module	N	221.9	
China Sunergy	CSUN 240M-250	250W Monocrystalline Module	N	226.6	
China Sunergy	CSUN 290M-265	265W Monocrystalline Module	N	239.5	
China Sunergy	CSUN 290M-270	270W Monocrystalline Module	N	244.1	
China Sunergy	CSUN 290M-275	275W Monocrystalline Module	N	248.7	
China Sunergy	CSUN 290M-280	280W Monocrystalline Module	N	253.4	
China Sunergy	CSUN 290M-285	285W Monocrystalline Module	N	258.0	
Chinalight Haoyu Photovoltaic Technology (Beijing)	CLS165P	165W Polycrystalline Module	N	146.0	

Chinalight Haoyu Photovoltaic Technology (Beijing)	CLS170P	170W Polycrystalline Module	N	150.5
Chinalight Haoyu Photovoltaic Technology (Beijing)	CLS175P	175W Polycrystalline Module	N	155.1
Chinalight Haoyu Photovoltaic Technology (Beijing)	CLS180P	180W Polycrystalline Module	N	159.6
Chinalight Haoyu Photovoltaic Technology (Beijing)	CLS185P	185W Polycrystalline Module	N	164.2
Chinalight Haoyu Photovoltaic Technology (Beijing)	CLS190P	190W Polycrystalline Module	N	168.8
Chinalight Haoyu Photovoltaic Technology (Beijing)	CLS215P	215W Polycrystalline Module	N	190.5
Chinalight Haoyu Photovoltaic Technology (Beijing)	CLS220P	220W Polycrystalline Module	N	195.0
Chinalight Haoyu Photovoltaic Technology (Beijing)	CLS225P	225W Polycrystalline Module	N	199.6
Chinalight Haoyu Photovoltaic Technology (Beijing)	CLS230P	230W Polycrystalline Module	N	204.2
Chinalight Haoyu Photovoltaic Technology (Beijing)	CLS235P	235W Polycrystalline Module	N	208.7
Chint Solar (Zhejiang)	CHSM5612M-175	175W Monocrystalline Module	N	153.8

(Continued)

MANUFACTURER NAME	MODULE MODEL NUMBER	DESCRIPTION	BIPV*	PTC**	NOTES
Chint Solar (Zhejiang)	CHSM6610P-215	215W Polycrystalline Module	N	194.8	
Chint Solar (Zhejiang)	CHSM6610M-220	220W Monocrystalline Module	N	195.3	
Chint Solar (Zhejiang)	CHSM6610P-220	220W Polycrystalline Module	N	199.5	
Chint Solar (Zhejiang)	CHSM6610M-225	225W Monocrystalline Module	N	199.9	
Chint Solar (Zhejiang)	CHSM6610P-225	225W Polycrystalline Module	N	204.1	
Chint Solar (Zhejiang)	CHSM6610M-230	230W Monocrystalline Module	N	204.4	
Chint Solar (Zhejiang)	CHSM6610P-230	230W Polycrystalline Module	N	208.8	
Chint Solar (Zhejiang)	CHSM6610P-235	235W Polycrystalline Module	N	213.4	
Clean Source & Energy	CSE115M-1	115W Monocrystalline Module	N	103.2	
Clean Source & Energy	CSE120M-1	120W Monocrystalline Module	N	107.8	
Clean Source & Energy	CSE125M-1	125W Monocrystalline Module	N	112.4	
Clean Source & Energy	CSE150M-2	150W Monocrystalline Module	N	132.0	
Clean Source & Energy	CSE155M-2	155W Monocrystalline Module	N	136.5	
Clean Source & Energy	CSE160M-1	160W Monocrystalline Module	N	143.5	
Clean Source & Energy	CSE160M-2	160W Monocrystalline Module	N	141.0	
Clean Source & Energy	CSE165M-1	165W Monocrystalline Module	N	148.0	
Clean Source & Energy	CSE165M-2	165W Monocrystalline Module	N	145.5	
Clean Source & Energy	CSE170M-1	170W Monocrystalline Module	N	152.6	
Clean Source & Energy	CSE170M-2	170W Monocrystalline Module	N	150.1	
Clean Source & Energy	CSE175M-1	175W Monocrystalline Module	N	156.1	
Clean Source & Energy	CSE175M-2	175W Monocrystalline Module	N	154.6	
Clean Source & Energy	CSE180M-1	180W Monocrystalline Module	N	160.7	
Clean Source & Energy	CSE180M-2	180W Monocrystalline Module	N	159.2	
Clean Source & Energy	CSE180M-3	180W Monocrystalline Module	N	163.4	
Clean Source & Energy	CSE185M-1	185W Monocrystalline Module	N	165.3	

Clean Source & Energy	CSE185M-2	185W Monocrystalline Module	N	163.7
Clean Source & Energy	CSE190P-3	190W Polycrystalline Module	N	170.5
Clean Source & Energy	CSE195P-3	195W Polycrystalline Module	N	175.1
Clean Source & Energy	CSE200P-1	200W Polycrystalline Module	N	179.8
Clean Source & Energy	CSE200P-3	200W Polycrystalline Module	N	179.7
Clean Source & Energy	CSE205P-1	205W Polycrystalline Module	N	184.4
Clean Source & Energy	CSE205P-3	205W Polycrystalline Module	N	184.3
Clean Source & Energy	CSE210P-1	210W Polycrystalline Module	N	189.0
Clean Source & Energy	CSE215P-1	215W Polycrystalline Module	N	193.6
Clean Source & Energy	CSE220P-1	220W Polycrystalline Module	N	198.2
Clean Source & Energy	CSE220P-3	220W Polycrystalline Module	N	198.5
Clean Source & Energy	CSE225P-3	225W Polycrystalline Module	N	203.1
Clean Source & Energy	CSE230P-3	230W Polycrystalline Module	N	207.7
Clean Source & Energy	CSE250P-3	250W Polycrystalline Module	N	224.3
Clean Source & Energy	CSE255P-3	255W Polycrystalline Module	N	228.9
Clean Source & Energy	CSE270P-3	270W Polycrystalline Module	N	242.7
Clean Source & Energy	CSE275P-3	275W Polycrystalline Module	N	247.3
Clean Source & Energy	CSE280P-3	280W Polycrystalline Module	N	252.0
Conergy	Conergy P 170M	170W Monocrystalline Module	N	151.8
Conergy	Conergy P 175M	175W Monocrystalline Module	N	156.4
Conergy	Conergy P 180M	180W Monocrystalline Module	N	161.0
Conergy	Conergy P 185M	185W Monocrystalline Module	N	165.6
Conergy	Conergy P 190P	190W Polycrystalline Module	N	168.6
Conergy	Conergy P 195P	195W Polycrystalline Module	N	173.1
Conergy	Conergy P 200P	200W Polycrystalline Module	N	177.7
Conergy	Conergy P 205P	205W Polycrystalline Module	N	182.3

(Continued)

MANUFACTURER NAME	MODULE MODEL NUMBER	DESCRIPTION	BIPV*	PTC**	NOTES
Conergy	Conergy P 210P	210W Polycrystalline Module	N	186.8	
Conergy	Conergy P 210PA	210W Polycrystalline Module	N	189.2	
Conergy	Conergy P 215PA	215W Polycrystalline Module	N	193.8	
Conergy	Conergy P 220PA	220W Polycrystalline Module	N	198.4	
Conergy	Conergy P 225PA	225W Polycrystalline Module	N	203.0	
Conergy	Conergy P 230PA	230W Polycrystalline Module	N	207.6	
Conergy	Conergy P 235PA	235W Polycrystalline Module	N	212.2	
Cosmos Energy	SL195-18	195W Polycrystalline Module	N	169.3	
Cosmos Energy	SL200-18	200W Polycrystalline Module	N	173.8	
Cosmos Energy	SL205-18	205W Polycrystalline Module	N	178.3	
Cosmos Energy	SL220-20	220W Polycrystalline Module	N	189.8	
Cosmos Energy	SL225-20	225W Polycrystalline Module	N	194.2	
Cosmos Energy	SL230-20	230W Polycrystalline Module	N	198.7	
CSG PVTech	CSG160M2-24	160W Polycrystalline Module	N	143.4	
CSG PVTech	CSG165M2-24	165W Polycrystalline Module	N	148.0	
CSG PVTech	CSG170M2-24	170W Polycrystalline Module	N	152.6	
CSG PVTech	CSG175M2-24	175W Polycrystalline Module	N	157.2	
CSG PVTech	CSG175S1-36	175W Monocrystalline Module	N	158.7	
CSG PVTech	CSG180M2-24	180W Polycrystalline Module	N	161.8	
CSG PVTech	CSG180S1-36	180W Monocrystalline Module	N	163.4	
CSG PVTech	CSG190M2-27	190W Polycrystalline Module	N	170.5	
CSG PVTech	CSG195M2-27	195W Polycrystalline Module	N	175.1	
CSG PVTech	CSG200M2-27	200W Polycrystalline Module	N	179.7	
CSG PVTech	CSG205M2-27	205W Polycrystalline Module	N	184.3	
CSG PVTech	CSG220M2-30	220W Polycrystalline Module	N	198.5	

CSG PVTech	CSG225M2-30	225W Polycrystalline Module	N	203.1
CSG PVTech	CSG230M2-30	230W Polycrystalline Module	N	207.7
CSG PVTech	CSG250M2-35	250W Polycrystalline Module	N	224.3
CSG PVTech	CSG255M2-35	255W Polycrystalline Module	N	228.9
CSG PVTech	CSG260M2-35	260W Polycrystalline Module	N	233.5
CSG PVTech	CSG265M2-35	265W Polycrystalline Module	N	238.1
CSG PVTech	CSG270M2-36	270W Polycrystalline Module	N	242.7
CSG PVTech	CSG275M2-36	275W Polycrystalline Module	N	247.3
CSG PVTech	CSG280M2-36	280W Polycrystalline Module	N	252.0
CSUN	SST275-72P	275W Polycrystalline Module	N	241.6
CSUN	SST280-72P	280W Polycrystalline Module	N	246.1
CSUN	SST285-72P	285W Polycrystalline Module	N	250.6
CSUN	SST290-72M	290W Monocrystalline Module	N	253.5
CSUN	SST290-72P	290W Polycrystalline Module	N	255.2
CSUN	SST295-72M	295W Monocrystalline Module	N	258.1
CSUN	SST300-72M	300W Monocrystalline Module	N	262.6
Cuantum Solar	SUNPORT 275P	275W Polycrystalline Module	N	248.5
Day4 Energy	Day448MC 155	155W Polycrystalline Module, Day4 Electrode Technology, Framed	N	138.7
Day4 Energy	Day448MC 160	160W Polycrystalline Module, Day4 Electrode Technology, Framed	N	143.3
Day4 Energy	Day448MC 165	165W Polycrystalline Module, Day4 Electrode Technology, Framed	N	147.9

(Continued)

MANUFACTURER NAME	MODULE MODEL NUMBER	DESCRIPTION	BIPV*	PTC**	NOTES
Day4 Energy	Day448MC 170	170W Polycrystalline Module, Day4 Electrode Technology, Framed	N	152.5	
Day4 Energy	Day448MC 175	175W Polycrystalline Module, Day4 Electrode Technology, Framed	N	157.1	
Day4 Energy	Day448MC 180	180W Polycrystalline Module, Day4 Electrode Technology, Framed	N	161.7	
Day4 Energy	Day448MC 185	185W Polycrystalline Module, Day4 Electrode Technology, Framed	N	166.3	
Day4 Energy	Day448MC 190	190W Polycrystalline Module, Day4 Electrode Technology, Framed	N	170.9	
DelSolar	D6P175A2E	175W Polycrystalline Module	N	156.0	
DelSolar	D6P180A2E	180W Polycrystalline Module	N	160.6	
DelSolar	D6P185A2E	185W Polycrystalline Module	N	165.2	
DelSolar	D6P210A3E	210W Polycrystalline Module	N	187.1	
DelSolar	D6P215A3E	215W Polycrystalline Module	N	191.6	
DelSolar	D6P220A3E	220W Polycrystalline Module	N	196.2	
DelSolar	D6P225A3E	225W Polycrystalline Module	N	200.8	
DelSolar	D6P230A3E	230W Polycrystalline Module	N	205.4	
DelSolar	D6P235A3E	235W Polycrystalline Module	N	210.0	
DelSolar	D6P240A3E	240W Polycrystalline Module	N	214.6	
Dmsolar	DM160M2-24	160W Polycrystalline Module	N	143.4	
Dmsolar	DM-M160	160W Monocrystalline Module	N	145.4	
Dmsolar	DM165M2-24	165W Polycrystalline Module	N	148.0	

Dmsolar	DM170M2-24	170W Polycrystalline Module	N	152.6
Dmsolar	DM-M170	170W Monocrystalline Module	N	154.7
Dmsolar	DM175M2-24	175W Polycrystalline Module	N	157.2
Dmsolar	DM175S1-36	175W Monocrystalline Module	N	158.7
Dmsolar	DM180M2-24	180W Polycrystalline Module	N	161.8
Dmsolar	DM180S1-36	180W Monocrystalline Module	N	163.4
Dmsolar	DM-M180	180W Monocrystalline Module	N	164.0
Dmsolar	DM190M2-27	190W Polycrystalline Module	N	170.5
Dmsolar	DM195M2-27	195W Polycrystalline Module	N	175.1
Dmsolar	DM200M2-27	200W Polycrystalline Module	N	179.7
Dmsolar	DM-P200	200W Polycrystalline Module	N	183.0
Dmsolar	DM205M2-27	205W Polycrystalline Module	N	184.3
Dmsolar	DM-P210	210W Polycrystalline Module	N	192.3
Dmsolar	DM220M2-30	220W Polycrystalline Module	N	198.5
Dmsolar	DM-M220	220W Monocrystalline Module	N	200.1
Dmsolar	DM-P220	220W Polycrystalline Module	N	201.6
Dmsolar	DM225M2-30	225W Polycrystalline Module	N	203.1
Dmsolar	DM230M2-30	230W Polycrystalline Module	N	207.7
Dmsolar	DM-M230	230W Monocrystalline Module	N	209.4
Dmsolar	DM-P230	230W Polycrystalline Module	N	211.0
Dmsolar	DM-M240	240W Monocrystalline Module	N	218.7
Dmsolar	DM250M2-35	250W Polycrystalline Module	N	224.3
Dmsolar	DM255M2-35	255W Polycrystalline Module	N	228.9
Dmsolar	DM260M2-35	260W Polycrystalline Module	N	233.5
Dmsolar	DM265M2-35	265W Polycrystalline Module	N	238.1

(Continued)

MANUFACTURER NAME	MODULE MODEL NUMBER	DESCRIPTION	BIPV*	PTC**	NOTES
Dmsolar	DM270M2-36	270W Polycrystalline Module	N	242.7	
Dmsolar	DM275M2-36	275W Polycrystalline Module	N	247.3	
Dmsolar	DM280M2-36	280W Polycrystalline Module	N	252.0	
Du Pont Apollo	DA095-A2	95W Thin Film a-Si Module	N	86.3	
Du Pont Apollo	DA100-A2	100W Thin Film a-Si Module	N	90.9	
Du Pont Apollo	DA105-A2	105W Thin Film a-Si Module	N	95.5	
Ecosolargy	TWES-(155)72M	155W Monocrystalline Module	N	135.1	
Ecosolargy	TWES-(160)72M	160W Monocrystalline Module	N	139.5	
Ecosolargy	SDM-170/(165)-72M	165W Monocrystalline Module	N	145.3	
Ecosolargy	TWES-(165)72M	165W Monocrystalline Module	N	144.0	
Ecosolargy	SDM-170/(170)-72M	170W Monocrystalline Module	N	149.8	
Ecosolargy	TWES-(170)72M	170W Monocrystalline Module	N	148.5	
Ecosolargy	SDM-170/(175)-72M	175W Monocrystalline Module	N	154.3	
Ecosolargy	TWES-(175)72M	175W Monocrystalline Module	N	153.0	
Ecosolargy	SDM-170/(180)-72M	180W Monocrystalline Module	N	158.9	
Ecosolargy	TWES-(180)72M	180W Monocrystalline Module	N	157.5	
Ecosolargy	SDM-170/(185)-72M	185W Monocrystalline Module	N	163.4	
Ecosolargy	TWES-(200)60P	200W Polycrystalline Module	N	173.4	
Ecosolargy	TWES-(205)60P	205W Polycrystalline Module	N	177.9	
Ecosolargy	TWES-(210)60P	210W Polycrystalline Module	N	182.4	
Ecosolargy	TWES-(215)60P	215W Polycrystalline Module	N	186.8	
Ecosolargy	TWES-(220)60P	220W Polycrystalline Module	N	191.3	
Ecosolargy	TWES-(225)60P	225W Polycrystalline Module	N	195.8	
Ecosolargy	TWES-(230)60P	230W Polycrystalline Module	N	200.3	
Ecosolargy	TWES-(235)60P	235W Polycrystalline Module	N	204.8	

ENN Solar Energy	EST-110	110W Thin Film Tandem Junction Module	N	100.9
ENN Solar Energy	EST-115A	114W Thin Film Tandem Junction Module	N	104.6
ENN Solar Energy	EST-115	115W Thin Film Tandem Junction Module	N	105.5
ENN Solar Energy	EST-120	120W Thin Film Tandem Junction Module	N	110.1
ENN Solar Energy	EST-220H	220W Thin Film Tandem Junction Module	N	201.7
ENN Solar Energy	EST-220V	220W Thin Film Tandem Junction Module	N	201.7
ENN Solar Energy	EST-228HA	229W Thin Film Tandem Junction Module	N	210.1
ENN Solar Energy	EST-228VA	229W Thin Film Tandem Junction Module	N	210.1
ENN Solar Energy	EST-230H	230W Thin Film Tandem Junction Module	N	211.0
ENN Solar Energy	EST-230V	230W Thin Film Tandem Junction Module	N	211.0
ENN Solar Energy	EST-240H	240W Thin Film Tandem Junction Module	N	220.3
ENN Solar Energy	EST-240V	240W Thin Film Tandem Junction Module	N	220.3
ENN Solar Energy	EST-440	440W Thin Film Tandem Junction Module	N	403.5
ENN Solar Energy	EST-460	458W Thin Film Tandem Junction Module	N	420.2

(Continued)

MANUFACTURER NAME	MODULE MODEL NUMBER	DESCRIPTION	BIPV*	PTC**	NOTES
ENN Solar Energy	EST-460A	458W Thin Film Tandem Junction Module	N	420.2	
ENN Solar Energy	EST-480	480W Thin Film Tandem Junction Module	N	440.6	
Enp Sonne Solar Technikk	ENP-M165MO	165W Monocrystalline Module	N	147.3	
Enp Sonne Solar Technikk	ENP-M170MO	170W Monocrystalline Module	N	151.8	
Enp Sonne Solar Technikk	ENP-M175MO	175W Monocrystalline Module	N	156.4	
Enp Sonne Solar Technikk	ENP-M180MO	180W Monocrystalline Module	N	161.0	
EPV SOLAR	EPV-40	40W Thin Film Dual Junction a-Si Module, Frameless	N	38.2	
EPV SOLAR	EPV-42	42W Thin Film Dual Junction a-Si Module, Frameless	N	40.1	
ET Solar Industry	ET-M572165	165W Monocrystalline Module	N	146.5	
ET Solar Industry	ET-M572170	170W Monocrystalline Module	N	151.1	
ET Solar Industry	ET-M572175	175W Monocrystalline Module	N	155.7	
ET Solar Industry	ET-M572175B	175W Monocrystalline Module, Black Frame, Black Backsheet	N	155.1	
ET Solar Industry	ET-M572180	180W Monocrystalline Module	N	157.7	
ET Solar Industry	ET-M572180B	180W Monocrystalline Module, Black Frame, Black Backsheet	N	159.6	
ET Solar Industry	ET-M572185	185W Monocrystalline Module	N	162.2	
ET Solar Industry	ET-M572185B	185W Monocrystalline Module	N	164.2	
ET Solar Industry	ET-P654190	190W Polycrystalline Module	N	167.9	

ET Solar Industry	ET-P654190B	190W Polycrystalline Module, Black Frame	N	167.9
ET Solar Industry	ET-P654195	195W Polycrystalline Module	N	172.4
ET Solar Industry	ET-P654195B	195W Polycrystalline Module, Black Frame	N	172.4
ET Solar Industry	ET-P654200	200W Polycrystalline Module	N	176.9
ET Solar Industry	ET-P654200B	200W Polycrystalline Module, Black Frame	N	176.9
ET Solar Industry	ET-P654205	205W Polycrystalline Module	N	181.5
ET Solar Industry	ET-P654205B	205W Polycrystalline Module, Black Frame	N	181.5
ET Solar Industry	ET-P654210	210W Polycrystalline Module	N	186.1
ET Solar Industry	ET-P654210B	210W Polycrystalline Module, Black Frame	N	186.1
ET Solar Industry	ET-P660220	220W Polycrystalline Module	N	198.2
ET Solar Industry	ET-P660220B	220W Polycrystalline Module, Black Frame	N	198.2
ET Solar Industry	ET-P660225	225W Polycrystalline Module	N	202.8
ET Solar Industry	ET-P660225B	225W Polycrystalline Module, Black Frame	N	202.8
ET Solar Industry	ET-P660230	230W Polycrystalline Module	N	207.5
ET Solar Industry	ET-P660230B	230W Polycrystalline Module, Black Frame	N	207.5
ET Solar Industry	ET-P660235	235W Polycrystalline Module	N	212.1
ET Solar Industry	ET-P660235B	235W Polycrystalline Module, Black Frame	N	212.1
ET Solar Industry	ET-P672250	250W Polycrystalline Module	N	222.0

(Continued)

MANUFACTURER NAME	MODULE MODEL NUMBER	DESCRIPTION	BIPV*	PTC**	NOTES
ET Solar Industry	ET-P672255	255W Polycrystalline Module	N	226.6	
ET Solar Industry	ET-P672260	260W Polycrystalline Module	N	231.2	
ET Solar Industry	ET-P672260B	260W Polycrystalline Module, Black Frame	N	231.2	
ET Solar Industry	ET-P672265	265W Polycrystalline Module	N	235.8	
ET Solar Industry	ET-P672265B	265W Polycrystalline Module, Black Frame	N	235.8	
ET Solar Industry	ET-P672270	270W Polycrystalline Module	N	237.2	
ET Solar Industry	ET-P672270B	270W Polycrystalline Module, Black Frame	N	237.2	
ET Solar Industry	ET-P672275	275W Polycrystalline Module	N	241.7	
ET Solar Industry	ET-P672275B	275W Polycrystalline Module, Black Frame	N	241.7	
ET Solar Industry	ET-P672280	280W Polycrystalline Module	N	246.3	
ET Solar Industry	ET-P672280B	280W Polycrystalline Module, Black Frame	N	246.3	
Evergreen Solar	ES-C-70-fa5	70W String Ribbon Module	N	62.4	
Evergreen Solar	ES-C-75-fa5	75W String Ribbon Module	N	66.9	
Evergreen Solar	ES-C-80-fa5	80W String Ribbon Module	N	71.5	
Evergreen Solar	ES-C-110-fa2	110W String Ribbon Module	N	96.6	
Evergreen Solar	ES-C-115-fa2	115W String Ribbon Module	N	101.1	
Evergreen Solar	ES-C-115-fa4	115W String Ribbon Module	N	102.2	
Evergreen Solar	ES-C-120-fa4	120W String Ribbon Module	N	106.7	
Evergreen Solar	ES-170-RL	170W Spruce Line Module, MC Connectors	N	147.1	
Evergreen Solar	ES-170-SL	170W Spruce Line Module, MC Connectors	N	147.1	

Evergreen Solar	ES-B-170-fa1	170W Glass-EVA-Tedlar Construction, MC Connectors	N	145.8
Evergreen Solar	ES-B-170-fb1	170W Glass-EVA-Tedlar Construction, MC Connectors	N	145.8
Evergreen Solar	ES-180-RL	180W Spruce Line Module, MC Connectors	N	158.7
Evergreen Solar	ES-180-SL	180W Spruce Line Module, MC Connectors	N	158.7
Evergreen Solar	ES-180-VL	180W Polycrystalline Module	N	156.8
Evergreen Solar	ES-B-180-fa1	180W Glass-EVA-Tedlar Construction, MC Connectors	N	154.6
Evergreen Solar	ES-B-180-fb1	180W Glass-EVA-Tedlar Construction, MC Connectors	N	154.6
Evergreen Solar	ES-190-RL	190W Spruce Line Module, MC Connectors	N	163.9
Evergreen Solar	ES-190-SL	190W Spruce Line Module, MC Connectors	N	163.9
Evergreen Solar	ES-A-190-fa2	190W String Ribbon Module	N	171.7
Evergreen Solar	ES-A-190-fa3	190W String Ribbon Module, Black Frame	N	171.7
Evergreen Solar	ES-B-190-fa1	190W Glass-EVA-Tedlar Construction, MC Connectors	N	163.4
Evergreen Solar	ES-B-190-fb1	190W Glass-EVA-Tedlar Construction, MC Connectors	N	163.4
Evergreen Solar	ES-195-RL	195W Spruce Line Module, MC Connectors	N	168.4
Evergreen Solar	ES-195-SL	195W Spruce Line Module, MC Connectors	N	168.4

(Continued)

MANUFACTURER NAME	MODULE MODEL NUMBER	DESCRIPTION	BIPV*	PTC**	NOTES
Evergreen Solar	ES-195-VL	195W Polycrystalline Module	N	168.9	
Evergreen Solar	ES-A-195-fa2	195W String Ribbon Module	N	176.1	
Evergreen Solar	ES-A-195-fa3	195W String Ribbon Module, Black Frame	N	176.1	
Evergreen Solar	ES-B-195-fa1	195W Glass-EVA-Tedlar Construction, MC Connectors	N	167.9	
Evergreen Solar	ES-B-195-fb1	195W Glass-EVA-Tedlar Construction, MC Connectors	N	167.9	
Evergreen Solar	ES-200-RL	200W Spruce Line Module, MC Connectors	N	171.7	
Evergreen Solar	ES-200-SL	200W Spruce Line Module, MC Connectors	N	171.7	
Evergreen Solar	ES-A-200-fa2	200W String Ribbon Module	N	180.7	
Evergreen Solar	ES-A-200-fa3	200W String Ribbon Module, Black Frame	N	180.7	
Evergreen Solar	ES-B-200-fa1	200W Glass-EVA-Tedlar Construction, MC Connectors	N	172.3	
Evergreen Solar	ES-B-200-fb1	200W Glass-EVA-Tedlar Construction, MC Connectors	N	172.3	
Evergreen Solar	ES-A-205-fa2	205W String Ribbon Module	N	185.4	
Evergreen Solar	ES-A-205-fa3	205W String Ribbon Module, Black Frame	N	185.4	
Evergreen Solar	ES-A-210-fa2	210W String Ribbon Module	N	190.0	
Evergreen Solar	ES-A-210-fa3	210W String Ribbon Module, Black Frame	N	190.0	
Evergreen Solar	ES-E-210-fc3	210W Polycrystalline Module	N	186.5	
Evergreen Solar	ES-A-215-fa2	215W Polycrystalline Module	N	192.7	
Evergreen Solar	ES-A-215-fa3	215W Polycrystalline Module	N	192.7	

Evergreen Solar	ES-E-215-fc3	215W Polycrystalline Module	N	191.9	
Evergreen Solar	ES-E-220-fc3	220W Polycrystalline Module	N	196.5	
Fire Energy	FE5A-160M	160W Monocrystalline Module	N	145.4	
Fire Energy	FE6A-160P	160W Polycrystalline Module	N	139.6	
Fire Energy	FE6A-160PE	160W Polycrystalline Module	N	141.7	
Fire Energy	FE5A-170M	170W Monocrystalline Module	N	154.7	
Fire Energy	FE6A-170P	170W Polycrystalline Module	N	148.6	
Fire Energy	FE6P-170PE	170W Polycrystalline Module	N	150.0	
Fire Energy	FE5A-180M	180W Monocrystalline Module	N	164.0	
Fire Energy	FE6A-180P	180W Polycrystalline Module	N	157.6	
Fire Energy	FE6P-180PE	180W Polycrystalline Module	N	159.0	
Fire Energy	FE5A-190M	190W Monocrystalline Module	N	172.3	
Fire Energy	FE6P-190PE	190W Polycrystalline Module	N	168.1	
Fire Energy	FE6P-200PE	200W Polycrystalline Module	N	177.1	
Fire Energy	FE6P-210P	210W Polycrystalline Module	N	192.3	
Fire Energy	FE5P-220M	220W Monocrystalline Module	N	200.1	
Fire Energy	FE6P-220P	220W Polycrystalline Module	N	201.6	
Fire Energy	FE5P-230M	230W Monocrystalline Module	N	209.4	
Fire Energy	FE6P-230P	230W Polycrystalline Module	N	211.0	
Fire Energy	FE5P-240M	240W Monocrystalline Module	N	218.7	
First Solar	FS-267	67.5W Thin Film CdTe Module	N	64.8	For use with First Solar mounting systems
First Solar	FS-270	70W Thin Film CdTe Module	N	67.2	For use with First Solar mounting systems

(Continued)

MANUFACTURER NAME	MODULE MODEL NUMBER	DESCRIPTION	BIPV*	PTC**	NOTES
First Solar	FS-272	72.5W Thin Film CdTe Module	N	69.6	For use with First Solar mounting systems
First Solar	FS-372	72.5W Thin Film CdTe Module	N	68.2	For use with First Solar mounting systems
First Solar	FS-275	75W Thin Film CdTe Module	N	72.0	For use with First Solar mounting systems
First Solar	FS-375	75W Thin Film CdTe Module	N	70.6	For use with First Solar mounting systems
First Solar	FS-377	77W Thin Film CdTe Module	N	70.8	For use with First Solar mounting systems
First Solar	FS-277	77.5W Thin Film CdTe Module	N	74.5	For use with First Solar mounting systems
First Solar	FS-380	80W Thin Film CdTe Module	N	73.6	For use with First Solar mounting systems
Fluitecnik	FTS210P	210W Polycrystalline Module	N	179.6	
Fluitecnik	FTS215P	215W Polycrystalline Module	N	184.0	
Fluitecnik	FTS220P	220W Polycrystalline Module	N	188.4	
Fluitecnik	FTS225P	225W Polycrystalline Module	N	190.9	
Fluitecnik	FTS230M	230W Monocrystalline Module	N	198.4	
Fluitecnik	FTS230P	230W Polycrystalline Module	N	195.3	
Fluitecnik	FTS235M	235W Monocrystalline Module	N	202.9	
Fluitecnik	FTS240M	240W Monocrystalline Module	N	207.4	

Fluitecnik	FTS245M	245W Monocrystalline Module	N	211.9
Fluitecnik	FTS250P	250W Polycrystalline Module	N	216.6
Fluitecnik	FTS260P	260W Polycrystalline Module	N	225.6
Fluitecnik	FTS270P	270W Polycrystalline Module	N	229.7
Fluitecnik	FTS280P	280W Polycrystalline Module	N	238.6
Future Solar Energy	FSM572165	165W Monocrystalline Module	N	146.5
Future Solar Energy	FSM572170	170W Monocrystalline Module	N	151.1
Future Solar Energy	FSM572175	175W Monocrystalline Module	N	155.7
Future Solar Energy	FSM572180	180W Monocrystalline Module	N	157.7
Future Solar Energy	FSM572185	185W Monocrystalline Module	N	162.2
Future Solar Energy	FSP654195	195W Polycrystalline Module	N	172.4
Future Solar Energy	FSP654200	200W Polycrystalline Module	N	176.9
Future Solar Energy	FSP654205	205W Polycrystalline Module	N	181.5
Future Solar Energy	FSP654210	210W Polycrystalline Module	N	186.1
Future Solar Energy	FSP672260	260W Polycrystalline Module	N	231.2
Future Solar Energy	FSP672265	265W Polycrystalline Module	N	235.8
Future Solar Energy	FSP672270	270W Polycrystalline Module	N	237.2
Future Solar Energy	FSP672275	275W Polycrystalline Module	N	241.7
Future Solar Energy	FSP672280	280W Polycrystalline Module	N	246.3
GE Energy	GEPVp-066-G	66W Polycrystalline Module	Y	52.1
GE Energy	GEPVp-200-M	200W Polycrystalline Module	N	173.1
GE Energy	GEPVp-205-M	205W Polycrystalline Module	N	177.6
GESOLAR	GES-M165	165W Monocrystalline Module	N	147.3
GESOLAR	GES-M170	170W Monocrystalline Module	N	151.8
GESOLAR	GES-M175	175W Monocrystalline Module	N	156.4

(Continued)

MANUFACTURER NAME	MODULE MODEL NUMBER	DESCRIPTION	BIPV*	PTC***	NOTES
GESOLAR	GES-M180	180W Monocrystalline Module	N	161.0	
GESOLAR	GES-P200	200W Polycrystalline Module	N	175.9	
GESOLAR	GES-P210	210W Polycrystalline Module	N	184.9	
GESOLAR	GES-P220	220W Polycrystalline Module	N	194.0	
GESOLAR	GES-P230	230W Polycrystalline Module	N	203.1	
GESOLAR	GES-P240	240W Polycrystalline Module	N	211.7	
GESOLAR	GES-P250	250W Polycrystalline Module	N	220.7	
GESOLAR	GES-P260	260W Polycrystalline Module	N	229.8	
GESOLAR	GES-P270	270W Polycrystalline Module	N	238.9	
GESOLAR	GES-P280	280W Polycrystalline Module	N	248.0	
Gloria Solar	GSS5-160A-E1	160W Monocrystalline Module	N	139.7	
Gloria Solar	GSS5-165A-E1	165W Monocrystalline Module	N	144.2	
Gloria Solar	GSS5-170A-E1	170W Monocrystalline Module	N	148.7	
Gloria Solar	GSS5-175A-E1	175W Monocrystalline PV Module	N	157.4	
Gloria Solar	GSS5-180A-E1	180W Monocrystalline PV Module	N	162.0	
Gloria Solar	GSM6-200D-E1	200W Polycrystalline Module	N	170.0	
Gloria Solar	GSM6-205D-E1	205W Polycrystalline Module	N	174.4	
Gloria Solar	GSM6-210D-E1	210W Polycrystalline Module	N	178.8	
Gloria Solar	GSM6-215D-E1	215W Polycrystalline Module	N	183.2	
Gloria Solar	GSM6-220D-E1	220W Polycrystalline Module	N	187.6	
Gloria Solar	GSM6-225D-E1	225W Polycrystalline PV Module	N	202.2	
Gloria Solar	GSM6-230D-E1	230W Polycrystalline PV Module	N	206.8	

Manufacturer	Model		Value	Part Number
Gloria Solar	235W Polycrystalline PV Module	N	211.4	GSM6-235D-E1
Gloria Solar	240W Polycrystalline PV Module	N	216.0	GSM6-240D-E1
Grape Solar	150W Monocrystalline Module	N	132.0	GS-S-150-SF
Grape Solar	155W Monocrystalline Module	N	136.5	GS-S-155-SF
Grape Solar	160W Monocrystalline Module	N	141.0	GS-S-160-SF
Grape Solar	165W Monocrystalline Module	N	147.3	CS-S-165-DJ
Grape Solar	165W Monocrystalline Module	N	145.5	GS-S-165-SF
Grape Solar	170W Monocrystalline Module	N	151.8	CS-S-170-DJ
Grape Solar	170W Monocrystalline Module	N	150.1	GS-S-170-SF
Grape Solar	175W Monocrystalline Module	N	156.4	CS-S-175-DJ
Grape Solar	175W Monocrystalline Module	N	154.6	GS-S-175-SF
Grape Solar	180W Monocrystalline Module	N	161.0	CS-S-180-DJ
Grape Solar	180W Monocrystalline Module	N	159.2	GS-S-180-SF
Grape Solar	185W Monocrystalline Module	N	163.7	GS-S-185-SF
Grape Solar	195W Polycrystalline Module	N	169.3	GS-P-190-SL6-18
Grape Solar	200W Polycrystalline Module	N	175.9	CS-P-200-DJ
Grape Solar	200W Polycrystalline Module	N	177.1	GS-P-200-CSPE
Grape Solar	200W Polycrystalline Module	N	173.8	GS-P-200-SL6-18
Grape Solar	205W Polycrystalline Module	N	178.3	GS-P-205-SL6-18
Grape Solar	210W Polycrystalline Module	N	184.9	CS-P-210-DJ
Grape Solar	220W Polycrystalline Module	N	194.0	CS-P-220-DJ
Grape Solar	220W Polycrystalline Module	N	201.6	GS-P-220-CS
Grape Solar	220W Polycrystalline Module	N	189.8	GS-P-220-SL6-20
Grape Solar	220W Monocrystalline Module	N	200.1	GS-S-220-CS

(Continued)

MANUFACTURER NAME	MODULE MODEL NUMBER	DESCRIPTION	BIPV*	PTC**	NOTES
Grape Solar	GS-P-225-SL6-20	225W Polycrystalline Module	N	194.2	
Grape Solar	CS-P-230-DJ	230W Polycrystalline Module	N	203.1	
Grape Solar	GS-P-230-CS	230W Polycrystalline Module	N	211.0	
Grape Solar	GS-P-230-SL6-20	230W Polycrystalline Module	N	198.7	
Grape Solar	GS-S-230-CS	230W Monocrystalline Module	N	209.4	
Grape Solar	CS-P-240-DJ	240W Polycrystalline Module	N	211.7	
Grape Solar	GS-S-240-CS	240W Monocrystalline Module	N	218.7	
Grape Solar	CS-P-250-DJ	250W Polycrystalline Module	N	220.7	
Grape Solar	CS-P-260-DJ	260W Polycrystalline Module	N	229.8	
Grape Solar	CS-P-270-DJ	270W Polycrystalline Module	N	238.9	
Grape Solar	CS-P-280-DJ	280W Polycrystalline Module	N	248.0	
Green Energy Solar & Wind	GESW1750 USA-170W	170W Monocrystalline Module	N	151.9	
Green Energy Solar & Wind	GESW1750 USA-175W	175W Monocrystalline Module	N	156.5	
Green Energy Solar & Wind	GESW1750 USA-180W	180W Monocyrstalline Module	N	161.1	
Green Energy Solar & Wind	GESW2100 USA-210W	210W Monocyrstalline Module	N	189.5	
Green Energy Solar & Wind	GESW2100 USA-215W	215W Monocrystalline Module	N	194.1	
Green Energy Solar & Wind	GESW2100 USA-220W	220W Monocrystalline Module	N	198.8	
Green Energy Technology	GET-085A	85W Thin Film Single Junction Amorphous Silicon Module	N	79.9	
Green Energy Technology	GET-090A	90W Thin Film Single Junction Amorphous Silicon Module	N	84.6	

Green Energy Technology	GET-170A	170W Thin Film Single Junction Amorphous Silicon Module	N	159.7
Green Energy Technology	GET-170B	170W Thin Film Single Junction Amorphous Silicon Module	N	159.7
Green Energy Technology	GET-180A	180W Thin Film Single Junction Amorphous Silicon Module	N	169.2
Green Energy Technology	GET-180B	180W Thin Film Single Junction Amorphous Silicon Module	N	169.2
Green Energy Technology	GET-340A	340W Thin Film Single Junction Amorphous Silicon Module	N	319.4
Green Energy Technology	GET-360A	360W Thin Film Single Junction Amorphous Silicon Module	N	338.3
Green Power	ET-M572165	165W Monocrystalline Module	N	146.5
Green Power	ET-M572170	170W Monocrystalline Module	N	151.1
Green Power	ET-M572175	175W Monocrystalline Module	N	155.7
Green Power	ET-M572175B	175W Monocrystalline Module, Black Frame, Black Backsheet	N	155.1
Green Power	ET-M572180	180W Monocrystalline Module	N	157.7
Green Power	ET-M572180B	180W Monocrystalline Module, Black Frame, Black Backsheet	N	159.6
Green Power	ET-M572185	185W Monocrystalline Module	N	162.2
Green Power	ET-M572185B	185W Monocrystalline Module	N	164.2

(Continued)

MANUFACTURER NAME	MODULE MODEL NUMBER	DESCRIPTION	BIPV*	PTC**	NOTES
Green Power	ET-P654190B	190W Polycrystalline Module, Black Frame	N	167.9	
Green Power	ET-P654195	195W Polycrystalline Module	N	172.4	
Green Power	ET-P654195B	195W Polycrystalline Module, Black Frame	N	172.4	
Green Power	ET-P654200	200W Polycrystalline Module	N	176.9	
Green Power	ET-P654200B	200W Polycrystalline Module, Black Frame	N	176.9	
Green Power	ET-P654205	205W Polycrystalline Module	N	181.5	
Green Power	ET-P654205B	205W Polycrystalline Module, Black Frame	N	181.5	
Green Power	ET-P654210	210W Polycrystalline Module	N	186.1	
Green Power	ET-P654210B	210W Polycrystalline Module, Black Frame	N	186.1	
Green Power	ET-P660220	220W Polycrystalline Module	N	198.2	
Green Power	ET-P660220B	220W Polycrystalline Module, Black Frame	N	198.2	
Green Power	ET-P660225	225W Polycrystalline Module	N	202.8	
Green Power	ET-P660225B	225W Polycrystalline Module, Black Frame	N	202.8	
Green Power	ET-P660230	230W Polycrystalline Module	N	207.5	
Green Power	ET-P660230B	230W Polycrystalline Module, Black Frame	N	207.5	
Green Power	ET-P660235	235W Polycrystalline Module	N	212.1	
Green Power	ET-P660235B	235W Polycrystalline Module, Black Frame	N	212.1	
Green Power	ET-P672260	260W Polycrystalline Module	N	231.2	

Green Power	ET-P672260B	260W Polycrystalline Module, Black Frame	N	231.2
Green Power	ET-P672265	265W Polycrystalline Module	N	235.8
Green Power	ET-P672265B	265W Polycrystalline Module, Black Frame	N	235.8
Green Power	ET-P672270	270W Polycrystalline Module	N	237.2
Green Power	ET-P672270B	270W Polycrystalline Module, Black Frame	N	237.2
Green Power	ET-P672275	275W Polycrystalline Module	N	241.7
Green Power	ET-P672275B	275W Polycrystalline Module, Black Frame	N	241.7
Green Power	ET-P672280	280W Polycrystalline Module	N	246.3
Green Power	ET-P672280B	280W Polycrystalline Module, Black Frame	N	246.3
Greenbrilliance Energy	GB54P6-200	200W Polycrystalline Module	N	181.1
Greenbrilliance Energy	GB60P6-210	210W Polycrystalline Module	N	189.9
Greenbrilliance Energy	GB60P6-215	215W Polycrystalline Module	N	194.5
Greenbrilliance Energy	GB54P6-220	220W Polycrystalline Module	N	197.9
Greenbrilliance Energy	GB60P6-225	225W Polycrystalline Module	N	202.5
Greenbrilliance Energy	GB60P6-230	230W Polycrystalline Module	N	207.2
Greenbrilliance Energy	GB72P6-270	270W Polycrystalline Module	N	245.2
Greenbrilliance Energy	GB72P6-275	275W Polycrystalline Module	N	247.7
Greenbrilliance Energy	GB72P6-280	280W Polycrystalline Module	N	252.3
Greenlite Solar Canada	GLS-200	200W Polycrystalline Module	N	180.0
Greenlite Solar Canada	GLS-220	220W Polycrystalline Module	N	197.9
Greenlite Solar Canada	GLS-270	270W Polycrystalline Module	N	243.1

(Continued)

MANUFACTURER NAME	MODULE MODEL NUMBER	DESCRIPTION	BIPV*	PTC**	NOTES
GS-Solar (Fujian)	GS-46SU	46W Thin Film Dual Junction a-Si Module	N	43.6	
Guangdong Fivestar Solar Energy	FS-M125-175W	175W Monocrystalline Module	N	156.4	
Guangdong Fivestar Solar Energy	FS-P156-210W	210W Polycrystalline Module	N	189.5	
Guangdong Fivestar Solar Energy	FS-P156-220W	220W Polycrystalline Module	N	198.7	
Guangdong Golden Glass Technologies	GG160M2-24/1324x992	160W Polycrystalline Module	N	143.4	
Guangdong Golden Glass Technologies	GG165M2-24/1324x992	165W Polycrystalline Module	N	148.0	
Guangdong Golden Glass Technologies	GG170M2-24/1324x992	170W Polycrystalline Module	N	152.6	
Guangdong Golden Glass Technologies	GG175M2-24/1324x992	175W Polycrystalline Module	N	157.2	
Guangdong Golden Glass Technologies	GG175S1-36/1589x807	175W Monocrystalline Module	N	158.7	
Guangdong Golden Glass Technologies	GG180M2-24/1324x992	180W Polycrystalline Module	N	161.8	
Guangdong Golden Glass Technologies	GG180S1-36/1589x807	180W Monocrystalline Module	N	163.4	
Guangdong Golden Glass Technologies	GG190M2-27/1482x992	190W Polycrystalline Module	N	170.5	
Guangdong Golden Glass Technologies	GG195M2-27/1482x992	195W Polycrystalline Module	N	175.1	
Guangdong Golden Glass Technologies	GG200M2-27/1482x992	200W Polycrystalline Module	N	179.7	

Guangdong Golden Glass Technologies	GG205M2-27/1482x992	205W Polycrystalline Module	N	184.3
Guangdong Golden Glass Technologies	GG220M2-30/1640x992	220W Polycrystalline Module	N	198.5
Guangdong Golden Glass Technologies	GG225M2-30/1640x992	225W Polycrystalline Module	N	203.1
Guangdong Golden Glass Technologies	GG230M2-30/1640x992	230W Polycrystalline Module	N	207.7
Guangdong Golden Glass Technologies	GG250M2-35/1956x992	250W Polycrystalline Module	N	224.3
Guangdong Golden Glass Technologies	GG255M2-35/1956x992	255W Polycrystalline Module	N	228.9
Guangdong Golden Glass Technologies	GG260M2-35/1956x992	260W Polycrystalline Module	N	233.5
Guangdong Golden Glass Technologies	GG270M2-36/1956x992	270W Polycrystalline Module	N	242.7
Guangdong Golden Glass Technologies	GG275M2-36/1956x992	275W Polycrystalline Module	N	247.3
Guangdong Golden Glass Technologies	GG280M2-36/1956x992	280W Polycrystalline Module	N	252.0
Hyundai Heavy Industries	HiS-M191SF	191W Polycrystalline Module	N	172.6
Hyundai Heavy Industries	HiS-M194SF	194W Polycrystalline Module	N	175.4
Hyundai Heavy Industries	HiS-M197SF	197W Polycrystalline Module	N	178.2
Hyundai Heavy Industries	HiS-M200SF	200W Polycrystalline Module	N	180.9

(Continued)

MANUFACTURER NAME	MODULE MODEL NUMBER	DESCRIPTION	BIPV*	PTC**	NOTES
Hyundai Heavy Industries	HiS-M203SF	203W Polycrystalline Module	N	183.7	
Hyundai Heavy Industries	HiS-M206SF	206W Polycrystalline Module	N	186.5	
Hyundai Heavy Industries	HiS-M209SF	209W Polycrystalline Module	N	189.3	
Hyundai Heavy Industries	HiS-M212SG	212W Polycrystalline Module	N	191.0	
Hyundai Heavy Industries	HiS-M215SG	215W Polycrystalline Module	N	193.7	
Hyundai Heavy Industries	HiS-M218SG	218W Polycrystalline Module	N	196.5	
Hyundai Heavy Industries	HiS-M221SG	221W Polycrystalline Module	N	199.3	
Hyundai Heavy Industries	HiS-M224SG	224W Polycrystalline Module	N	202.0	
Hyundai Heavy Industries	HiS-M227SG	227W Polycrystalline Module	N	204.8	
Hyundai Heavy Industries	HiS-M230SG	230W Polycrystalline Module	N	207.6	
Innotech Solar	ITS210XXXX	210W Polycrystalline Module	N	187.4	
Innotech Solar	ITS215XXXX	215W Polycrystalline Module	N	192.0	
Innotech Solar	ITS220XXXX	220W Polycrystalline Module	N	196.6	
Innotech Solar	ITS225XXXX	225W Polycrystalline Module	N	201.1	
Innotech Solar	ITS230XXXX	230W Polycrystalline Module	N	205.7	
Innovosolar International	TEM165M	165W Monocrystalline Module	N	147.3	
Innovosolar International	TEM170M	170W Monocrystalline Module	N	151.8	

Innovosolar International	TEM175M	175W Monocrystalline Module	N	156.4
Innovosolar International	TEM180M	180W Monocrystalline Module	N	161.0
Instalaciones Pevafersa	IP-VA180	180W Monocrystalline Module	N	159.2
Instalaciones Pevafersa	IP-VAP210	210W Polycrystalline Module	N	183.7
Instalaciones Pevafersa	IP-VAP230	230W Polycrystalline Module	N	201.0
Isofoton	IS-210/32	210W Monocrystalline Module	N	182.8
Isofoton	IS-215/32	215W Monocrystalline Module	N	187.3
Isofoton	IS-220/32	220W Monocrystalline Module	N	191.8
Jiangsu Aide Solar Energy Technology	XZST-160W	160W Monocrystalline Module	N	138.3
Jiangsu Aide Solar Energy Technology	XZST-165W	165W Monocrystalline Module	N	142.8
Jiangsu Aide Solar Energy Technology	XZST-170W	170W Monocrystalline Module	N	146.1
Jiangsu Aide Solar Energy Technology	XZST-175W	175W Monocrystalline Module	N	150.5
Jiangsu Aide Solar Energy Technology	XZST-180W	180W Monocrystalline Module	N	155.0
Jiangsu Aide Solar Energy Technology	XZST-190W	190W Polycrystalline Module	N	163.9
Jiangsu Aide Solar Energy Technology	XZST-195W	195W Polycrystalline Module	N	168.3
Jiangsu Aide Solar Energy Technology	XZST-200W	200W Polycrystalline Module	N	172.8
Jiangsu Aide Solar Energy Technology	XZST-205W	205W Polycrystalline Module	N	177.2

(Continued)

517

MANUFACTURER NAME	MODULE MODEL NUMBER	DESCRIPTION	BIPV*	PTC**	NOTES
Jiangsu Aide Solar Energy Technology	XZST-210W	210W Polycrystalline Module	N	181.7	
Jiangsu Aide Solar Energy Technology	XZST-215W	215W Polycrystalline Module	N	185.5	
Jiangsu Aide Solar Energy Technology	XZST-220W	220W Polycrystalline Module	N	190.0	
Jiangsu Aide Solar Energy Technology	XZST-225W	225W Polycrystalline Module	N	194.0	
Jiangsu Aide Solar Energy Technology	XZST-230W	230W Polycrystalline Module	N	198.5	
Jiangsu Best Solar High-Tech	BEST-170D-24	170W Monocrystalline Module	N	148.2	
Jiangsu Best Solar High-Tech	BEST-175D-24	175W Monocrystalline Module	N	152.7	
Jiangsu Best Solar High-Tech	BEST-180D-24	180W Monocrystalline Module	N	157.2	
Jiangsu Best Solar High-Tech	BEST-190P-18	190W Polycrystalline Module	N	166.2	
Jiangsu Best Solar High-Tech	BEST-200P-18	200W Polycrystalline Module	N	175.2	
Jiangsu Best Solar High-Tech	BEST-210P-18	210W Polycrystalline Module	N	184.3	
Jiangsu Best Solar High-Tech	BEST-210P-20	210W Polycrystalline Module	N	183.0	
Jiangsu Best Solar High-Tech	BEST-220D-20	220W Monocrystalline Module	N	189.7	
Jiangsu Best Solar High-Tech	BEST-220P-20	220W Polycrystalline Module	N	192.0	

Jiangsu Best Solar High-Tech	BEST-230D-20	230W Monocrystalline Module	N	198.6
Jiangsu Best Solar High-Tech	BEST-230P-20	230W Polycrystalline Module	N	201.0
Jiangsu Best Solar High-Tech	BEST-240D-20	240W Monocrystalline Module	N	207.6
Jiangsu Best Solar High-Tech	BEST-260P-24	260W Polycrystalline Module	N	228.0
Jiangsu Best Solar High-Tech	BEST-270P-24	270W Polycrystalline Module	N	237.0
Jiangsu Best Solar High-Tech	BEST-280P-24	280W Polycrystalline Module	N	246.1
Jiangsu Sainty Machinery	GES-M165	165W Monocrystalline Module	N	147.3
Jiangsu Sainty Machinery	GES-M170	170W Monocrystalline Module	N	151.8
Jiangsu Sainty Machinery	GES-M175	175W Monocrystalline Module	N	156.4
Jiangsu Sainty Machinery	GES-M180	180W Monocrystalline Module	N	161.0
Jiangsu Sainty Machinery	GES-P200	200W Polycrystalline Module	N	175.9
Jiangsu Sainty Machinery	GES-P210	210W Polycrystalline Module	N	184.9
Jiangsu Sainty Machinery	GES-P220	220W Polycrystalline Module	N	194.0
Jiangsu Sainty Machinery	GES-P230	230W Polycrystalline Module	N	203.1

(Continued)

MANUFACTURER NAME	MODULE MODEL NUMBER	DESCRIPTION	BIPV*	PTC**	NOTES
Jiangsu Sainty Machinery	GES-P240	240W Polycrystalline Module	N	211.7	
Jiangsu Sainty Machinery	GES-P250	250W Polycrystalline Module	N	220.7	
Jiangsu Sainty Machinery	GES-P260	260W Polycrystalline Module	N	229.8	
Jiangsu Sainty Machinery	GES-P270	270W Polycrystalline Module	N	238.9	
Jiangsu Sainty Machinery	GES-P280	280W Polycrystalline Module	N	248.0	
Jiangsu Shunda PV-Tech	SDM-170/(165)-72M	165W Monocrystalline Module	N	145.3	
Jiangsu Shunda PV-Tech	SDM-170/(170)-72M	170W Monocrystalline Module	N	149.8	
Jiangsu Shunda PV-Tech	SDM-170/(175)-72M	175W Monocrystalline Module	N	154.3	
Jiangsu Shunda PV-Tech	SDM-170/(180)-72M	180W Monocrystalline Module	N	158.9	
Jiangsu Shunda PV-Tech	SDM-170/(185)-72M	185W Monocrystalline Module	N	163.4	
Jiangxi Best Solar High-Tech	BEST-170D-24	170W Monocrystalline Module	N	148.2	
Jiangxi Best Solar High-Tech	BEST-175D-24	175W Monocrystalline Module	N	152.7	
Jiangxi Best Solar High-Tech	BEST-180D-24	180W Monocrystalline Module	N	157.2	
Jiangxi Best Solar High-Tech	BEST-190P-18	190W Polycrystalline Module	N	166.2	

Jiangxi Best Solar High-Tech	BEST-200P-18	200W Polycrystalline Module	N	175.2
Jiangxi Best Solar High-Tech	BEST-210P-18	210W Polycrystalline Module	N	184.3
Jiangxi Best Solar High-Tech	BEST-210P-20	210W Polycrystalline Module	N	183.0
Jiangxi Best Solar High-Tech	BEST-220D-20	220W Monocrystalline Module	N	189.7
Jiangxi Best Solar High-Tech	BEST-220P-20	220W Polycrystalline Module	N	192.0
Jiangxi Best Solar High-Tech	BEST-230D-20	230W Monocrystalline Module	N	198.6
Jiangxi Best Solar High-Tech	BEST-230P-20	230W Polycrystalline Module	N	201.0
Jiangxi Best Solar High-Tech	BEST-240D-20	240W Monocrystalline Module	N	207.6
Jiangxi Best Solar High-Tech	BEST-260P-24	260W Polycrystalline Module	N	228.0
Jiangxi Best Solar High-Tech	BEST-270P-24	270W Polycrystalline Module	N	237.0
Jiangxi Best Solar High-Tech	BEST-280P-24	280W Polycrystalline Module	N	246.1
Jiangyin Hareon Power	HR-115W/17V	115W Monocrystalline Module	N	103.2
Jiangyin Hareon Power	HR-120W/17V	120W Monocrystalline Module	N	107.8
Jiangyin Hareon Power	HR-125W/17V	125W Monocrystalline Module	N	112.4
Jiangyin Hareon Power	HR-160W/24V	160W Monocrystalline Module	N	143.5
Jiangyin Hareon Power	HR-165W/24V	165W Monocrystalline Module	N	148.0
Jiangyin Hareon Power	HR-170W/24V	170W Monocrystalline Module	N	152.6

(Continued)

MANUFACTURER NAME	MODULE MODEL NUMBER	DESCRIPTION	BIPV*	PTC**	NOTES
Jiangyin Hareon Power	HR-175W/24V	175W Monocrystalline Module	N	156.1	
Jiangyin Hareon Power	HR-180W/24V	180W Monocrystalline Module	N	160.7	
Jiangyin Hareon Power	HR-185W/24V	185W Monocrystalline Module	N	165.3	
Jiangyin Hareon Power	HR-200W/18V	200W Polycrystalline Module	N	179.8	
Jiangyin Hareon Power	HR-205W/18V	205W Polycrystalline Module	N	184.4	
Jiangyin Hareon Power	HR-210W/18V	210W Polycrystalline Module	N	189.0	
Jiangyin Hareon Power	HR-215W/18V	215W Polycrystalline Module	N	193.6	
Jiangyin Hareon Power	HR-220W/18V	220W Polycrystalline Module	N	198.2	
Jiangyin Jetion Science and Technology	JT175(35)S1580x808	175W Monocrystalline Module	N	153.9	
Jiangyin Jetion Science and Technology	JT180(36)S1580x808	180W Monocrystalline Module	N	158.5	
Jiangyin Jetion Science and Technology	JT185(36)S1580x808	185W Monocrystalline Module	N	163.0	
Jiangyin Jetion Science and Technology	JT210(29)P1655x992	210W Polycrystalline Module	N	184.1	
Jiangyin Jetion Science and Technology	JT215(30)P1655x992	215W Polycrystalline Module	N	188.6	
Jiangyin Jetion Science and Technology	JT220(30)P1655x992	220W Polycrystalline Module	N	193.2	
Jiangyin Jetion Science and Technology	JT225(30)P1655x992	225W Polycrystalline Module	N	197.7	
Jiangyin Jetion Science and Technology	JT230(30)P1655x992	230W Polycrystalline Module	N	202.2	
Jingao Solar	JAM5(L)-175	175W Monocrystalline Module	N	153.4	
Jingao Solar	JAM5(L)-180	180W Monocrystalline Module	N	158.0	
Jingao Solar	JAM5(L)-185	185W Monocrystalline Module	N	162.5	

Jingao Solar	JAM6-200	200W Monocrystalline Module	N	179.5
Jingao Solar	JAP6-200	200W Polycrystalline Module	N	178.8
Jingao Solar	JAM6-205	205W Monocrystalline Module	N	184.1
Jingao Solar	JAP6-205	205W Polycrystalline Module	N	183.4
Jingao Solar	JAM6-210	210W Monocrystalline Module	N	188.7
Jingao Solar	JAP6-210	210W Polycrystalline Module	N	187.9
Jingao Solar	JAM6-215	215W Monocrystalline Module	N	193.3
Jingao Solar	JAP6-215	215W Polycrystalline Module	N	192.5
Jingao Solar	JAM6-220	220W Monocrystalline Module	N	197.9
Jingao Solar	JAP6-220	220W Polycrystalline Module	N	197.1
Jingao Solar	JAM6-225	225W Monocrystalline Module	N	202.5
Jingao Solar	JAP6-225	225W Polycrystalline Module	N	201.7
Jingao Solar	JAM6-230	230W Monocrystalline Module	N	207.1
Jingao Solar	JAP6-230	230W Polycrystalline Module	N	206.3
Jingao Solar	JAM6-235	235W Monocrystalline Module	N	211.7
Jingao Solar	JAP6-235	235W Polycrystalline Module	N	210.9
Jingao Solar	JAM6-240	240W Monocrystalline Module	N	216.3
Jingao Solar	JAP6-240	240W Polycrystalline Module	N	215.5
Juli New Energy	JLS155M	155W Monocrystalline Module	N	132.6
Juli New Energy	JLS160M	160W Monocrystalline Module	N	137.0
Juli New Energy	JLS165M	165W Monocrystalline Module	N	141.5
Juli New Energy	JLS170M	170W Monocrystalline Module	N	148.6
Juli New Energy	JLS175M	175W Monocrystalline Module	N	153.1
Juli New Energy	JLS180M	180W Monocrystalline Module	N	157.6
Juli New Energy	JLS200M	200W Monocrystalline Module	N	177.8

(Continued)

MANUFACTURER NAME	MODULE MODEL NUMBER	DESCRIPTION	BIPV*	PTC**	NOTES
Juli New Energy	JLS205M	205W Monocrystalline Module	N	182.4	
Juli New Energy	JLS210M	210W Monocrystalline Module	N	187.0	
Juli New Energy	JLS215P	215W Polycrystalline Module	N	188.6	
Juli New Energy	JLS220P	220W Polycrystalline Module	N	191.5	
Juli New Energy	JLS225P	225W Polycrystalline Module	N	196.0	
Juli New Energy	JLS230P	230W Polycrystalline Module	N	200.5	
Jumao Photonic (Xiamen)	JMP-85W-S5-G	85W Monocrystalline Module	N	72.6	
Jumao Photonic (Xiamen)	JMP-160W-S5-G	160W Monocrystalline Module	N	134.8	
Jumao Photonic (Xiamen)	JMP-170W-S5-G	170W Monocrystalline Module	N	143.5	
Jumao Photonic (Xiamen)	JMP-210W-M6-G	210W Polycrystalline Module	N	187.4	
Jumao Photonic (Xiamen)	JMP-215W-M6-G	215W Polycrystalline Module	N	192.0	
Jumao Photonic (Xiamen)	JMP-220W-M6-G	220W Polycrystalline Module	N	196.6	
Jumao Photonic (Xiamen)	JMP-225W-M6-G	225W Polycrystalline Module	N	201.1	
Jumao Photonic (Xiamen)	JMP-230W-M6-G	230W Polycrystalline Module	N	205.7	
Kaneka	G-SA060	60W a-Si Module	N	57.0	
Kaneka	GSA221	60W a-Si Module	N	57.0	
KD Solar	KD-20012A2	200W Polycrystalline Module	N	176.5	
KD Solar	KD-20013A2	200W Polycrystalline Module	N	175.4	
KD Solar	KD-20512A2	205W Polycrystalline Module	N	181.1	

KD Solar	KD-20513A2	205W Polycrystalline Module	N	180.0
KD Solar	KD-21012A2	210W Polycrystalline Module	N	185.6
KD Solar	KD-21013A2	210W Polycrystalline Module	N	184.5
KD Solar	KD-21512A2	215W Polycrystalline Module	N	190.2
KD Solar	KD-21513A2	215W Polycrystalline Module	N	189.0
KD Solar	KD-22012A2	220W Polycrystalline Module	N	195.8
KD Solar	KD-22013A2	220W Polycrystalline Module	N	192.5
KD Solar	KD-22512A2	225W Polycrystalline Module	N	200.3
KD Solar	KD-22513A2	225W Polycrystalline Module	N	197.0
KD Solar	KD-23012A2	230W Polycrystalline Module	N	204.9
KD Solar	KD-23013A2	230W Polycrystalline Module	N	201.6
KD Solar	KD-23512A2	235W Polycrystalline Module	N	209.5
KD Solar	KD-24012A2	240W Polycrystalline Module	N	214.1
Kenmec Mechanical Engineering	TKSE-13001	130W Polycrystalline Module	N	116.5
Kenmec Mechanical Engineering	TKSE-13501	135W Polycrystalline Module	N	121.1
Kenmec Mechanical Engineering	TKSE-14001	140W Polycrystalline Module	N	125.7
Kenmec Mechanical Engineering	TKSE-14501	145W Polycrystalline Module	N	130.3
Kenmec Mechanical Engineering	TKSD-15001	150W Polycrystalline Module	N	134.5
Kenmec Mechanical Engineering	TKSD-15501	155W Polycrystalline Module	N	139.1
Kenmec Mechanical Engineering	TKSD-16001	160W Polycrystalline Module	N	143.7

(Continued)

MANUFACTURER NAME	MODULE MODEL NUMBER	DESCRIPTION	BIPV*	PTC**	NOTES
Kenmec Mechanical Engineering	TKSD-16501	165W Polycrystalline Module	N	148.3	
Kenmec Mechanical Engineering	TKSD-17001	170W Polycrystalline Module	N	152.9	
Kenmec Mechanical Engineering	TKSC-17501	175W Polycrystalline Module	N	157.0	
Kenmec Mechanical Engineering	TKSC-18001	180W Polycrystalline Module	N	161.6	
Kenmec Mechanical Engineering	TKSC-18501	185W Polycrystalline Module	N	166.2	
Kenmec Mechanical Engineering	TKSC-19001	190W Polycrystalline Module	N	170.8	
Kenmec Mechanical Engineering	TKSC-19501	195W Polycrystalline Module	N	175.4	
Kenmec Mechanical Engineering	TKSB-20001	200W Polycrystalline Module	N	179.5	
Kenmec Mechanical Engineering	TKSB-20501	205W Polycrystalline Module	N	184.1	
Kenmec Mechanical Engineering	TKSB-21001	210W Polycrystalline Module	N	188.7	
Kenmec Mechanical Engineering	TKSB-21501	215W Polycrystalline Module	N	193.3	
Kenmec Mechanical Engineering	TKSB-22001	220W Polycrystalline Module	N	197.9	
Kenmec Mechanical Engineering	TKSA-22501	225W Polycrystalline Module	N	202.0	
Kenmec Mechanical Engineering	TKSA-23001	230W Polycrystalline Module	N	206.6	

Kenmec Mechanical Engineering	TKSA-23501	235W Polycrystalline Module	N	211.2
Kenmec Mechanical Engineering	TKSA-24001	240W Polycrystalline Module	N	215.9
Kenmos Photovoltaic	aTT-50W-02	50W Thin Film Dual Junction a-Si Module	N	44.3
Kinmac Solar	LPS-6P6A-220-B	220W Polycrystalline Module	N	197.7
Kinmac Solar	LPS-6P6A-225-B	225W Polycrystalline Module	N	202.3
Kinmac Solar	LPS-6P6A-230-B	230W Polycrystalline Module	N	207.0
Kinmac Solar	LPS-6P6A-235-B	235W Polycrystalline Module	N	211.6
Kinmac Solar	LPS-6P6A-240-B	240W Polycrystalline Module	N	216.2
Kinve Solar Power	KV170-72M	170W Monocrystalline Module	N	147.3
Kinve Solar Power	KV175-72M	175W Monocrystalline Module	N	151.8
Kinve Solar Power	KV180-72M	180W Monocrystalline Module	N	156.2
Kyocera Solar	SU53BU	53W Polycrystalline Module, Deep Blue	Y	41.6
Kyocera Solar	KC130GT	130W Polycrystalline Module, High Efficiency, Deep Blue	N	114.2
Kyocera Solar	KC130TM	130W Polycrystalline Module, High Efficiency, Deep Blue	N	114.2
Kyocera Solar	KD130GX-L	130W Polycrystalline Module, High Efficiency, Deep Blue	N	117.5
Kyocera Solar	KD130GX-LP	130W Polycrystalline Module, High Efficiency, Deep Blue	N	117.5
Kyocera Solar	KD135GX-L	135W Polycrystalline Module, High Efficiency, Deep Blue	N	122.2
Kyocera Solar	KD135GX-LP	135W Polycrystalline Module, High Efficiency, Deep Blue	N	122.2

(Continued)

MANUFACTURER NAME	MODULE MODEL NUMBER	DESCRIPTION	BIPV*	PTC**	NOTES
Kyocera Solar	KD135GX-LPU	135W Polycrystalline Module, High Efficiency, Deep Blue	N	122.2	
Kyocera Solar	KC175GT	175W Polycrystalline Module, High Efficiency, Deep Blue	N	153.6	
Kyocera Solar	KD180GX-L	180W Polycrystalline Module, High Efficiency, Deep Blue	N	162.5	
Kyocera Solar	KD180GX-LP	180W Polycrystalline Module, High Efficiency, Deep Blue	N	162.5	
Kyocera Solar	KD185GX-LPU	185W Polycrystalline Module, High Efficiency, Deep Blue	N	167.1	
Kyocera Solar	KC200GT	200W Polycrystalline Module, High Efficiency, Deep Blue	N	175.7	
Kyocera Solar	KD200GX-LPU	200W Polycrystalline Module, High Efficiency, Deep Blue	N	180.5	
Kyocera Solar	KD205GX-L	205W Polycrystalline Module, High Efficiency, Deep Blue	N	185.2	
Kyocera Solar	KD205GX-LFBS	205W Polycrystalline Module	N	185.3	
Kyocera Solar	KD205GX-LP	205W Polycrystalline Module, High Efficiency, Deep Blue	N	185.2	
Kyocera Solar	KD205GX-LPU	205W Polycrystalline Module, High Efficiency, Deep Blue	N	185.2	
Kyocera Solar	KD210GX-L	210W Polycrystalline Module, High Efficiency, Deep Blue	N	189.8	
Kyocera Solar	KD210GX-LFBS	210W Polycrystalline Module	N	190.0	
Kyocera Solar	KD210GX-LP	210W Polycrystalline Module, High Efficiency, Deep Blue	N	189.8	
Kyocera Solar	KD210GX-LPU	210W Polycrystalline Module, High Efficiency, Deep Blue	N	189.8	

Manufacturer	Model	Description		Rating
Kyocera Solar	KD215GX-LFBS	215W Polycrystalline Module	N	194.6
Kyocera Solar	KD215GX-LPU	215W Polycrystalline Module, High Efficiency, Deep Blue	N	194.4
Kyocera Solar	KD225GX-LFB	225W Polycrystalline Module	N	203.3
Kyocera Solar	KD225GX-LPB	225W Polycrystalline Module	N	200.2
Kyocera Solar	KD230GX-LFB	230W Polycrystalline Module	N	208.0
Kyocera Solar	KD230GX-LPB	230W Polycrystalline Module	N	204.8
Kyocera Solar	KD235GX-LFB	235W Polycrystalline Module	N	212.6
Kyocera Solar	KD235GX-LPB	235W Polycrystalline Module	N	209.4
Kyocera Solar	KD240GX-LFB	240W Polycrystalline Module	N	217.3
LDK Solar	LDK-170D-24	170W Monocrystalline Module	N	148.2
LDK Solar	LDK-175D-24	175W Monocrystalline Module	N	152.7
LDK Solar	LDK-180D-24	180W Monocrystalline Module	N	157.2
LDK Solar	LDK-190P-18	190W Polycrystalline Module	N	166.2
LDK Solar	LDK-200P-18	200W Polycrystalline Module	N	175.2
LDK Solar	LDK-210P-18	210W Polycrystalline Module	N	184.3
LDK Solar	LDK-210P-20	210W Polycrystalline Module	N	183.0
LDK Solar	LDK-220D-20	220W Monocrystalline Module	N	189.7
LDK Solar	LDK-220P-20	220W Polycrystalline Module	N	192.0
LDK Solar	LDK-230D-20	230W Monocrystalline Module	N	198.6
LDK Solar	LDK-230P-20	230W Polycrystalline Module	N	201.0
LDK Solar	LDK-240D-20	240W Monocrystalline Module	N	207.6
LDK Solar	LDK-260P-24	260W Polycrystalline Module	N	228.0
LDK Solar	LDK-270P-24	270W Polycrystalline Module	N	237.0
LDK Solar	LDK-280P-24	280W Polycrystalline Module	N	246.1

(Continued)

MANUFACTURER NAME	MODULE MODEL NUMBER	DESCRIPTION	BIPV*	PTC**	NOTES
LG Electronics Solar Cell Division	LG210P1W-G1	210W Polycrystalline Module	N	188.9	
LG Electronics Solar Cell Division	LG215P1W	215W Polycrystalline Module	N	191.0	
LG Electronics Solar Cell Division	LG215P1W-G1	215W Polycrystalline Module	N	193.5	
LG Electronics Solar Cell Division	LG220P1W	220W Polycrystalline Module	N	195.5	
LG Electronics Solar Cell Division	LG220P1W-G1	220W Polycrystalline Module	N	198.1	
LG Electronics Solar Cell Division	LG225P1W	225W Polycrystalline Module	N	200.1	
LG Electronics Solar Cell Division	LG225P1W-G1	225W Polycrystalline Module	N	202.8	
LG Electronics Solar Cell Division	LG230P1W	230W Polycrystalline Module	N	204.7	
LG Electronics Solar Cell Division	LG230P1W-G1	230W Polycrystalline Module	N	207.4	
LG Electronics Solar Cell Division	LG235P1W	235W Polycrystalline Module	N	209.2	
Lifeline Energy	LF-235WP-US	235W Monocrystalline Module	N	205.1	
Lifeline Energy	LF-240WP-US	240W Monocrystalline Module	N	209.6	
Lifeline Energy	LF-245WP-US	245W Monocrystalline Module	N	214.1	
Lifeline Energy	LF-250WP-US	250W Monocrystalline Module	N	218.6	
Lifeline Energy International	LF-165WP	165W Monocrystalline Module	N	147.3	
Lifeline Energy International	LF-170WP	170W Monocrystalline Module	N	151.8	

Manufacturer	Model	Description		
Lifeline Energy International	LF-175WP	175W Monocrystalline Module	N	156.4
Lifeline Energy International	LF-180WP	180W Monocrystalline Module	N	161.0
Lifeline Energy International	LF-200WP	200W Polycrystalline Module	N	175.9
Lifeline Energy International	LF-210WP	210W Polycrystalline Module	N	184.9
Lifeline Energy International	LF-220WP	220W Polycrystalline Module	N	194.0
Lifeline Energy International	LF-230WP	230W Polycrystalline Module	N	203.1
Lifeline Energy International	LF-240WP	240W Polycrystalline Module	N	211.7
Lifeline Energy International	LF-250WP	250W Polycrystalline Module	N	220.7
Lifeline Energy International	LF-260WP	260W Polycrystalline Module	N	229.8
Lifeline Energy International	LF-270WP	270W Polycrystalline Module	N	238.9
Lifeline Energy International	LF-280WP	280W Polycrystalline Module	N	248.0
Lifeline Energy USA	LFUS-205-p	200W Polycrystalline Module	N	173.1
Lifeline Energy USA	LFUS-200-p	205W Polycrystalline Module	N	177.6
Ligitek Photovoltaic	LG052BB00	52W Polycrystalline Module	Y	41.2
Ligitek Photovoltaic	LM175AA00	175W Monocrystalline Module	N	157.5
Ligitek Photovoltaic	LM175BB00	175W Polycrystalline Module	N	156.0
Ligitek Photovoltaic	LM180AA00	180W Monocrystalline Module	N	162.1
Ligitek Photovoltaic	LM180BB00	180W Polycrystalline Module	N	160.6

(Continued)

MANUFACTURER NAME	MODULE MODEL NUMBER	DESCRIPTION	BIPV*	PTC**	NOTES
Ligitek Photovoltaic	LM185AA00	185W Monocrystalline Module	N	166.7	
Ligitek Photovoltaic	LM185BB00	185W Polycrystalline Module	N	165.2	
Ligitek Photovoltaic	LM210BB00	210W Polycrystalline Module	N	187.1	
Ligitek Photovoltaic	LM215BB00	215W Polycrystalline Module	N	191.6	
Ligitek Photovoltaic	LM220BB00	220W Polycrystalline Module	N	196.2	
Ligitek Photovoltaic	LM225BB00	225W Polycrystalline Module	N	200.8	
Ligitek Photovoltaic	LM230BB00	230W Polycrystalline Module	N	205.4	
Ligitek Photovoltaic	LM235BB00	235W Polycrystalline Module	N	210.0	
Ligitek Photovoltaic	LM240BB00	240W Polycrystalline Module	N	214.6	
Ligitek Photovoltaic	LM265BB02	265W Polycrystalline Module	N	236.3	
Ligitek Photovoltaic	LM270BB02	270W Polycrystalline Module	N	240.9	
Ligitek Photovoltaic	LM275BB02	275W Polycrystalline Module	N	245.5	
Ligitek Photovoltaic	LM280BB02	280W Polycrystalline Module	N	250.1	
Ligitek Photovoltaic	LM285BB02	285W Polycrystalline Module	N	254.7	
Lumos	LS185-72M	185W Monocrystalline Module	N	165.6	
Lumos	LS220-60M	220W Monocrystalline Module	N	198.8	
Lumos	LS225-60M	225W Monocrystalline Module	N	203.4	
Lumos	LS230-60M	230W Monocrystalline Module	N	208.0	
Lumos	LS235-60M	235W Monocrystalline Module	N	212.6	
Lumos	LS240-60M	240W Monocrystalline Module	N	217.3	
Lumos	LS245-60M	245W Monocrystalline Module	N	221.9	
Lumos	LS250-60M	250W Monocrystalline Module	N	226.6	
Lumos	LS265-72M	265W Monocrystalline Module	N	239.5	
Lumos	LS270-72M	270W Monocrystalline Module	N	244.1	
Lumos	LS275-72M	275W Monocrystalline Module	N	248.7	

Lumos	SST275-72P	275W Polycrystalline Module	N	241.6
Lumos	LS280-72M	280W Monocrystalline Module	N	253.4
Lumos	SST280-72P	280W Polycrystalline Module	N	246.1
Lumos	LS285-72M	285W Monocrystalline Module	N	258.0
Lumos	SST285-72P	285W Polycrystalline Module	N	250.6
Lumos	SST290-72M	290W Monocrystalline Module	N	253.5
Lumos	SST290-72P	290W Polycrystalline Module	N	255.2
Lumos	SST295-72M	295W Monocrystalline Module	N	258.1
Lumos	SST300-72M	300W Monocrystalline Module	N	262.6
MAGE Solar	Powertec Plus 170/5 ME	170W Monocrystalline Module	N	151.0
MAGE Solar	Powertec Plus 170/5 MH	170W Monocrystalline Module	N	152.6
MAGE Solar	Powertec Plus 175/5 ME	175W Monocrystalline Module	N	155.6
MAGE Solar	Powertec Plus 175/5 MH	175W Monocrystalline Module	N	156.1
MAGE Solar	Powertec Plus 175/5MJ	175W Monocrystalline Module	N	153.9
MAGE Solar	Powertec Plus 180/5 ME	180W Monocrystalline Module	N	160.1
MAGE Solar	Powertec Plus 180/5 MH	180W Monocrystalline Module	N	160.7
MAGE Solar	Powertec Plus 180/5MJ	180W Monocrystalline Module	N	158.5
MAGE Solar	Powertec Plus 185/5 MH	185W Monocrystalline Module	N	165.3
MAGE Solar	Powertec Plus 185/5MJ	185W Monocrystalline Module	N	163.0
MAGE Solar	Powertec Plus 220/6 PH	220W Polycrystalline Module	N	198.2
MAGE Solar	Powertec Plus 220/6PJ	220W Polycrystalline Module	N	193.2
MAGE Solar	Powertec Plus 225/6PJ	225W Polycrystalline Module	N	197.7
MAGE Solar	Powertec Plus 230/6PJ	230W Polycrystalline Module	N	202.2
MiaSole	MR-107	107W Thin Film CIGS Module	N	93.5
Mitsubishi Electric	PV-EE120MF5F	120W Polycrystalline Module, Lead-Free Solder	N	108.7

(Continued)

MANUFACTURER NAME	MODULE MODEL NUMBER	DESCRIPTION	BIPV*	PTC**	NOTES
Mitsubishi Electric	PV-EE125MF5F	125W Polycrystalline Module, Lead-Free Solder	N	112.4	
Mitsubishi Electric	PV-EE130MF5F	130W Polycrystalline Module, Lead-Free Solder	N	117.0	
Mitsubishi Electric	PV-MF160EB4	160W Polycrystalline Module, Lead-Free Solder, MC Connectors	N	141.6	
Mitsubishi Electric	PV-MF165EB4	165W Polycrystalline Module, Lead-Free Solder, MC Connectors	N	146.1	
Mitsubishi Electric	PV-MF170EB3	170W Polycrystalline Module, Lead-Free Solder, MC Connectors	N	150.7	
Mitsubishi Electric	PV-MF170EB4	170W Polycrystalline Module, Lead-Free Solder, MC Connectors	N	150.7	
Mitsubishi Electric	PV-MF170UD4	170W Polycrystalline Module, Lead-Free Solder, MC Connectors	N	151.6	
Mitsubishi Electric	PV-UD170MF5	170W Polycrystalline Module, Lead-Free Solder, MC Connectors	N	150.9	
Mitsubishi Electric	PV-MF175UD4	175W Polycrystalline Module, Lead-Free Solder, MC Connectors	N	156.1	
Mitsubishi Electric	PV-UD175MF5	175W Polycrystalline Module, Lead-Free Solder, MC Connectors	N	157.0	
Mitsubishi Electric	PV-MF180UD4	180W Polycrystalline Module, Lead-Free Solder, MC Connectors	N	160.7	

Mitsubishi Electric	PV-UD180MF5	180W Polycrystalline Module, Lead-Free Solder, MC Connectors	N	161.6
Mitsubishi Electric	PV-UD185MF5	185W Polycrystalline Module, Lead-Free Solder, MC Connectors	N	166.2
Mitsubishi Electric	PV-UD190MF5	190W Polycrystalline Module, Lead-Free Solder, MC Connectors	N	169.8
Mitsubishi Electric	PV-UJ212G6	212W Polycrystalline Modue, Lead-Free Solder, MC Connectors	N	194.5
Mitsubishi Electric	PV-UJ218G6	218W Polycrystalline Modue, Lead-Free Solder, MC Connectors	N	200.1
Mitsubishi Electric	PV-UJ220GA6	220W Polycrystalline Modue, Lead-Free Solder, MC Connectors	N	198.6
Mitsubishi Electric	PV-UJ224G6	224W Polycrystalline Modue, Lead-Free Solder, MC Connectors	N	205.7
Mitsubishi Electric	PV-UJ225GA6	225W Polycrystalline Modue, Lead-Free Solder, MC Connectors	N	203.2
Mitsubishi Electric	PV-UJ230GA6	230W Polycrystalline Modue, Lead-Free Solder, MC Connectors	N	207.8
Mitsubishi Electric	PV-UJ235GA6	235W Polycrystalline Modue, Lead-Free Solder, MC Connectors	N	212.4

(Continued)

MANUFACTURER NAME	MODULE MODEL NUMBER	DESCRIPTION	BIPV*	PTC**	NOTES
Moser Baer Photovoltaic	MBPV CAAP 210	210W Polycrystalline Module	N	186.5	
Moser Baer Photovoltaic	MBPV CAAP BB 210W	210W Polycrystalline Module	N	188.3	
Moser Baer Photovoltaic	MBPV CAAP BB 215W	215W Polycrystalline Module	N	192.9	
Moser Baer Photovoltaic	MBPV CAAP 220	220W Polycrystalline Module	N	195.6	
Moser Baer Photovoltaic	MBPV CAAP BB 220W	220W Polycrystalline Module	N	197.5	
Moser Baer Photovoltaic	MBPV CAAP BB 225W	225W Polycrystalline Module	N	202.1	
Moser Baer Photovoltaic	MBPV CAAP BB 230W	230W Polycrystalline Module	N	206.7	
Moser Baer Photovoltaic	MBPV CAAP 215W	215W Polycrystalline Module	N	191.4	
Motech Americas	GEPVp-200-M	200W Polycrystalline Module	N	173.1	
Motech Americas	GEPVp-205-M	205W Polycrystalline Module	N	177.6	
Motech Americas	GEPVp-210-M	210W Polycrystalline Module	N	184.3	
Nexpower Technology	NH-100UT_3A	95W Thin Film a-Si Module	N	90.0	
NexPower Technology	NH-100UX 3A	95W Thin Film a-Si Module	N	90.7	
Nexpower Technology	NH-100UT_4A	100W Thin Film a-Si Module	N	94.8	
NexPower Technology	NH-100UX 4A	100W Thin Film a-Si Module	N	95.5	
Nexpower Technology	NH-100UT_5A	105W Thin Film a-Si Module	N	99.5	
NexPower Technology	NH-100UX 5A	105W Thin Film a-Si Module	N	100.3	
Nexpower Technology	NH-100UT_6A	110W Thin Film a-Si Module	N	104.3	
Nexpower Technology	NT-125UX	125W Thin Film a-Si/uc-Si Module	N	117.5	

Manufacturer	Model	Description		Rating
Nexpower Technology	NT-130UX	130W Thin Film a-Si/uc-Si Module	N	122.3
Nexpower Technology	NT-135UX	135W Thin Film a-Si/uc-Si Module	N	127.0
Next Solar Systems	NXT18m	180W Monocrystalline Module, Black Frame, Black Backsheet	N	164.0
Next Solar Systems	NXT20e	200W Polycrystalline Module, Black Frame	N	177.1
Ningbo Maxsolar	MP-150WP	150W Monocrystalline Module	N	131.7
Ningbo Maxsolar	MP-155WP	155W Monocrystalline Module	N	136.2
Ningbo Maxsolar	MP-160WP	160W Monocrystalline Module	N	140.7
Ningbo Maxsolar	MP-165WP	165W Monocrystalline Module	N	145.2
Ningbo Maxsolar	MP-170WP	170W Monocrystalline Module	N	149.7
Ningbo Maxsolar	MP-175WP	175W Monocrystalline Module	N	154.2
Ningbo Maxsolar	MP-180WP	180W Monocrystalline Module	N	158.8
Ningbo Qixin Solar Electrical Appliance	SL160UL-36M	160W Monocrystalline Module	N	143.3
Ningbo Qixin Solar Electrical Appliance	SL165UL-36M	165W Monocrystalline Module	N	147.9
Ningbo Qixin Solar Electrical Appliance	SL170UL-36M	170W Monocrystalline Module	N	152.5
Ningbo Qixin Solar Electrical Appliance	SL180UL-36M	180W Monocrystalline Module	N	161.7
Ningbo Qixin Solar Electrical Appliance	SL185UL-36M	185W Monocrystalline Module	N	166.3
Ningbo Qixin Solar Electrical Appliance	SL210UL-30P	210W Polycrystalline Module	N	188.1
Ningbo Qixin Solar Electrical Appliance	SL215UL-30P	215W Polycrystalline Module	N	192.6

(Continued)

MANUFACTURER NAME	MODULE MODEL NUMBER	DESCRIPTION	BIPV*	PTC**	NOTES
Ningbo Qixin Solar Electrical Appliance	SL220UL-30P	220W Polycrystalline Module	N	197.2	
Ningbo Qixin Solar Electrical Appliance	SL220UL-48M	220W Monocrystalline Module	N	195.5	
Ningbo Qixin Solar Electrical Appliance	SL225UL-30P	225W Polycrystalline Module	N	201.8	
Ningbo Qixin Solar Electrical Appliance	SL225UL-48M	225W Monocrystalline Module	N	200.1	
Ningbo Qixin Solar Electrical Appliance	SL230UL-30P	230W Polycrystalline Module	N	206.4	
Ningbo Qixin Solar Electrical Appliance	SL230UL-48M	230W Monocrystalline Module	N	204.7	
Ningbo Qixin Solar Electrical Appliance	SL235UL-30P	235W Polycrystalline Module	N	211.0	
Ningbo Qixin Solar Electrical Appliance	SL235UL-48M	235W Monocrystalline Module	N	209.2	
Ningbo Qixin Solar Electrical Appliance	SL240UL-48M	240W Monocrystalline Module	N	213.8	
Ningbo Qixin Solar Electrical Appliance	SL245UL-48M	245W Monocrystalline Module	N	218.4	
Ningbo Qixin Solar Electrical Appliance	SL250UL-36M	250W Monocrystalline Module	N	222.1	
Ningbo Qixin Solar Electrical Appliance	SL250UL-48M	250W Monocrystalline Module	N	223.0	
Ningbo Qixin Solar Electrical Appliance	SL255UL-36M	255W Monocrystalline Module	N	226.7	
Ningbo Qixin Solar Electrical Appliance	SL260UL-36M	260W Monocrystalline Module	N	231.6	
Ningbo Qixin Solar Electrical Appliance	SL265UL-36M	265W Monocrystalline Module	N	236.1	

Ningbo Qixin Solar Electrical Appliance	SL270UL-36M	270W Monocrystalline Module	N	240.7
Ningbo Qixin Solar Electrical Appliance	SL275UL-36M	275W Monocrystalline Module	N	245.3
Ningbo Qixin Solar Electrical Appliance	SL280UL-36M	280W Monocrystalline Module	N	249.9
Ningbo Solar Electric Power	TDB125x125-72-P 150W	150W Monocrystalline Module	N	133.2
Ningbo Solar Electric Power	TPB125x125-72-P 150W	150W Polycrystalline Module	N	135.1
Ningbo Solar Electric Power	TDB125x125-72-P 155W	155W Monocrystalline Module	N	137.7
Ningbo Solar Electric Power	TPB125x125-72-P 155W	155W Polycrystalline Module	N	139.7
Ningbo Solar Electric Power	TDB125x125-72-P 160W	160W Monocrystalline Module	N	142.3
Ningbo Solar Electric Power	TPB125x125-72-P 160W	160W Polycrystalline Module	N	144.3
Ningbo Solar Electric Power	TDB125x125-72-P 165W	165W Monocrystalline Module	N	146.8
Ningbo Solar Electric Power	TPB125x125-72-P 165W	165W Polycrystalline Module	N	149.0
Ningbo Solar Electric Power	TDB125x125-72-P 170W	170W Monocrystalline Module	N	151.4
Ningbo Solar Electric Power	TPB125x125-72-P 170W	170W Polycrystalline Module	N	153.6
Ningbo Solar Electric Power	TDB125x125-72-P 175W	175W Monocrystalline Module	N	155.9
Ningbo Solar Electric Power	TPB125x125-72-P 175W	175W Polycrystalline Module	N	158.2

(Continued)

MANUFACTURER NAME	MODULE MODEL NUMBER	DESCRIPTION	BIPV*	PTC**	NOTES
Ningbo Solar Electric Power	TDB125x125-72-P 180W	180W Monocrystalline Module	N	160.5	
Ningbo Solar Electric Power	TPB125x125-72-P 180W	180W Polycrystalline Module	N	162.8	
Ningbo Solar Electric Power	TPB156x156-54-P 180W	180W Polycrystalline Module	N	160.6	
Ningbo Solar Electric Power	TPB156x156-54-P 185W	185W Polycrystalline Module	N	165.2	
Ningbo Solar Electric Power	TPB156x156-54-P 190W	190W Polycrystalline Module	N	169.8	
Ningbo Solar Electric Power	TPB156x156-54-P 195W	195W Polycrystalline Module	N	174.3	
Ningbo Solar Electric Power	TPB125x125-96-P 200W	200W Polycrystalline Module	N	175.7	
Ningbo Solar Electric Power	TPB156x156-54-P 200W	200W Polycrystalline Module	N	178.9	
Ningbo Solar Electric Power	TPB156x156-60-P 200W	200W Polycrystalline Module	N	178.5	
Ningbo Solar Electric Power	TPB156x156-60-P 205W	205W Polycrystalline Module	N	183.0	
Ningbo Solar Electric Power	TPB125x125-96-P 210W	210W Polycrystalline Module	N	184.7	
Ningbo Solar Electric Power	TPB156x156-60-P 210W	210W Polycrystalline Module	N	187.6	
Ningbo Solar Electric Power	TPB156x156-60-P 215W	215W Polycrystalline Module	N	192.2	
Ningbo Solar Electric Power	TPB125x125-96-P 220W	220W Polycrystalline Module	N	193.7	
Ningbo Solar Electric Power	TPB156x156-60-P 220W	220W Polycrystalline Module	N	196.8	

Ningbo Solar Electric Power	TPB156x156-60-P 225W	225W Polycrystalline Module	N	201.4
Ningbo Solar Electric Power	TPB125x125-96-P 230W	230W Polycrystalline Module	N	202.8
Ningbo Solar Electric Power	TPB156x156-60-P 230W	230W Polycrystalline Module	N	206.0
Ningbo Solar Electric Power	TPB156x156-72-P 230W	230W Polycrystalline Module	N	205.1
Ningbo Solar Electric Power	TPB125x125-96-P 240W	240W Polycrystalline Module	N	211.9
Ningbo Solar Electric Power	TPB156x156-72-P 240W	240W Polycrystalline Module	N	214.2
Ningbo Solar Electric Power	TPB156x156-72-P 250W	250W Polycrystalline Module	N	223.3
Ningbo Solar Electric Power	TPB156x156-72-P 255W	255W Polycrystalline Module	N	227.9
Ningbo Solar Electric Power	TPB156x156-72-P 260W	260W Polycrystalline Module	N	232.5
Ningbo Solar Electric Power	TPB156x156-72-P 270W	270W Polycrystalline Module	N	241.7
Ningbo Solar Electric Power	TPB156x156-72-P 280W	280W Polycrystalline Module	N	250.9
Ningbo Ulica Solar Science & Technology	165(35) D UL811x1581	165W Monocrystalline Module	N	142.7
Ningbo Ulica Solar Science & Technology	230(35) D UL997x1953	230W Monocrystalline Module	N	201.8
Perfect Source Technology	PST 212GT	212W Polycrystalline Module	N	191.0
Perfect Source Technology	PST 218GT	218W Polycrystalline Module	N	196.6

(Continued)

MANUFACTURER NAME	MODULE MODEL NUMBER	DESCRIPTION	BIPV*	PTC***	NOTES
Perfect Source Technology	PST 224GT	224W Polycrystalline Module	N	202.1	
Perfect Source Technology	PST 230GT	230W Polycrystalline Module	N	207.6	
Perfect Source Technology	PST 236GT	236W Polycrystalline Module	N	213.2	
Perlight Solar	PLM-5B-175M	175W Monocrystalline Module	N	149.8	
Perlight Solar	PLM-5B-185M	185W Monocrystalline Module	N	161.5	
Perlight Solar	PLM-6B-220P	220W Polycrystalline Module	N	191.5	
Perlight Solar	PLM-6B-230P	230W Polycrystalline Module	N	199.2	
Phono Solar Technology	PS160M-24/F	160W Monocrystalline Module	N	145.1	
Phono Solar Technology	PS165M-24/F	165W Monocrystalline Module	N	149.7	
Phono Solar Technology	PS170M-24/F	170W Monocrystalline Module	N	154.4	
Phono Solar Technology	PS180M-24/F	180W Monocrystalline Module	N	163.7	
Phono Solar Technology	PS185M-24/F	185W Monocrystalline Module	N	168.3	
Phono Solar Technology	PS190M-24/F	190W Monocrystalline Module	N	173.0	
Phono Solar Technology	PS210M-20/U	210W Monocrystalline Module	N	189.7	
Phono Solar Technology	PS210P-20/U	210W Polycrystalline Module	N	190.1	
Phono Solar Technology	PS215M-20/U	215W Monocrystalline Module	N	194.3	

Phono Solar Technology	PS215P-20/U	215W Polycrystalline Module	N	194.7
Phono Solar Technology	PS220M-20/U	220W Monocrystalline Module	N	199.0
Phono Solar Technology	PS220P-20/U	220W Polycrystalline Module	N	199.4
Phono Solar Technology	PS225M-20/U	225W Monocrystalline Module	N	203.6
Phono Solar Technology	PS225P-20/U	225W Polycrystalline Module	N	204.0
Phono Solar Technology	PS230M-20/U	230W Monocrystalline Module	N	208.2
Phono Solar Technology	PS230P-20/U	230W Polycrystalline Module	N	208.6
Phono Solar Technology	PS235M-20/U	235W Monocrystalline Module	N	212.8
Phono Solar Technology	PS235P-20/U	235W Polycrystalline Module	N	213.3
Phono Solar Technology	PS240M-20/U	240W Monocrystalline Module	N	217.5
Phono Solar Technology	PS240P-20/U	240W Polycrystalline Module	N	217.9
Polar Photovoltaics	TFSM-T-0	42W Thin Film Dual Junction a-Si Module	N	38.7
Polar Photovoltaics	TFSM-T-1	44W Thin Film Dual Junction a-Si Module	N	40.6
Polar Photovoltaics	TFSM-T-2	46W Thin Film Dual Junction a-Si Module	N	42.4
Polar Photovoltaics	TFSM-T-3	48W Thin Film Dual Junction a-Si Module	N	44.3

(Continued)

MANUFACTURER NAME	MODULE MODEL NUMBER	DESCRIPTION	BIPV*	PTC**	NOTES
Polar Photovoltaics	TFSM-T-4	50W Thin Film Dual Junction a-Si Module	N	46.1	
Polar Photovoltaics	TFSM-T-5	52W Thin Film Dual Junction a-Si Module	N	48.0	
Powercom	PPV-130M6L	126W Polycrystalline PV Module	N	110.1	
Powercom	PPV-180M6L	178W Polycrystalline PV Module	N	155.9	
Powercom	PPV-210M6L	210W Polycrystalline PV Module	N	182.9	
Powercom	PPV-230M6L	223W Polycrystalline PV Module	N	194.2	
Quantum Technologies	QS 215W/60-156 AP 2BB	215W Polycrystalline Module	N	187.6	
Quantum Technologies	QS 220W/60-156 AP 2BB	220W Polycrystalline Module	N	192.1	
Quantum Technologies	QS 220W/60-156 SP 2BB	220W Polycrystalline Module	N	189.1	
Quantum Technologies	QS 225W/60-156 AP 2BB	225W Polycrystalline Module	N	196.6	
Quantum Technologies	QS 225W/60-156 SP 2BB	225W Polycyrystalline Module	N	193.5	
Quantum Technologies	QS 225W/60-256 EM 3BB	225W Monocrystalline Module	N	192.4	
Quantum Technologies	QS 230W/60-156 SP 2BB	230W Polycrystalline Module	N	198.0	
REC ScanModule	SCM205	205W Polycrystalline Module	N	178.6	
REC ScanModule	SCM210	210W Polycrystalline Module	N	183.0	
REC ScanModule	SCM215	215W Polycrystalline Module	N	187.5	
REC ScanModule	SCM220	220W Polycrystalline Module	N	192.0	
REC ScanModule	SCM225	225W Polycrystalline Module	N	196.5	
REC Solar	REC205AE-US	205W Polycrystalline Module, High Performance	N	178.2	
REC Solar	REC205AE-US (BLK)	205W Polycrystalline Module, High Performance	N	178.2	

REC Solar	REC210AE-US	210W Polycrystalline Module, High Performance	N	182.7
REC Solar	REC210AE-US (BLK)	210W Polycrystalline Module, High Performance	N	182.7
REC Solar	REC215AE-US	215W Polycrystalline Module, High Performance	N	187.2
REC Solar	REC215AE-US (BLK)	215W Polycrystalline Module, High Performance	N	187.2
REC Solar	REC215PE	215W Polycrystalline Module	N	190.6
REC Solar	REC215PE-US	215W Polycrystalline Module	N	190.6
REC Solar	REC220AE-US	220W Polycrystalline Module, High Performance	N	191.7
REC Solar	REC220AE-US (BLK)	220W Polycrystalline Module, High Performance	N	191.7
REC Solar	REC220PE	220W Polycrystalline Module	N	195.2
REC Solar	REC220PE-US	220W Polycrystalline Module	N	195.2
REC Solar	REC225AE-US	225W Polycrystalline Module, High Performance	N	196.2
REC Solar	REC225AE-US (BLK)	225W Polycrystalline Module, High Performance	N	196.2
REC Solar	REC225PE	225W Polycrystalline Module	N	199.7
REC Solar	REC225PE-US	225W Polycrystalline Module	N	199.7
REC Solar	REC230AE-US	230W Polycrystalline Module, High Performance	N	200.7
REC Solar	REC230AE-US (BLK)	230W Polycrystalline Module, High Performance	N	200.7
REC Solar	REC230PE	230W Polycrystalline Module	N	204.3
REC Solar	REC230PE-US	230W Polycrystalline Module	N	204.3

(Continued)

MANUFACTURER NAME	MODULE MODEL NUMBER	DESCRIPTION	BIPV*	PTC**	NOTES
REC Solar	REC235PE	235W Polycrystalline Module	N	208.8	
REC Solar	REC235PE-US	235W Polycrystalline Module	N	208.8	
Risen Energy	SYP-85S	85W Monocrystalline Module	N	72.5	
Risen Energy	SYP-110S	110W Monocrystalline Module	N	94.8	
Risen Energy	SYP-160S	160W Monocrystalline Module	N	137.8	
Risen Energy	SYP-165S	165W Monocrystalline Module	N	142.2	
Risen Energy	SYP-170S	170W Monocrystalline Module	N	146.6	
Risen Energy	SYP-175S	175W Monocrystalline Module	N	151.0	
Risen Energy	SYP-180S	180W Monocrystalline Module	N	155.5	
Risen Energy	SYP-200S	200W Monocrystalline Module	N	169.0	
Risen Energy	SYP-210S	210W Monocrystalline Module	N	177.7	
Ritek	PM130	130W Polycrystalline Module	N	115.4	
Ritek	MM135	135W Monocrystalline Module	N	120.9	
Ritek	PM135	135W Polycrystalline Module	N	119.9	
Ritek	MM140	140W Monocrystalline Module	N	125.4	
Ritek	PM140	140W Polycrystalline Module	N	124.5	
Ritek	MM145	145W Monocrystalline Module	N	130.0	
Ritek	PM175	175W Polycrystalline Module	N	155.4	
Ritek	MM180	180W Monocrystalline Module	N	161.2	
Ritek	PM180	180W Polycrystalline Module	N	160.0	
Ritek	MM185	185W Monocrystalline Module	N	165.8	
Ritek	PM185	185W Polycrystalline Module	N	164.5	
Ritek	MM190	190W Monocrystalline Module	N	170.4	
Ritek	PM190	190W Polycrystalline Module	N	168.6	
Ritek	MM195	195W Monocrystalline Module	N	174.5	

Ritek	PM195	195W Polycrystalline Module	N	173.1
Ritek	MM200	200W Monocrystalline Module	N	179.1
Ritek	PM200	200W Polycrystalline Module	N	177.7
Ritek	MM205	205W Monocrystalline Module	N	183.7
Ritek	PM205	205W Polycrystalline Module	N	182.3
Ritek	MM210	210W Monocrystalline Module	N	188.3
Ritek	PM210	210W Polycrystalline Module	N	186.3
Ritek	MM215	215W Monocrystalline Module	N	192.9
Ritek	PM215	215W Polycrystalline Module	N	190.9
Ritek	MM220	220W Monocrystalline Module	N	197.0
Ritek	PM220	220W Polycrystalline Module	N	195.4
Ritek	MM225	225W Monocrystalline Module	N	201.6
Ritek	PM225	225W Polycrystalline Module	N	200.0
Ritek	MM230	230W Monocrystalline Module	N	206.1
Ritek	PM230	230W Polycrystalline Module	N	204.6
Ritek	MM235	235W Monocrystalline Module	N	210.7
Ritek	MM240	240W Monocrystalline Module	N	215.4
Robert Bosch	c-Si M 60-225-16	225W Monocrystalline Module	N	194.4
Robert Bosch	c-Si M 60-230-16	230W Monocrystalline Module	N	198.9
Robert Bosch	c-Si M 60-235-16	235W Monocrystalline Module	N	203.4
Robert Bosch	c-Si M 60-240-16	240W Monocrystalline Module	N	207.9
Samsung Electronics	LPC232SM-02	232W Monocrystalline Module	N	205.1
Samsung Electronics	LPC235SM-02	235W Monocrystalline Module	N	207.8
Samsung Electronics	LPC238SM-02	238W Monocrystalline Module	N	210.6
Samsung Electronics	LPC241SM-02	241W Monocrystalline Module	N	213.3

(Continued)

MANUFACTURER NAME	MODULE MODEL NUMBER	DESCRIPTION	BIPV*	PTC**	NOTES
Samsung Electronics	LPC244SM-02	244W Monocrystalline Module	N	216.0	
Sanyo Electric	HIP-180BA19	180W HIT Power Hybrid Amorphous/Monocrystalline Module	N	166.9	
Sanyo Electric	HIP-180BA20	180W HIT Power Hybrid Amorphous/Monocrystalline Module	N	166.9	
Sanyo Electric	HIP-180DA3	180W HIT Power Hybrid Amorphous/Monocrystalline Module	N	167.3	
Sanyo Electric	HIP-186BA19	186W HIT Power Hybrid Amorphous/Monocrystalline Module	N	172.6	
Sanyo Electric	HIP-186BA20	186W HIT Power Hybrid Amorphous/Monocrystalline Module	N	172.6	
Sanyo Electric	HIP-186BA3	186W HIT Power Hybrid Amorphous/Monocrystalline Module	N	167.5	
Sanyo Electric	HIP-186DA3	186W HIT Power Hybrid Amorphous/Monocrystalline Module	N	173.0	
Sanyo Electric	HIP-190BA19	190W HIT Power Hybrid Amorphous/Monocrystalline Module	N	176.4	
Sanyo Electric	HIP-190BA20	190W HIT Power Hybrid Amorphous/Monocrystalline Module	N	176.4	
Sanyo Electric	HIP-190BA3	190W HIT Power Hybrid Amorphous/Monocrystalline Module	N	171.2	

Sanyo Electric	HIP-190DA3	190W HIT Power Hybrid Amorphous/Monocrystalline Module	N	176.8
Sanyo Electric	HIP-195BA19	195W HIT Power Hybrid Amorphous/Monocrystalline Module	N	181.1
Sanyo Electric	HIP-195BA20	195W HIT Power Hybrid Amorphous/Monocrystalline Module	N	181.1
Sanyo Electric	HIP-195BA3	195W HIT Power Hybrid Amorphous/Monocrystalline Module	N	179.8
Sanyo Electric	HIP-195DA3	195W HIT Power Hybrid Amorphous/Monocrystalline Module	N	181.5
Sanyo Electric	HIP-195NKHA1	195W HIT Power N Hybrid Amorphous/Monocrystalline Module	N	180.7
Sanyo Electric	HIP-195NKHA5	195W HIT Power N Hybrid Amorphous/Monocrystalline Module	N	180.7
Sanyo Electric	HIP-195NKHA6	195W HIT Power N Hybrid Amorphous/Monocrystalline Module	N	180.7
Sanyo Electric	HIT-N195A01	195W HIT Hybrid Amorphous/Monocrystalline Module	N	180.7
Sanyo Electric	HIP-200BA19	200W HIT Power Hybrid Amorphous/Monocrystalline Module	N	185.9
Sanyo Electric	HIP-200BA20	200W HIT Power Hybrid Amorphous/Monocrystalline Module	N	185.9

MANUFACTURER NAME	MODULE MODEL NUMBER	DESCRIPTION	BIPV*	PTC**	NOTES
Sanyo Electric	HIP-200BA3	200W HIT Power Hybrid Amorphous/Monocrystalline Module	N	184.5	
Sanyo Electric	HIP-200DA3	200W HIT Power Hybrid Amorphous/Monocrystalline Module	N	186.2	
Sanyo Electric	HIP-200NKHA1	200W HIT Power N Hybrid Amorphous/Monocrystalline Module	N	185.4	
Sanyo Electric	HIP-200NKHA5	200W HIT Power N Hybrid Amorphous/Monocrystalline Module	N	185.4	
Sanyo Electric	HIP-200NKHA6	200W HIT Power N Hybrid Amorphous/Monocrystalline Module	N	185.4	
Sanyo Electric	HIT-N200A01	200W HIT Hybrid Amorphous/ Monocrystalline Module	N	185.4	
Sanyo Electric	HIP-205BA19	205W HIT Power Hybrid Amorphous/Monocrystalline Module	N	190.7	
Sanyo Electric	HIP-205BA20	205W HIT Power Hybrid Amorphous/Monocrystalline Module	N	190.7	
Sanyo Electric	HIP-205BA3	205W HIT Power Hybrid Amorphous/Monocrystalline Module	N	185.1	
Sanyo Electric	HIP-205NKHA1	205W HIT Power N Hybrid Amorphous/Monocrystalline Module	N	190.2	

Sanyo Electric	HIP-205NKHA5	205W HIT Power N Hybrid Amorphous/Monocrystalline Module	N	190.2
Sanyo Electric	HIP-205NKHA6	205W HIT Power N Hybrid Amorphous/Monocrystalline Module	N	190.2
Sanyo Electric	HIT-N205A01	205W HIT Hybrid Amorphous/Monocrystalline Module	N	190.2
Sanyo Electric	HIP-210NKHA1	210W HIT Power N Hybrid Amorphous/Monocrystalline Module	N	194.9
Sanyo Electric	HIP-210NKHA5	210W HIT Power N Hybrid Amorphous/Monocrystalline Module	N	194.9
Sanyo Electric	HIP-210NKHA6	210W HIT Power N Hybrid Amorphous/Monocrystalline Module	N	194.9
Sanyo Electric	HIT-N210A01	210W HIT Hybrid Amorphous/Monocrystalline Module	N	194.9
Sanyo Electric	HIP-215NKHA1	215W HIT Power N Hybrid Amorphous/Monocrystalline Module	N	199.6
Sanyo Electric	HIP-215NKHA5	215W HIT Power N Hybrid Amorphous/Monocrystalline Module	N	199.6
Sanyo Electric	HIP-215NKHA6	215W HIT Power N Hybrid Amorphous/Monocrystalline Module	N	199.6
Sanyo Electric	HIT-N215A01	215W HIT Hybrid Amorphous/Monocrystalline Module	N	199.6

(Continued)

MANUFACTURER NAME	MODULE MODEL NUMBER	DESCRIPTION	BIPV*	PTC**	NOTES
Sanyo Electric	HIT-N220A01	220W HIT Hybrid Amorphous/ Monocrystalline Module	N	204.4	
Scheuten Solar USA	P6-54 185W	185W Polycrystalline Module	N	165.9	
Scheuten Solar USA	P6-54 190W	190W Polycrystalline Module	N	170.5	
Scheuten Solar USA	P6-54 195W	195W Polycrystalline Module	N	175.1	
Scheuten Solar USA	P6-54 200W	200W Polycrystalline Module	N	179.7	
Scheuten Solar USA	P6-54 205W	205W Polycrystalline Module	N	184.3	
Schott Solar	ASE-300-DGF/25-145	145W Polycrystalline Module, 25V, Double Glass, Framed	N	131.8	
Schott Solar	SAPC-165	165W Polycrystalline Module	N	144.6	
Schott Solar	SAPC-170	170W Polycrystalline Module	N	149.1	
Schott Solar	SAPC-175	175W Monocrystalline Module	N	151.8	
Schott Solar	ASE-300-DGF/34-195	195W Polycrystalline Module, 34V, Double Glass, Framed	N	178.7	
Schott Solar	Poly 202	202W Polycrystalline Module	N	177.6	
Schott Solar	Poly 210	210W Polycrystalline Module	N	184.8	
Schott Solar	Poly 217	217W Polycrystalline Module	N	191.2	
Schott Solar	Poly 220	220W Polycrystalline Module	N	193.9	
Schott Solar	Poly 225	225W Polycrystalline Module	N	198.4	
Schott Solar	Poly 230	230W Polycrystalline Module	N	198.9	
Schott Solar	Poly 235	235W Polycrystalline Module	N	203.4	
Schott Solar	ASE-300-DGF/42-240	240W Polycrystalline Module	N	212.5	
Schott Solar	Poly 240	240W Polycrystalline Module	N	207.9	
Schott Solar	ASE-300-DGF/50-250	250W Polycrystalline Module, 50V, Double Glass, Framed	N	221.6	
Schott Solar	ASE-300-DGF/50-260	260W Polycrystalline Module	N	230.6	

Schott Solar	ASE-300-DGF/50-270	270W Polycrystalline Module	N	239.7
Schott Solar	ASE-300-DGF/50-280	280W Polycrystalline Module 50V, Double Glass, Framed	N	248.8
Schott Solar	ASE-300-DGF/50-290	290W Polycrystalline Module	N	257.9
Schott Solar	ASE-300-DGF/50-300	300W Polycrystalline Module	N	267.0
Schott Solar	ASE-300-DGF/50-310	310W Polycrystalline Module, 50V, Double Glass, Framed	N	276.2
Schuco International	MPE 310 MP 02	310W Monocrystalline Module, Premium, Dark Frame	N	274.9
Schuco International	MPE 320 MP 02	320W Monocrystalline Module, Premium, Dark Frame	N	284.0
Schuco USA	S 165-SPU	165W Polycrystalline Module, Lead-Free Solder, MC Connectors	N	146.1
Schuco USA	S 165-SPU-4	165W Polycrystalline Module, Lead-Free Solder, MC Connectors	N	146.4
Schuco USA	MPE 170 MS 05	170W Monocrystalline Module, Black Backsheet	N	153.3
Schuco USA	S 170-SPU	170W Polycrystalline Module, Lead-Free Solder, MC Connectors	N	150.7
Schuco USA	S 170-SPU-4	170W Polycrystalline Module, Lead-Free Solder, MC Connectors	N	150.9
Schuco USA	MPE 175 MS 05	175W Monocrystalline Module, Black Backsheet	N	158.0
Schuco USA	S 175-SPU-4	175W Polycrystalline Module, Lead-Free Solder, MC Connectors	N	157.0

(Continued)

MANUFACTURER NAME	MODULE MODEL NUMBER	DESCRIPTION	BIPV*	PTC**	NOTES
Schuco USA	MPE 180 MS 05	180W Monocrystalline Module, Black Backsheet	N	162.6	
Schuco USA	S 180-SPU-4	180W Polycrystalline Module, Lead-Free Solder, MC Connectors	N	161.6	
Schuco USA	MPE 185 MS 05	185W Monocrystalline Module, Black Backsheet	N	167.2	
Schuco USA	MPE 200 PS 04	200W Polycrystalline Module	N	183.0	
Schuco USA	SPV 200 SMAU-1	200W Monocrystalline Module, Black Frame	N	179.0	
Schuco USA	MPE 205 PS 04	205W Polycrystalline Module	N	187.6	
Schuco USA	MPE 210 PS 04	210W Polycrystalline Module	N	192.3	
Schuco USA	SPV 210 SMAU-1	210W Monocrystalline Module, Black Frame	N	188.2	
Schuco USA	MPE 215 PS 04	215W Polycrystalline Module	N	197.0	
Schuco USA	MPE 220 PS 04	220W Polycrystalline Module	N	201.6	
Schuco USA	MPE 220 PS 09	220W Polycrystalline Module	N	199.5	
Schuco USA	MPE 225 PS 04	225W Polycrystalline Module	N	206.3	
Schuco USA	MPE 225 PS 09	225W Polycrystalline Module	N	204.1	
Schuco USA	MPE 230 PS 04	230W Polycrystalline Module	N	211.0	
Schuco USA	MPE 230 PS 09	230W Polycrystalline Module	N	208.8	
S-Energy	SM-200MC1	200W Polycrystalline Module	N	178.0	
S-Energy	SM-205MC1	205W Polycrystalline Module	N	182.6	
S-Energy	SM-210MC1	210W Polycrystalline Module	N	187.1	
S-Energy	SM-215MC1	215W Polycrystalline Module	N	193.2	
S-Energy	SM-215PC5	215W Polycrystalline Module	N	188.6	
S-Energy	SM-220MC1	220W Polycrystalline Module	N	197.8	

S-Energy	SM-220PC5	220W Polycrystalline Module	N	193.2
S-Energy	SM-225MC1	225W Polycrystalline Module	N	202.4
S-Energy	SM-225PC5	225W Polycrystalline Module	N	197.7
S-Energy	SM-230MC1	230W Polycrystalline Module	N	207.0
S-Energy	SM-230PC5	230W Polycrystalline Module	N	202.2
S-Energy	SM-235MC1	235W Polycrystalline Module	N	211.6
S-Energy	SM-235PC5	235W Polycrystalline Module	N	206.7
S-Energy	SM-265PC5	265W Polycrystalline Module	N	234.1
S-Energy	SM-270PC5	270W Polycrystalline Module	N	238.7
S-Energy	SM-275PC5	275W Polycrystalline Module	N	243.2
S-Energy	SM-280PC5	280W Polycrystalline Module	N	247.8
S-Energy	SM-285PC5	285W Polycrystalline Module	N	252.3
SET-Solar	SETA-125-150W-72M	150W Monocrystalline Module	N	132.0
SET-Solar	SETA-125-155W-72M	155W Monocrystalline Module	N	136.5
SET-Solar	SETA-125-160W-72M	160W Monocrystalline Module	N	141.0
SET-Solar	SDM-170/(165)-72M	165W Monocrystalline Module	N	145.3
SET-Solar	SETA-125-165W-72M	165W Monocrystalline Module	N	145.5
SET-Solar	SDM-170/(170)-72M	170W Monocrystalline Module	N	149.8
SET-Solar	SETA-125-170W-72M	170W Monocrystalline Module	N	150.1
SET-Solar	SDM-170/(175)-72M	175W Monocrystalline Module	N	154.3
SET-Solar	SETA-125-175W-72M	175W Monocrystalline Module	N	154.6
SET-Solar	SETB-125-175W-72M	175W Monocrystalline Module	N	150.8
SET-Solar	SDM-170/(180)-72M	180W Monocrystalline Module	N	158.9
SET-Solar	SETA-125-180W-72M	180W Monocrystalline Module	N	159.2
SET-Solar	SETB-125-180W-72M	180W Monocrystalline Module	N	155.3
SET-Solar	SDM-170/(185)-72M	185W Monocrystalline Module	N	163.4

(Continued)

MANUFACTURER NAME	MODULE MODEL NUMBER	DESCRIPTION	BIPV*	PTC**	NOTES
SET-Solar	SETA-125-185W-72M	185W Monocrystalline Module	N	163.7	
SET-Solar	SETB-125-185W-72M	185W Monocrystalline Module	N	159.8	
SET-Solar	SETB-156-190W-54P	190W Polycrystalline Module	N	165.4	
SET-Solar	SETB-156-195W-54P	195W Polycrystalline Module	N	169.9	
SET-Solar	SETB-156-200W-54P	200W Polycrystalline Module	N	174.4	
SET-Solar	SETB-156-205W-54P	205W Polycrystalline Module	N	178.9	
SET-Solar	SETB-156-210W-54P	210W Polycrystalline Module	N	183.4	
Shandong Linuo Photovoltaic Hi-Tech	LN180(36)M-175	175W Monocrystalline Module	N	152.6	
Shandong Linuo Photovoltaic Hi-Tech	LN180(36)M-180	180W Monocrystalline Module	N	157.1	
Shandong Linuo Photovoltaic Hi-Tech	LN180(36)M-185	185W Monocrystalline Module	N	161.7	
Shandong Linuo Photovoltaic Hi-Tech	LN180(36)M-190	190W Monocrystalline Module	N	166.2	
Shandong Linuo Photovoltaic Hi-Tech	LN240(30)P-220	220W Polycrystalline Module	N	191.5	
Shandong Linuo Photovoltaic Hi-Tech	LN240(30)P-225	225W Polycrystalline Module	N	196.0	
Shandong Linuo Photovoltaic Hi-Tech	LN240(30)P-230	230W Polycrystalline Module	N	200.5	
Shandong Linuo Photovoltaic Hi-Tech	LN240(30)P-235	235W Polycrystalline Module	N	205.0	
Shandong Linuo Photovoltaic Hi-Tech	LN240(30)P-240	240W Polycrystalline Module	N	209.5	
Shanghai Alex Solar Energy Science & Technology	ALM-170D-24	170W Monocrystalline Module	N	153.1	

Shanghai Alex Solar Energy Science & Technology	ALM-175D-24	175W Monocrystalline Module	N	157.8
Shanghai Alex Solar Energy Science & Technology	ALM-180D-24	180W Monocrystalline Module	N	162.4
Shanghai Chaori Solar Energy Science & Technology	CRM60S125S	60W Monocrystalline Module	N	53.6
Shanghai Chaori Solar Energy Science & Technology	CRM85S125S	85W Monocrystalline Module	N	76.6
Shanghai Chaori Solar Energy Science & Technology	CRM115S125S	115W Monocrystalline Module	N	102.2
Shanghai Chaori Solar Energy Science & Technology	CRM145S125S	145W Monocrystalline Module	N	126.5
Shanghai Chaori Solar Energy Science & Technology	CRM175S125S	175W Monocrystalline Module	N	155.1
Shanghai Chaori Solar Energy Science & Technology	CRM185S156P-54	185W Polycrystalline Module	N	163.5
Shanghai Chaori Solar Energy Science & Technology	CRM190S156P-54	190W Polycrystalline Module	N	168.1
Shanghai Chaori Solar Energy Science & Technology	CRM195S125S	195W Monocrystsalline Module	N	171.7

(*Continued*)

MANUFACTURER NAME	MODULE MODEL NUMBER	DESCRIPTION	BIPV*	PTC**	NOTES
Shanghai Chaori Solar Energy Science & Technology	CRM195S156P-54	195W Polycrystalline Module	N	172.6	
Shanghai Chaori Solar Energy Science & Technology	CRM200S156P-54	200W Polycrystalline Module	N	176.1	
Shanghai Chaori Solar Energy Science & Technology	CRM200S156P-60	200W Polycrystalline Module	N	180.1	
Shanghai Chaori Solar Energy Science & Technology	CRM205S156P-54	205W Polycrystalline Module	N	180.6	
Shanghai Chaori Solar Energy Science & Technology	CRM205S156P-60	205W Polycrystalline Module	N	184.7	
Shanghai Chaori Solar Energy Science & Technology	CRM210S156P-54	210W Polycrystalline Module	N	185.3	
Shanghai Chaori Solar Energy Science & Technology	CRM210S156P-60	210W Polycrystalline Module	N	189.3	
Shanghai Chaori Solar Energy Science & Technology	CRM215S156P-54	215W Polycrystalline Module	N	189.9	
Shanghai Chaori Solar Energy Science & Technology	CRM215S156P-60	215W Polycrystalline Module	N	193.9	
Shanghai Chaori Solar Energy Science & Technology	CRM220S156P-54	220W Polycrystalline Module	N	194.4	

Shanghai Chaori Solar Energy & Technology	CRM220S156P-60	220W Polycrystalline Module	N	195.8
Shanghai Chaori Solar Energy & Technology	CRM225S156P-54	225W Polycrystalline Module	N	199.0
Shanghai Chaori Solar Energy & Technology	CRM225S156P-60	225W Polycrystalline Module	N	200.3
Shanghai Chaori Solar Energy & Technology	CRM230S156P-54	230W Polycrystalline Module	N	203.6
Shanghai Chaori Solar Energy & Technology	CRM230S156P-60	230W Polycrystalline Module	N	201.4
Shanghai Chaori Solar Energy & Technology	CRM235S125S	235W Monocrystalline Module	N	209.3
Shanghai Chaori Solar Energy & Technology	CRM235S156P-60	235W Polycrystalline Module	N	206.0
Shanghai Chaori Solar Energy & Technology	CRM240S156P-60	240W Polycrystalline Module	N	210.5
Shanghai Chaori Solar Energy & Technology	CRM240S156P-72	240W Polycrystalline Module	N	211.3
Shanghai Chaori Solar Energy & Technology	CRM245S156P-60	245W Polycrystalline Module	N	215.0

(Continued)

MANUFACTURER NAME	MODULE MODEL NUMBER	DESCRIPTION	BIPV*	PTC**	NOTES
Shanghai Chaori Solar Energy Science & Technology	CRM245S156P-72	245W Polycrystalline Module	N	215.8	
Shanghai Chaori Solar Energy Science & Technology	CRM250S156P-60	250W Polycrystalline Module	N	224.8	
Shanghai Chaori Solar Energy Science & Technology	CRM250S156P-72	250W Polycrystalline Module	N	220.4	
Shanghai Chaori Solar Energy Science & Technology	CRM255S156P-60	255W Polycrystalline Module	N	229.4	
Shanghai Chaori Solar Energy Science & Technology	CRM255S156P-72	255W Polycrystalline Module	N	224.9	
Shanghai Chaori Solar Energy Science & Technology	CRM260S156P-60	260W Polycrystalline Module	N	234.1	
Shanghai Chaori Solar Energy Science & Technology	CRM260S156P-72	260W Polycrystalline Module	N	230.5	
Shanghai Chaori Solar Energy Science & Technology	CRM265S156P-72	265W Polycrystalline Module	N	235.1	
Shanghai Chaori Solar Energy Science & Technology	CRM270S156P-72	270W Polycrystalline Module	N	239.6	
Shanghai Chaori Solar Energy Science & Technology	CRM275S156P-72	275W Polycrystalline Module	N	244.2	

Shanghai Chaori Solar Energy Science & Technology	CRM280S156P-72	280W Polycrystalline Module	N	248.8
Shanghai Chaori Solar Energy Science & Technology	CRM285S156P-72	285W Polycrystalline Module	N	253.4
Shanghai Chaori Solar Energy Science & Technology	CRM290S156P-72	290W Polycrystalline Module	N	257.9
Shanghai Chaori Solar Energy Science & Technology	CRM295S156P-72	295W Polycrystalline Module	N	262.5
Shanghai Chaori Solar Energy Science & Technology	CRM300S156P-72	300W Polycrystalline Module	N	267.1
Shanghai Jinglong Solar Energy	JLM160S-I	160W Monocrystalline Module	N	143.0
Shanghai Jinglong Solar Energy	JLM165S-I	165W Monocrystalline Module	N	147.6
Shanghai Jinglong Solar Energy	JLM170S-I	170W Monocrystalline Module	N	152.2
Shanghai Jinglong Solar Energy	JLM175S-I	175W Monocrystalline Module	N	156.8
Shanghai Jinglong Solar Energy	JLM180S-I	180W Monocrystalline Module	N	161.4
Shanghai Jinglong Solar Energy	JLM185S-I	185W Monocrystalline Module	N	166.0
Shanghai Pubsolar	GYS-180C	180W Monocrystalline Module	N	162.9
Shanghai Pubsolar	GYS-185C	185W Monocrystalline Module	N	167.6
Shanghai Pubsolar	GYS-190C	190W Monocrystalline Module	N	172.2

(Continued)

MANUFACTURER NAME	MODULE MODEL NUMBER	DESCRIPTION	BIPV*	PTC**	NOTES
Shanghai Solar Energy S&T	S-75D-1	75W Polycrystalline Module	N	64.1	
Shanghai Solar Energy S&T	S-90C-1	90W Monocrystalline Module	N	78.5	
Shanghai Solar Energy S&T	S-125D-1	125W Polycrystalline Module	N	105.5	
Shanghai Solar Energy S&T	S-165D-1	165W Polycrystalline Module	N	139.3	
Shanghai Solar Energy S&T	S-165D-A-1	165W Polycrystalline Module	N	146.3	
Shanghai Solar Energy S&T	S-175C	175W Monocrystalline Module	N	155.2	
Shanghai Solar Energy S&T	S-180C	180W Monocrystalline Module	N	159.7	
Shanghai Solar Energy S&T	S-185C	185W Monocrystalline Module	N	164.3	
Shanghai Solar Energy S&T	S-205C	205W Monocrystalline Module	N	182.6	
Shanghai Solar Energy S&T	S-215C	215W Monocrystalline Module	N	191.7	
Shanghai Solar Energy S&T	S-220D	220W Polycrystalline Module	N	196.7	
Shanghai Solar Energy S&T	S-225C	225W Monocrystalline Module	N	200.9	
Shanghai Solar Energy S&T	S-235C	235W Monocrystalline Module	N	210.1	
Shanghai Solar Energy S&T	S-245C	245W Monocrystalline Module	N	219.3	

Shanghai Topsolar Green Energy	TSM72-125M 160W	160W Monocrystalline Module	N	143.9
Shanghai Topsolar Green Energy	TSM72-125M 165W	165W Monocrystalline Module	N	148.5
Shanghai Topsolar Green Energy	TSM72-125M 170W	170W Monocrystalline Module	N	153.1
Shanghai Topsolar Green Energy	TSM72-125M 175W	175W Monocrystalline Module	N	157.8
Shanghai Topsolar Green Energy	TSM72-125M 180W	180W Monocrystalline Module	N	162.4
Sharp	ND-62RU1	62W Polycrystalline Roof Module	Y	51.2
Sharp	ND-62RU1F	62W Polycrystalline Roof Module	Y	51.2
Sharp	ND-62RU2	62W Polycrystalline Roof Module	Y	51.2
Sharp	ND-62RUC1	62W Polycrystalline Roof Module, Locking Connector	Y	51.2
Sharp	ND-65RU1F	65W Polycrystalline Roof Module	Y	53.0
Sharp	ND-65RUC1	65W Polycrystalline Roof Module, Locking Connector	Y	53.0
Sharp	ND-72ELU	72W Polycrystalline Triangle Module, Left Side	N	63.2
Sharp	ND-72ELUC	72W Polycrystalline Triangle Module, Left Side, Locking Connector	N	63.2
Sharp	ND-72ELUF	72W Polycrystalline Triangle Module, Left Side	N	63.2

(Continued)

MANUFACTURER NAME	MODULE MODEL NUMBER	DESCRIPTION	BIPV*	PTC**	NOTES
Sharp	ND-72ERU	72W Polycrystalline Triangle Module, Right Side	N	63.2	
Sharp	ND-72ERUC	72W Polycrystalline Triangle Module, Right Side, Locking Connector	N	63.2	
Sharp	ND-72ERUF	72W Polycrystalline Triangle Module, Right Side	N	63.2	
Sharp	ND-V075CL	75W Polycrystalline Triangle Module, Left Side, Locking Connector	N	65.9	
Sharp	ND-V075CR	75W Polycrystalline Triangle Module, Right Side, Locking Connector	N	65.9	
Sharp	NA-V115H1	115W Thin Film Module, Locking Connector	N	108.4	
Sharp	NA-V121H1	121W Thin Film Module, Locking Connector	N	114.1	
Sharp	ND-123UJF	123W Polycrystalline Module	N	108.2	
Sharp	NA-V128H1	128W Thin Film Module, Locking Connector	N	119.2	
Sharp	ND-130UJF	130W Polycrystalline Module	N	113.8	
Sharp	NA-V135H1	135W Thin Film Module, Locking Connector	N	125.8	
Sharp	ND-N2ECU	142W Polycrystalline Module	N	124.1	
Sharp	ND-N2ECUC	142W Polycrystalline Module, Locking Connector	N	124.1	
Sharp	ND-N2ECUF	142W Polycrystalline Module	N	124.1	
Sharp	NE-160U1	160W Polycrystalline Module, Flat Screw	N	140.1	

Sharp	ND-162U1F	162W Polycrystalline Module	N	142.8
Sharp	ND-162U1Y	162W Polycrystalline Module	N	142.8
Sharp	ND-162U2	162W Polycrystalline Module	N	142.8
Sharp	NE-165U1	165W Polycrystalline Module, Flat Screw	N	144.6
Sharp	NE-165U5	165W Polycrystalline Module	N	144.6
Sharp	NE-165UC1	165W Polycrystalline Module, Locking Connector	N	144.6
Sharp	ND-167U1	167W Polycrystalline Module	N	147.3
Sharp	ND-167U1F	167W Polycrystalline Module	N	147.3
Sharp	ND-167U1Y	167W Polycrystalline Module	N	147.3
Sharp	ND-167U2	167W Polycrystalline Module	N	147.3
Sharp	ND-167U3A	167W Polycrystalline Module	N	147.3
Sharp	ND-167UC1	167W Polycrystalline Module, Locking Connector	N	147.3
Sharp	NE-170U1	170W Polycrystalline Module	N	149.1
Sharp	NE-170UC1	170W Polycrystalline Module, Locking Connector	N	149.1
Sharp	NT-170U1	170W Monocrystalline Module	N	147.3
Sharp	NT-170UC1	170W Monocrystalline Module, Locking Connector	N	147.3
Sharp	NT-175U1	175W Monocrystalline Module	N	151.8
Sharp	NT-175UC1	175W Monocrystalline Module, Locking Connector	N	151.8
Sharp	ND-176U1Y	176W Polycrystalline Module	N	152.4
Sharp	ND-176UC1	176W Polycrystalline Module, Locking Connector	N	152.4

(Continued)

MANUFACTURER NAME	MODULE MODEL NUMBER	DESCRIPTION	BIPV*	PTC**	NOTES
Sharp	NT-180U1	180W Monocrystalline Module	N	156.3	
Sharp	NU-U180FC	180W Monocrystalline Module, Locking Connector	N	161.9	
Sharp	ND-181U1	181W Polycrystalline Module	N	159.3	
Sharp	ND-181U1F	181W Polycrystalline Module	N	159.3	
Sharp	ND-181U2	181W Polycrystalline Module	N	159.3	
Sharp	ND-187U1	187W Polycrystalline Module	N	164.7	
Sharp	ND-187U1F	187W Polycrystalline Module	N	164.7	
Sharp	ND-187U2	187W Polycrystalline Module	N	164.7	
Sharp	ND-187UC1	187W Polycrystalline Module, Locking Connector	N	164.7	
Sharp	ND-198U1F	198W Polycrystalline Module	N	170.5	
Sharp	ND-198UC1	198W Polycrystalline Module, Locking Connector	N	170.5	
Sharp	ND-200U1	200W Polycrystalline Module	N	173.0	
Sharp	ND-200U1F	200W Polycrystalline Module	N	173.0	
Sharp	ND-200U2	200W Polycrystalline Module	N	173.0	
Sharp	ND-200UC1	200W Polycrystalline Module, Locking Connector	N	173.0	
Sharp	ND-208U1	208W Polycrystalline Module	N	180.1	
Sharp	ND-208U1F	208W Polycrystalline Module	N	180.1	
Sharp	ND-208U2	208W Polycrystalline Module	N	180.1	
Sharp	ND-208UC1	208W Polycrystalline Module, Locking Connector	N	180.1	
Sharp	NU-U208FC	208W Monocrystalline Module, Locking Connector	N	187.2	
Sharp	ND-216U1F	216W Polycrystalline Module	N	190.4	

Sharp	ND-216U2	216W Polycrystalline Module	N	187.3
Sharp	ND-216UC1	216W Polycrystalline Module, Locking Connector	N	190.4
Sharp	ND-U216C1	216W Polycrystalline Module, Locking Connector	N	190.4
Sharp	ND-220U1F	220W Polycrystalline Module	N	194.0
Sharp	ND-220UC1	220W Polycrystalline Module, Locking Connector	N	194.0
Sharp	ND-224U1F	224W Polycrystalline Module	N	197.6
Sharp	ND-224UC1	224W Polycrystalline Module, Locking Connector	N	197.6
Sharp	ND-U224C1	224W Polycrystalline Module, Locking Connector	N	197.6
Sharp	ND-U230C1	230W Polycrystalline Module	N	203.1
Sharp	ND-V230A1	230W Polycrystalline Module	N	203.1
Sharp	NU-U230F3	230W Monocrystalline Module, Locking Connector, Black Frame	N	207.1
Sharp	NU-U235F1	235W Monocrystalline Module, Locking Connector, Silver Frame	N	211.7
Sharp	NU-U235F3	235W Monocrystalline Module, Locking Connector, Black Frame	N	211.7
Sharp	NU-U240F1	240W Monocrystalline Module, Locking Connector, Silver Frame	N	216.3
Shenzhen Zhongjing Solar	PWM-170W	170W Monocrystalline Module	N	147.3

(Continued)

MANUFACTURER NAME	MODULE MODEL NUMBER	DESCRIPTION	BIPV*	PTC***	NOTES
Shenzhen Zhongjing Solar	PWM-175W	175W Monocrystalline Module	N	151.8	
Shenzhen Zhongjing Solar	PWM-180W	180W Monocrystalline Module	N	156.2	
Signet Solar	S4	86W Thin Film Single Junction Amorphous Silicon Module	N	80.8	
Signet Solar	S2	172W Thin Film Single Junction Amorphous Silicon Module	N	161.6	
Signet Solar	S3	172W Thin Film Single Junction Amorphous Silicon Module	N	161.6	
Signet Solar	S1	343W Thin Film Single Junction Amorphous Silicon Module	N	322.3	
Silicon Energy	SiE-165	165W Glass-Glass Polycrystalline Module	N	142.8	
Silicon Energy	SiE-170	170W Glass-Glass Polycrystalline Module	N	147.3	
Silicon Energy	SiE-175	175W Glass-Glass Polycrystalline Module	N	151.7	
Silicon Energy	SiE-180	180W Glass-Glass Polycrystalline Module	N	156.2	
Siliken	SLK60P6L BLK/WHT 205Wp	205W Polycrystalline Module	N	182.7	
Siliken	SLK60P6L SLV/WHT 205Wp	205W Polycrystalline Module, Silver Frame, White Backsheet	N	182.7	
Siliken	SLK60P6L BLK/WHT 210Wp	210W Polycrystalline Module	N	187.2	

Siliken	SLK60P6L SLV/WHT 210Wp	210W Polycrystalline Module, Silver Frame, White Backsheet	N	187.2
Siliken	SLK60P6L BLK/BLK 215Wp	215W Polycrystalline Module, Black Frame, Black Backsheet	N	191.8
Siliken	SLK60P6L BLK/WHT 215Wp	215W Polycrystalline Module	N	191.8
Siliken	SLK60P6L SLV/WHT 215Wp	215W Polycrystalline Module, Silver Frame, White Backsheet	N	192.4
Siliken	SLK60P6L BLK/BLK 220Wp	220W Polycrystalline Module, Black Frame, Black Backsheet	N	196.4
Siliken	SLK60P6L BLK/WHT 220Wp	220W Polycrystalline Module	N	196.4
Siliken	SLK60P6L SLV/WHT 220Wp	220W Polycrystalline Module, Silver Frame, White Backsheet	N	197.0
Siliken	SLK60P6L BLK/BLK 225Wp	225W Polycrystalline Module, Black Frame, Black Backsheet	N	199.8
Siliken	SLK60P6L BLK/WHT 225Wp	225W Polycrystalline Module	N	200.9
Siliken	SLK60P6L SLV/WHT 225Wp	225W Polycrystalline Module, Silver Frame, White Backsheet	N	201.5
Siliken	SLK60P6L BLK/BLK 230Wp	230W Polycrystalline Module, Black Frame, Black Backsheet	N	204.4
Siliken	SLK60P6L BLK/WHT 230Wp	230W Polycrystalline Module	N	202.0
Siliken	SLK60P6L SLV/WHT 230Wp	230W Polycrystalline Module, Silver Frame, White Backsheet	N	206.1
Siliken	SLK60P6L BLK/BLK 235Wp	235W Polycrystalline Module, Black Frame, Black Backsheet	N	209.0
Siliken	SLK60P6L BLK/WHT 235Wp	235W Polycrystalline Module	N	206.6

(Continued)

MANUFACTURER NAME	MODULE MODEL NUMBER	DESCRIPTION	BIPV*	PTC**	NOTES
Siliken	SLK60P6L SLV/WHT 235Wp	235W Polycrystalline Module, Silver Frame, White Backsheet	N	210.7	
Siliken	SLK60P6L BLK/WHT 240Wp	240W Polycrystalline Module	N	211.1	
Siliken	SLK60P6L SLV/WHT 240Wp	240W Polycrystalline Module, Silver Frame, White Backsheet	N	211.1	
Siliken California	SLK60P6L BLK/WHT 205Wp	205W Polycrystalline Module	N	182.7	
Siliken California	SLK60P6L SLV/WHT 205Wp	205W Polycrystalline Module, Silver Frame, White Backsheet	N	182.7	
Siliken California	SLK60P6L BLK/WHT 210Wp	210W Polycrystalline Module	N	187.2	
Siliken California	SLK60P6L SLV/WHT 210Wp	210W Polycrystalline Module, Silver Frame, White Backsheet	N	187.2	
Siliken California	SLK60P6L BLK/BLK 215Wp	215W Polycrystalline Module, Black Frame, Black Backsheet	N	191.8	
Siliken California	SLK60P6L BLK/WHT 215Wp	215W Polycrystalline Module	N	191.8	
Siliken California	SLK60P6L SLV/WHT 215Wp	215W Polycrystalline Module, Silver Frame, White Backsheet	N	192.4	
Siliken California	SLK60P6L BLK/BLK 220Wp	220W Polycrystalline Module, Black Frame, Black Backsheet	N	196.4	
Siliken California	SLK60P6L BLK/WHT 220Wp	220W Polycrystalline Module	N	196.4	
Siliken California	SLK60P6L SLV/WHT 220Wp	220W Polycrystalline Module, Silver Frame, White Backsheet	N	197.0	
Siliken California	SLK60P6L BLK/BLK 225Wp	225W Polycrystalline Module, Black Frame, Black Backsheet	N	199.8	

Manufacturer	Model	Description		
Siliken California	SLK60P6L BLK/WHT 225Wp	225W Polycrystalline Module	N	200.9
Siliken California	SLK60P6L SLV/WHT 225Wp	225W Polycrystalline Module, Silver Frame, White Backsheet	N	201.5
Siliken California	SLK60P6L BLK/BLK 230Wp	230W Polycrystalline Module, Black Frame, Black Backsheet	N	204.4
Siliken California	SLK60P6L BLK/WHT 230Wp	230W Polycrystalline Module	N	202.0
Siliken California	SLK60P6L SLV/WHT 230Wp	230W Polycrystalline Module, Silver Frame, White Backsheet	N	206.1
Siliken California	SLK60P6L BLK/BLK 235Wp	235W Polycrystalline Module, Black Frame, Black Backsheet	N	209.0
Siliken California	SLK60P6L BLK/WHT 235Wp	235W Polycrystalline Module	N	206.6
Siliken California	SLK60P6L SLV/WHT 235Wp	235W Polycrystalline Module, Silver Frame, White Backsheet	N	210.7
Siliken California	SLK60P6L BLK/WHT 240Wp	240W Polycrystalline Module	N	211.1
Siliken California	SLK60P6L SLV/WHT 240Wp	240W Polycrystalline Module, Silver Frame, White Backsheet	N	211.1
Silray	SR-180	180W Monocrystalline Module	N	159.7
Sintek Photronic	STK-S6C2PE-A	175W Polycrystalline Module	N	155.9
Sintek Photronic	STK-S6C1ME-A	180W Monocrystalline Module	N	159.7
Sintek Photronic	STK-S6B1PE-A	200W Polycrystalline Module	N	176.9
Sintek Photronic	STK-S6B1ME-A	205W Monocrystalline Module	N	183.6
Solar Enertech (Shanghai)	SE175-72M	175W Monocrystalline Module	N	150.8
Solar Enertech (Shanghai)	SE180-72M	180W Monocrystalline Module	N	155.3

(Continued)

MANUFACTURER NAME	MODULE MODEL NUMBER	DESCRIPTION	BIPV*	PTC**	NOTES
Solar Enertech (Shanghai)	SE185-72M	185W Monocrystalline Module	N	159.8	
Solar Enertech (Shanghai)	SE195-54P	195W Polycrystalline Module	N	169.9	
Solar Enertech (Shanghai)	SE200-54P	200W Polycrystalline Module	N	174.4	
Solar Enertech (Shanghai)	SE205-54P	205W Polycrystalline Module	N	178.9	
Solar Enertech (Shanghai)	SE210-54P	210W Polycrystalline Module	N	183.4	
Solar Frontier	SC70-US-P	70W Thin Film Module	N	60.6	
Solar Frontier	SF70-US-B	70W Thin Film Module	N	60.6	
Solar Frontier	SF70-US-P	70W Thin Film Module	N	60.6	
Solar Frontier	SC75-US-P	75W Thin Film Module	N	64.9	
Solar Frontier	SF75-US-B	75W Thin Film Module	N	64.9	
Solar Frontier	SF75-US-P	75W Thin Film Module	N	64.9	
Solar Frontier	SC80-US-P	80W Thin Film Module	N	71.1	
Solar Frontier	SF80-US-B	80W Thin Film Module	N	71.1	
Solar Frontier	SF80-US-P	80W Thin Film Module	N	71.1	
Solar Frontier	SF85-US-B	85W Thin Film Module	N	76.7	
Solar Frontier	SF85-US-P	85W Thin Film Module	N	76.7	
Solar Integrated Technologies	SI-G1 288	288W Thin Film Roofing Membrane	Y	260.0	
Solar Integrated Technologies	SI-T1 288	288W Thin Film Roofing Membrane	Y	260.0	
Solar Integrated Technologies	SI-G1 576	576W Thin Film Roofing Membrane	Y	519.9	

Manufacturer	Model	Description		
Solar Integrated Technologies	SI-T1 576	576W Thin Film Roofing Membrane	Y	519.9
Solar Integrated Technologies	SI-G1 720	720W Thin Film Roofing Membrane	Y	649.9
Solar Integrated Technologies	SI-G1 864	864W Thin Film Roofing Membrane	Y	779.8
Solar Power (SPI)	SP165FM52	165W Monocrystalline Module	N	149.8
Solar Power (SPI)	SP170FM12	170W Monocrystalline Module	N	152.0
Solar Power (SPI)	SP170FM22	170W Monocrystalline Module	N	146.6
Solar Power (SPI)	SP170FM52	170W Monocrystalline Module	N	154.5
Solar Power (SPI)	SP175FM12	175W Monocrystalline Module	N	156.6
Solar Power (SPI)	SP175FM52	175W Monocrystalline Module	N	159.1
Solar Power (SPI)	SP180FM12	180W Monocrystalline Module	N	161.2
Solar Power (SPI)	SP180FM52	180W Monocrystalline Module	N	163.7
Solar Power (SPI)	SP190FM52	190W Monocrystalline Module	N	171.8
Solar Power (SPI)	SP190FP12	190W Polycrystalline Module	N	173.8
Solar Power (SPI)	SP190FP52	190W Polycrystalline Module	N	170.4
Solar Power (SPI)	SP195FM12	195W Monocrystalline Module	N	173.2
Solar Power (SPI)	SP195FM52	195W Monocrystalline Module	N	176.4
Solar Power (SPI)	SP195FP12	195W Polycrystalline Module	N	178.5
Solar Power (SPI)	SP195FP42	195W Polycrystalline Module	N	175.4
Solar Power (SPI)	SP195FP52	195W Polycrystalline Module	N	175.0
Solar Power (SPI)	SP195FP82	195W Polycrystalline Module	N	172.0
Solar Power (SPI)	SP200FM12	200W Monocrystalline Module	N	177.8
Solar Power (SPI)	SP200FM52	200W Monocrystalline Module	N	181.0
Solar Power (SPI)	SP200FP12	200W Polycrystalline Module	N	183.1
Solar Power (SPI)	SP200FP42	200W Polycrystalline Module	N	180.0

(Continued)

MANUFACTURER NAME	MODULE MODEL NUMBER	DESCRIPTION	BIPV*	PTC**	NOTES
Solar Power (SPI)	SP200FP52	200W Polycrystalline Module	N	179.6	
Solar Power (SPI)	SP200FP82	200W Polycrystalline Module	N	176.5	
Solar Power (SPI)	SP200FPA2	200W Polycrystalline Module	N	177.8	
Solar Power (SPI)	SP205FM12	205W Monocrystalline Module	N	182.4	
Solar Power (SPI)	SP205FM52	205W Monocrystalline Module	N	185.7	
Solar Power (SPI)	SP205FP12	205W Polycrystalline Module	N	187.8	
Solar Power (SPI)	SP205FP42	205W Polycrystalline Module	N	184.6	
Solar Power (SPI)	SP205FP52	205W Polycrystalline Module	N	184.2	
Solar Power (SPI)	SP205FP82	205W Polycrystalline Module	N	181.1	
Solar Power (SPI)	SP205FPA2	205W Polycrystalline Module	N	182.4	
Solar Power (SPI)	SP210FM12	210W Monocrystalline Module	N	187.0	
Solar Power (SPI)	SP210FM52	210W Monocrystalline Module	N	190.3	
Solar Power (SPI)	SP210FP12	210W Polycrystalline Module	N	192.5	
Solar Power (SPI)	SP210FP42	210W Polycrystalline Module	N	189.2	
Solar Power (SPI)	SP210FP52	210W Polycrystalline Module	N	188.8	
Solar Power (SPI)	SP210FP82	210W Polycrystalline Module	N	185.6	
Solar Power (SPI)	SP210FPA2	210W Polycrystalline Module	N	187.0	
Solar Power (SPI)	SP220FP12	220W Polycrystalline Module	N	197.4	
Solar Power (SPI)	SP220FP52	220W Polycrystalline Module	N	198.3	
Solar Power (SPI)	SP220FP82	220W Polycrystalline Module	N	197.3	
Solar Power (SPI)	SP220FPA2	220W Polycrystalline Module	N	195.2	
Solar Power (SPI)	SP225FP12	225W Polycrystalline Module	N	202.0	
Solar Power (SPI)	SP225FP52	225W Polycrystalline Module	N	202.9	
Solar Power (SPI)	SP225FP82	225W Polycrystalline Module	N	201.9	
Solar Power (SPI)	SP225FPA2	225W Polycrystalline Module	N	199.8	

Solar Power (SPI)	SP230FP12	230W Polycrystalline Module	N	206.6
Solar Power (SPI)	SP230FP52	230W Polycrystalline Module	N	207.5
Solar Power (SPI)	SP230FP82	230W Polycrystalline Module	N	206.5
Solar Power (SPI)	SP230FPA2	230W Polycrystalline Module	N	204.4
Solar Power Industries	SPI-M160-60	160W Polycrystalline Module	N	141.0
Solar Power Industries	SPI-M170-60	170W Polycrystalline Module	N	150.0
Solar Power Industries	SPI-M180-60	180W Polycrystalline Module	N	159.0
Solar Power Industries	SPI-M190-60	190W Polycrystalline Module	N	168.1
Solar Power Industries	SPI-156-200W	200W Polycrystalline Module	N	180.2
Solar Power Industries	SPI-M200-60	200W Polycrystalline Module	N	177.1
Solar Power Industries	SPI-M210-60	210W Polycrystalline Module	N	192.3
Solar Power Industries	SPI-M215-60	215W Polycrystalline Module	N	197.0
Solar Power Industries	SPI-M220-60	220W Polycrystalline Module	N	201.6
Solar Power Industries	SPI-M225-60	225W Polycrystalline Module	N	206.3
Solar Power Industries	SPI-M230-60	230W Polycrystalline Module	N	211.0
Solar Semiconductor	SSI-S6-125	125W Monocrystalline Module	N	106.3
Solar Semiconductor	SSI-S6-130	130W Monocrystalline Module	N	110.7
Solar Semiconductor	SSI-S6-135	135W Monocrystalline Module	N	115.1
Solar Semiconductor	SSI-S6-140	140W Monocrystalline Module	N	119.5
Solar Semiconductor	SSI-S6-145	145W Monocrystalline Module	N	124.0
Solar Semiconductor	SSI-S6-170	170W Monocrystalline Module	N	145.5
Solar Semiconductor	SSI-S6-175	175W Monocrystalline Module	N	149.9
Solar Semiconductor	SSI-S6-180	180W Monocrystalline Module	N	154.3
Solar Semiconductor	SSI-S6-185	185W Monocrystalline Module	N	158.8
Solar Semiconductor	SSI-S6-190	190W Monocrystalline Module	N	166.5
Solar Semiconductor	SSI-S6-195	195W Monocrystalline Module	N	171.0

(Continued)

MANUFACTURER NAME	MODULE MODEL NUMBER	DESCRIPTION	BIPV*	PTC**	NOTES
Solar Semiconductor	SSI-S6-200	200W Monocrystalline Module	N	175.5	
Solar Semiconductor	SSI-S6-205	205W Monocrystalline Module	N	180.0	
Solar Semiconductor	SSI-S6-210	210W Monocrystalline Module	N	184.5	
Solar Semiconductor	SSI-3M6-215	215W Polycrystalline Module	N	193.5	
Solar Semiconductor	SSI-M6-215	215W Polycrystalline Module	N	190.2	
Solar Semiconductor	SSI-S6-215	215W Monocrystalline Module	N	185.1	
Solar Semiconductor	SSI-3M6-220	220W Polycrystalline Module	N	198.1	
Solar Semiconductor	SSI-M6-220	220W Polycrystalline Module	N	194.8	
Solar Semiconductor	SSI-S6-220	220W Monocrystalline Module	N	189.6	
Solar Semiconductor	SSI-3M6-225	225W Polycrystalline Module	N	202.7	
Solar Semiconductor	SSI-M6-225	225W Polycrystalline Module	N	199.3	
Solar Semiconductor	SSI-S6-225	225W Monocrystalline Module	N	194.1	
Solar Semiconductor	SSI-3M6-230	230W Polycrystalline Module	N	207.3	
Solar Semiconductor	SSI-M6-230	230W Polycrystalline Module	N	203.9	
Solar Semiconductor	SSI-S6-230	230W Monocrystalline Module	N	198.5	
Solar Semiconductor	SSI-3M6-235	235W Polycrystalline Module	N	212.0	
Solar Semiconductor	SSI-M6-235	235W Polycrystalline Module	N	208.4	
Solar Semiconductor	SSI-S6-235	235W Monocrystalline Module	N	203.0	
Solar Semiconductor	SSI-S6-260	260W Monocrystalline Module	N	217.1	
Solar Semiconductor	SSI-S6-265	265W Monocrystalline Module	N	221.5	
Solar Semiconductor	SSI-S6-270	270W Monocrystalline Module	N	225.8	
Solar Semiconductor	SSI-S6-275	275W Monocrystalline Module	N	230.2	
Solar Semiconductor	SSI-S6-280	280W Monocrystalline Module	N	234.6	
Solarfun Power	SF160-24-M150	150W Monocrystalline Module	N	133.6	
Solarfun Power	SF160-24-M155	155W Monocrystalline Module	N	138.1	

Solarfun Power	SF160-24-M160	160W Monocrystalline Module	N	142.7
Solarfun Power	SF160-24-M160B	160W Monocrystalline Module, Black Backsheet	N	144.1
Solarfun Power	SF160-24-M165	165W Monocrystalline Module	N	147.2
Solarfun Power	SF160-24-M165B	165W Monocrystalline Module, Black Backsheet	N	148.7
Solarfun Power	SF160-24-P165	165W Polycrystalline Module	N	145.5
Solarfun Power	SF160-24-M170	170W Monocrystalline Module	N	151.8
Solarfun Power	SF160-24-M170B	170W Monocrystalline Module, Black Backsheet	N	153.3
Solarfun Power	SF160-24-P170	170W Polycrystalline Module	N	150.0
Solarfun Power	SF190-27-P170	170W Polycrystalline Module	N	150.4
Solarfun Power	SF160-24-M175	175W Monocrystalline Module	N	156.4
Solarfun Power	SF160-24-M175B	175W Monocrystalline Module, Black Backsheet	N	158.0
Solarfun Power	SF160-24-P175	175W Polycrystalline Module	N	154.5
Solarfun Power	SF190-27-P175	175W Polycrystalline Module	N	155.0
Solarfun Power	SF160-24-M180	180W Monocrystalline Module	N	161.0
Solarfun Power	SF160-24-M180B	180W Monocrystalline Module, Black Backsheet	N	162.6
Solarfun Power	SF190-27-M180B	180W Monocrystalline Module, Black Backsheet	N	159.4
Solarfun Power	SF190-27-P180	180W Polycrystalline Module	N	159.5
Solarfun Power	SF160-24-M185	185W Monocrystalline Module	N	165.6
Solarfun Power	SF160-24-M185B	185W Monocrystalline Module, Black Backsheet	N	167.2
Solarfun Power	SF190-27-M185B	185W Monocrystalline Module, Black Backsheet	N	164.0

(Continued)

MANUFACTURER NAME	MODULE MODEL NUMBER	DESCRIPTION	BIPV*	PTC**	NOTES
Solarfun Power	SF190-27-P185	185W Polycrystalline Module	N	164.0	
Solarfun Power	SF190-27-M190B	190W Monocrystalline Module, Black Backsheet	N	168.5	
Solarfun Power	SF190-27-P190	190W Polycrystalline Module	N	168.6	
Solarfun Power	SF190-27-M195B	195W Monocrystalline Module, Black Backsheet	N	173.1	
Solarfun Power	SF190-27-P195	195W Polycrystalline Module	N	173.1	
Solarfun Power	SF190-27-M200B	200W Monocrystalline Module, Black Backsheet	N	177.6	
Solarfun Power	SF190-27-P200	200W Polycrystalline Module	N	177.7	
Solarfun Power	SF220-30-M200	200W Monocrystalline Module	N	179.9	
Solarfun Power	SF220-30-M200B	200W Monocrystalline Module	N	178.9	
Solarfun Power	SF220-30-P200	200W Polycrystalline Module	N	180.0	
Solarfun Power	SF220-30-P200B	200W Polycrystalline Module	N	178.9	
Solarfun Power	SF190-27-M205B	205W Monocrystalline Module, Black Backsheet	N	182.2	
Solarfun Power	SF190-27-P205	205W Polycrystalline Module	N	182.3	
Solarfun Power	SF220-30-M205	205W Monocrystalline Module	N	184.5	
Solarfun Power	SF220-30-M205B	205W Monocrystalline Module	N	183.5	
Solarfun Power	SF220-30-P205	205W Polycrystalline Module	N	184.6	
Solarfun Power	SF220-30-P205B	205W Polycrystalline Module	N	183.4	
Solarfun Power	SF190-27-M210B	210W Monocrystalline Module, Black Backsheet	N	186.7	
Solarfun Power	SF190-27-P210	210W Polycrystalline Module	N	186.8	
Solarfun Power	SF220-30-M210	210W Monocrystalline Module	N	189.1	
Solarfun Power	SF220-30-M210B	210W Monocrystalline Module	N	188.1	
Solarfun Power	SF220-30-P210	210W Polycrystalline Module	N	189.2	

Solarfun Power	SF220-30-P210B	210W Polycrystalline Module	N	188.0
Solarfun Power	SF220-30-M215	215W Monocrystalline Module	N	193.7
Solarfun Power	SF220-30-M215B	215W Monocrystalline Module	N	192.6
Solarfun Power	SF220-30-P215	215W Polycrystalline Module	N	193.8
Solarfun Power	SF220-30-P215B	215W Polycrystalline Module	N	192.6
Solarfun Power	SF220-30-M220	220W Monocrystalline Module	N	198.3
Solarfun Power	SF220-30-M220B	220W Monocrystalline Module	N	197.2
Solarfun Power	SF220-30-P220	220W Polycrystalline Module	N	198.4
Solarfun Power	SF220-30-P220B	220W Polycrystalline Module	N	197.2
Solarfun Power	SF220-30-M225	225W Monocrystalline Module	N	202.9
Solarfun Power	SF220-30-M225B	225W Monocrystalline Module	N	201.8
Solarfun Power	SF220-30-P225	225W Polycrystalline Module	N	203.0
Solarfun Power	SF220-30-P225B	225W Polycrystalline Module	N	201.8
Solarfun Power	SF220-30-M230	230W Monocrystalline Module	N	207.5
Solarfun Power	SF220-30-M230B	230W Monocrystalline Module	N	206.4
Solarfun Power	SF220-30-P230	230W Polycrystalline Module	N	207.6
Solarfun Power	SF220-30-P230B	230W Polycrystalline Module	N	206.4
Solarfun Power	SF220-30-M235	235W Monocrystalline Module	N	212.2
Solarfun Power	SF220-30-M235B	235W Monocrystalline Module	N	211.0
Solarfun Power	SF220-30-P235	235W Polycrystalline Module	N	212.2
Solarfun Power	SF220-30-P235B	235W Polycrystalline Module	N	211.0
Solarfun Power	SF220-30-M240	240W Monocrystalline Module	N	216.8
Solarfun Power	SF220-30-M240B	240W Monocrystalline Module	N	215.6
Solarfun Power	SF220-30-P240	240W Polycrystalline Module	N	216.9
Solarfun Power	SF220-30-P240B	240W Polycrystalline Module	N	215.6

(Continued)

MANUFACTURER NAME	MODULE MODEL NUMBER	DESCRIPTION	BIPV*	PTC***	NOTES
Solarfun Power	SF220-30-M245	245W Monocrystalline Module	N	221.4	
Solarfun Power	SF220-30-M245B	245W Monocrystalline Module	N	220.3	
Solarfun Power	SF220-30-P245	245W Polycrystalline Module	N	221.5	
Solarfun Power	SF220-30-P245B	245W Polycrystalline Module	N	220.2	
Solarland USA	SLP115S-17H	115W Monocrystalline Module	N	103.2	
Solarland USA	SLP120S-17H	120W Monocrystalline Module	N	107.8	
Solarland USA	SLP125S-17H	125W Monocrystalline Module	N	112.4	
Solarland USA	SLP160S-24H	160W Monocrystalline Module	N	143.5	
Solarland USA	SLP165S-24H	165W Monocrystalline Module	N	148.0	
Solarland USA	SLP170S-24H	170W Monocrystalline Module	N	152.6	
Solarland USA	SLP175S-24H	175W Monocrystalline Module	N	156.1	
Solarland USA	SLP175S-24J	175W Monocrystalline Module	N	153.4	
Solarland USA	SLP180S-24H	180W Monocrystalline Module	N	160.7	
Solarland USA	SLP180S-24J	180W Monocrystalline Module	N	158.0	
Solarland USA	SLP185S-24H	185W Monocrystalline Module	N	165.3	
Solarland USA	SLP185S-24J	185W Monocrystalline Module	N	162.5	
Solarland USA	SLP200-18H	200W Polycrystalline Module	N	179.8	
Solarland USA	SLP205-18H	205W Polycrystalline Module	N	184.4	
Solarland USA	SLP210-18H	210W Polycrystalline Module	N	189.0	
Solarland USA	SLP215-18H	215W Polycrystalline Module	N	193.6	
Solarland USA	SLP220-18H	220W Polycrystalline Module	N	198.2	
Solartec USA	SDM-170/(165)-72M	165W Monocrystalline Module	N	145.3	
Solartec USA	SDM-170/(170)-72M	170W Monocrystalline Module	N	149.8	
Solartec USA	SDM-170/(175)-72M	175W Monocrystalline Module	N	154.3	
Solartec USA	SDM-170/(180)-72M	180W Monocrystalline Module	N	158.9	

Solartec USA	SDM-170/(185)-72M	185W Monocrystalline Module	N	163.4
Solartech Energy	SEC-210W-M6-G	210W Polycrystalline Module	N	187.4
Solartech Energy	SEC-215W-M6-G	215W Polycrystalline Module	N	192.0
Solartech Energy	SEC-220W-M6-G	220W Polycrystalline Module	N	196.6
Solartech Energy	SEC-225W-M6-G	225W Polycrystalline Module	N	201.1
Solartech Energy	SEC-230W-M6-G	230W Polycrystalline Module	N	205.7
Solartech Power	SPM190P	190W Polycrystalline Module, Framed	N	159.1
Solartech Power	SPM195P	195W Polycrystalline Module, Framed	N	163.5
Solartech Power	SPM200P	200W Polycrystalline Module, Framed	N	167.8
Solartech Power	SPM205P	205W Polycrystalline Module, Framed	N	172.1
Solartech Power	SPM210P	210W Polycrystalline Module, Framed	N	176.5
Solartech Power	SPM220P	220W Polycrystalline Module	N	196.2
Solartech Power	SPM225P	225W Polycrystalline Module	N	200.8
Solartech Power	SPM230P	230W Polycrystalline Module	N	205.4
Solartech Power	SPM235P	235W Polycrystalline Module	N	210.0
Solartech Power	SPM240P	240W Polycrystalline Module	N	214.6
SolarWorld	SW165 mono	165W Monocrystalline Module	N	147.4
SolarWorld	SW175 mono	175W Monocrystalline Module	N	156.6
SolarWorld	SW220 mono	220W Monocrystalline Module	N	196.5
SolarWorld	SW220 mono black	220W Monocrystalline Module	N	195.9
SolarWorld	SW225 mono	225W Monocrystalline Module	N	201.2
SolarWorld	SW225 mono black	225W Monocrystalline Module	N	200.4

(Continued)

MANUFACTURER NAME	MODULE MODEL NUMBER	DESCRIPTION	BIPV*	PTC**	NOTES
SolarWorld	SW230 mono	230W Monocrystalline Module	N	205.8	
SolarWorld	SW230 mono black	230W Monocrystalline Module, Black Backsheet	N	205.0	
SolarWorld	SW235 mono	235W Monocrystalline Module	N	210.4	
SolarWorld	SW240 mono	240W Monocrystalline Module	N	215.0	
SolarWorld	SW245 mono	245W Monocrystalline Module	N	219.6	
SOLON	P220/6+/01 215 Wp	215W Polycrystalline Module	N	189.8	
SOLON	P220/6+/01 220 Wp	220W Polycrystalline Module	N	194.3	
SOLON	M 230/15/01 225WP	225W Monocrystalline Module	N	191.8	
SOLON	P220/6+/01 225 Wp	225W Polycrystalline Module	N	198.9	
SOLON	SOLON Black 230/15/01 225WP	225W Monocrystalline Module	N	191.8	
SOLON	SOLON Black 230/6+/01 225 Wp	225W Monocrystalline Module	N	194.4	
SOLON	M 230/15/01 230WP	230W Monocrystalline Module	N	196.2	
SOLON	P220/6+/01 230 Wp	230W Polycrystalline Module	N	203.4	
SOLON	SOLON Black 230/15/01 230WP	230W Monocrystalline Module	N	196.2	
SOLON	SOLON Black 230/6+/01 230 Wp	230W Monocrystalline Module	N	198.9	
SOLON	M 230/15/01 235WP	235W Monocrystalline Module	N	200.7	
SOLON	P220/6+/01 235 Wp	235W Polycrystalline Module	N	208.0	
SOLON	SOLON Black 230/15/01 235WP	235W Monocrystalline Module	N	200.7	
SOLON	SOLON Black 230/6+/01 235 Wp	235W Monocrystalline Module	N	203.4	
SOLON	M 230/15/01 240WP	240W Monocrystalline Module	N	205.1	

SOLON	SOLON Black 230/15/01 240WP	240W Monocrystalline Module	N	205.1
SOLON	SOLON Black 230/6+/01 240 Wp	240W Monocrystalline Module	N	207.9
SOLON	SOLON Blue 270/09/01 250 Wp	250W Polycrystalline Module	N	220.8
SOLON	SOLON Blue 270/09/01 255 Wp	255W Polycrystalline Module	N	225.4
SOLON	SOLON Blue 270/09/01 260 Wp	260W Polycrystalline Module	N	229.9
SOLON	SOLON Black 280/09/01 265 Wp	265W Monocrystalline Module	N	229.2
SOLON	SOLON Blue 270/09/01 265 Wp	265W Polycrystalline Module	N	234.5
SOLON	SOLON Black 280/09/01 270 Wp	270W Monocrystalline Module	N	233.7
SOLON	SOLON Blue 270/09/01 270 Wp	270W Polycrystalline Module	N	239.0
SOLON	SOLON Black 280/09/01 275 Wp	275W Monocrystalline Module	N	238.2
SOLON	SOLON Blue 270/09/01 275 Wp	275W Polycrystalline Module	N	243.1
SOLON	SOLON Black 280/09/01 280 Wp	280W Monocrystalline Module	N	242.7
SOLON	SOLON Blue 270/09/01 280 Wp	280W Polycrystalline Module	N	247.6
SOLON	SOLON Black 280/09/01 285 Wp	285W Monocrystalline Module	N	247.2
SOLON	SOLON Blue 270/09/01 285 Wp	285W Polycrystalline Module	N	252.2

(Continued)

MANUFACTURER NAME	MODULE MODEL NUMBER	DESCRIPTION	BIPV*	PTC**	NOTES
SOLON	SOLON Black 280/09/01 290 Wp	290W Monocrystalline Module	N	250.3	
SOLON	SOLON Blue 270/09/01 290 Wp	290W Polycrystalline Module	N	256.7	
SOLON	SOLON Black 280/09/01 295 Wp	295W Monocrystalline Module	N	254.8	
SOLON	SOLON Black 280/09/01 300 Wp	300W Monocrystalline Module	N	259.3	
Solyndra	SL-001-135	135W CIGS Thin Film Module	N	126.7	
Solyndra	SL-001-135N	135W CIGS Thin Film Module	N	126.7	
Solyndra	SL-001-135U	135W CIGS Thin Film Module	N	126.7	
Solyndra	SL-001-150	150W CIGS Thin Film Module	N	141.0	
Solyndra	SL-001-150N	150W CIGS Thin Film Module	N	141.0	
Solyndra	SL-001-150U	150W CIGS Thin Film Module	N	141.0	
Solyndra	SL-001-157	157W CIGS Thin Film Module	N	147.6	
Solyndra	SL-001-157N	157W CIGS Thin Film Module	N	147.6	
Solyndra	SL-001-157U	157W CIGS Thin Film Module	N	147.6	
Solyndra	SL-150-157	157W CIGS Thin Film Module	N	147.1	
Solyndra	SL-001-165	165W CIGS Thin Film Module	N	155.2	
Solyndra	SL-001-165N	165W CIGS Thin Film Module	N	155.2	
Solyndra	SL-001-165U	165W CIGS Thin Film Module	N	155.2	
Solyndra	SL-150-165	165W CIGS Thin Film Module	N	154.7	
Solyndra	SL-001-173	173W CIGS Thin Film Module	N	162.8	
Solyndra	SL-001-173N	173W CIGS Thin Film Module	N	162.8	
Solyndra	SL-001-173U	173W CIGS Thin Film Module	N	162.8	
Solyndra	SL-150-173	173W CIGS Thin Film Module	N	162.2	

Solyndra	SL-200-173	173W CIGS Thin Film Module	N	159.7
Solyndra	SL-001-182	182W CIGS Thin Film Module	N	171.4
Solyndra	SL-001-182N	182W CIGS Thin Film Module	N	171.4
Solyndra	SL-001-182U	182W CIGS Thin Film Module	N	171.4
Solyndra	SL-150-182	182W CIGS Thin Film Module	N	170.7
Solyndra	SL-200-182	182W CIGS Thin Film Module	N	168.0
Solyndra	SL-001-191	191W CIGS Thin Film Module	N	179.9
Solyndra	SL-001-191N	191W CIGS Thin Film Module	N	179.9
Solyndra	SL-001-191U	191W CIGS Thin Film Module	N	179.9
Solyndra	SL-150-191	191W CIGS Thin Film Module	N	179.3
Solyndra	SL-200-191	191W CIGS Thin Film Module	N	176.4
Solyndra	SL-001-200	200W CIGS Thin Film Module	N	188.5
Solyndra	SL-001-200N	200W CIGS Thin Film Module	N	188.5
Solyndra	SL-001-200U	200W CIGS Thin Film Module	N	188.5
Solyndra	SL-150-200	200W CIGS Thin Film Module	N	187.8
Solyndra	SL-200-200	200W CIGS Thin Film Module	N	184.8
Solyndra	SL-200-210	210W CIGS Thin Film Module	N	194.2
Solyndra	SL-200-220	220W CIGS Thin Film Module	N	203.5
Sonali Energees USA	SS1750-170W	170W Monocyrstalline Module	N	151.9
Sonali Energees USA	SS1750-175W	175W Monocyrstalline Module	N	156.5
Sonali Energees USA	SS1750-180W	180W Monocyrstalline Module	N	161.1
Sonali Energees USA	SS2100-210W	210W Monocyrstalline Module	N	189.5
Sonali Energees USA	SS2100-215W	215W Monocyrstalline Module	N	194.1
Sonali Energees USA	SS2100-220W	220W Monocyrstalline Module	N	198.8
SRS Energy	SPT16	15.75W Thin Film 3-a-Si Module	Y	14.8

(Continued)

MANUFACTURER NAME	MODULE MODEL NUMBER	DESCRIPTION	BIPV*	PTC**	NOTES
Sun Energy Engineering	SE14-50	50W Polycrystalline Waterproof Integrated PV Roof Tile	Y	38.9	
Sun Energy Engineering	SE14-52	52W Polycrystalline Waterproof Integrated PV Roof Tile	Y	41.2	
SUNGEN International	SGM-165D	165W Monocrystalline Module	N	147.3	
SUNGEN International	SGM-170D	170W Monocrystalline Module	N	151.8	
SUNGEN International	SGM-175D	175W Monocrystalline Module	N	156.4	
SUNGEN International	SGM-180D	180W Monocrystalline Module	N	161.0	
SUNGEN International	SGM-200P	200W Polycrystalline Module	N	175.9	
SUNGEN International	SGM-210P	210W Polycrystalline Module	N	184.9	
SUNGEN International	SGM-220P	220W Polycrystalline Module	N	194.0	
SUNGEN International	SGM-230P	230W Polycrystalline Module	N	203.1	
SUNGEN International	SGM-240P	240W Polycrystalline Module	N	211.7	
SUNGEN International	SGM-250P	250W Polycrystalline Module	N	220.7	
SUNGEN International	SGM-260P	260W Polycrystalline Module	N	229.8	
SUNGEN International	SGM-270P	270W Polycrystalline Module	N	238.9	
SUNGEN International	SGM-280P	280W Polycrystalline Module	N	248.0	
Suniva	Suniva Art 235-60-2	235W Monocrystalline Module	N	205.3	
Suniva	Suniva Art 240-60-2	240W Monocrystalline Module	N	209.8	
Suniva	Suniva Art 245-60-2	245W Monocrystalline Module	N	214.3	
Sunperfect Solar	CRM60S125S	60W Monocrystalline Module	N	53.6	
Sunperfect Solar	CRM85S125S	85W Monocrystalline Module	N	76.6	
Sunperfect Solar	CRM115S125S	115W Monocrystalline Module	N	102.2	
Sunperfect Solar	CRM145S125S	145W Monocrystalline Module	N	126.5	

Sunperfect Solar	CRM175S125S	175W Monocrystalline Module	N	155.1
Sunperfect Solar	CRM185S156P-54	185W Polycrystalline Module	N	163.5
Sunperfect Solar	CRM190S156P-54	190W Polycrystalline Module	N	168.1
Sunperfect Solar	CRM195S125S	195W Monocrystsalline Module	N	171.7
Sunperfect Solar	CRM195S156P-54	195W Polycrystalline Module	N	172.6
Sunperfect Solar	CRM200S156P-54	200W Polycrystalline Module	N	176.1
Sunperfect Solar	CRM200S156P-60	200W Polycrystalline Module	N	180.1
Sunperfect Solar	CRM205S156P-54	205W Polycrystalline Module	N	180.6
Sunperfect Solar	CRM205S156P-60	205W Polycrystalline Module	N	184.7
Sunperfect Solar	CRM210S156P-54	210W Polycrystalline Module	N	185.3
Sunperfect Solar	CRM210S156P-60	210W Polycrystalline Module	N	189.3
Sunperfect Solar	CRM215S156P-54	215W Polycrystalline Module	N	189.9
Sunperfect Solar	CRM215S156P-60	215W Polycrystalline Module	N	193.9
Sunperfect Solar	CRM220S156P-54	220W Polycrystalline Module	N	194.4
Sunperfect Solar	CRM220S156P-60	220W Polycrystalline Module	N	195.8
Sunperfect Solar	CRM225S156P-54	225W Polycrystalline Module	N	199.0
Sunperfect Solar	CRM225S156P-60	225W Polycrystalline Module	N	200.3
Sunperfect Solar	CRM230S156P-54	230W Polycrystalline Module	N	203.6
Sunperfect Solar	CRM230S156P-60	230W Polycrystalline Module	N	201.4
Sunperfect Solar	CRM235S125S	235W Monocrystalline Module	N	209.3
Sunperfect Solar	CRM235S156P-60	235W Polycrystalline Module	N	206.0
Sunperfect Solar	CRM240S156P-60	240W Polycrystalline Module	N	210.5
Sunperfect Solar	CRM240S156P-72	240W Polycrystalline Module	N	211.3
Sunperfect Solar	CRM245S156P-60	245W Polycrystalline Module	N	215.0
Sunperfect Solar	CRM245S156P-72	245W Polycrystalline Module	N	215.8

(Continued)

MANUFACTURER NAME	MODULE MODEL NUMBER	DESCRIPTION	BIPV*	PTC**	NOTES
Sunperfect Solar	CRM250S156P-60	250W Polycrystalline Module	N	224.8	
Sunperfect Solar	CRM250S156P-72	250W Polycrystalline Module	N	220.4	
Sunperfect Solar	CRM255S156P-60	255W Polycrystalline Module	N	229.4	
Sunperfect Solar	CRM255S156P-72	255W Polycrystalline Module	N	224.9	
Sunperfect Solar	CRM260S156P-60	260W Polycrystalline Module	N	234.1	
Sunperfect Solar	CRM260S156P-72	260W Polycrystalline Module	N	230.5	
Sunperfect Solar	CRM265S156P-72	265W Polycrystalline Module	N	235.1	
Sunperfect Solar	CRM270S156P-72	270W Polycrystalline Module	N	239.6	
Sunperfect Solar	CRM275S156P-72	275W Polycrystalline Module	N	244.2	
Sunperfect Solar	CRM280S156P-72	280W Polycrystalline Module	N	248.8	
Sunperfect Solar	CRM285S156P-72	285W Polycrystalline Module	N	253.4	
Sunperfect Solar	CRM290S156P-72	290W Polycrystalline Module	N	257.9	
Sunperfect Solar	CRM295S156P-72	295W Polycrystalline Module	N	262.5	
Sunperfect Solar	CRM300S156P-72	300W Polycrystalline Module	N	267.1	
SunPower	PL-PLT-63L-BLK-U	63W Monocrystalline Module, SunTile	Y	55.1	
SunPower	SPR-76R-BLK-U	76W Monocrystalline Module, SunTile	Y	64.9	
SunPower	SPR-76RE-BLK-U	76W Monocrystalline Module, SunTile	Y	64.9	
SunPower	SPR-200-BLK-U	200W Monocrystalline Module, Framed, Grid Connect, Black Backsheet	N	179.7	
SunPower	SPR-200-WHT-U	200W Monocrystalline Module, Framed, Grid Connect, White Backsheet	N	181.5	

SunPower	SPR-205-BLK-U	205W Monocrystalline Module, Framed, Grid Connect, Black Backsheet	N	184.3
SunPower	SPR-208-WHT-U	208W Monocrystalline Module, Framed, Grid Connect, White Backsheet	N	188.9
SunPower	SPR-210-BLK-U	210W Monocrystalline Module, Framed, Grid Connect, Black Backsheet	N	188.9
SunPower	SPR-210-WHT-U	210W Monocrystalline Module, Framed, Grid Connect, White Backsheet	N	193.5
SunPower	SPR-215-WHT-U	215W Monocrystalline Module, Framed, Grid Connect, White Backsheet	N	198.3
SunPower	SPR-217-WHT-U	217W Monocrystalline Module, Framed, Grid Connect, White Backsheet	N	197.4
SunPower	SER-220P	220W Polycrystalline Module	N	194.8
SunPower	SPR-220-BLK-U	220W Monocrystalline Module, Framed, Grid Connect, Black Backsheet	N	198.2
SunPower	SPR-220-WHT-U	220W Monocrystalline Module, Framed, Grid Connect, White Backsheet	N	200.2
SunPower	SPR-225-BLK-U	225W Monocrystalline Module, Framed, Grid Connect, Black Backsheet	N	205.0
SunPower	SER-228P	228W Polycrystalline Module	N	202.0

(Continued)

MANUFACTURER NAME	MODULE MODEL NUMBER	DESCRIPTION	BIPV*	PTC**	NOTES
SunPower	SPR-230-WHT-U	230W Monocrystalline Module, Framed, Grid Connect, White Backsheet	N	212.3	
SunPower	PL-SUNP-SPR-290	290W Monocrystalline Module, Laminate, Grid Connect, White Backsheet	N	266.4	
SunPower	SPR-290-WHT-U	290W Monocrystalline Module, Framed, Grid Connect, White Backsheet	N	266.4	
SunPower	T5-SPR-290	290W Monocrystalline Module, Integrated Laminate/Frame/ Mounting System, Grid Connect, White Backsheet	N	266.4	
SunPower	SPR-295E-WHT-U	295W Monocrystalline Module, Framed, Grid Connect, White Backsheet	N	271.1	
SunPower	PL-SUNP-SPR-305	305W Monocrystalline Module, Laminate, Grid Connect, White Backsheet	N	280.6	
SunPower	SPR-305E-WHT-U	305W Monocrystalline Module, Framed, Grid Connect, White Backsheet	N	280.6	
SunPower	SPR-305-WHT-U	305W Monocrystalline Module, Framed, Grid Connect, White Backsheet	N	280.6	
SunPower	T5-SPR-305	305W Monocrystalline Module, Integrated Laminate/Frame/ Mounting System, Grid Connect, White Backsheet	N	280.6	

SunPower	T5-SPR-305E	305W Monocrystalline Module, Integrated Laminate/Frame/Mounting System, Grid Connect, White Backsheet	N	280.6
SunPower	PL-SUNP-SPR-310	310W Monocrystalline Module, Laminate, Grid Connect, White Backsheet	N	285.3
SunPower	SPR-310E-WHT-U	310W Monocrystalline Module, Framed, Grid Connect, White Backsheet	N	285.3
SunPower	SPR-310-WHT-U	310W Monocrystalline Module, Framed, Grid Connect, White Backsheet	N	285.3
SunPower	T5-SPR-310	310W Monocrystalline Module, Integrated Laminate/Frame/Mounting System, Grid Connect, White Backsheet	N	285.3
SunPower	PL-SUNP-SPR-315E	315W Monocrystalline Module, Laminate, Grid Connect, White Backsheet	N	290.0
SunPower	SPR-315E-WHT-D	315W Monocrystalline Module, Framed, Grid Connect, White Backsheet	N	290.0
SunPower	SPR-315E-WHT-U	315W Monocrystalline Module, Framed, Grid Connect, White Backsheet	N	290.0
SunPower	T5-SPR-315	315W Monocrystalline Module, Integrated Laminate/Frame/Mounting System, Grid Connect, White Backsheet	N	290.0

(Continued)

MANUFACTURER NAME	MODULE MODEL NUMBER	DESCRIPTION	BIPV*	PTC***	NOTES
SunPower	T5-SPR-315E	315W Monocrystalline Module, Integrated Laminate/Frame/ Mounting System, Grid Connect, White Backsheet	N	290.0	
SunPower	PL-SUNP-SPR-318E	318W Monocrystalline Module, Laminate, Grid Connect, White Backsheet	N	292.9	
SunPower	SPR-318E-WHT-D	318W Monocrystalline Module	N	292.9	
SunPower	T5-SPR-318E	318W Monocrystalline Module, Integrated Laminate/Frame/ Mounting System, Grid Connect, White Backsheet	N	292.9	
SunPower	SPR-390E-WHT-D	390W Monocrystalline Module	N	361.4	
SunPower	SPR-400E-WHT-D	400W Monocrystalline Module	N	370.9	
SunPower	SPR-415E-WHT-D	415W Monocrystalline Module	N	385.2	
Suntech Power	STP050D-5/ZCB	50W Polycrystalline Waterproof Built-in PV Roof Tile	Y	44.0	
Suntech Power	STP050D-5/ZCF	50W Polycrystalline Waterproof Built-in PV Roof Tile	Y	44.0	
Suntech Power	STP050D-5/ZCG	50W Polycrystalline Waterproof Built-in PV Roof Tile	Y	44.0	
Suntech Power	STP120D-12/VEC	120W Polycrystalline Module, MC Connectors, Transparent Backsheet	N	106.9	
Suntech Power	STP130D-12/VEC	130W Polycrystalline Module, MC Connectors, Transparent Backsheet	N	113.7	

Suntech Power	STP135D-12/VEC	135W Polycrystalline Module, MC Connectors, Transparent Backsheet	N	118.2
Suntech Power	STP160S-24/Ab-1 Black	160W Monocrystalline Module, MC Connectors, Black Backsheet	N	138.4
Suntech Power	STP165S-24/Ab-1 Black	165W Monocrystalline Module, MC Connectors, Black Backsheet	N	142.8
Suntech Power	STP170S-24/Ab-1	170W Monocrystalline Module, MC Connectors	N	151.4
Suntech Power	STP170S-24/Ab-1 Black	170W Monocrystalline Module, MC Connectors, Black Backsheet	N	152.9
Suntech Power	STP170S-24/Ad+	170W Monocrystalline Module, MC Connectors, H4 Connectors	N	151.4
Suntech Power	STP170S-24/Adb+	170W Monocrystalline Module, MC Connectors, Black Backsheet, H4 Connectors	N	152.9
Suntech Power	MSZ175B-F	175W Monocrystalline Module, MC Connectors, Black Backsheet	N	157.5
Suntech Power	MSZ175J-F	175W Monocrystalline Module, MC Connectors, Black Backsheet	N	157.5
Suntech Power	STP175S-24/Ab-1	175W Monocrystalline Module, MC Connectors	N	156.0
Suntech Power	STP175S-24/Ab-1 Black	175W Monocrystalline Module, MC Connectors, Black Backsheet	N	157.5

(Continued)

MANUFACTURER NAME	MODULE MODEL NUMBER	DESCRIPTION	BIPV*	PTC**	NOTES
Suntech Power	STP175S-24/Ad+	175W Monocrystalline Module, MC Connectors, H4 Connectors	N	156.0	
Suntech Power	STP175S-24/Adb+	175W Monocrystalline Module, MC Connectors, Black Backsheet, H4 Connectors	N	157.5	
Suntech Power	MSZ180B-F	180W Monocrystalline Module, MC Connectors, Black Backsheet	N	162.1	
Suntech Power	MSZ180J-F	180W Monocrystalline Module, MC Connectors, Black Backsheet	N	162.1	
Suntech Power	STP180S-24/Ab-1	180W Monocrystalline Module, MC Connectors	N	160.5	
Suntech Power	STP180S-24/Ab-1 Black	180W Monocrystalline Module, MC Connectors, Black Backsheet	N	162.1	
Suntech Power	STP180S-24/Ad+	180W Monocrystalline Module, MC Connectors, H4 Connectors	N	160.5	
Suntech Power	STP180S-24/Adb+	180W Monocrystalline Module, MC Connectors, Black Backsheet, H4 Connectors	N	162.1	
Suntech Power	STP185S-24/Ab-1	185W Monocrystalline Module, MC Connectors	N	165.1	
Suntech Power	STP185S-24/Ad+	185W Monocrystalline Module, MC Connectors, H4 Connectors	N	165.1	
Suntech Power	STP185S-24/Adb	185W Monocrystalline Module, Black Backsheet	N	167.2	

Suntech Power	STP185S-24/Adb+	185W Monocrystalline Module, Black Backsheet	N	167.2
Suntech Power	PLUTO190-Ada	190W Monocrystalline Module, MC Connectors	N	173.2
Suntech Power	PLUTO190-Adb	190W Monocrystalline Module, MC Connectors, Black Backsheet	N	173.2
Suntech Power	PLUTO190-Ade	190W Monocrystalline Module, MC Connectors	N	173.2
Suntech Power	PLUTO190-Adf	190W Monocrystalline Module, MC Connectors, Black Backsheet	N	173.2
Suntech Power	STP190-18/UB-1	190W Polycrystalline Module, MC Connectors	N	172.0
Suntech Power	STP190S-24/Ad	190W Monocrystalline Module	N	171.5
Suntech Power	STP190S-24/Ad+	190W Monocrystalline Module	N	171.5
Suntech Power	STP190S-24/Adb	190W Monocrystalline Module, Black Backsheet	N	171.8
Suntech Power	STP190S-24/Adb+	190W Monocrystalline Module, Black Backsheet	N	171.8
Suntech Power	PLUTO195-Ada	195W Monocrystalline Module, MC Connectors	N	177.9
Suntech Power	PLUTO195-Adb	195W Monocrystalline Module, MC Connectors, Black Backsheet	N	177.8
Suntech Power	PLUTO195-Ade	195W Monocrystalline Module, MC Connectors	N	177.9
Suntech Power	PLUTO195-Adf	195W Monocrystalline Module, MC Connectors, Black Backsheet	N	177.8

(Continued)

MANUFACTURER NAME	MODULE MODEL NUMBER	DESCRIPTION	BIPV*	PTC**	NOTES
Suntech Power	STP195S-24/Ad	195W Monocrystalline Module	N	176.2	
Suntech Power	STP195S-24/Ad+	195W Monocrystalline Module	N	176.2	
Suntech Power	STP195S-24/Adb	195W Monocrystalline Module, Black Backsheet	N	176.5	
Suntech Power	STP195S-24/Adb+	195W Monocrystalline Module, Black Backsheet	N	176.5	
Suntech Power	PLUTO200-Ada	200W Monocrystalline Module, MC Connectors	N	182.6	
Suntech Power	PLUTO200-Adb	200W Monocrystalline Module, MC Connectors, Black Backsheet	N	182.5	
Suntech Power	PLUTO200-Ade	200W Monocrystalline Module, MC Connectors	N	182.6	
Suntech Power	PLUTO200-Adf	200W Monocrystalline Module, MC Connectors, Black Backsheet	N	182.5	
Suntech Power	STP200-18/UB-1	200W Polycrystalline Module, MC Connectors	N	181.3	
Suntech Power	STP200-18/Ud	200W Polycrystalline Module, MC Connectors	N	181.3	
Suntech Power	PLUTO205-Ada	205W Monocrystalline Module	N	188.3	
Suntech Power	PLUTO205-Adb	205W Monocrystalline Module	N	187.5	
Suntech Power	PLUTO205-Ade	205W Monocrystalline Module	N	188.3	
Suntech Power	PLUTO205-Adf	205W Monocrystalline Module	N	187.5	
Suntech Power	STP205-18/Ud	205W Polycrystalline Module, MC Connectors	N	185.9	
Suntech Power	PLUTO210-Ada	210W Monocrystalline Module	N	193.0	
Suntech Power	PLUTO210-Adb	210W Monocrystalline Module	N	192.2	

Suntech Power	PLUTO210-Ade	210W Monocrystalline Module	N	193.0
Suntech Power	PLUTO210-Adf	210W Monocrystalline Module	N	192.2
Suntech Power	PLUTO210-Udm	210W Polycrystalline Module	N	191.6
Suntech Power	STP210-18/UB-1	210W Polycrystalline Module, MC Connectors	N	190.5
Suntech Power	STP210-18/Ud	210W Polycrystalline Module, MC Connectors	N	190.5
Suntech Power	STP210-20/Wd	210W Polycrystalline Module	N	190.0
Suntech Power	PLUTO215-Udm	215W Polycrystalline Module	N	196.3
Suntech Power	STP215-18/Ud	215W Polycrystalline Module	N	195.2
Suntech Power	STP215-20/Wd	215W Polycrystalline Module	N	194.7
Suntech Power	PLUTO220-Udm	220W Polycrystalline Module	N	200.9
Suntech Power	STP220-18/Ud	220W Polycrystalline Module	N	199.8
Suntech Power	STP220-20/Wd	220W Polycrystalline Module	N	199.3
Suntech Power	PLUTO225-Udm	225W Polycrystalline Module	N	205.6
Suntech Power	STP225-18/Ud	225W Polycrystalline Module	N	204.5
Suntech Power	STP225-20/Wd	225W Polycrystalline Module	N	203.9
Suntech Power	PLUTO230-Udm	230W Polycrystalline Module	N	210.3
Suntech Power	STP230-20/Wd	230W Polycrystalline Module	N	208.6
Suntech Power	STP235-20/Wd	235W Polycrystalline Module	N	213.2
Suntech Power	STP240-20/Wd	240W Polycrystalline Module	N	217.8
Suntech Power	STP250D-24/VEC	250W Polycrystalline Module, MC Connectors, Transparent Backsheet	N	219.6
Suntech Power	STP260-24/Vb-1	260W Polycrystalline Module, MC Connectors	N	235.4

(Continued)

MANUFACTURER NAME	MODULE MODEL NUMBER	DESCRIPTION	BIPV*	PTC**	NOTES
Suntech Power	STP260D-24/VEC	260W Polycrystalline Module, MC Connectors, Transparent Backsheet	N	226.1	
Suntech Power	STP260-VRM-1	260W Monocrystalline Module	N	235.4	
Suntech Power	STP270-24/Vb-1	270W Polycrystalline Module, MC Connectors	N	244.7	
Suntech Power	STP270-24/Vd	270W Polycrystalline Module, MC Connectors	N	244.7	
Suntech Power	STP270D-24/VEC	270W Polycrystalline Module, MC Connectors, Transparent Backsheet	N	235.1	
Suntech Power	STP270-VRM-1	270W Monocrystalline Module	N	244.7	
Suntech Power	STP275-24/Vd	275W Polycrystalline Module, MC Connectors	N	249.3	
Suntech Power	STP275-VRM-1	275W Monocrystalline Module	N	249.3	
Suntech Power	PLUTO280-Vdm	280W Polycrystalline Module	N	255.5	
Suntech Power	STP280-24/Vb-1	280W Polycrystalline Module, MC Connectors	N	254.0	
Suntech Power	STP280-24/Vd	280W Polycrystalline Module, MC Connectors	N	254.0	
Suntech Power	STP280-VRM-1	280W Monocrystalline Module	N	254.0	
Suntech Power	PLUTO285-Vdm	285W Polycrystalline Module	N	260.2	
Suntech Power	STP285-24/Vd	285W Polycrystalline Module, MC Connectors	N	258.6	
Suntech Power	STP285-VRM-1	285W Monocrystalline Module	N	258.6	
Suntech Power	PLUTO290-Vdm	290W Polycrystalline Module	N	264.9	
Suntech Power	STP290-24/Vd	290W Polycrystalline Module, MC Connectors	N	263.3	

Suntech Power	STP290-VRM-1	290W Monocrystalline Module	N	263.3
Suntech Power	PLUTO295-Vdm	295W Polycrystalline Module	N	269.5
Suntech Power	PLUTO300-Vdm	300W Polycrystalline Module	N	274.2
Suntech Power	PLUTO305-Vdm	305W Polycrystalline Module	N	278.9
Suntech Power	PLUTO310-Vdm	310W Polycrystalline Module	N	283.6
SunWize Technologies	SW170	170W Monocrystalline Module	N	151.9
SunWize Technologies	SW175	175W Monocrystalline Module	N	156.5
SunWize Technologies	SW180	180W Monocrystalline Module	N	161.1
Supreme Solar	SS-165M	165W Monocrystalline Module	N	147.3
Supreme Solar	SS-170M	170W Monocrystalline Module	N	151.8
Supreme Solar	SS-175M	175W Monocrystalline Module	N	156.4
Supreme Solar	SS-180M	180W Monocrystalline Module	N	161.0
Supreme Solar	SS-200P	200W Polycrystalline Module	N	175.9
Supreme Solar	SS-210P	210W Polycrystalline Module	N	184.9
Supreme Solar	SS-220P	220W Polycrystalline Module	N	194.0
Supreme Solar	SS-230P	230W Polycrystalline Module	N	203.1
Supreme Solar	SS-240P	240W Polycrystalline Module	N	211.7
Supreme Solar	SS-250P	250W Polycrystalline Module	N	220.7
Supreme Solar	SS-260P	260W Polycrystalline Module	N	229.8
Supreme Solar	SS-270P	270W Polycrystalline Module	N	238.9
Supreme Solar	SS-280P	280W Polycrystalline Module	N	248.0
Suzhou Shenglong PV-Tech	SLSM-175D	175W Monocrystalline Module	N	151.6
Suzhou Shenglong PV-Tech	SLSM-180D	180W Monocrystalline Module	N	156.1
Suzhou Shenglong PV-Tech	SLSM-185D	185W Monocrystalline Module	N	160.6

(Continued)

MANUFACTURER NAME	MODULE MODEL NUMBER	DESCRIPTION	BIPV*	PTC**	NOTES
Suzhou Shenglong PV-Tech	SLSM-220P	220W Polycrystalline Module	N	191.7	
Suzhou Shenglong PV-Tech	SLSM-225P	225W Polycrystalline Module	N	196.2	
Suzhou Shenglong PV-Tech	SLSM-230P	230W Polycrystalline Module	N	200.7	
Symphony Energy	SE-M215Gy	215W Polycrystalline Module	N	195.4	
Symphony Energy	SE-M216Gy	216W Polycrystalline Module	N	196.4	
Symphony Energy	SE-M217Gy	217W Polycrystalline Module	N	197.3	
Symphony Energy	SE-M218Gy	218W Polycrystalline Module	N	198.2	
Symphony Energy	SE-M219Gy	219W Polycrystalline Module	N	199.2	
Symphony Energy	SE-M220Gy	220W Polycrystalline Module	N	200.1	
Symphony Energy	SE-M221Gy	221W Polycrystalline Module	N	201.0	
Symphony Energy	SE-M222Gy	222W Polycrystalline Module	N	202.0	
Symphony Energy	SE-M223Gy	223W Polycrystalline Module	N	202.9	
Symphony Energy	SE-M224Gy	224W Polycrystalline Module	N	203.8	
Symphony Energy	SE-M225Gy	225W Polycrystalline Module	N	204.7	
Symphony Energy	SE-M226Gy	226W Polycrystalline Module	N	205.7	
Symphony Energy	SE-M227Gy	227W Polycrystalline Module	N	206.6	
Symphony Energy	SE-M228Gy	228W Polycrystalline Module	N	207.5	
Symphony Energy	SE-M229Gy	229W Polycrystalline Module	N	208.5	
Symphony Energy	SE-M230Gy	230W Polycrystalline Module	N	209.4	
Symphony Energy	SE-M231Gy	231W Polycrystalline Module	N	210.3	
Symphony Energy	SE-M232Gy	232W Polycrystalline Module	N	211.3	
Symphony Energy	SE-M233Gy	233W Polycrystalline Module	N	212.2	
Symphony Energy	SE-M234Gy	234W Polycrystalline Module	N	213.1	

Symphony Energy	SE-M235Gy	235W Polycrystalline Module	N	214.1
Taiwan Semiconductor Manufacturing	TS-230P4-AD	230W Polycrystalline Module	N	203.5
Taiwan Semiconductor Manufacturing	TS-235P4-AD	235W Polycrystalline Module	N	208.1
Taiwan Semiconductor Manufacturing	TS-240P4-AD	240W Polycrystalline Module	N	212.7
Taiwan Semiconductor Manufacturing	TS-245P4-AD	245W Polycrystalline Module	N	217.2
Taiwan Semiconductor Manufacturing	TS-250P4-AD	250W Polycrystalline Module	N	221.8
Tianwei Ecosolargy	TWES-(155)72M	155W Monocrystalline Module	N	135.1
Tianwei Ecosolargy	TWES-(160)72M	160W Monocrystalline Module	N	139.5
Tianwei Ecosolargy	TWES-(165)72M	165W Monocrystalline Module	N	144.0
Tianwei Ecosolargy	TWES-(170)72M	170W Monocrystalline Module	N	148.5
Tianwei Ecosolargy	TWES-(175)72M	175W Monocrystalline Module	N	153.0
Tianwei Ecosolargy	TWES-(180)72M	180W Monocrystalline Module	N	157.5
Tianwei Ecosolargy	TWES-(200)60P	200W Polycrystalline Module	N	173.4
Tianwei Ecosolargy	TWES-(205)60P	205W Polycrystalline Module	N	177.9
Tianwei Ecosolargy	TWES-(210)60P	210W Polycrystalline Module	N	182.4
Tianwei Ecosolargy	TWES-(215)60P	215W Polycrystalline Module	N	186.8
Tianwei Ecosolargy	TWES-(220)60P	220W Polycrystalline Module	N	191.3
Tianwei Ecosolargy	TWES-(225)60P	225W Polycrystalline Module	N	195.8
Tianwei Ecosolargy	TWES-(230)60P	230W Polycrystalline Module	N	200.3
Tianwei Ecosolargy	TWES-(235)60P	235W Polycrystalline Module	N	204.8
Tianwei New Energy (Chengdu) PV Module	TW155(35)D	155W Monocrystalline Module	N	135.1

(Continued)

MANUFACTURER NAME	MODULE MODEL NUMBER	DESCRIPTION	BiPV*	PTC**	NOTES
Tianwei New Energy (Chengdu) PV Module	TW160(35)D	160W Monocrystalline Module	N	139.5	
Tianwei New Energy (Chengdu) PV Module	TW165(35)D	165W Monocrystalline Module	N	144.0	
Tianwei New Energy (Chengdu) PV Module	TW170(35)D	170W Monocrystalline Module	N	148.5	
Tianwei New Energy (Chengdu) PV Module	TW175(35)D	175W Monocrystalline Module	N	153.0	
Tianwei New Energy (Chengdu) PV Module	TW180(28)P	180W Polycrystalline Module	N	155.6	
Tianwei New Energy (Chengdu) PV Module	TW180(35)D	180W Monocrystalline Module	N	157.5	
Tianwei New Energy (Chengdu) PV Module	TW185(28)P	185W Polycrystalline Module	N	160.1	
Tianwei New Energy (Chengdu) PV Module	TW185(35)D	185W Monocrystalline Module	N	162.1	
Tianwei New Energy (Chengdu) PV Module	TW190(28)P	190W Polycrystalline Module	N	164.5	
Tianwei New Energy (Chengdu) PV Module	TW195(28)P	195W Polycrystalline Module	N	169.0	
Tianwei New Energy (Chengdu) PV Module	TW200(28)P	200W Polycrystalline Module	N	173.4	
Tianwei New Energy (Chengdu) PV Module	TW205(28)P	205W Polycrystalline Module	N	177.9	
Tianwei New Energy (Chengdu) PV Module	TW210(28)P	210W Polycrystalline Module	N	182.4	
Tianwei New Energy (Chengdu) PV Module	TW215(28)P	215W Polycrystalline Module	N	186.8	

Tianwei New Energy (Chengdu) PV Module	TW220(28)P	220W Polycrystalline Module	N	191.3
Tianwei New Energy (Chengdu) PV Module	TW225(28)P	225W Polycrystalline Module	N	195.8
Tianwei New Energy (Chengdu) PV Module	TW230(28)P	230W Polycrystalline Module	N	200.3
Tianwei New Energy (Chengdu) PV Module	TW235(28)P	235W Polycrystalline Module	N	204.8
Titan Energy Systems	Titan S6-60	210W Monocrystalline Module	N	186.6
Titan Energy Sytems	Titan S6-60-2-235	235W Monocrystalline Module	N	205.3
Titan Energy Sytems	Titan S6-60-2-240	240W Monocrystalline Module	N	209.8
Titan Energy Sytems	Titan S6-60-2-245	245W Monocrystalline Module	N	214.3
Top Solar	TSC-200W-E	200W Polycrystalline Module	N	177.1
Trina Solar	TSM-165DA01	165W Monocrystalline Module	N	146.0
Trina Solar	TSM-170D	170W Monocrystalline Module	N	152.1
Trina Solar	TSM-170DA01	170W Monocrystalline Module	N	150.6
Trina Solar	TSM-170DA03	170W Monocrystalline Module	N	150.5
Trina Solar	TSM-170PA03	170W Polycrystalline Module	N	148.7
Trina Solar	TSM-175D	175W Monocrystalline Module	N	156.7
Trina Solar	TSM-175DA01	175W Monocrystalline Module	N	155.1
Trina Solar	TSM-175DA03	175W Monocrystalline Module	N	155.0
Trina Solar	TSM-175PA03	175W Polycrystalline Module	N	153.2
Trina Solar	TSM-180D	180W Monocrystalline Module	N	161.3
Trina Solar	TSM-180DA01	180W Monocrystalline Module	N	159.7
Trina Solar	TSM-180DA03	180W Monocrystalline Module	N	159.6
Trina Solar	TSM-180PA03	180W Polycrystalline Module	N	157.7

(Continued)

MANUFACTURER NAME	MODULE MODEL NUMBER	DESCRIPTION	BIPV*	PTC**	NOTES
Trina Solar	TSM-185DA01	185W Monocrystalline Module	N	164.3	
Trina Solar	TSM-185DA03	185W Monocrystalline Module	N	164.2	
Trina Solar	TSM-185PA03	185W Polycrystalline Module	N	162.3	
Trina Solar	TSM-190DA03	190W Monocrystalline Module	N	168.7	
Trina Solar	TSM-190PA03	190W Polycrystalline Module	N	166.8	
Trina Solar	TSM-220DA05	220W Monocrystalline Module	N	193.6	
Trina Solar	TSM-220PA05	220W Polycrystalline Module	N	192.9	
Trina Solar	TSM-225PA05	225W Polycrystalline Module	N	197.9	
Trina Solar	TSM-230DA05	230W Monocrystalline Module	N	202.7	
Trina Solar	TSM-230PA05	230W Polycrystalline Module	N	202.0	
Trina Solar	TSM-235PA05	235W Polycrystalline Module	N	207.0	
Trina Solar	TSM-240DA05	240W Monocrystalline Module	N	211.8	
Trina Solar	TSM-240PA05	240W Polycrystalline Module	N	211.0	
TynSolar	TYN-168P6	165W Polycrystalline Module	N	142.1	
TynSolar	TYN-170S5	170W Monocrystalline Module	N	149.2	
TynSolar	TYN-180P6	175W Polycrystalline Module	N	151.0	
TynSolar	TYN-180S5	175W Monocrystalline Module	N	153.7	
UE Solar	ZHM175	175W Monocrystalline Module	N	154.2	
UE Solar	ZHM180	180W Monocrystalline Module	N	158.8	
UE Solar	ZHM185	185W Monocrystalline Module	N	163.3	
UE Solar	ZHM220P	220W Polycrystalline Module	N	193.3	
UE Solar	ZHM225P	225W Polycrystalline Module	N	197.8	
UE Solar	ZHM230P	230W Polycrystalline Module	N	202.3	
Unicor Federal Prison Industries	PVM220PS-Q6LTT200	200W Polycrystalline Module	N	179.3	

Unicor Federal Prison Industries	PVM220PS-Q6LTT205	205W Polycrystalline Module	N	183.9
Unicor Federal Prison Industries	PVM220PS-Q6LTT210	210W Polycrystalline Module	N	188.5
Unicor Federal Prison Industries	PVM220PS-Q6LTT215	215W Polycrystalline Module	N	193.1
Unicor Federal Prison Industries	PVM220PS-Q6LTT220	220W Polycrystalline Module	N	197.7
United Solar Ovonic	PT-29	29W 3-a-Si Thin Film Module	Y	26.1
United Solar Ovonic	PVL-29	29W Field Applied 3J a-Si Laminate	Y	26.1
United Solar Ovonic	PT-31	31W 3-a-Si Thin Film Module	Y	28.0
United Solar Ovonic	PVL-31	31W Field Applied 3J a-Si Laminate	Y	28.0
United Solar Ovonic	PT-33	33W 3-a-Si Thin Film Module	Y	29.8
United Solar Ovonic	PVL-33	33W Field Applied 3J a-Si Laminate	Y	29.8
United Solar Ovonic	PT-58	58W 3-a-Si Thin Film Module	Y	52.3
United Solar Ovonic	PVL-58	58W Field Applied 3J a-Si Laminate	Y	52.3
United Solar Ovonic	PT-62	62W 3-a-Si Thin Film Module	Y	56.0
United Solar Ovonic	PVL-62	62W Field Applied 3J a-Si Laminate	Y	56.0
United Solar Ovonic	PT-64	64W 3-a-Si Thin Film Module	Y	57.7
United Solar Ovonic	PVL-64	64W Field Applied 3J a-Si Laminate	Y	57.7
United Solar Ovonic	PT-66	66W 3-a-Si Thin Film Module	Y	59.6
United Solar Ovonic	PVL-66	66W Field Applied 3J a-Si Laminate	Y	59.6

(Continued)

MANUFACTURER NAME	MODULE MODEL NUMBER	DESCRIPTION	BIPV*	PTC**	NOTES
United Solar Ovonic	PT-68	68W 3-a-Si Thin Film Module	Y	61.4	
United Solar Ovonic	PVL-68	68W Field Applied 3J a-Si Laminate	Y	61.4	
United Solar Ovonic	PT-72	72W 3-a-Si Thin Film Module	Y	65.0	
United Solar Ovonic	PVL-72	72W Field Applied 3J a-Si Laminate	Y	65.0	
United Solar Ovonic	PT-116	116W 3-a-Si Thin Film Module	Y	104.7	
United Solar Ovonic	PVL-116	116W Field Applied 3J a-Si Laminate	Y	104.7	
United Solar Ovonic	PT-124	124W 3-a-Si Thin Film Module	Y	112.0	
United Solar Ovonic	PVL-124	124W Field Applied 3J a-Si Laminate	Y	112.0	
United Solar Ovonic	PT-128	128W 3-a-Si Thin Film Module	Y	115.5	
United Solar Ovonic	PVL-128	128W Field Applied 3J a-Si Laminate	Y	115.5	
United Solar Ovonic	PT-131	131W 3-a-Si Thin Film Module	Y	118.2	
United Solar Ovonic	PVL-131	131W Field Applied 3J a-Si Laminate	Y	118.2	
United Solar Ovonic	PT-136	136W 3-a-Si Thin Film Module	Y	122.8	
United Solar Ovonic	PVL-136	136W Field Applied 3J a-Si Laminate	Y	122.8	
United Solar Ovonic	PT-144	144W 3-a-Si Thin Film Module	Y	130.1	
United Solar Ovonic	PVL-144	144W Field Applied 3J a-Si Laminate	Y	130.1	
Upsolar	UP-M155M	155W Monocrystalline Module	N	135.1	
Upsolar	UP-M160M	160W Monocrystalline Module	N	139.5	
Upsolar	UP-M165M	165W Monocrystalline Module	N	147.3	

Upsolar	UP-M170M	170W Monocrystalline Module	N	151.8
Upsolar	UP-M175M	175W Monocrystalline Module	N	156.4
Upsolar	UP-M180M	180W Monocrystalline Module	N	161.0
Upsolar	UP-M180P	180W Polycrystalline Module	N	155.6
Upsolar	UP-M185P	185W Polycrystalline Module	N	160.1
Upsolar	UP-M190P	190W Polycrystalline Module	N	164.5
Upsolar	UP-M195P	195W Polycrystalline Module	N	169.0
Upsolar	UP-M200P	200W Polycrystalline Module	N	173.4
Upsolar	UP-M205P	205W Polycrystalline Module	N	177.9
Upsolar	UP-M210P	210W Polycrystalline Module	N	182.4
Upsolar	UP-M215P	215W Polycrystalline Module	N	186.8
Upsolar	UP-M220P	220W Polycrystalline Module	N	191.3
Upsolar	UP-M225P	225W Polycrystalline Module	N	195.8
Upsolar	UP-M230P	230W Polycrystalline Module	N	200.3
Upsolar	UP-M235P	235W Polycrystalline Module	N	204.8
Waaree Energies	WU-120	120W Polycrystalline Module	N	106.4
Waaree Energies	WU-125	125W Polycrystalline Module	N	111.0
Waaree Energies	WU-130	130W Polycrystalline Module	N	115.5
Waaree Energies	WU-140	140W Polycrystalline Module	N	124.7
Waaree Energies	WU-175	175W Polycrystalline Module	N	155.1
Waaree Energies	WU-180	180W Polycrystalline Module	N	159.7
Waaree Energies	WU-190	190W Polycrystalline Module	N	168.8
Waaree Energies	WU-200	200W Polycrystalline Module	N	177.9
Waaree Energies	WU-210	210W Polycrystalline Module	N	186.4
Waaree Energies	WU-220	220W Polycrystalline Module	N	195.6
Waaree Energies	WU-230	230W Polycrystalline Module	N	204.7

(Continued)

MANUFACTURER NAME	MODULE MODEL NUMBER	DESCRIPTION	BIPV*	PTC***	NOTES
Waaree Energies	WU-240	240W Polycrystalline Module	N	213.8	
Wanxiang New Energy	WXS175S-US	175W Monocrystalline Module	N	157.6	
Wanxiang New Energy	WXS180P-US	180W Polycrystalline Module	N	162.3	
Wanxiang New Energy	WXS180S-US	180W Monocrystalline Module	N	162.3	
Wanxiang New Energy	WXS185P-US	185W Polycrystalline Module	N	167.0	
Wanxiang New Energy	WXS185S-US	185W Monocrystalline Module	N	166.9	
Wanxiang New Energy	WXS190P-US	190W Polycrystalline Module	N	171.6	
Wanxiang New Energy	WXS200P-US	200W Polycrystalline Module	N	179.8	
Wanxiang New Energy	WXS205P-US	205W Polycrystalline Module	N	184.4	
Wanxiang New Energy	WXS210P-US	210W Polycrystalline Module	N	189.0	
Wanxiang New Energy	WXS215P-US	215W Polycrystalline Module	N	193.6	
Wanxiang New Energy	WXS220P-US	220W Polycrystalline Module	N	198.2	
Wanxiang New Energy	WXS225P-US	225W Polycrystalline Module	N	202.9	
Wanxiang New Energy	WXS230P-US	230W Polycrystalline Module	N	206.5	
Wanxiang New Energy	WXS235P-US	235W Polycrystalline Module	N	211.1	
Wanxiang New Energy	WXS240P-US	240W Polycrystalline Module	N	215.7	
Wanxiang New Energy	WXS245P-US	245W Polycrystalline Module	N	220.3	
Wanxiang New Energy	WXS250P-US	250W Polycrystalline Module	N	224.9	
Wanxiang New Energy	WXS255P-US	255W Polycrystalline Module	N	229.6	
Wanxiang New Energy	WXS260P-US	260W Polycrystalline Module	N	234.2	
Wanxiang New Energy	WXS265P-US	265W Polycrystalline Module	N	238.8	
Wanxiang New Energy	WXS270P-US	270W Polycrystalline Module	N	243.4	
Wanxiang New Energy	WXS275P-US	275W Polycrystalline Module	N	248.1	
Wanxiang New Energy	WXS280P-US	280W Polycrystalline Module	N	252.7	
Webel SL Energy Systems	W 1750-170 Wp	170W Monocrystalline Module	N	151.9	

Manufacturer	Model	Description		
Webel SL Energy Systems	W 1750-175 Wp	175W Monocrystalline Module	N	156.5
Webel SL Energy Systems	W 1750-180 Wp	180W Monocrystalline Module	N	161.1
Webel SL Energy Systems	W 2100-210 Wp	210W Monocrystalline Module	N	189.5
Webel SL Energy Systems	W 2100-215 Wp	215W Monocrystalline Module	N	194.1
Webel SL Energy Systems	W 2100-220 Wp	220W Monocrystalline Module	N	198.8
Webel SL Energy Systems	W 2300-230	230W Monocrystalline Module	N	200.4
Webel SL Energy Systems	W 2300-235 Wp	235W Monocrystalline Module	N	204.9
Webel SL Energy Systems	W 2300-240 Wp	240W Monocrystalline Module	N	209.4
Websol Energy Systems	W2500-250	250W Monocrystalline Module	N	213.4
Websol Energy Systems	W2800-275	275W Monocrystalline Module	N	228.1
Websol Energy Systems	W2800-280	280W Monocrystalline Module	N	232.5
Websol Energy Systems	W2800-285	285W Monocrystalline Module	N	236.8
Westinghouse Solar	WS-165-1-DC0-0-A	165W Monocrystalline Module	N	142.8
Westinghouse Solar	WS-170-1-DC0-0-A	170W Monocrystalline Module	N	152.9
Westinghouse Solar	WS-175-1-DC0-0-A	175W Monocrystalline Module	N	157.5
Westinghouse Solar	WS-180-1-DC0-0-A	180W Monocrystalline Module	N	162.1

(Continued)

MANUFACTURER NAME	MODULE MODEL NUMBER	DESCRIPTION	BIPV*	PTC**	NOTES
Worldwide Energy & Mfg USA	AS-5M36-165W	165W Monocrystalline Module	N	146.0	
Worldwide Energy & Mfg USA	AS-5M-170W	170W Monocrystalline Module	N	152.1	
Worldwide Energy & Mfg USA	AS-5M36-170W	170W Monocrystalline Module	N	150.6	
Worldwide Energy & Mfg USA	AS-6M24-170W	170W Monocrystalline Module	N	150.5	
Worldwide Energy & Mfg USA	AS-6P24-170W	170W Polycrystalline Module	N	148.7	
Worldwide Energy & Mfg USA	AS-5M-175W	175W Monocrystalline Module	N	156.7	
Worldwide Energy & Mfg USA	AS-5M36-175W	175W Monocrystalline Module	N	155.1	
Worldwide Energy & Mfg USA	AS-6M24-175W	175W Monocrystalline Module	N	155.0	
Worldwide Energy & Mfg USA	AS-6P24-175W	175W Polycrystalline Module	N	153.2	
Worldwide Energy & Mfg USA	AS-5M-180W	180W Monocrystalline Module	N	161.3	
Worldwide Energy & Mfg USA	AS-5M36-180W	180W Monocrystalline Module	N	159.7	
Worldwide Energy & Mfg USA	AS-6M24-180W	180W Monocrystalline Module	N	159.6	
Worldwide Energy & Mfg USA	AS-6P24-180W	180W Polycrystalline Module	N	157.7	
Worldwide Energy & Mfg USA	AS-5M36-185W	185W Monocrystalline Module	N	164.3	

Manufacturer	Model	Description		
Worldwide Energy & Mfg USA	AS-6M24-185W	185W Monocrystalline Module	N	164.2
Worldwide Energy & Mfg USA	AS-6P24-185W	185W Polycrystalline Module	N	162.3
Worldwide Energy & Mfg USA	AS-6M24-190W	190W Monocrystalline Module	N	168.7
Worldwide Energy & Mfg USA	AS-6P24-190W	190W Polycrystalline Module	N	166.8
Worldwide Energy & Mfg USA	AS-6M30-220W	220W Monocrystalline Module	N	193.6
Worldwide Energy & Mfg USA	AS-6P30-220W	220W Polycrystalline Module	N	192.9
Worldwide Energy & Mfg USA	AS-6M30-230W	230W Monocrystalline Module	N	202.7
Worldwide Energy & Mfg USA	AS-6P30-230W	230W Polycrystalline Module	N	202.0
Worldwide Energy & Mfg USA	AS-6M30-240W	240W Monocrystalline Module	N	211.8
Worldwide Energy & Mfg USA	AS-6P30-240W	240W Polycrystalline Module	N	211.0
Wuxi Shangpin Solar Energy Science & Technology	SPSM-165D	165W Monocrystalline Module	N	145.0
Wuxi Shangpin Solar Energy Science & Technology	SPSM-170D	170W Monocrystalline Module	N	149.5
Wuxi Shangpin Solar Energy Science & Technology	SPSM-175D	175W Monocrystalline Module	N	154.1

(Continued)

MANUFACTURER NAME	MODULE MODEL NUMBER	DESCRIPTION	BIPV*	PTC**	NOTES
Wuxi Shangpin Solar Energy Science & Technology	SPSM-180D	180W Monocrystalline Module	N	158.6	
Wuxi Shangpin Solar Energy Science & Technology	SPSM-250P	250W Polycrystalline Module	N	227.8	
Wuxi Shangpin Solar Energy Science & Technology	SPSM-260P	260W Polycrystalline Module	N	237.1	
Wuxi Shangpin Solar Energy Science & Technology	SPSM-270P	270W Polycrystalline Module	N	246.4	
XL Telecom & Energy	XL 30225	225W Polycrystalline Module	N	198.5	
Xunlight	XR 36-264	264W 3-a-Si Thin Film Module	Y	243.5	
Yes! Solar	ES190W-NW	190W Polycrystalline Module	N	173.8	
Yes! Solar	ES190W-RW	190W Polycrystalline Module	N	170.4	
Yes! Solar	ES195W-NW	195W Polycrystalline Module	N	178.5	
Yes! Solar	ES195W-RW	195W Polycrystalline Module	N	175.0	
Yes! Solar	ES200W-NW	200W Polycrystalline Module	N	183.1	
Yes! Solar	ES200W-RW	200W Polycrystalline Module	N	179.6	
Yes! Solar	ES205W-NW	205W Polycrystalline Module	N	187.8	
Yes! Solar	ES205W-RW	205W Polycrystalline Module	N	184.2	
Yingli Green Energy	YL170P-23b	170W Polycrystalline Module, High Transmission and Textured Glass	N	154.2	
Yingli Green Energy	YL175(156)	175W Polycrystalline Module, High Transmission and Textured Glass	N	148.9	

Yingli Green Energy	YL175P-23b	175W Polycrystalline Module, High Transmission and Textured Glass	N	158.9
Yingli Green Energy	YL180P-23b	180W Polycrystalline Module, High Transmission and Textured Glass	N	163.5
Yingli Green Energy	YL185P-23b	185W Polycrystalline Module, High Transmission and Textured Glass	N	168.2
Yingli Green Energy	YL195P-26b	195W Polycrystalline Module, High Transmission and Textured Glass	N	173.0
Yingli Green Energy	YL200P-26b	200W Polycrystalline Module, High Transmission and Textured Glass	N	177.5
Yingli Green Energy	YL205P-26b	205W Polycrystalline Module, High Transmission and Textured Glass	N	182.1
Yingli Green Energy	YL210P-26b	210W Polycrystalline Module, High Transmission and Textured Glass	N	186.7
Yingli Green Energy	YL220P-29b	220W Polycrystalline Module, High Transmission and Textured Glass	N	197.4
Yingli Green Energy	YL225P-29b	225W Polycrystalline Module, High Transmission and Textured Glass	N	202.0
Yingli Green Energy	YL230P-29b	230W Polycrystalline Module, High Transmission and Textured Glass	N	206.6

(Continued)

MANUFACTURER NAME	MODULE MODEL NUMBER	DESCRIPTION	BIPV*	PTC**	NOTES
Yingli Green Energy	YL235P-29b	235W Polycrystalline Module, High Transmission and Textured Glass	N	211.2	
Yingli Green Energy	YL240P-32b	240W Polycrystalline Module, High Transmission and Textured Glass	N	211.5	
Yingli Green Energy	YL245P-32b	245W Polycrystalline Module, High Transmission and Textured Glass	N	216.0	
Yingli Green Energy	YL250P-32b	250W Polycrystalline Module, High Transmission and Textured Glass	N	220.6	
Yingli Green Energy	YL255P-32b	255W Polycrystalline Module, High Transmission and Textured Glass	N	225.1	
Yingli Green Energy	YL260P-35b	260W Polycrystalline Module, High Transmission and Textured Glass	N	232.5	
Yingli Green Energy	YL265P-35b	265W Polycrystalline Module, High Transmission and Textured Glass	N	237.0	
Yingli Green Energy	YL270P-35b	270W Polycrystalline Module, High Transmission and Textured Glass	N	241.6	
Yingli Green Energy	YL275P-35b	275W Polycrystalline Module, High Transmission and Textured Glass	N	246.2	
Yingli Green Energy	YL280P-35b	280W Polycrystalline Module, High Transmission and Textured Glass	N	250.8	

Yunnan Tianda Photovoltaic	TD160M5	160W Monocrystalline Module	N	139.0
Yunnan Tianda Photovoltaic	TD165M5	165W Monocrystalline Module	N	143.5
Yunnan Tianda Photovoltaic	TD170M5	170W Monocrystalline Module	N	148.0
Yunnan Tianda Photovoltaic	TD175M5	175W Monocrystalline Module	N	152.5
Yunnan Tianda Photovoltaic	TD180M5	180W Monocrystalline Module	N	157.0
Zerquest Solar	Zp170	170W Monocrystalline Module	N	143.5
Zhangjiagang City Sunlink PV	SL195-18	195W Polycrystalline Module	N	169.3
Zhangjiagang City Sunlink PV	SL200-18	200W Polycrystalline Module	N	173.8
Zhangjiagang City Sunlink PV	SL205-18	205W Polycrystalline Module	N	178.3
Zhangjiagang City Sunlink PV	SL220-20	220W Polycrystalline Module	N	189.8
Zhangjiagang City Sunlink PV	SL225-20	225W Polycrystalline Module	N	194.2
Zhangjiagang City Sunlink PV	SL230-20	230W Polycrystalline Module	N	198.7
Zhejiang ERA Solar Technology	ESPSA-170	170W Monocrystalline Module	N	154.7
Zhejiang ERA Solar Technology	ESPSA-175	175W Monocrystalline Module	N	159.4
Zhejiang ERA Solar Technology	ESPSA-180	180W Monocrystalline Module	N	164.1

(Continued)

MANUFACTURER NAME	MODULE MODEL NUMBER	DESCRIPTION	BIPV*	PTC**	NOTES
Zhejiang Global Photovoltaic Technology	GSM-155	155W Monocrystalline Module	N	135.4	
Zhejiang Global Photovoltaic Technology	GSM-160	160W Monocrystalline Module	N	139.9	
Zhejiang Global Photovoltaic Technology	GSM-165	165W Monocrystalline Module	N	144.4	
Zhejiang Global Photovoltaic Technology	GSM-170	170W Monocrystalline Module	N	148.9	
Zhejiang Global Photovoltaic Technology	GSM-175	175W Monocrystalline Module	N	153.4	
Zhejiang Global Photovoltaic Technology	GSM-180	180W Monocrystalline Module	N	157.9	
Zhejiang Sunflower Light Energy Science & Technology	SF125x125-72-M-150W	150W Monocrystalline Module	N	132.0	
Zhejiang Sunflower Light Energy Science & Technology	SF125x125-72-M-155W	155W Monocrystalline Module	N	136.5	
Zhejiang Sunflower Light Energy Science & Technology	SF125x125-72-M-160W	160W Monocrystalline Module	N	141.0	
Zhejiang Sunflower Light Energy Science & Technology	SF125x125-72-M-165W	165W Monocrystalline Module	N	145.5	

Zhejiang Sunflower Light Energy Science & Technology	SF125x125-72-M-170W	170W Monocrystalline Module	N	150.1
Zhejiang Sunflower Light Energy Science & Technology	SF125x125-72-M-175W	175W Monocrystalline Module	N	154.6
Zhejiang Sunflower Light Energy Science & Technology	SF125x125-72-M-180W	180W Monocrystalline Module	N	159.2
Zhejiang Sunflower Light Energy Science & Technology	SF125x125-72-M-185W	185W Monocrystalline Module	N	163.7
Zhejiang Wanxiang Solar	WXS175P	175W Polycrystalline Module	N	157.7
Zhejiang Wanxiang Solar	WXS175S	175W Monocrystalline Module	N	157.6
Zhejiang Wanxiang Solar	WXS180P	180W Polycrystalline Module	N	162.3
Zhejiang Wanxiang Solar	WXS180S	180W Monocrystalline Module	N	162.3
Zhejiang Wanxiang Solar	WXS185P	185W Polycrystalline Module	N	167.0
Zhejiang Wanxiang Solar	WXS185S	185W Monocrystalline Module	N	166.9
Zhejiang Wanxiang Solar	WXS190P	190W Polycrystalline Module	N	171.6
Zhejiang Wanxiang Solar	WXS200P	200W Polycrystalline Module	N	179.8
Zhejiang Wanxiang Solar	WXS205P	205W Polycrystalline Module	N	184.4

(Continued)

617

MANUFACTURER NAME	MODULE MODEL NUMBER	DESCRIPTION	BIPV*	PTC**	NOTES
Zhejiang Wanxiang Solar	WXS210P	210W Polycrystalline Module	N	189.0	
Zhejiang Wanxiang Solar	WXS215P	215W Polycrystalline Module	N	193.6	
Zhejiang Wanxiang Solar	WXS220P	220W Polycrystalline Module	N	198.2	
Zhejiang Wanxiang Solar	WXS225P	225W Polycrystalline Module	N	202.9	
Zhejiang Wanxiang Solar	WXS230P	230W Polycrystalline Module	N	206.5	
Zhejiang Wanxiang Solar	WXS235P	235W Polycrystalline Module	N	211.1	
Zhejiang Wanxiang Solar	WXS240P	240W Polycrystalline Module	N	215.7	
Zhejiang Wanxiang Solar	WXS245P	245W Polycrystalline Module	N	220.3	
Zhejiang Wanxiang Solar	WXS250P	250W Polycrystalline Module	N	224.9	
Zhejiang Wanxiang Solar	WXS255P	255W Polycrystalline Module	N	229.6	
Zhejiang Wanxiang Solar	WXS260P	260W Polycrystalline Module	N	234.2	
Zhejiang Wanxiang Solar	WXS265P	265W Polycrystalline Module	N	238.8	
Zhejiang Wanxiang Solar	WXS270P	270W Polycrystalline Module	N	243.4	
Zhejiang Wanxiang Solar	WXS275P	275W Polycrystalline Module	N	248.1	
Zhejiang Wanxiang Solar	WXS280P	280W Polycrystalline Module	N	252.7	

* BIPV = Building Integrated Photovoltaics

** In the New Solar Homes Partnership (NSHP), the incentive is based on expected kWh generation with a time-dependent weighting, via the CECPV Calculator. The CECPV Calculator shall be used to determine the incentive under NSHP, as the PTC rating is not used in the NSHP incentive calculation.

PTC refers to PVUSA Test Conditions, which were developed to test and compare PV systems as part of the PVUSA (Photovoltaics for Utility Scale Applications) project. PTC are 1,000 Watts per square meter solar irradiance, 20 degrees C air temperature, and wind speed of 1 meter per second at 10 meters above ground level. PV manufacturers use Standard Test Conditions, or STC, to rate their PV products. STC are 1,000 Watts per square meter solar irradiance, 25 degrees C cell temperature, air mass equal to 1.5, and ASTM G173-03 standard spectrum. The PTC rating, which is lower than the STC rating, is generally recognized as a more realistic measure of PV output because the test conditions better reflect "real-world" solar and climatic conditions, compared to the STC rating. All ratings in the list are DC (direct current) watts.

Neither PTC nor STC account for all "real-world" losses. Actual solar systems will produce lower outputs due to soiling, shading, module mismatch, wire losses, inverter and transformer losses, shortfalls in actual nameplate ratings, panel degradation over time, and high-temperature losses for arrays mounted close to or integrated within a roofline. These loss factors can vary by season, geographic location, mounting technique, azimuth, and array tilt. Examples of estimated losses from varying factors can be found at: http://www.nrel.gov/rredc/pvwatts.

List of Eligible Inverters per SB1 Guidelines

Guideline for the use of the Performance Test Protocol for Evaluating Inverters Used in Grid-Connected Photovoltaic Systems

Performance Test Protocol for Evaluating Inverters Used in Grid-Connected Photovoltaic Systems

View performance test results for inverters performed by a nationally recognized testing laboratory in accordance with the protocols adopted by the Energy Commission

MANUFACTURER NAME	INVERTER MODEL NUMBER	DESCRIPTION	POWER RATING (WATTS)	WEIGHTED EFFICIENCY	APPROVED BUILT-IN METER	NOTES
Advanced Energy Industries	Solaron 250kW (3159200-XXXX)	250 kW 480 Vac Three Phase Utility-Interactive Inverter	250000	97.5	No	N/A
Advanced Energy Industries	Solaron 333kW (3159000-XXXX)	333 kW 480 Vac Three Phase Utility-Interactive Inverter	333000	97.5	No	NA
Advanced Energy Industries	Solaron 500kW (3159500-XXXX)	500 kW 480 Vac Three Phase Utility-Interactive Inverter	500000	97.5	No	NA
Alpha Technologies	Solaris 3500 XP	3.5 kW, 240 Vac, NEMA-3R, Grid Interactive PV Inverter & Batt. Backed UPS, MPPT	3500	90.5	Yes	NA
Ballard Power Systems	EPC-PV-208-30kW	Utility Interactive 208V 30kW PV Power Converter System	30000	93	Yes	NA
Ballard Power Systems	EPC-PV-208-75kW	Utility Interactive 75kW PV Power Converter System	75000	91.5	Yes	NA
Ballard Power Systems	EPC-PV-480-30kW	Utility Interactive 480V 30kW PV Power Converter System	30000	93	Yes	NA
Ballard Power Systems	EPC-PV-480-75KW	Utility Interactive 480V 75kW PV Power Converter System	75000	91.5	Yes	NA
Beacon Power	M4	4 kW Power Conversion System	4000	90	No	NA
Beacon Power	M4 Plus	4 kW Power Conversion System with Meter	4000	91.5	Yes	NA
Beacon Power	M5	5 kW Power Conversion System	5000	89	No	NA
Beacon Power	M5 Plus	5 kW Power Conversion System with Meter	5000	90	Yes	NA
Bergey Windpower	Gridtek 10	10kW, 240Vac split-phase, utility interactive inverter	10000	91	No	NA
Beyond Building Group	SR1500TLI (208V)	1.5kW (208Vac) Utility-Interactive Inverter	1500	94.5	Yes	NA
Beyond Building Group	SR1500TLI (240)	1.5kW (240Vac) Utility-Interactive Inverter	1500	94.5	Yes	NA
Beyond Building Group	SR2000TLI (208V)	2kW (208Vac) Utility-Interactive Inverter	2000	95.5	Yes	NA

MANUFACTURER NAME	INVERTER MODEL NUMBER	DESCRIPTION	POWER RATING (WATTS)	WEIGHTED EFFICIENCY	APPROVED BUILT-IN METER	NOTES
Beyond Building Group	SR2000TLI (240)	2kW (240Vac) Utility-Interactive Inverter	2000	95	Yes	NA
Beyond Building Group	SR3000TLI (208V)	3kW (208Vac) Utility-Interactive Inverter	3000	95.5	Yes	NA
Beyond Building Group	SR3000TLI (240)	3kW (240Vac) Utility-Interactive Inverter	3000	95.5	Yes	NA
Beyond Building Group	SR4000TLI (208V)	4kW (208Vac) Utility-Interactive Inverter	4000	95.5	Yes	NA
Beyond Building Group	SR4000TLI (240)	4kW (240Vac) Utility-Interactive Inverter	4000	96	Yes	NA
Bloom Energy	INV-500	25 kW 480 Vac Three Phase Utility-Interactive Inverter for Fuel Cell	25000	93.5	No	NA
CFM Equipment Distributors	Green Power 1100	1100W (120Vac) Utility Interactive Inverter	1100	90.5	Yes	NA
CFM Equipment Distributors	Green Power 2000	2000W (240Vac) Utility Interactive Inverter	2000	92	Yes	NA
CFM Equipment Distributors	Green Power 2500	2500W (240Vac) Utility Interactive Inverter	2500	94.5	Yes	NA
CFM Equipment Distributors	Green Power 3000	3000W (240Vac) Utility Interactive Inverter	3000	93.5	Yes	NA
CFM Equipment Distributors	Green Power 3500	3500W (240Vac) Utility Interactive Inverter	3500	95.5	Yes	NA
CFM Equipment Distributors	Green Power 4600	4600W (208Vac) Utility Interactive Inverter	4600	95.5	Yes	NA
CFM Equipment Distributors	Green Power 4800	4800W (240Vac) Utility Interactive Inverter	4800	96	Yes	NA
CFM Equipment Distributors	Green Power 5200	5200W (240Vac) Utility Interactive Inverter	5200	96	Yes	NA
Connect Renewable Energy	CE 4000	3000W Grid Tied, 240Vac, Low DC Voltage Inverter	2850	93	Yes	NA

Manufacturer	Model	Description				
Delta Energy Systems	SI 1800 US (208V)	1800W 208Vac Grid-Tied Solar PV Inverter	1799	92.5	Yes	NA
Delta Energy Systems	SI 1800 US (240V)	1800W 240Vac Grid-Tied Solar PV Inverter	1759	92.5	Yes	NA
Delta Energy Systems	SI 2500 US (208V)	2500W 208Vac Grid-Tied Solar PV Inverter	2485	92	Yes	NA
Delta Energy Systems	SI 2500 US (240V)	2500W 240Vac Grid-Tied Solar PV Inverter	2498	93	Yes	NA
Diehl AKO Stiftung & Co. KG	4301 S-A (208V)	3.68kW (208Vac) Utility-Interactive Inverter	3680	94	Yes	NA
Diehl AKO Stiftung & Co. KG	4301 S-A (240V)	3.68kW (240Vac) Utility-Interactive Inverter	3680	94.5	Yes	NA
Diehl AKO Stiftung & Co. KG	Platinum 100 CS-A 208	100kW 208Vac Three phase Utility Interactive inverter	100000	94.5	Yes	NA
Diehl AKO Stiftung & Co. KG	Platinum 100 CS-A 480	100kW 480Vac Three phase Utility Interactive inverter	100000	94.5	Yes	NA
Diehl AKO Stiftung & Co. KG	Platinum 100 CS-A HE 208	100kW 208Vac Three phase Utility Interactive inverter	100000	95	Yes	NA
Diehl AKO Stiftung & Co. KG	Platinum 100 CS-A HE 480	100kW 480Vac Three phase Utility Interactive inverter	100000	95.5	Yes	NA
Enphase Energy	D380-72-208-S1x	380 W, 208Vac, utility-interactive modular inverter, 1x are connector types	380	95	No	NA
Enphase Energy	D380-72-240-S1x	380 W, 240Vac, utility-interactive modular inverter, 1x are connector types	380	95	No	NA
Enphase Energy	M175-24-208-Sxx	175 W, 24Vdc, 208Vac, utility-interactive modular inverter, xx is 01 or 02.	175	94.5	No	NA
Enphase Energy	M175-24-240-Sxx	175 W, 24Vdc, 240Vac, utility-interactive modular inverter, xx is 01 or 02.	175	94.5	No	NA
Enphase Energy	M190-72-208-Sxx	190 W, 32Vdc, 208Vac, utility-interactive modular inverter, xx is 11, 12, or 13.	190	95	No	NA

(*Continued*)

MANUFACTURER NAME	INVERTER MODEL NUMBER	DESCRIPTION	POWER RATING (WATTS)	WEIGHTED EFFICIENCY	APPROVED BUILT-IN METER	NOTES
Enphase Energy	M190-72-240-Sxx	190 W, 32Vdc, 240Vac, utility-interactive modular inverter, xx is 11, 12, or 13.	190	95	No	NA
Enphase Energy	M200-32-208-Sxx	200 W, 52.5Vdc, 208Vac, utility-interactive modular inverter, xx is 01 or 02.	200	95	No	NA
Enphase Energy	M200-32-240-Sxx	200 W, 52.5Vdc, 240Vac, utility-interactive modular inverter, xx is 01 or 02.	200	95	No	NA
Enphase Energy	M210-84-208-Sxx	210 W, 40.5Vdc, 208Vac, utility-interactive modular inverter, xx is 11, 12, or 13.	210	95.5	No	NA
Enphase Energy	M210-84-240-Sxx	210 W, 40.5Vdc, 240Vac, utility-interactive modular inverter, xx is 11, 12, or 13.	210	95.5	No	NA
E-Village Solar	EVS1100EVR	1100W (120Vac) Utility Interactive Inverter	1100	90.5	Yes	NA
E-Village Solar	EVS2000EVR	2000W (240Vac) Utility Interactive Inverter	2000	92	Yes	NA
E-Village Solar	EVS2500	2500W (240Vac) Utility Interactive Inverter	2500	94.5	Yes	NA
E-Village Solar	EVS3000	3000W (240Vac) Utility Interactive Inverter	3000	93.5	Yes	NA
E-Village Solar	EVS3500	3500W (240Vac) Utility Interactive Inverter	3500	95.5	Yes	NA
E-Village Solar	EVS4600	4600W (208Vac) Utility Interactive Inverter	4600	95.5	Yes	NA
E-Village Solar	EVS4800	4800W (240Vac) Utility Interactive Inverter	4800	96	Yes	NA
E-Village Solar	EVS5200	5200W (240Vac) Utility Interactive Inverter	5200	96	Yes	NA
Exeltech	AC2-1-B-6-5	212W (240Vac) Utility-Interactive Inverter	212	94.5	Yes	NA
Exeltech	XLGT18A60 & XLGT18A60-01	1.8kW (120Vac) Utility-Interactive Inverter (-01 for disconnect)	1800	96.5	Yes	NA
Fronius USA	Fronius CL 33.3 Delta (208V)	33.3 kW, 208Vac Delta Utility Interactive Inverter	33300	94.5	Yes	NA
Fronius USA	Fronius CL 33.3 Delta (240V)	33.3 kW, 240Vac Delta Utility Interactive Inverter	33300	95	Yes	NA
Fronius USA	Fronius CL 36.0 Wye (277V)	36 kW, 277Vac Wye Utility Interactive Inverter	36000	95.5	Yes	NA

Fronius USA	Fronius CL 44.4 Delta (208V)	44.4 kW, 208Vac Delta Utility Interactive Inverter	44400	94.5	Yes	NA
Fronius USA	Fronius CL 44.4 Delta (240V)	44.4 kW, 240Vac Delta Utility Interactive Inverter	44400	95	Yes	NA
Fronius USA	Fronius CL 48.0 Wye (277V)	48 kW, 277Vac Wye Utility Interactive Inverter	48000	95.5	Yes	NA
Fronius USA	Fronius CL 55.5 Delta (208V)	55.5 kW, 208Vac Delta Utility Interactive Inverter	55500	94.5	Yes	NA
Fronius USA	Fronius CL 55.5 Delta (240V)	55.5 kW, 240Vac Delta Utility Interactive Inverter	55500	95	Yes	NA
Fronius USA	Fronius CL 60.0 Wye (277V)	60 kW, 277Vac Wye Utility Interactive Inverter	60000	95.5	Yes	NA
Fronius USA	IG 2000 NEG	2000W Grid-tied unit with Integrated Disconnects and Performance Meter	2000	93.5	Yes	NA
Fronius USA	IG 2000 POS	2000W Grid-tied unit with Integrated Disconnects and Performance Meter	2000	93.5	Yes	NA
Fronius USA	IG 2500-LV NEG	2350W Grid-tied unit with Integrated Disconnects and Performance Meter	2350	93	Yes	NA
Fronius USA	IG 2500-LV POS	2350W Grid-tied unit with Integrated Disconnects and Performance Meter	2350	93	Yes	NA
Fronius USA	IG 3000 NEG	2700W Grid-tied unit with Integrated Disconnects and Performance Meter	2700	94	Yes	NA
Fronius USA	IG 3000 POS	2700W Grid-tied unit with Integrated Disconnects and Performance Meter	2700	94	Yes	NA
Fronius USA	IG 4000 NEG	4000W Grid-tied unit with Integrated Disconnects and Performance Meter	4000	94	Yes	NA
Fronius USA	IG 4000 POS	4000W Grid-tied unit with Integrated Disconnects and Performance Meter	4000	94	Yes	NA
Fronius USA	IG 4500-LV NEG	4500W Grid-tied unit with Integrated Disconnects and Performance Meter	4500	93.5	Yes	NA

(*Continued*)

MANUFACTURER NAME	INVERTER MODEL NUMBER	DESCRIPTION	POWER RATING (WATTS)	WEIGHTED EFFICIENCY	APPROVED BUILT-IN METER	NOTES
Fronius USA	IG 4500-LV POS	4500W Grid-tied unit with Integrated Disconnects and Performance Meter	4500	93.5	Yes	NA
Fronius USA	IG 5100 NEG	5100W Grid-tied unit with Integrated Disconnects and Performance Meter	5100	94.5	Yes	NA
Fronius USA	IG 5100 POS	5100W Grid-tied unit with Integrated Disconnects and Performance Meter	5100	94.5	Yes	NA
Fronius USA	IG Plus 10.0-208	10000W (208Vac) Utility Interactive Inverter	10000	95	Yes	NA
Fronius USA	IG Plus 10.0-240	10000W (240Vac) Utility Interactive Inverter	10000	95.5	Yes	NA
Fronius USA	IG Plus 10.0-277	10000W (277Vac) Utility Interactive Inverter	10000	96	Yes	NA
Fronius USA	IG Plus 11.4-208	11400W (208Vac) Utility Interactive Inverter	11400	95.5	Yes	NA
Fronius USA	IG Plus 11.4-240	11400W (240Vac) Utility Interactive Inverter	11400	96	Yes	NA
Fronius USA	IG Plus 11.4-277	11400W (277Vac) Utility Interactive Inverter	11400	96	Yes	NA
Fronius USA	IG Plus 11.4-3 Delta-208	11400W (208Vac) 3-Phase Utility Interactive Inverter	11400	95	Yes	NA
Fronius USA	IG Plus 11.4-3 Delta-240	11400W (240Vac) 3-Phase Utility Interactive Inverter	11400	95.5	Yes	NA
Fronius USA	IG Plus 12.0-3WYE277	12000W (3-Phase WYE 277Vac) Utility Interactive Inverter	12000	96	Yes	NA
Fronius USA	IG Plus 3.0-208	3000W (208Vac) Utility Interactive Inverter	3000	95	Yes	NA
Fronius USA	IG Plus 3.0-240	3000W (240Vac) Utility Interactive Inverter	3000	95.5	Yes	NA
Fronius USA	IG Plus 3.0-277	3000W (277Vac) Utility Interactive Inverter	3000	95.5	Yes	NA

Manufacturer	Model	Description	Power (W)	Efficiency		
Fronius USA	IG Plus 3.8-208	3800W (208Vac) Utility Interactive Inverter	3800	95	Yes	NA
Fronius USA	IG Plus 3.8-240	3800W (240Vac) Utility Interactive Inverter	3800	95.5	Yes	NA
Fronius USA	IG Plus 3.8-277	3800W (277Vac) Utility Interactive Inverter	3800	95.5	Yes	NA
Fronius USA	IG Plus 5.0-208	5000W (208Vac) Utility Interactive Inverter	5000	95.5	Yes	NA
Fronius USA	IG Plus 5.0-240	5000W (240Vac) Utility Interactive Inverter	5000	95.5	Yes	NA
Fronius USA	IG Plus 5.0-277	5000W (277Vac) Utility Interactive Inverter	5000	96	Yes	NA
Fronius USA	IG Plus 6.0-208	6000W (208Vac) Utility Interactive Inverter	6000	95.5	Yes	NA
Fronius USA	IG Plus 6.0-240	6000W (240Vac) Utility Interactive Inverter	6000	96	Yes	NA
Fronius USA	IG Plus 6.0-277	6000W (277Vac) Utility Interactive Inverter	6000	96	Yes	NA
Fronius USA	IG Plus 7.5-208	7500W (208Vac) Utility Interactive Inverter	7500	95	Yes	NA
Fronius USA	IG Plus 7.5-240	7500W (240Vac) Utility Interactive Inverter	7500	95.5	Yes	NA
Fronius USA	IG Plus 7.5-277	7500W (277Vac) Utility Interactive Inverter	7500	96	Yes	NA
Fronius USA	IG Plus V 10.0-208	10000W (208Vac) Utility Interactive Inverter	10000	95	Yes	NA
Fronius USA	IG Plus V 10.0-240	10000W (240Vac) Utility Interactive Inverter	10000	95.5	Yes	NA
Fronius USA	IG Plus V 10.0-277	10000W (277Vac) Utility Interactive Inverter	10000	96	Yes	NA
Fronius USA	IG Plus V 11.4-208	11400W (208Vac) Utility Interactive Inverter	11400	95.5	Yes	NA
Fronius USA	IG Plus V 11.4-240	11400W (240Vac) Utility Interactive Inverter	11400	95.5	Yes	NA
Fronius USA	IG Plus V 11.4-277	11400W (277Vac) Utility Interactive Inverter	11400	96	Yes	NA
Fronius USA	IG Plus V 11.4-3-208 Delta	11400W (208Vac) 3-Phase Utility Interactive Inverter	11400	95	Yes	NA
Fronius USA	IG Plus V 11.4-3-240 Delta	11400W (240Vac) 3-Phase Utility Interactive Inverter	11400	96	Yes	NA

(Continued)

MANUFACTURER NAME	INVERTER MODEL NUMBER	DESCRIPTION	POWER RATING (WATTS)	WEIGHTED EFFICIENCY	APPROVED BUILT-IN METER	NOTES
Fronius USA	IG Plus V 12.0-3WYE277	12000W (3-Phase WYE 277Vac) Utility Interactive Inverter	12000	96	Yes	NA
Fronius USA	IG Plus V 3.0-208	3000W (208Vac) Utility Interactive Inverter	3000	95	Yes	NA
Fronius USA	IG Plus V 3.0-240	3000W (240Vac) Utility Interactive Inverter	3000	95.5	Yes	NA
Fronius USA	IG Plus V 3.0-277	3000W (277Vac) Utility Interactive Inverter	3000	96	Yes	NA
Fronius USA	IG Plus V 3.8-208	3800W (208Vac) Utility Interactive Inverter	3800	95	Yes	NA
Fronius USA	IG Plus V 3.8-240	3800W (240Vac) Utility Interactive Inverter	3800	95.5	Yes	NA
Fronius USA	IG Plus V 3.8-277	3800W (277Vac) Utility Interactive Inverter	3800	96	Yes	NA
Fronius USA	IG Plus V 5.0-208	5000W (208Vac) Utility Interactive Inverter	5000	95.5	Yes	NA
Fronius USA	IG Plus V 5.0-240	5000W (240Vac) Utility Interactive Inverter	5000	95.5	Yes	NA
Fronius USA	IG Plus V 5.0-277	5000W (277Vac) Utility Interactive Inverter	5000	96	Yes	NA
Fronius USA	IG Plus V 6.0-208	6000W (208Vac) Utility Interactive Inverter	6000	95.5	Yes	NA
Fronius USA	IG Plus V 6.0-240	6000W (240Vac) Utility Interactive Inverter	6000	96	Yes	NA
Fronius USA	IG Plus V 6.0-277	6000W (277Vac) Utility Interactive Inverter	6000	96	Yes	NA
Fronius USA	IG Plus V 7.5-208	7500W (208Vac) Utility Interactive Inverter	7500	95	Yes	NA
Fronius USA	IG Plus V 7.5-240	7500W (240Vac) Utility Interactive Inverter	7500	95.5	Yes	NA
Fronius USA	IG Plus V 7.5-277	7500W (277Vac) Utility Interactive Inverter	7500	96	Yes	NA
GE Energy	GEPVb-2500-NA-240	2.5 kW Brilliance-DC Disconnect-Display	2500	94	Yes	NA
GE Energy	GEPVb-2800-NA-240/208-02 (208V)	2.7 kW, 208 Vac, 195-600Vdc Grid Tie Inverter	2700	93.5	Yes	NA
GE Energy	GEPVb-2800-NA-240/208-02 (240V)	2.8 kW, 240 Vac, 195-600Vdc Grid Tie Inverter	2800	94	Yes	NA
GE Energy	GEPVb-3000-NA-240	3.0 kW Brilliance-DC Disconnect-Display	3000	94.5	Yes	NA

GE Energy	GEPVb-3300-NA-208	3.3 kW Brilliance-DC Disconnect-Display	3300	94	Yes	NA
GE Energy	GEPVb-3300-NA-240	3.3 kW Brilliance-DC Disconnect-Display	3300	94.5	Yes	NA
GE Energy	GEPVb-3300-NA-240/208-02 (208V)	3.1 kW, 208 Vac, 195-600Vdc Grid Tie Inverter	3100	95	Yes	NA
GE Energy	GEPVb-3300-NA-240/208-02 (240V)	3.3 kW, 240 Vac, 195-600Vdc Grid Tie Inverter	3300	95.5	Yes	NA
GE Energy	GEPVb-3800-NA-240	3.8 kW Brilliance-DC Disconnect-Display	3800	95	Yes	NA
GE Energy	GEPVb-4000-NA-240/208-02 (208V)	3.8 kW, 208 Vac, 195-600Vdc Grid Tie Inverter	3800	95	Yes	NA
GE Energy	GEPVb-4000-NA-240/208-02 (240V)	4.0 kW, 240 Vac, 195-600Vdc Grid Tie Inverter	4000	95.5	Yes	NA
GE Energy	GEPVb-5000-NA-240/208-02 (208V)	5.0 kW, 208 Vac, 195-600Vdc Grid Tie Inverter	4500	95	Yes	NA
GE Energy	GEPVb-5000-NA-240/208-02 (240V)	5.0 kW, 240 Vac, 195-600Vdc Grid Tie Inverter	5000	95.5	Yes	NA
GE Energy	GEPVb-5000-NA-240-01	5.0 kW, 240 Vac, 195-600Vdc Grid Tie Inverter	5000	95.5	Yes	NA
GE Energy	GEPVe-1100-NA-120	1100W (120Vac) Utility Interactive Inverter with PV System Disconnect	1100	90.5	Yes	NA
GE Energy	GEPVe-2000-NA-240	2000W (240Vac) Utility Interactive Inverter with PV System Disconnect	2000	92.5	Yes	NA
GE Energy	GEPVe-2500-NA-240	2500W (240Vac) Utility Interactive Inverter with PV System Disconnect	2500	94.5	Yes	NA

(Continued)

MANUFACTURER NAME	INVERTER MODEL NUMBER	DESCRIPTION	POWER RATING (WATTS)	WEIGHTED EFFICIENCY	APPROVED BUILT-IN METER	NOTES
GE Energy	GEPVe-2800-NA-208	2800W (208Vac) Utility Interactive Inverter with PV System Disconnect	2800	94	Yes	NA
GE Energy	GEPVe-3000-NA-240	3000W (240Vac) Utility Interactive Inverter with PV System Disconnect	3000	93.5	Yes	NA
GE Energy	GEPVe-3500-NA-240	3500W (240Vac) Utility Interactive Inverter with PV System Disconnect	3500	95.5	Yes	NA
GE Energy	GEPVe-4600-NA-240	4600W (208Vac) Utility Interactive Inverter with PV System Disconnect	4600	95.5	Yes	NA
GE Energy	GEPVe-4800-NA-240	4800W (240Vac) Utility Interactive Inverter with PV System Disconnect	4800	96	Yes	NA
GE Energy	GEPVe-5200-NA-240	5200W (240Vac) Utility Interactive Inverter with PV System Disconnect	5200	96	Yes	NA
GridPoint	Connect C36	3600 W Utility Interactive (w/ battery backup) 48Vdc Inverter	3037	91	No	NA
Home Director	HD-SUN-INV1100-EVR	1100W (120Vac) Utility Interactive Inverter	1100	90.5	Yes	NA
Home Director	HD-SUN-INV2000-EVR	2000W (240Vac) Utility Interactive Inverter	2000	92	Yes	NA
Home Director	HD-SUN-INV2500	2500W (240Vac) Utility Interactive Inverter	2500	94.5	Yes	NA
Home Director	HD-SUN-INV3000	3000W (240Vac) Utility Interactive Inverter	3000	93.5	Yes	NA
Home Director	HD-SUN-INV3500	3500W (240Vac) Utility Interactive Inverter	3500	95.5	Yes	NA
Home Director	HD-SUN-INV4600	4600W (208Vac) Utility Interactive Inverter	4600	95.5	Yes	NA
Home Director	HD-SUN-INV4800	4800W (240Vac) Utility Interactive Inverter	4800	96	Yes	NA

Home Director	HD-SUN-INV5200	5200W (240Vac) Utility Interactive Inverter	5200	96	Yes	NA
Ingeteam	INGECON SUN 100U 208V	100kW 208Vac Three phase Utility Interactive inverter	100000	94.5	Yes	NA
Ingeteam	INGECON SUN 100U 480V	100kW 480Vac Three phase Utility Interactive inverter	100000	94.5	Yes	NA
Ingeteam	INGECON SUN 100UP 208V	100kW 208Vac Three phase Utility Interactive inverter	100000	95	Yes	NA
Ingeteam	INGECON SUN 100UP 480V	100kW 480Vac Three phase Utility Interactive inverter	100000	95.5	Yes	NA
Ingeteam	INGECON SUN 15U 208V	15kW 208Vac Three phase Utility Interactive inverter	15000	95	Yes	NA
Ingeteam	INGECON SUN 15U 480V	15kW 480Vac Three phase Utility Interactive inverter	15000	95	Yes	NA
Ingeteam	INGECON SUN 25U 208V	25kW 208Vac Three phase Utility Interactive inverter	25000	95.5	Yes	NA
Ingeteam	INGECON SUN 25U 480V	25kW 480Vac Three phase Utility Interactive inverter	25000	95.5	Yes	NA
iPower	SHO-1.1	1100W (120Vac) Utility Interactive Inverter	1100	90.5	Yes	NA
iPower	SHO-2.0	2000W (240Vac) Utility Interactive Inverter	2000	92	Yes	NA
iPower	SHO-2.5	2500W (240Vac) Utility Interactive Inverter	2500	94.5	Yes	NA
iPower	SHO-3.0	3000W (240Vac) Utility Interactive Inverter	3000	93.5	Yes	NA
iPower	SHO-3.5	3500W (240Vac) Utility Interactive Inverter	3500	95.5	Yes	NA
iPower	SHO-4.6	4600W (208Vac) Utility Interactive Inverter	4600	95.5	Yes	NA
iPower	SHO-4.8	4800W (240Vac) Utility Interactive Inverter	4800	96	Yes	NA
iPower	SHO-5.2	5200W (240Vac) Utility Interactive Inverter	5200	96	Yes	NA
KACO	blueplanet 1501xi	1.5 kW, 240 Vac, Utility Interactive Inverter	1500	94	Yes	NA
KACO	blueplanet 1502xi (208V)	1.5 kW, 208 Vac, Utility Interactive Inverter	1500	95	Yes	NA

(Continued)

MANUFACTURER NAME	INVERTER MODEL NUMBER	DESCRIPTION	POWER RATING (WATTS)	WEIGHTED EFFICIENCY	APPROVED BUILT-IN METER	NOTES
KACO	blueplanet 1502xi (240V)	1.5 kW, 240 Vac, Utility Interactive Inverter	1500	95.5	Yes	NA
KACO	blueplanet 2502xi (208V)	2.5 kW, 208 Vac, Utility Interactive Inverter	2500	95	Yes	NA
KACO	blueplanet 2502xi (240V)	2.5 kW, 240 Vac, Utility Interactive Inverter	2500	95.5	Yes	NA
KACO	blueplanet 2901xi	2.9 kW, 240 Vac, Utility Interactive Inverter	2864	94	Yes	NA
KACO	blueplanet 3502xi (208V)	3.5 kW, 208 Vac, Utility Interactive Inverter	3500	95.5	Yes	NA
KACO	blueplanet 3502xi (240V)	3.5 kW, 240 Vac, Utility Interactive Inverter	3500	95.5	Yes	NA
KACO	blueplanet 3601xi	3.6 kW, 240 Vac, Utility Interactive Inverter	3600	93.5	Yes	NA
KACO	blueplanet 5002xi (208V)	5 kW, 208 Vac, Utility Interactive Inverter	5000	95	Yes	NA
KACO	blueplanet 5002xi (240V)	5 kW, 240 Vac, Utility Interactive Inverter	5000	95.5	Yes	NA
KACO	XP100U-H2	100 kW 480Vac Three Phase Utility Interactive Inverter	100000	95.5	Yes	NA
KACO	XP100U-H4	100 kW 480Vac Three Phase Utility Interactive Inverter	100000	96	Yes	NA
Mariah Power	Windspire 1.2G Inverter 800021	1.2 kW, 120Vac Utility-Interactive Integrated Wind Turbine Inverter	1000	84.5	No	NA
Mohr Power	MPS1100EVR	1100W (120Vac) Utility Interactive Inverter	1100	90.5	Yes	NA
Mohr Power	MPS2000EVR	2000W (240Vac) Utility Interactive Inverter	2000	92	Yes	NA
Mohr Power	MPS2500	2500W (240Vac) Utility Interactive Inverter	2500	94.5	Yes	NA

Mohr Power	MPS3000	3000W (240Vac) Utility Interactive Inverter	3000	93.5	Yes	NA
Mohr Power	MPS3500	3500W (240Vac) Utility Interactive Inverter	3500	95.5	Yes	NA
Mohr Power	MPS4600	4600W (208Vac) Utility Interactive Inverter	4600	95.5	Yes	NA
Mohr Power	MPS4800	4800W (240Vac) Utility Interactive Inverter	4800	96	Yes	NA
Mohr Power	MPS5200	5200W (240Vac) Utility Interactive Inverter	5200	96	Yes	NA
Motech Industries	PVMate 2900U (208V)	2.7kW (208Vac) Utility-Interactive Inverter	2700	95.5	Yes	NA
Motech Industries	PVMate 2900U (240V)	2.9kW (240Vac) Utility-Interactive Inverter	2900	96	Yes	NA
Motech Industries	PVMate 2900U-PG (208V)	2.7kW (208Vac) Utility-Interactive Inverter	2700	95.5	Yes	NA
Motech Industries	PVMate 2900U-PG (240V)	2.9kW (240Vac) Utility-Interactive Inverter	2900	96	Yes	NA
Motech Industries	PVMate 3840U (208V)	Same unit as PVMate 3900U with max power for use with 20-A breaker	3320	95.5	Yes	NA
Motech Industries	PVMate 3840U (240V)	Same unit as PVMate 3900U with max power for use with 20-A breaker	3840	96	Yes	NA
Motech Industries	PVMate 3840U-PG (208V)	Same unit as PVMate 3900U-PG with max power for use with 20-A breaker	3320	95.5	Yes	NA
Motech Industries	PVMate 3840U-PG (240V)	Same unit as PVMate 3900U-PG with max power for use with 20-A breaker	3840	96	Yes	NA
Motech Industries	PVMate 3900U (208V)	3.4kW (208Vac) Utility-Interactive Inverter	3400	95.5	Yes	NA
Motech Industries	PVMate 3900U (240V)	3.9kW (240Vac) Utility-Interactive Inverter	3900	96	Yes	NA

(Continued)

MANUFACTURER NAME	INVERTER MODEL NUMBER	DESCRIPTION	POWER RATING (WATTS)	WEIGHTED EFFICIENCY	APPROVED BUILT-IN METER	NOTES
Motech Industries	PVMate 3900U-PG (208V)	3.4kW (208Vac) Utility-Interactive Inverter	3400	95.5	Yes	NA
Motech Industries	PVMate 3900U-PG (240V)	3.9kW (240Vac) Utility-Interactive Inverter	3900	96	Yes	NA
Motech Industries	PVMate 4900U (208V)	4.3kW (208Vac) Utility-Interactive Inverter	4300	96	Yes	NA
Motech Industries	PVMate 4900U (240V)	4.9kW (240Vac) Utility-Interactive Inverter	4900	96	Yes	NA
Motech Industries	PVMate 4900U-PG (208V)	4.3kW (208Vac) Utility-Interactive Inverter	4300	96	Yes	NA
Motech Industries	PVMate 4900U-PG (240V)	4.9kW (240Vac) Utility-Interactive Inverter	4900	96	Yes	NA
Motech Industries	PVMate 5300U (208V)	4.6kW (208Vac) Utility-Interactive Inverter	4600	95.5	Yes	NA
Motech Industries	PVMate 5300U (240V)	5.3kW (240Vac) Utility-Interactive Inverter	5300	96	Yes	NA
Motech Industries	PVMate 5300U-PG (208V)	4.6kW (208Vac) Utility-Interactive Inverter	4600	95.5	Yes	NA
Motech Industries	PVMate 5300U-PG (240V)	5.3kW (240Vac) Utility-Interactive Inverter	5300	96	Yes	NA
Ningbo Ginlong Technologies	GCI-2K (240V)	2 kW 240Vac, Utility-Interactive Inverter	2000	94	Yes	N/A
Open Energy	Solar Save 1100	1100W (240Vac) Utility Interactive Inverter	1100	90.5	Yes	NA
Open Energy	Solar Save 2000	2000W (240Vac) Utility Interactive Inverter	2000	92	Yes	NA
Open Energy	Solar Save 2500	2500W (240Vac) Utility Interactive Inverter	2500	94.5	Yes	NA
Open Energy	Solar Save 3500	3500W (240Vac) Utility Interactive Inverter	3500	95.5	Yes	NA
Open Energy	Solar Save 4600	4600W (208Vac) Utility Interactive Inverter	4600	95.5	Yes	NA

Open Energy	Solar Save 4800	4800W (240Vac) Utility Interactive Inverter	4800	96	Yes	NA
Open Energy	Solar Save 5200	5200W (240Vac) Utility Interactive Inverter	5200	96	Yes	NA
OPTI International	GT 1500 (208V)	1.5kW (208Vac) Utility-Interactive Inverter	1500	94.5	Yes	NA
OPTI International	GT 1500 (240V)	1.5kW (240Vac) Utility-Interactive Inverter	1500	94.5	Yes	NA
OPTI International	GT 2000 (208V)	2kW (208Vac) Utility-Interactive Inverter	2000	95.5	Yes	NA
OPTI International	GT 2000 (240V)	2kW (240Vac) Utility-Interactive Inverter	2000	95	Yes	NA
OPTI International	GT 3000 (208V)	3kW (208Vac) Utility-Interactive Inverter	3000	95.5	Yes	NA
OPTI International	GT 3000 (240V)	3kW (240Vac) Utility-Interactive Inverter	3000	95.5	Yes	NA
OutBack Power Systems	GVFX 3048	3000 W Utility Interactive (w/battery backup) 48Vdc Inverter	2560	91	No	NA
OutBack Power Systems	GVFX 3648	3600 W Utility Interactive (w/battery backup) 48vdc Inverter	3037	91	No	NA
Phoenixtec Power	PS240US3R (208V)	4000W (208Vac) Utility Interactive Inverter	4000	95.5	Yes	NA
Phoenixtec Power	PS240US3R (240V)	4000W (240Vac) Utility Interactive Inverter	4000	95.5	Yes	NA
Powercom	SLK-1500 (208V)	1.5kW (208Vac) Utility-Interactive Inverter	1500	94.5	Yes	N/A
Powercom	SLK-1500 (240V)	1.5kW (240Vac) Utility-Interactive Inverter	1500	94.5	Yes	N/A
Powercom	SLK-2000 (208V)	2kW (208Vac) Utility-Interactive Inverter	2000	95.5	Yes	N/A
Powercom	SLK-2000 (240V)	2kW (240Vac) Utility-Interactive Inverter	2000	95	Yes	N/A
Powercom	SLK-3000 (208V)	3kW (208Vac) Utility-Interactive Inverter	3000	95.5	Yes	N/A
Powercom	SLK-3000 (240V)	3kW (240Vac) Utility-Interactive Inverter	3000	95.5	Yes	N/A
Powercom	SLK-4000 (208V)	4kW (208Vac) Utility-Interactive Inverter	4000	95.5	Yes	NA
Powercom	SLK-4000 (240V)	4kW (240Vac) Utility-Interactive Inverter	4000	96	Yes	NA
Power-One	PVI-3.0-OUTD-S-US (208 V)	3 kW (208Vac) Utility Interactive Inverter	3000	96	Yes	NA
Power-One	PVI-3.0-OUTD-S-US (240 V)	3 kW (240Vac) Utility Interactive Inverter	3000	96	Yes	NA

(Continued)

MANUFACTURER NAME	INVERTER MODEL NUMBER	DESCRIPTION	POWER RATING (WATTS)	WEIGHTED EFFICIENCY	APPROVED BUILT-IN METER	NOTES
Power-One	PVI-3.0-OUTD-S-US (277 V)	3 kW (277Vac) Utility Interactive Inverter	3000	96	Yes	NA
Power-One	PVI-3.6-OUTD-S-US (208 V)	3.6 kW (208Vac) Utility Interactive Inverter	3600	96	Yes	NA
Power-One	PVI-3.6-OUTD-S-US (240 V)	3.6 kW (240Vac) Utility Interactive Inverter	3600	96	Yes	NA
Power-One	PVI-3.6-OUTD-S-US (277 V)	3.6 kW (277Vac) Utility Interactive Inverter	3600	96	Yes	NA
Power-One	PVI-3000-I-OUTD-US	3 kW, 150-600 VDC Utility Interactive Inverter	2520	92	Yes	NA
Power-One	PVI-3000-I-OUTD-US (208V)	3 kW, 150-600 VDC Utility Interactive Inverter	2500	90.5	Yes	NA
Power-One	PVI-3000-I-OUTD-US-F (208V)	3 kW, 150-600 VDC Utility Interactive Inverter	2700	90.5	Yes	NA
Power-One	PVI-3000-I-OUTD-US-F (240V)	3 kW, 150-600 VDC Utility Interactive Inverter	3000	91	Yes	NA
Power-One	PVI-3600-OUTD-US-F (208V)	3.3 kW, 150-600 VDC Utility Interactive Inverter	3300	94.5	Yes	NA
Power-One	PVI-3600-OUTD-US-F (240V)	3.6 kW, 150-600 VDC Utility Interactive Inverter	3600	95	Yes	NA
Power-One	PVI-3600-US (208V)	3.3 kW, 150-600 VDC Utility Interactive Inverter	3300	94	Yes	NA
Power-One	PVI-3600-US (240V)	3.6 kW, 150-600 VDC Utility Interactive Inverter	3600	94.5	Yes	NA
Power-One	PVI-4.2-OUTD-S-US (208 V)	4.2 kW (208Vac) Utility Interactive Inverter	4200	96	Yes	NA
Power-One	PVI-4.2-OUTD-S-US (240 V)	4.2 kW (240Vac) Utility Interactive Inverter	4200	96	Yes	NA

Power-One	PVI-4.2-OUTD-S-US (277 V)	4.2 kW (277Vac) Utility Interactive Inverter	4200	96	Yes	NA
Power-One	PVI-5000-OUTD-US (208 V)	5 kW (208Vac) Utility Interactive Inverter	5000	96	Yes	NA
Power-One	PVI-5000-OUTD-US (240 V)	5 kW (240Vac) Utility Interactive Inverter	5000	96.5	Yes	NA
Power-One	PVI-5000-OUTD-US (277 V)	5 kW (277Vac) Utility Interactive Inverter	5000	96.5	Yes	NA
Power-One	PVI-6000-OUTD-US (208 V)	6 kW (208Vac) Utility Interactive Inverter	6000	96	Yes	NA
Power-One	PVI-6000-OUTD-US (240 V)	6 kW (240Vac) Utility Interactive Inverter	6000	96.5	Yes	NA
Power-One	PVI-6000-OUTD-US (277 V)	6 kW (277Vac) Utility Interactive Inverter	6000	96.5	Yes	NA
Power-One	PVI-6000-OUTD-US-W (208 V)	6 kW (208Vac) Utility Interactive Inverter for Wind Turbines	6000	96	Yes	NA
Power-One	PVI-6000-OUTD-US-W (240 V)	6 kW (240Vac) Utility Interactive Inverter for Wind Turbines	6000	96.5	Yes	NA
Power-One	PVI-6000-OUTD-US-W (277 V)	6 kW (277Vac) Utility Interactive Inverter for Wind Turbines	6000	96.5	Yes	NA
Power-One	PVI-CENTRAL-100-US (208 V)	100 kW (208Vac) Utility Interactive Inverter	100000	95	Yes	NA
Power-One	PVI-CENTRAL-100-US (480 V)	100 kW (480Vac) Utility Interactive Inverter	100000	95	Yes	NA
Power-One	PVI-CENTRAL-50-US (208 V)	50 kW (208Vac) Utility Interactive Inverter	50000	95	Yes	NA
Power-One	PVI-CENTRAL-50-US (480 V)	50 kW (480Vac) Utility Interactive Inverter	50000	95	Yes	NA
Princeton Power Systems	GTIB-480-100-xxxx(480Vdc Max)	100 kW, 480Vac, Utility Interactive Inverter (480Vdc Max)	100000	94.5	No	NA

(Continued)

MANUFACTURER NAME	INVERTER MODEL NUMBER	DESCRIPTION	POWER RATING (WATTS)	WEIGHTED EFFICIENCY	APPROVED BUILT-IN METER	NOTES
Princeton Power Systems	GTIB-480-100-xxxx(600Vdc Max)	100 kW, 480Vac, Utility Interactive Inverter (600Vdc Max)	100000	95	No	NA
PV Powered	PVP100KW-208	100kW (208Vac) 3-Phase Utility Interactive Inverter w/ 295-600Vdc input	100000	95.5	Yes	NA
PV Powered	PVP100KW-480	100kW (480Vac) 3-Phase Utility Interactive Inverter w/ 295-600Vdc input	100000	96	Yes	NA
PV Powered	PVP1100	1100W (120Vac) Utility Interactive Inverter	1100	91.5	Yes	NA
PV Powered	PVP1100EVR	1100W (120Vac) Utility Interactive Inverter w/ extended voltage range	1100	90.5	Yes	NA
PV Powered	PVP2000	2000W (240Vac) Utility Interactive Inverter	2000	92.5	Yes	NA
PV Powered	PVP2000EVR	2000W (240Vac) Utility Interactive Inverter w/ extended voltage range	2000	92	Yes	NA
PV Powered	PVP2500	2500W (240Vac) Utility Interactive Inverter	2500	94.5	Yes	NA
PV Powered	PVP260KW	260kW (480Vac) 3-Phase Utility Interactive Inverter w/ 295-600Vdc input	260000	97	Yes	NA
PV Powered	PVP260KW-LV	260kW (480Vac) 3-Phase Utility Interactive Inverter w/ 265-600Vdc input	260000	96.5	Yes	NA
PV Powered	PVP2800	2800W (208Vac) Utility Interactive Inverter	2800	92	Yes	NA
PV Powered	PVP3000	3000W (240Vac) Utility Interactive Inverter	3000	93.5	Yes	NA
PV Powered	PVP30KW-208	30kW (208Vac) 3-Phase Utility Interactive Inverter w/ 330-600Vdc input	30000	94	Yes	NA
PV Powered	PVP30KW-480	30kW (480Vac) 3-Phase Utility Interactive Inverter w/ 330-600Vdc input	30000	94	Yes	NA
PV Powered	PVP30KW-LV-208	30kW (208Vac) 3-Phase Utility Interactive Inverter w/ 295-600Vdc input	30000	93	Yes	NA
PV Powered	PVP30KW-LV-480	30kW (480Vac) 3-Phase Utility Interactive Inverter w/ 295-600Vdc input	30000	93.5	Yes	NA

Manufacturer	Model	Description				
PV Powered	PVP3500	3500W (240Vac) Utility Interactive Inverter	3500	95.5	Yes	NA
PV Powered	PVP4600	4600W (208Vac) Utility Interactive Inverter	4600	95.5	Yes	NA
PV Powered	PVP4800	4800W (240Vac) Utility Interactive Inverter	4800	96	Yes	NA
PV Powered	PVP5200	5200W (240Vac) Utility Interactive Inverter	5200	96	Yes	NA
PV Powered	PVP75KW-208	75kW (208Vac) 3-Phase Utility Interactive Inverter w/ 295-600Vdc input	75000	95.5	Yes	NA
PV Powered	PVP75KW-480	75kW (480Vac) 3-Phase Utility Interactive Inverter w/ 295-600Vdc input	75000	95.5	Yes	NA
Renergy	RS-1500 (208V)	1.5kW (208Vac) Utility-Interactive Inverter	1500	94.5	Yes	NA
Renergy	RS-1500 (240V)	1.5kW (240Vac) Utility-Interactive Inverter	1500	94.5	Yes	NA
Renergy	RS-2000 (208V)	2kW (208Vac) Utility-Interactive Inverter	2000	95.5	Yes	NA
Renergy	RS-2000 (240V)	2kW (240Vac) Utility-Interactive Inverter	2000	95	Yes	NA
Renergy	RS-3000 (208V)	3kW (208Vac) Utility-Interactive Inverter	3000	95.5	Yes	NA
Renergy	RS-3000 (240V)	3kW (240Vac) Utility-Interactive Inverter	3000	95.5	Yes	NA
Renergy	RS-4000 (208V)	4kW (208Vac) Utility-Interactive Inverter	4000	95.5	Yes	NA
Renergy	RS-4000 (240V)	4kW (240Vac) Utility-Interactive Inverter	4000	96	Yes	NA
SatCon Technology	AE-100-60-PV-A	100kW 480Vac Three phase Utility Interactive inverter	100000	94.5	Yes	NA
SatCon Technology	AE-100-60-PV-A-HE	100kW 480Vac Three phase Utility Interactive inverter	100000	95.5	Yes	NA
SatCon Technology	AE-100-60-PV-D	100kW 208Vac Three phase Utility Interactive inverter	100000	94.5	Yes	NA
SatCon Technology	AE-100-60-PV-F	100kW 240Vac Three phase Utility Interactive inverter	100000	95	Yes	NA
SatCon Technology	AE-135-60-PV-A	135 kW 480Vac Three Phase Utility Interactive Inverter	135000	95.5	Yes	NA
SatCon Technology	AE-135-60-PV-D	135 kW 208Vac Three Phase Utility Interactive Inverter	135000	95	Yes	NA

(Continued)

MANUFACTURER NAME	INVERTER MODEL NUMBER	DESCRIPTION	POWER RATING (WATTS)	WEIGHTED EFFICIENCY	APPROVED BUILT-IN METER	NOTES
SatCon Technology	AE-225-60-PV-A	225kW 480Vac Three Phase Utility Interactive Inverter	225000	94.5	Yes	NA
SatCon Technology	AE-225-60-PV-D	225 kW 208Vac Three Phase Utility Interactive Inverter	225000	95.5	Yes	NA
SatCon Technology	AE-30-60-PV-A	30 kW 480Vac Three Phase Utility Interactive Inverter	30000	93	Yes	NA
SatCon Technology	AE-30-60-PV-D	30 kW 208Vac Three Phase Utility Interactive Inverter	30000	92.5	Yes	NA
SatCon Technology	AE-30-60-PV-E	30 kW 240Vac Single Phase Utility Interactive Inverter	30000	93	Yes	NA
SatCon Technology	AE-30-60-PV-F	30 kW 240Vac Three Phase Utility Interactive Inverter	30000	94.5	Yes	NA
SatCon Technology	AE-500-60-PV-A	500 kW 480Vac Three Phase Utility Interactive Inverter	500000	95	Yes	NA
SatCon Technology	AE-50-60-PV-A	50 kW 480Vac Three Phase Utility Interactive Inverter	50000	93	Yes	NA
SatCon Technology	AE-50-60-PV-D	50 kW 208Vac Three Phase Utility Interactive Inverter	50000	94.5	Yes	NA
SatCon Technology	AE-50-60-PV-F	50 kW 240Vac Three Phase Utility Interactive Inverter	50000	93	Yes	NA
SatCon Technology	AE-75-60-PV-A	75 kW 480Vac Three Phase Utility Interactive Inverter	75000	95.5	Yes	NA
SatCon Technology	AE-75-60-PV-D	75 kW 208Vac Three Phase Utility Interactive Inverter	75000	95.5	Yes	NA
SatCon Technology	PVS-100 (208 V)	100 kW 208Vac Three Phase Utility Interactive Inverter	100000	96	Yes	NA
SatCon Technology	PVS-100 (240 V)	100 kW 240Vac Three Phase Utility Interactive Inverter	100000	96	Yes	NA

SatCon Technology	PVS-100 (480 V)	100 kW 480Vac Three Phase Utility Interactive Inverter	100000	96	Yes	NA
SatCon Technology	PVS-1000 (MVT)	1000 kW Three Phase Inverter for Medium Voltage Transformer	1000000	96	Yes	NA
Satcon Technology	PVS-110-S-MT (208V)	110 kW 208Vac Three Phase Utility Interactive Inverter	110000	95.5	Yes	NA
Satcon Technology	PVS-110-S-MT (240V)	110 kW 240Vac Three Phase Utility Interactive Inverter	110000	95.5	Yes	NA
Satcon Technology	PVS-110-S-MT (480V)	110 kW 480Vac Three Phase Utility Interactive Inverter	110000	95	Yes	NA
SatCon Technology	PVS-135 (208 V)	135 kW 208Vac Three Phase Utility Interactive Inverter	135000	96	Yes	NA
SatCon Technology	PVS-135 (240 V)	135 kW 240Vac Three Phase Utility Interactive Inverter	135000	96	Yes	NA
SatCon Technology	PVS-135 (480 V)	135 kW 480Vac Three Phase Utility Interactive Inverter	135000	96	Yes	NA
Satcon Technology	PVS-210-S (208 V)	210 kW 208Vac Three Phase Utility Interactive Inverter	210000	95.5	Yes	NA
Satcon Technology	PVS-210-S (240 V)	210 kW 240Vac Three Phase Utility Interactive Inverter	210000	95.5	Yes	NA
Satcon Technology	PVS-210-S (480 V)	210 kW 480Vac Three Phase Utility Interactive Inverter	210000	95.5	Yes	NA
SatCon Technology	PVS-250 (208 V)	250 kW 208Vac Three Phase Utility Interactive Inverter	250000	96	Yes	NA
SatCon Technology	PVS-250 (240 V)	250 kW 240Vac Three Phase Utility Interactive Inverter	250000	96	Yes	NA
SatCon Technology	PVS-250 (480 V)	250 kW 480Vac Three Phase Utility Interactive Inverter	250000	96	Yes	NA
SatCon Technology	PVS-250 (MVT)	250 kW Three Phase Inverter for Medium Voltage Transformer	250000	95.5	Yes	NA

(Continued)

MANUFACTURER NAME	INVERTER MODEL NUMBER	DESCRIPTION	POWER RATING (WATTS)	WEIGHTED EFFICIENCY	APPROVED BUILT-IN METER	NOTES
SatCon Technology	PVS-30 (208 V)	30 kW 208Vac Three Phase Utility Interactive Inverter	30000	95	Yes	NA
SatCon Technology	PVS-30 (240 V)	30 kW 240Vac Three Phase Utility Interactive Inverter	30000	95	Yes	NA
SatCon Technology	PVS-30 (480 V)	30 kW 480Vac Three Phase Utility Interactive Inverter	30000	95	Yes	NA
SatCon Technology	PVS-375 (480 V)	375 kW 480Vac Three Phase Utility Interactive Inverter	375000	95.5	Yes	NA
SatCon Technology	PVS-50 (208 V)	50 kW 208Vac Three Phase Utility Interactive Inverter	50000	95.5	Yes	NA
SatCon Technology	PVS-50 (240 V)	50 kW 240Vac Three Phase Utility Interactive Inverter	50000	95.5	Yes	NA
SatCon Technology	PVS-50 (480 V)	50 kW 480Vac Three Phase Utility Interactive Inverter	50000	95.5	Yes	NA
SatCon Technology	PVS-500 (480 V)	500 kW 480Vac Three Phase Utility Interactive Inverter	500000	96	Yes	NA
SatCon Technology	PVS-500 (MVT)	500 kW Three Phase Inverter for Medium Voltage Transformer	500000	95.5	Yes	NA
Satcon Technology	PVS-50-S-MT(208V)	50 kW 208Vac Three Phase Utility Interactive Inverter	50000	95.5	Yes	NA
Satcon Technology	PVS-50-S-MT(240V)	50 kW 240Vac Three Phase Utility Interactive Inverter	50000	95.5	Yes	NA
Satcon Technology	PVS-50-S-MT(480V)	50 kW 480Vac Three Phase Utility Interactive Inverter	50000	95.5	Yes	NA
SatCon Technology	PVS-75 (208 V)	75 kW 208Vac Three Phase Utility Interactive Inverter	75000	96	Yes	NA
SatCon Technology	PVS-75 (240 V)	75 kW 240Vac Three Phase Utility Interactive Inverter	75000	96	Yes	NA

Manufacturer	Model	Description			
SatCon Technology	PVS-75 (480 V)	75 kW 480Vac Three Phase Utility Interactive Inverter	75000	96	Yes	NA
Satcon Technology	SDMS0100208LNIU	100 kW 208Vac Three Phase Inverter with Combiner	100000	95	Yes	NA
Satcon Technology	SDMS0100240LNIU	100 kW 240Vac Three Phase Inverter with Combiner	100000	95	Yes	NA
Satcon Technology	SDMS0100480LNIU	100 kW 480Vac Three Phase Inverter with Combiner	100000	95	Yes	NA
Schuco USA	SB3000US (208V)	3kW, 208Vac Sunny Boy Utility Interactive Inverter with display	3000	95	Yes	NA
Schuco USA	SB3000US (240V)	3kW, 240Vac Sunny Boy Utility Interactive Inverter with display	3000	95.5	Yes	NA
Schuco USA	SB3300U (240V)	3.3kW, 240Vac Sunny Boy Utility Interactive Inverter with display	3300	94.5	Yes	NA
Schuco USA	SB3800U (208V)	3.5kW, 208Vac Sunny Boy Utility Interactive Inverter with display	3500	94	Yes	NA
Schuco USA	SB3800U (240V)	3.8kW, 240Vac Sunny Boy Utility Interactive Inverter with display	3800	94.5	Yes	NA
Schuco USA	SB4000US (208V)	4kW, 208Vac Sunny Boy Utility Interactive Inverter with display	4000	95.5	Yes	NA
Schuco USA	SB4000US (240V)	4kW, 240Vac Sunny Boy Utility Interactive Inverter with display	4000	96	Yes	NA
Schuco USA	SB5000US (208V)	5kW, 208Vac Sunny Boy Utility Interactive Inverter with display	5000	95.5	Yes	NA
Schuco USA	SB5000US (240V)	5kW, 240Vac Sunny Boy Utility Interactive Inverter with display	5000	95.5	Yes	NA
Schuco USA	SB5000US (277V)	5kW, 277Vac Sunny Boy Utility Interactive Inverter with display	5000	95.5	Yes	NA
Schuco USA	SB6000U (208V)	6kW, 208Vac Sunny Boy Utility Interactive Inverter with display	5200	94.5	Yes	NA

(Continued)

MANUFACTURER NAME	INVERTER MODEL NUMBER	DESCRIPTION	POWER RATING (WATTS)	WEIGHTED EFFICIENCY	APPROVED BUILT-IN METER	NOTES
Schuco USA	SB6000U (240V)	6kW, 240Vac Sunny Boy Utility Interactive Inverter with display	6000	94.5	Yes	NA
Schuco USA	SB6000U (277V)	6kW, 277Vac Sunny Boy Utility Interactive Inverter with display	6000	94.5	Yes	NA
Schuco USA	SB6000US (208V)	6kW, 208Vac Sunny Boy Utility Interactive Inverter with display	6000	95.5	Yes	NA
Schuco USA	SB6000US (240V)	6kW, 240Vac Sunny Boy Utility Interactive Inverter with display	6000	95.5	Yes	NA
Schuco USA	SB6000US (277V)	6kW, 277Vac Sunny Boy Utility Interactive Inverter with display	6000	96	Yes	NA
Schuco USA	SB7000US (208V)	7kW, 208Vac Sunny Boy Utility Interactive Inverter with display	7000	95.5	Yes	NA
Schuco USA	SB7000US (240V)	7kW, 240Vac Sunny Boy Utility Interactive Inverter with display	7000	96	Yes	NA
Schuco USA	SB7000US (277V)	7kW, 277Vac Sunny Boy Utility Interactive Inverter with display	7000	96	Yes	NA
Schuco USA	ST42 (208V)	42kW, 208Vac Sunny Tower Utility Interactive Inverter with display	42000	95.5	Yes	NA
Schuco USA	ST42 (277V)	42kW, 277Vac Sunny Tower Utility Interactive Inverter with display	42000	96	Yes	NA
Schuco USA	SWR1800U	1.8kW, 120Vac Sunny Boy String Inverter	1800	91.5	Yes	NA
Schuco USA	SWR1800U-SBD	1.8kW, 120Vac Sunny Boy String Inverter with display	1800	91.5	Yes	NA
Schuco USA	SWR2100U	2.1kW, 240Vac Sunny Boy String Inverter	2100	93	Yes	NA
Schuco USA	SWR2500U (208V)	2.1kW, 208Vac Sunny Boy String Inverter	2100	92.5	Yes	NA
Schuco USA	SWR2500U (240V)	2.5kW, 240Vac Sunny Boy String Inverter	2500	93	Yes	NA

Schuco USA	SWR2500U-SBD (208V)	2.1kW, 208Vac Sunny Boy String Inverter with display	2100	92.5	Yes	NA
Schuco USA	SWR2500U-SBD (240V)	2.5kW, 240Vac Sunny Boy String Inverter with display	2500	93	Yes	NA
Sharp	JH-3500U	Utility Interactive Inverter 240Vac L-L, 3.5kW	3500	91	Yes	NA
Siemens Industry	SINVERT PVS1051 UL	1050 kW 480Vac Three Phase Inverter (Master Unit, 2 Slave Units)	1050000	96	Yes	NA
Siemens Industry	SINVERT PVS1401 UL	1400 kW 480Vac Three Phase Inverter (Master Unit, 3 Slave Units)	1400000	-96	Yes	NA
Siemens Industry	SINVERT PVS351 UL	350 kW 480Vac Three Phase Inverter (Master Unit)	350000	96	Yes	NA
Siemens Industry	SINVERT PVS701 UL	700 kW 480Vac Three Phase Inverter (Master Unit, 1 Slave Unit)	700000	96	Yes	NA
SMA America	SB1100U	1100W, 240Vac Sunny Boy String Inverter	1100	91	No	NA
SMA America	SB1100U-SBD	1100W, 240Vac Sunny Boy String Inverter	1100	91	Yes	NA
SMA America	SB3000US (208V)	3kW, 208Vac Sunny Boy Utility Interactive Inverter with display	3000	95	Yes	NA
SMA America	SB3000US (240V)	3kW, 240Vac Sunny Boy Utility Interactive Inverter with display	3000	95.5	Yes	NA
SMA America	SB3300U	3.3kW, 240Vac Sunny Boy Utility Interactive Inverter with display	3300	94.5	Yes	NA
SMA America	SB3800U (208V)	3.5kW, 208Vac Sunny Boy Utility Interactive Inverter with display	3500	94	Yes	NA
SMA America	SB3800U (240V)	3.8kW, 240Vac Sunny Boy Utility Interactive Inverter with display	3800	94.5	Yes	NA
SMA America	SB4000US (208V)	4kW, 208Vac Sunny Boy Utility Interactive Inverter with display	3500	95.5	Yes	NA
SMA America	SB4000US (240V)	4kW, 240Vac Sunny Boy Utility Interactive Inverter with display	4000	96	Yes	NA

(Continued)

MANUFACTURER NAME	INVERTER MODEL NUMBER	DESCRIPTION	POWER RATING (WATTS)	WEIGHTED EFFICIENCY	APPROVED BUILT-IN METER	NOTES
SMA America	SB4000US (CL) (208V)	4kW, 208Vac-Current Limited Utility Interactive Inverter with display	3300	95.5	Yes	NA
SMA America	SB4000US (CL) (240V)	4kW, 240Vac-Current Limited Utility Interactive Inverter with display	3800	96	Yes	NA
SMA America	SB5000US (208V)	5kW, 208Vac Sunny Boy Utility Interactive Inverter with display	5000	95.5	Yes	NA
SMA America	SB5000US (240V)	5kW, 240Vac Sunny Boy Utility Interactive Inverter with display	5000	95.5	Yes	NA
SMA America	SB5000US (277V)	5kW, 277Vac Sunny Boy Utility Interactive Inverter with display	5000	95.5	Yes	NA
SMA America	SB6000U (208V)	6kW, 208Vac Sunny Boy Utility Interactive Inverter with display	5200	94.5	Yes	NA
SMA America	SB6000U (240V)	6kW, 240Vac Sunny Boy Utility Interactive Inverter with display	6000	94.5	Yes	NA
SMA America	SB6000U (277V)	6kW, 277Vac Sunny Boy Utility Interactive Inverter with display	6000	94.5	Yes	NA
SMA America	SB6000US (208V)	6kW, 208Vac Sunny Boy Utility Interactive Inverter with display	6000	95.5	Yes	NA
SMA America	SB6000US (240V)	6kW, 240Vac Sunny Boy Utility Interactive Inverter with display	6000	95.5	Yes	NA
SMA America	SB6000US (277V)	6kW, 277Vac Sunny Boy Utility Interactive Inverter with display	6000	96	Yes	NA
SMA America	SB7000US (208V)	7kW, 208Vac Sunny Boy Utility Interactive Inverter with display	7000	95.5	Yes	NA
SMA America	SB7000US (240V)	7kW, 240Vac Sunny Boy Utility Interactive Inverter with display	7000	96	Yes	NA
SMA America	SB7000US (277V)	7kW, 277Vac Sunny Boy Utility Interactive Inverter with display	7000	96	Yes	NA

SMA America	SB700U	700W, 120Vac Sunny Boy String Inverter	696	91.5	No	NA
SMA America	SB700U-SBD	700W, 120Vac Sunny Boy String Inverter with display	696	91.5	Yes	NA
SMA America	SB8000US (240V)	8kW, 240Vac Sunny Boy Utility Interactive Inverter with display	7600	96	Yes	NA
SMA America	SB8000US (277V)	8kW, 277Vac Sunny Boy Utility Interactive Inverter with display	8000	96	Yes	NA
SMA America	SC125U (208V)	125 kW 3-phase 208Vac, 275-600Vdc, Utility Interactive Inverter	125000	93.5	Yes	NA
SMA America	SC125U (480V)	125 kW 3-phase 480Vac, 275-600Vdc, Utility Interactive Inverter	125000	94	Yes	NA
SMA America	SC250U (480V)	250 kW 3-phase 480Vac, Utility Interactive Inverter	250000	97	Yes	NA
SMA America	SC500HE-US-MV (w/ Cooper xfmr)	500 kW 3-phase, Util. Interactive Inverter with Medium Voltage Cooper Xfmr	500000	97	Yes	N/A
SMA America	SC500HE-US-MV (w/ TP1 xfmr)	500 kW 3-phase, Utility Interactive Inverter with Medium Voltage TP1 Xfmr	500000	96.5	Yes	N/A
SMA America	SC500U	500 kW 3-phase 480Vac, Utility Interactive Inverter	500000	97	Yes	NA
SMA America	ST36 (208V)	36kW, 208Vac Sunny Tower Utility Interactive Inverter with display	36000	95.5	Yes	NA
SMA America	ST36 (240V)	36kW, 240Vac Sunny Tower Utility Interactive Inverter with display	36000	95.5	Yes	NA
SMA America	ST36 (277V)	36kW, 277Vac Sunny Tower Utility Interactive Inverter with display	36000	96	Yes	NA
SMA America	ST42 (208V)	42kW, 208Vac Sunny Tower Utility Interactive Inverter with display	42000	95.5	Yes	NA
SMA America	ST42 (240V)	42kW, 240Vac Sunny Tower Utility Interactive Inverter with display	42000	96	Yes	NA

(Continued)

647

MANUFACTURER NAME	INVERTER MODEL NUMBER	DESCRIPTION	POWER RATING (WATTS)	WEIGHTED EFFICIENCY	APPROVED BUILT-IN METER	NOTES
SMA America	ST42 (277V)	42kW, 277Vac Sunny Tower Utility Interactive Inverter with display	42000	96	Yes	NA
SMA America	ST48 (240V)	45.6kW, 240Vac Sunny Tower Utility Interactive Inverter with display	45600	96	Yes	NA
SMA America	ST48 (277V)	48kW, 240Vac Sunny Tower Utility Interactive Inverter with display	48000	96	Yes	NA
SMA America	SWR1800U	1.8kW, 120Vac Sunny Boy String Inverter	1800	91.5	No	NA
SMA America	SWR1800U-SBD	1.8kW, 120Vac Sunny Boy String Inverter with display	1800	91.5	Yes	NA
SMA America	SWR2100U	2.1kW, 240Vac Sunny Boy String Inverter	2100	93	No	NA
SMA America	SWR2100U-SBD	2.1 kW, 240Vac Sunny Boy String Inverter	2100	93	Yes	NA
SMA America	SWR2500U (208V)	2.1kW, 208Vac Sunny Boy String Inverter	2100	92.5	No	NA
SMA America	SWR2500U (240V)	2.5kW, 240Vac Sunny Boy String Inverter	2500	93	No	NA
SMA America	SWR2500U-SBD (208V)	2.1kW, 208Vac Sunny Boy String Inverter with display	2100	92.5	Yes	NA
SMA America	SWR2500U-SBD (240V)	2.5kW, 240Vac Sunny Boy String Inverter with display	2500	93	Yes	NA
SolarEdge Technologies	SE3300 (208V)	3.3 kW 208Vac, Utility-Interactive Inverter	3300	97	Yes	N/A
SolarEdge Technologies	SE3300 (240V)	3.3 kW 240Vac, Utility-Interactive Inverter	3300	97.5	Yes	N/A
SolarEdge Technologies	SE4000 (208V)	4 kW 208Vac, Utility-Interactive Inverter	4000	97	Yes	N/A
SolarEdge Technologies	SE4000 (240V)	4 kW 240Vac, Utility-Interactive Inverter	4000	97.5	Yes	N/A
SolarEdge Technologies	SE5000 (208V)	5 kW 208Vac, Utility-Interactive Inverter	5000	97	Yes	N/A
SolarEdge Technologies	SE5000 (240V)	5 kW 240Vac, Utility-Interactive Inverter	5000	97.5	Yes	N/A
SolarEdge Technologies	SE6000 (208V)	5.3 kW 208Vac, Utility-Interactive Inverter	5300	97	Yes	N/A

SolarEdge Technologies	SE6000 (240V)	6 kW 240Vac, Utility-Interactive Inverter	6000	97.5	Yes	N/A
Solectria Renewables	PVI 3000 (208V)	2.7kW (208Vac) Utility-Interactive Inverter	2700	95.5	Yes	NA
Solectria Renewables	PVI 3000 (240V)	2.9kW (240Vac) Utility-Interactive Inverter	2900	96	Yes	NA
Solectria Renewables	PVI 3000-P (208V)	2.7kW (208Vac) Utility-Interactive Inverter	2700	95.5	Yes	NA
Solectria Renewables	PVI 3000-P (240V)	2.9kW (240Vac) Utility-Interactive Inverter	2900	96	Yes	NA
Solectria Renewables	PVI 4000 (208V)	3.4kW (208Vac) Utility-Interactive Inverter	3400	95.5	Yes	NA
Solectria Renewables	PVI 4000 (240V)	3.9kW (240Vac) Utility-Interactive Inverter	3900	96	Yes	NA
Solectria Renewables	PVI 4000-P (208V)	3.4kW (208Vac) Utility-Interactive Inverter	3400	95.5	Yes	NA
Solectria Renewables	PVI 4000-P (240V)	3.9kW (240Vac) Utility-Interactive Inverter	3900	96	Yes	NA
Solectria Renewables	PVI 5000 (208V)	4.3kW (208Vac) Utility-Interactive Inverter	4300	96	Yes	NA
Solectria Renewables	PVI 5000 (240V)	4.9kW (240Vac) Utility-Interactive Inverter	4900	96	Yes	NA
Solectria Renewables	PVI 5000-P (208V)	4.3kW (208Vac) Utility-Interactive Inverter	4300	96	Yes	NA
Solectria Renewables	PVI 5000-P (240V)	4.9kW (240Vac) Utility-Interactive Inverter	4900	96	Yes	NA
Solectria Renewables	PVI 5300 (208V)	4.6kW (208Vac) Utility-Interactive Inverter	4600	95.5	Yes	NA
Solectria Renewables	PVI 5300 (240V)	5.3kW (240Vac) Utility-Interactive Inverter	5300	96	Yes	NA
Solectria Renewables	PVI 5300-P (208V)	4.6kW (208Vac) Utility-Interactive Inverter	4600	95.5	Yes	NA
Solectria Renewables	PVI 5300-P (240V)	5.3kW (240Vac) Utility-Interactive Inverter	5300	96	Yes	NA
Solectria Renewables	PVI13kW-208	13 kW 208Vac Commercial Grid-Tied Solar PV Inverter	13200	94	No	NA
Solectria Renewables	PVI13kW-480	13 kW 480Vac Commercial Grid-Tied Solar PV Inverter	13200	94.5	No	NA

(Continued)

MANUFACTURER NAME	INVERTER MODEL NUMBER	DESCRIPTION	POWER RATING (WATTS)	WEIGHTED EFFICIENCY	APPROVED BUILT-IN METER	NOTES
Solectria Renewables	PVI15kW-208	15 kW 208Vac Commercial Grid-Tied Solar PV Inverter	15000	94	No	NA
Solectria Renewables	PVI15kW-480	15 kW 480Vac Commercial Grid-Tied Solar PV Inverter	15000	94.5	No	NA
Solectria Renewables	PVI1800-208	1800W 208Vac Grid-Tied Solar PV Inverter	1799	92.5	Yes	NA
Solectria Renewables	PVI1800-240	1800W 240Vac Grid-Tied Solar PV Inverter	1759	92.5	Yes	NA
Solectria Renewables	PVI2500-208	2500W 208Vac Grid-Tied Solar PV Inverter	2485	92	Yes	NA
Solectria Renewables	PVI2500-240	2500W 240Vac Grid-Tied Solar PV Inverter	2498	93	Yes	NA
Solectria Renewables	PVI60kW-208	60 kW 208Vac Commercial Grid-Tied Solar PV Inverter	60000	94	No	NA
Solectria Renewables	PVI60kW-480	60 kW 480Vac Commercial Grid-Tied Solar PV Inverter	60000	95.5	No	NA
Solectria Renewables	PVI82kW-208	82 kW 208Vac Commercial Grid-Tied Solar PV Inverter	82000	94.5	No	NA
Solectria Renewables	PVI82kW-480	82 kW 480Vac Commercial Grid-Tied Solar PV Inverter	82000	95.5	No	NA
Solectria Renewables	PVI95kW-208	95 kW 208Vac Commercial Grid-Tied Solar PV Inverter	95000	94.5	No	NA
Solectria Renewables	PVI95kW-480	95 kW 480Vac Commercial Grid-Tied Solar PV Inverter	95000	95.5	No	NA
Sonnetek	Sonnetek 2000	2000W (120Vac) Utility Interactive Inverter	2000	91.5	Yes	NA
Southwest Windpower	Skystream 3.7 inverter (208V)	1.8 kW, 208Vac Utility-Interactive Inverter integrated into turbine	1800	92.5	No	NA
Southwest Windpower	Skystream 3.7 inverter (240V)	1.8 kW, 240Vac Utility-Interactive Inverter integrated into turbine	1800	93	No	NA
Sungrow Power Supply	SG100KU	100kW (480Vac) Utility-Interactive Inverter	100000	96.5	Yes	NA
SunPower	SPR 11401f	11400W (240Vac) Utility Interactive Inverter	11400	95.5	Yes	NA

SunPower	SPR 11401f	11400W (208Vac) Utility Interactive Inverter	11400	95	Yes	NA
SunPower	SPR 12000f (208V)	11400W (208Vac) 3-Phase Utility Interactive Inverter	11400	95	Yes	NA
SunPower	SPR 12000f (240V)	12000W (240Vac) 3-Phase Utility Interactive Inverter	12000	95.5	Yes	NA
SunPower	SPR-10000f-208	10000W (208Vac) Utility Interactive Inverter	10000	95	Yes	N/A
SunPower	SPR-10000f-240	10000W (240Vac) Utility Interactive Inverter	10000	95.5	Yes	N/A
SunPower	SPR-10000f-277	10000W (277Vac) Utility Interactive Inverter	10000	96	Yes	N/A
SunPower	SPR-10001f-208	10000W (208Vac) Utility Interactive Inverter	10000	95	Yes	NA
SunPower	SPR-10001f-240	10000W (240Vac) Utility Interactive Inverter	10000	95.5	Yes	NA
SunPower	SPR-10001f-277	10000W (277Vac) Utility Interactive Inverter	10000	96	Yes	NA
SunPower	SPR-11400f-3-208	11400W (208Vac) 3-phase Utility Interactive Inverter	11400	95	Yes	N/A
SunPower	SPR-11400f-3-240	11400W (240Vac) 3-phase Utility Interactive Inverter	11400	95.5	Yes	N/A
SunPower	SPR-11401f-3-208 Delta	11400W (208Vac) 3-phase Utility Interactive Inverter	11400	95	Yes	NA
SunPower	SPR-11401f-3-240 Delta	11400W (240Vac) 3-phase Utility Interactive Inverter	11400	96	Yes	NA
SunPower	SPR-12000f-3-277	12000W (277Vac) 3-phase Utility Interactive Inverter	12000	96	Yes	N/A
SunPower	SPR-12001f-3-277WYE	12000W (277Vac) 3-phase Utility Interactive Inverter	12000	96	Yes	NA
SunPower	SPR-2500	2500W (240Vac) Utility Interactive Inverter	2500	94.5	Yes	NA

(*Continued*)

MANUFACTURER NAME	INVERTER MODEL NUMBER	DESCRIPTION	POWER RATING (WATTS)	WEIGHTED EFFICIENCY	APPROVED BUILT-IN METER	NOTES
SunPower	SPR-2500x	2.5 kW, 240Vac, 195-600Vdc Grid Tie Inverter	2500	94	Yes	NA
SunPower	SPR-2800x (208V)	2.7 kW, 208 Vac, 195-600Vdc Grid Tie Inverter	2700	93.5	Yes	NA
SunPower	SPR-2800x (240V)	2.8 kW, 240 Vac, 195-600Vdc Grid Tie Inverter	2800	94	Yes	NA
SunPower	SPR-3000m (208V)	3kW, 208Vac Utility Interactive Inverter with display	3000	95	Yes	NA
SunPower	SPR-3000m (240V)	3kW, 240Vac Utility Interactive Inverter with display	3000	95.5	Yes	NA
SunPower	SPR-3300f-208	3300W (208Vac) Utility Interactive Inverter	3300	95	Yes	NA
SunPower	SPR-3300f-240	3300W (240Vac) Utility Interactive Inverter	3300	95.5	Yes	NA
SunPower	SPR-3300f-277	3300W (277Vac) Utility Interactive Inverter	3300	95.5	Yes	NA
SunPower	SPR-3300x (208V)	3.1 kW, 208 Vac, 195-600Vdc Grid Tie Inverter	3100	95	Yes	NA
SunPower	SPR-3300x (240V)	3.3 kW, 240 Vac, 195-600Vdc Grid Tie Inverter	3300	95.5	Yes	NA
SunPower	SPR-3301f-208	3300W (208Vac) Utility Interactive Inverter	3300	95.5	Yes	NA
SunPower	SPR-3301f-240	3300W (240Vac) Utility Interactive Inverter	3300	95.5	Yes	NA
SunPower	SPR-3301f-277	3300W (277Vac) Utility Interactive Inverter	3300	96	Yes	NA
SunPower	SPR-3500	3500W (240Vac) Utility Interactive Inverter	3500	95.5	Yes	NA
SunPower	SPR-3801f-208	3800W (208Vac) Utility Interactive Inverter	3800	95	Yes	NA
SunPower	SPR-3801f-240	3800W (240Vac) Utility Interactive Inverter	3800	95.5	Yes	NA
SunPower	SPR-3801f-277	3800W (277Vac) Utility Interactive Inverter	3800	96	Yes	NA
SunPower	SPR-4000f-208	4000W (208Vac) Utility Interactive Inverter	4000	95	Yes	NA
SunPower	SPR-4000f-240	4000W (240Vac) Utility Interactive Inverter	4000	95.5	Yes	NA

SunPower	SPR-4000f-277	4000W (277Vac) Utility Interactive Inverter	4000	95.5	Yes	NA
SunPower	SPR-4000m (208V)	4kW, 208Vac Utility Interactive Inverter with display	4000	95.5	Yes	NA
SunPower	SPR-4000m (240V)	4kW, 240Vac Utility Interactive Inverter with display	4000	96	Yes	NA
SunPower	SPR-4000x (208V)	3.8 kW, 208 Vac, 195-600Vdc Grid Tie Inverter	3800	95	Yes	NA
SunPower	SPR-4000x (240V)	4.0 kW, 240 Vac, 195-600Vdc Grid Tie Inverter	4000	95.5	Yes	NA
SunPower	SPR-4600	4600W (208Vac) Utility Interactive Inverter	4600	95.5	Yes	NA
SunPower	SPR-5000m (208V)	5kW, 208Vac Utility Interactive Inverter with display	5000	95.5	Yes	NA
SunPower	SPR-5000m (240V)	5kW, 240Vac Utility Interactive Inverter with display	5000	95.5	Yes	NA
SunPower	SPR-5000m (277V)	5kW, 277Vac Utility Interactive Inverter with display	5000	95.5	Yes	NA
SunPower	SPR-5000x (208V)	4.5 kW, 208 Vac, 195-600Vdc Grid Tie Inverter	4500	95	Yes	NA
SunPower	SPR-5000x (240V)	5.0 kW, 240 Vac, 195-600Vdc Grid Tie Inverter	5000	95.5	Yes	NA
SunPower	SPR-5200	5200W (240Vac) Utility Interactive Inverter	5200	96	Yes	NA
SunPower	SPR-6000m (208V)	6kW, 208Vac Utility Interactive Inverter with display	6000	95.5	Yes	NA
SunPower	SPR-6000m (240V)	6kW, 240Vac Utility Interactive Inverter with display	6000	95.5	Yes	NA
SunPower	SPR-6000m (277V)	6kW, 277Vac Utility Interactive Inverter with display	6000	96	Yes	NA
SunPower	SPR-6500f-208	6500W (208Vac) Utility Interactive Inverter	6500	95	Yes	NA
SunPower	SPR-6500f-240	6500W (240Vac) Utility Interactive Inverter	6500	95.5	Yes	NA

(Continued)

MANUFACTURER NAME	INVERTER MODEL NUMBER	DESCRIPTION	POWER RATING (WATTS)	WEIGHTED EFFICIENCY	APPROVED BUILT-IN METER	NOTES
SunPower	SPR-6500f-277	6500W (277Vac) Utility Interactive Inverter	6500	96	Yes	NA
SunPower	SPR-6501f-208	6500W (208Vac) Utility Interactive Inverter	6500	95	Yes	NA
SunPower	SPR-6501f-240	6500W (240Vac) Utility Interactive Inverter	6500	95.5	Yes	NA
SunPower	SPR-6501f-277	6500W (277Vac) Utility Interactive Inverter	6500	96	Yes	NA
SunPower	SPR-7000m (208V)	7kW, 208Vac Utility Interactive Inverter with display	7000	95.5	Yes	NA
SunPower	SPR-7000m (240V)	7kW, 240Vac Utility Interactive Inverter with display	7000	96	Yes	NA
SunPower	SPR-7000m (277V)	7kW, 277Vac Utility Interactive Inverter with display	7000	96	Yes	NA
SunPower	SPR-7501f-208	7500W (208Vac) Utility Interactive Inverter	7500	95	Yes	NA
SunPower	SPR-7501f-240	7500W (240Vac) Utility Interactive Inverter	7500	95.5	Yes	NA
SunPower	SPR-7501f-277	7500W (277Vac) Utility Interactive Inverter	7500	96	Yes	NA
SunPower	SPR-8000f-208	7600W (208Vac) Utility Interactive Inverter	7600	95	Yes	NA
SunPower	SPR-8000f-240	8000W (240Vac) Utility Interactive Inverter	8000	95.5	Yes	NA
SunPower	SPR-8000f-277	8000W (277Vac) Utility Interactive Inverter	8000	96	Yes	NA
Sunset	SUNstring 4000	2.9 kW, 240 Vac, 125-400Vdc Utility Interactive Inverter	2864	94	Yes	NA
Sunset	SUNstring 5000	3.6 kW, 240 Vac, 125-400Vdc Utility Interactive Inverter	3600	93.5	Yes	NA
Sysgration	Soleil 2000-120	2000W (120Vac) Utility Interactive Inverter	2000	91.5	Yes	NA
Windterra Systems	ECO1200	1200W (120Vac) Utility-Interactive Inverter for ECO1200 Wind Turbine	1200	85.5	Yes	NA
Xantrex Technology	GT100-208	100 kW 208 Vac Three Phase Utility Interactive Inverter	100000	95	Yes	NA
Xantrex Technology	GT100-208-PG	100 kW 208 Vac Three Phase Utility Interactive Inverter, Positive Ground	100000	95	Yes	NA

Xantrex Technology	GT100-480	100 kW 480 Vac Three Phase Utility Interactive Inverter	100000	96	Yes	NA
Xantrex Technology	GT100-480-PG	100 kW 480 Vac Three Phase Utility Interactive Inverter, Positive Ground	100000	96	Yes	NA
Xantrex Technology	GT2.5-NA-DS-240	2.5 kW, 240 Vac, 195-600Vdc Grid Tie Inverter	2500	94	Yes	NA
Xantrex Technology	GT2.5-NA-DS-240-P	2.5 kW, 240 Vac, 195-600Vdc Grid Tie Inverter (positive ground)	2500	94	Yes	NA
Xantrex Technology	GT2.8-NA-240/208 (208V)	2.7 kW, 208 Vac, 195-600Vdc Grid Tie Inverter	2700	93.5	Yes	NA
Xantrex Technology	GT2.8-NA-240/208 (240V)	2.8 kW, 240 Vac, 195-600Vdc Grid Tie Inverter	2800	94	Yes	NA
Xantrex Technology	GT250-480	250 kW 480 Vac Three Phase Utility Interactive Inverter	250000	96	Yes	NA
Xantrex Technology	GT250-480-PG	250 kW 480 Vac Three Phase Utility Interactive Inverter, Positive Ground	250000	96	Yes	NA
Xantrex Technology	GT3.0-NA-DS-240	3.0 kW, 240 Vac, 195-600Vdc Grid Tie Inverter	3000	94.5	Yes	NA
Xantrex Technology	GT3.0-NA-DS-240-P	3.0 kW, 240 Vac, 195-600Vdc Grid Tie Inverter	3000	94.5	Yes	NA
Xantrex Technology	GT3.3-NA-DS-208	3.3 kW, 208 Vac, 195-600Vdc Grid Tie Inverter	3300	94	Yes	NA
Xantrex Technology	GT3.3-NA-DS-208-P or POS	3.3 kW, 208 Vac, 195-600Vdc Grid Tie Inverter	3300	94	Yes	NA
Xantrex Technology	GT3.3-NA-DS-240	3.3 kW, 240 Vac, 195-600Vdc Grid Tie Inverter	3300	94.5	Yes	NA
Xantrex Technology	GT3.3-NA-DS-240-P or POS	3.3 kW, 240 Vac, 195-600Vdc Grid Tie Inverter	3300	94.5	Yes	NA
Xantrex Technology	GT3.3N-NA-240/208 (208V)	3.1 kW, 208 Vac, 195-600Vdc Grid Tie Inverter	3100	95	Yes	NA
Xantrex Technology	GT3.3N-NA-240/208 (240V)	3.3 kW, 240 Vac, 195-600Vdc Grid Tie Inverter	3300	95.5	Yes	NA

(Continued)

MANUFACTURER NAME	INVERTER MODEL NUMBER	DESCRIPTION	POWER RATING (WATTS)	WEIGHTED EFFICIENCY	APPROVED BUILT-IN METER	NOTES
Xantrex Technology	GT3.8-NA-240/208 UL-05 (208Vac)	3.5 kW, 208 Vac, 195-600Vdc Grid Tie Inverter	3500	95	Yes	NA
Xantrex Technology	GT3.8-NA-240/208 UL-05 (240Vac)	3.8 kW, 240 Vac, 195-600Vdc Grid Tie Inverter	3800	95	Yes	NA
Xantrex Technology	GT3.8-NA-DS-240	3.8 kW, 240 Vac, 195-600Vdc Grid Tie Inverter	3800	95	Yes	NA
Xantrex Technology	GT3.8-NA-DS-240-P or POS	3.8 kW, 240 Vac, 195-600Vdc Grid Tie Inverter	3800	95	Yes	NA
Xantrex Technology	GT30-208	30 kW 208 Vac Three Phase Utility Interactive Inverter	28800	96	Yes	NA
Xantrex Technology	GT4.0N-NA-240/208 (208V)	3.8 kW, 208 Vac, 195-600Vdc Grid Tie Inverter	3800	95	Yes	NA
Xantrex Technology	GT4.0N-NA-240/208 (240V)	4.0 kW, 240 Vac, 195-600Vdc Grid Tie Inverter	4000	95.5	Yes	NA
Xantrex Technology	GT5.0-NA-240/208 (208V)	5.0 kW, 208 Vac, 195-600Vdc Grid Tie Inverter	4500	95	Yes	NA
Xantrex Technology	GT5.0-NA-240/208 (240V)	5.0 kW, 240 Vac, 195-600Vdc Grid Tie Inverter	5000	95.5	Yes	NA
Xantrex Technology	GT5.0-NA-DS-240	5.0 kW, 240 Vac, 195-600Vdc Grid Tie Inverter	5000	95.5	Yes	NA
Xantrex Technology	GT500-MVX	500 kW Three Phase Inverter for Medium Voltage Applications	500000	95.5	Yes	NA
Xantrex Technology	PV100S-208	100kW 208Vac, 330-600Vdc Inverter System with Automatic Transformer Disconnect	100000	92.5	Yes	NA

Xantrex Technology	PV100S-208-HE	100kW 208Vac, 330–600Vdc Inverter System with Automatic Transformer Disconnect	100000	94.5	Yes	NA
Xantrex Technology	PV100S-480	100kW 480Vac, 330–600Vdc Inverter System with Automatic Transformer Disconnect	100000	93	Yes	NA
Xantrex Technology	PV100S-480-HE	100kW 480Vac, 330–600Vdc Inverter System with Automatic Transformer Disconnect	100000	95	Yes	NA
Xantrex Technology	PV10-208	10kW, 208Vac/3phase Utility Interactive Inverter with 208Vac Transformer	10000	90	Yes	NA
Xantrex Technology	PV10-480	10kW, 208Vac/3phase Utility Interactive Inverter with 480Vac Transformer	10000	92	Yes	NA
Xantrex Technology	PV15-208	15kW, 208Vac/3phase Utility Interactive Inverter with 208Vac Transformer	15000	91.5	Yes	NA
Xantrex Technology	PV15-480	15kW, 208Vac/3phase Utility Interactive Inverter with 480Vac Transformer	15000	91.5	Yes	NA
Xantrex Technology	PV20-208	20kW, 208Vac/3phase Utility Interactive Inverter with 208Vac Transformer	20000	92	Yes	NA
Xantrex Technology	PV20-480	20kW, 208Vac/3phase Utility Interactive Inverter with 480Vac Transformer	20000	92	Yes	NA
Xantrex Technology	PV225S-480	225kW 480Vac, 330–600Vdc Inverter System with Automatic Transformer Disconnect	225000	93.5	Yes	NA
Xantrex Technology	PV225S-480-P	225kW 480Vac, 330–600Vdc Inverter System with Automatic Transformer Disconnect	225000	94.5	Yes	NA
Xantrex Technology	PV30-208	30kW, 208Vac/3phase Utility Interactive Inverter with 208Vac Transformer	30000	92	Yes	NA
Xantrex Technology	PV30-480	30kW, 208Vac/3phase Utility Interactive Inverter with 480Vac Transformer	30000	92	Yes	NA

(Continued)

MANUFACTURER NAME	INVERTER MODEL NUMBER	DESCRIPTION	POWER RATING (WATTS)	WEIGHTED EFFICIENCY	APPROVED BUILT-IN METER	NOTES
Xantrex Technology	PV45-208	45kW 208Vac/3phase Utility Interactive Inverter with 208Vac Transformer	45000	92.5	Yes	NA
Xantrex Technology	PV45-480	45kW 208Vac/3phase Utility Interactive Inverter with 480Vac Transformer	45000	92	Yes	NA
Xantrex Technology	SW4024	4.0 kVA, 24Vdc, 120Vac, Trace Engr. batt bkp, sine wave inverter (w/GTI)	3297	89	No	NA
Xantrex Technology	SW4048	4.0 kVA, 48Vdc, 120Vac, Trace Engr. batt bkp, sine wave inverter (w/GTI)	3211	90.5	No	NA
Xantrex Technology	SW5548	5.5 kVA, 48Vdc, 120Vac, Trace Engr. batt bkp, sine wave inverter (w/GTI)	4264	91	No	NA
Xantrex Technology	XW4024-120/240-60	4 kVA, 24Vdc, 120/240Vac, battery backup, utility-interactive inverter	4000	91	No	NA
Xantrex Technology	XW4548-120/240-60	4.5 kVA, 48Vdc, 120/240Vac, battery backup, utility-interactive inverter	4400	93	No	NA
Xantrex Technology	XW6048-120/240	6 kVA, 48Vdc, 120/240Vac, battery backup, utility-interactive inverter	6000	92.5	No	NA
Xantrex Technology	XW6048-120-60	6 kVA, 48Vdc, 120Vac, battery backup, utility-interactive inverter	5760	92.5	No	NA
Xslent Energy Technologies	XPX A1000	180 W 120Vac Single Phase Utility-Interactive Smart Grid Inverter	180	89	No	NA
Yes! Solar	ES3000 (208V)	2.7kW (208Vac) Utility-Interactive Inverter	2700	95.5	Yes	NA
Yes! Solar	ES3000 (240V)	2.9kW (240Vac) Utility-Interactive Inverter	2900	96	Yes	NA
Yes! Solar	ES3000P (208V)	2.7kW (208Vac) Utility-Interactive Inverter	2700	95.5	Yes	NA
Yes! Solar	ES3000P (240V)	2.9kW (240Vac) Utility-Interactive Inverter	2900	96	Yes	NA
Yes! Solar	ES4000 (208V)	3.4kW (208Vac) Utility-Interactive Inverter	3400	95.5	Yes	NA
Yes! Solar	ES4000 (240V)	3.9kW (240Vac) Utility-Interactive Inverter	3900	96	Yes	NA

Yes! Solar	ES4000P (208V)	3.4kW (208Vac) Utility-Interactive Inverter	3400	95.5	Yes	NA
Yes! Solar	ES4000P (240V)	3.9kW (240Vac) Utility-Interactive Inverter	3900	96	Yes	NA
Yes! Solar	ES5000 (208V)	4.3kW (208Vac) Utility-Interactive Inverter	4300	96	Yes	NA
Yes! Solar	ES5000 (240V)	4.9kW (240Vac) Utility-Interactive Inverter	4900	96	Yes	NA
Yes! Solar	ES5000P (208V)	4.3kW (208Vac) Utility-Interactive Inverter	4300	96	Yes	NA
Yes! Solar	ES5000P (240V)	4.9kW (240Vac) Utility-Interactive Inverter	4900	96	Yes	NA
Yes! Solar	ES5400 (208V)	4.6kW (208Vac) Utility-Interactive Inverter	4600	95.5	Yes	NA
Yes! Solar	ES5400 (240V)	5.3kW (240Vac) Utility-Interactive Inverter	5300	96	Yes	NA
Yes! Solar	ES5400P (208V)	4.6kW (208Vac) Utility-Interactive Inverter	4600	95.5	Yes	NA
Yes! Solar	ES5400P (240V)	5.3kW (240Vac) Utility-Interactive Inverter				NA

INDEX